MEMBRANES WITH FUNCTIONALIZED NANOMATERIALS

MEMBRANES WITH FUNCTIONALIZED NANOMATERIALS

Current & Emerging Research Trends in Membrane Technology

Edited by

SUMAN DUTTA

Department of Chemical Engineering, IIT(ISM), Dhanbad, India

CHAUDHERY MUSTANSAR HUSSAIN

Department of Chemistry and Environmental Sciences, New Jersey Institute of Technology (NJIT), Newark, NJ, United States

Elsevier
Radarweg 29, PO Box 211, 1000 AE Amsterdam, Netherlands
The Boulevard, Langford Lane, Kidlington, Oxford OX5 1GB, United Kingdom
50 Hampshire Street, 5th Floor, Cambridge, MA 02139, United States

Notices
Knowledge and best practice in this field are constantly changing. As new research and experience broaden our understanding, changes in research methods, professional practices, or medical treatment may become necessary.

Practitioners and researchers must always rely on their own experience and knowledge in evaluating and using any information, methods, compounds, or experiments described herein. In using such information or methods they should be mindful of their own safety and the safety of others, including parties for whom they have a professional responsibility.

To the fullest extent of the law, neither the Publisher nor the authors, contributors, or editors, assume any liability for any injury and/or damage to persons or property as a matter of products liability, negligence or otherwise, or from any use or operation of any methods, products, instructions, or ideas contained in the material herein.

ISBN: 978-0-323-85946-2

For information on all Elsevier publications
visit our website at https://www.elsevier.com/books-and-journals

Publisher: Susan Dennis
Acquisitions Editor: Anita A Koch
Editorial Project Manager: Andrea Dulberger
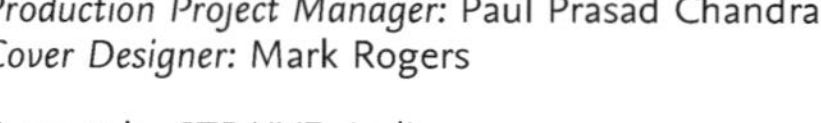
Production Project Manager: Paul Prasad Chandramohan
Cover Designer: Mark Rogers

Typeset by STRAIVE, India

Contents

Contributors

A.A. Abuhabi
Water Technology PhD Joint Programme Between Islamic University of Gaza IUG & Al-Azhar University of Gaza AUG, Gaza, Palestine

Nur Hashimah Alias
Department of Oil and Gas Engineering, School of Chemical Engineering, College of Engineering, Universiti Teknologi MARA, Shah Alam, Selangor, Malaysia

Abtin Ebadi Amooghin
Department of Chemical Engineering, Faculty of Engineering, Arak University, Arak, Iran

Sanjib Barma
Department of Chemical Technology, University of Calcutta, Kolkata, West Bengal, India

Zinnia Chowdhury
Department of Chemical Technology, University of Calcutta, Kolkata, West Bengal, India

Maher Darwish
Department of Pharmaceutical Chemistry and Drug Control, Faculty of Pharmacy, Wadi International University, Homs, Syria

Bimal Das
Department of Chemical Engineering, National Institute of Technology Durgapur, Durgapur, India

Krishna Priyadarshini Das
Department of Material Science and Engineering, Indian Institute of Technology, Delhi, India

P. Das
CSIR-Central Institute of Mining and Fuel Research, Dhanbad, Jharkhand, India

Deepshikha Datta
Department of Chemical Engineering, GMR Institute of Technology, Rajam, India

K.S. Deepak
Department of Chemical Engineering, National Institute of Technology Durgapur, Durgapur, India

Suman Dutta
Department of Chemical Engineering, IIT(ISM), Dhanbad, India

Abtin Ebadi Amooghin
Department of Chemical Engineering, Faculty of Engineering, Arak University, Arak, Iran

Jasir Jawad
Centre for Advanced Materials, Qatar University, Doha, Qatar

G.T.M. Kadja
Division of Inorganic and Physical Chemistry, Faculty of Mathematics and Natural Sciences, Institut Teknologi Bandung, Bandung, Indonesia

K. Khoiruddin
Department of Chemical Engineering, Faculty of Industrial Technology, Institut Teknologi Bandung, Bandung, Indonesia

Woei Jye Lau
Advanced Membrane Technology Research Centre (AMTEC), School of Chemical and Energy Engineering, Faculty of Engineering, Universiti Teknologi Malaysia, Skudai, Johor, Malaysia

Fauziah Marpani
Department of Oil and Gas Engineering, School of Chemical Engineering, College of Engineering, Universiti Teknologi MARA, Shah Alam, Selangor, Malaysia

Hanan Mohammad
Department of Pharmaceutical Chemistry and Drug Control, Faculty of Pharmacy, Wadi International University, Homs, Syria

Mohamad Nor Nor Azureen
Department of Chemical Sciences, Faculty of Science & Technology, The National University of Malaysia (UKM), Bangi, Selangor, Malaysia

Nur Hidayati Othman
Department of Oil and Gas Engineering, School of Chemical Engineering, College of Engineering, Universiti Teknologi MARA, Shah Alam, Selangor, Malaysia

Hamidreza Sanaeepur
Department of Chemical Engineering, Faculty of Engineering, Arak University, Arak, Iran

Aparna Ray Sarkar
Department of Chemical Engineering, Heritage Institute of Technology, Kolkata, West Bengal, India

Dwaipayan Sen
Department of Chemical Engineering, Heritage Institute of Technology, Kolkata, West Bengal, India

Munawar Zaman Shahruddin
Department of Oil and Gas Engineering, School of Chemical Engineering, College of Engineering, Universiti Teknologi MARA, Shah Alam, Selangor, Malaysia

Putu Doddy Sutrisna
Department of Chemical Engineering, University of Surabaya (UBAYA), Surabaya, Indonesia

I.G. Wenten
Department of Chemical Engineering, Faculty of Industrial Technology, Institut Teknologi Bandung, Bandung, Indonesia

Syed Javaid Zaidi
Centre for Advanced Materials, Qatar University, Doha, Qatar

Mehrzad Zandieh
Department of Chemical Engineering, Faculty of Engineering, Arak University, Arak, Iran

CHAPTER 1

Modern perspective in membrane technologies—Sustainable membranes with FNMs

Mehrzad Zandieh, Abtin Ebadi Amooghin, and Hamidreza Sanaeepur
Department of Chemical Engineering, Faculty of Engineering, Arak University, Arak, Iran

1.1 Introduction

Separation technology has acquired significance in different industrial usages [1]. Membrane technology is favored over other purification methods such as disinfection or distillation owing to the absence of chemical additives, energy yield, and the convenience of the technology [2]. In other words, membranes suggest energy-effective processes versus available separation processes [1]. Membrane technology renders a preferable operating efficiency along with small energy and performance costs. A lot of effort has been focused on making unique materials or structures with a higher capability to improve the membrane performance [3]. To attain membranes with the best performance, different kinds of materials have been applied. Among them, polymeric membranes suggest an extensive scope of molecular transfer features with more straightforward methods and lower membrane production costs [1]. Commercial membranes are mainly manufactured from polymeric materials containing polysulfone and polyamides [2]. Nevertheless, polymeric membranes involve a challenge of choosing between selectivity and permeability of penetrants. The focus to eliminate this challenge is to combine excellent-performance particles like zeolites, metal-organic frameworks, and nanoparticles such as porous layers or nanotubes in the membrane; the presence of different nanoparticles augments the performance of separation as well as physicochemical features of the membrane [1,4]. A novel approach in the development of membranes is incorporating nanomaterials that can improve the porosity, reactivity, and permeability of the membranes. Considering this approach, various types of membranes can be constructed, such as thin-film nanocomposite (TFN) membranes or mixed matrix membranes (MMMs) [5]. Embedding nanomaterials in

Membranes with Functionalized Nanomaterials
https://doi.org/10.1016/B978-0-323-85946-2.00009-6

thin-film layer, high-performance TFN membranes have developed with excellent physicochemical properties [6].

On the other hand, thin-film composite (TFC) membranes have excellent potential due to their specific structure, a thin selective layer created on the top of a porous supporting layer. These membranes are very versatile, as they have an acceptable permeability (flux) simultaneous with excellent selectivity (rejection). In addition, incorporating the nanoparticles in the TFC membrane structure is a prevalent approach. Different nanoparticles, like zeolites, titanium dioxide, and graphene oxide, have been effectively embedded into the separating layer of TFC membranes [7]. Moreover, porous organic polymers (POPs)—as a novel class of nanomaterials—can be excellent candidates for making TFC with high performance [5]. On the other hand, mixed matrix membranes (MMMs)—as another form of nanostructured membranes—have superior selectivity and/or permeability and excellent capability to overcome restrictions of polymeric membranes as well as inorganic membranes [8]. The application of advanced functional materials is developing to enhance MMMs' performance as energy-efficient substitutes for common membranes. Enhancing the facilitated transfer and selective adsorption of a specified molecule via MMMs are the original purposes of designing and developing functional materials for the preparation of these membranes. Excellent inherent permeability and selectivity of advanced nanomaterials simultaneously with well mechanical features of polymers result in excellent separation performance of MMMs compared to a pristine polymer [9]. Today, modern functionalized MMMs are the great candidates for the specific separations [10].

1.2 Nanostructured membranes

The manipulation of atomic and molecular interactions results in a correction of macroscopic properties, leading to a fundamental reconstruction of material features like optical, mechanical, and transfer phenomena. Macroscopic adjustments can be ascribed to small size effects, interface effects, quantum size effects, quantum tunneling effects, and so on. A nanostructured membrane represents the internal and external nanostructures together since nano-based characteristics can be in bulk or all over the membrane's surface (Fig. 1.1) [11].

Pores with sizes ranging from 1 to 100 nm that permit particles and compounds are separated according to their sizes are common structural elements that enable membranes to be categorized as nanostructures. In addition, the incorporation of nanofillers in polymer matrixes puts the

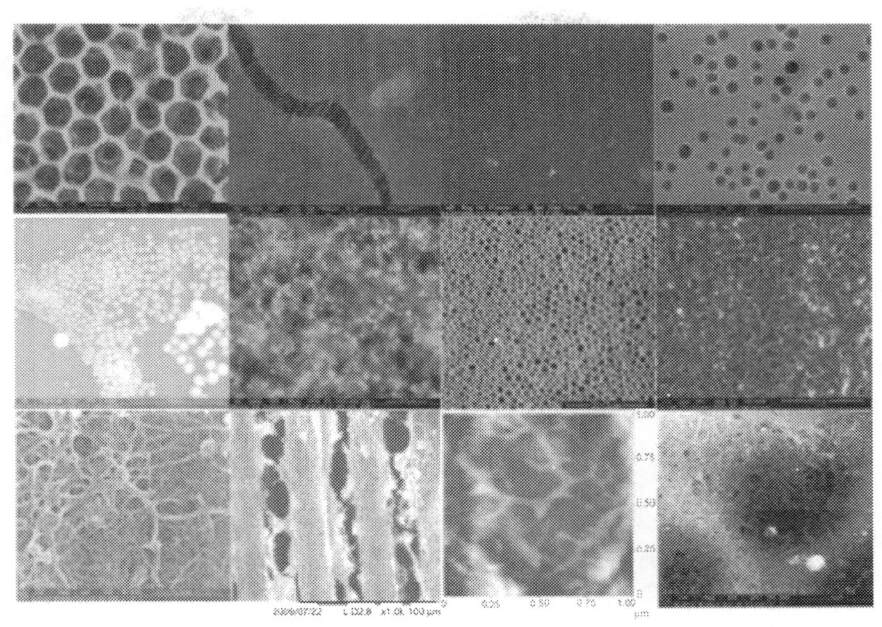

Fig. 1.1 Examples of tailor-made nanostructured membranes. *(From E. Drioli, L. Giorno, A. Gugliuzza, An introduction to nanostructured membranes for advanced applications in strategic fields, in: E. Drioli, L. Giorno, A. Gugliuzza (Eds.), Functional Nanostructured Membranes, first ed., Jenny Stanford Publishing, 2019, pp. 1–11. https://doi.org/10.1201/9781351135115.)*

membranes in the classification of nanostructures. The hierarchical texture on the membrane surfaces is another reason to categorize it in the nanostructures. Block copolymers in combination with nanotubes, catalytic nanoparticles, zeolites, or bio-combined compounds can be incorporated in the polymer matrix or deposited onto the relevant surface to enhance the separation performance as well as decreasing energy and capital costs (Fig. 1.1) [11]. The substantial allocation of nano-objects and complementary components in confined spaces yields new tunable functions and advanced performances that depend on intrinsic relationships between local structure, transport, and surface properties. The source of transport phenomena controlling the membrane permeability and selectivity depends on the activity at nano levels. It is essential to enhance the structural complexity of nanostructured membranes since many species need to be arranged in complex organizations that depend on the previous level of assembly. This applies organizational principles and accurate control on various length scales [8]. In nano-porous membranes, the pores are considered structural constituent sectors of the entire framework. This suggests that nano-porous membranes are not related to inherent risks, such as toxicity, that arise in conventional nanomaterials [11].

1.3 Nano-engineered membranes

In nanostructured membranes (NSMs), the word "nano" refers to the membrane's interior configuration. Nanoscience has created a way to manufacture nano-added or nano-engineered membranes (NEMs), where the word "nano" signifies NM-based membranes. Various kinds of nanomaterials (NMs) that have been investigated in preparing membranes are illustrated in Fig. 1.2. NEMs are applied to make reasonable adjustments to the morphology, permeability, selectivity, physical, and chemical properties of membranes to supply the following practical items [12]:

- enhanced selectivity via improving the membrane structure and surface features;
- extensive usage of membranes in different separation processes;
- improved mechanical and thermal consistency NEMs, which can be categorized as follows:
  - **(1)** nanostructured membranes are prepared from a layer of NMs on top of porous membrane substrate or fabricated as a freestanding NM film; and
  - **(2)** nanocomposite membranes (NCMs), in which NMs are embedded through a polymer binder (Fig. 1.3).

Moreover, NCMs are subclassified into:

(a) (1) thin film nanocomposite membranes (TFNCMs), where particles are immobilized/entrapped within the top layer of the membrane; (2)surface-coated TFNCMs, where NMs are coated onto the top surface of an NCM; and (3) substrate-coated TFNCMs, where NMs are located on a TFNCM support membrane; and

(b) MMMs in which NMs are embedded in the membrane matrix. TMNCMs can be fabricated by employing an MMM or a conventional nanostructured UF/MF membrane as the support [12].

TFC membranes are made by applying an ultra-thin selective layer on top of a microporous substrate with the support of nonwoven fabric. The structure of a totally fibrous membrane is known as a membrane of thin-film nanofibrous composite (TFNC). Applying nanofibrous scaffolds presents a novel combination of interlinked pore structures with high porosity that makes TFNC membranes more permeable than standard TFC membranes [13].

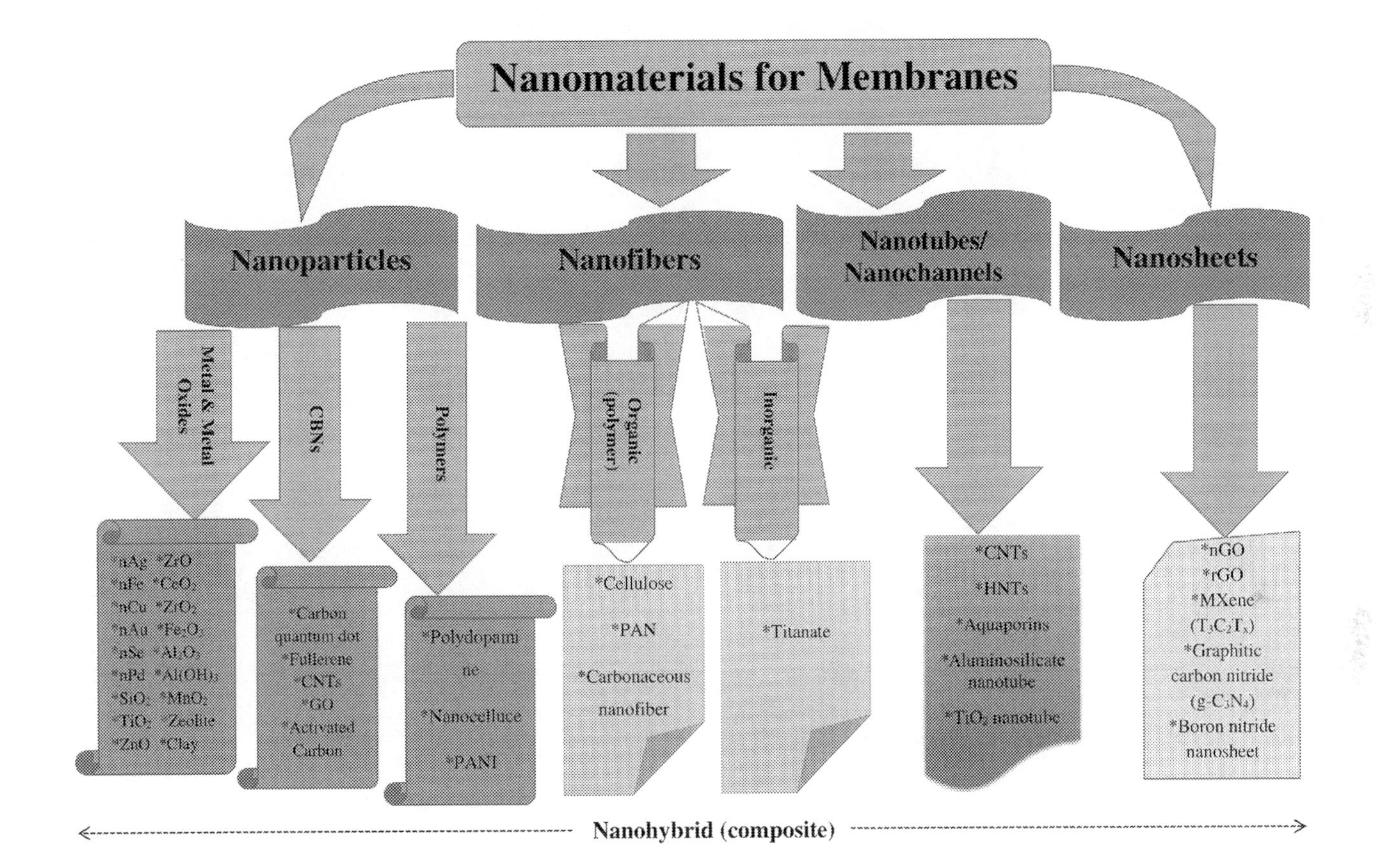

Fig. 1.2 Various kinds of nanomaterials for preparation of nano-engineered membranes. *(From J.M. Gohil, R.R. Choudhury, S. Thomas, D. Pasquini, S.-Y. Leu, D.A. Gopakumar, Introduction to nanostructured and nano-enhanced polymeric membranes: preparation, function, and application for water purification, in: Micro and Nano Technologies, Elsevier, 2019, pp. 25–57. https://doi.org/10.1016/B978-0-12-813926-4.00038-0. (Chapter 2))*

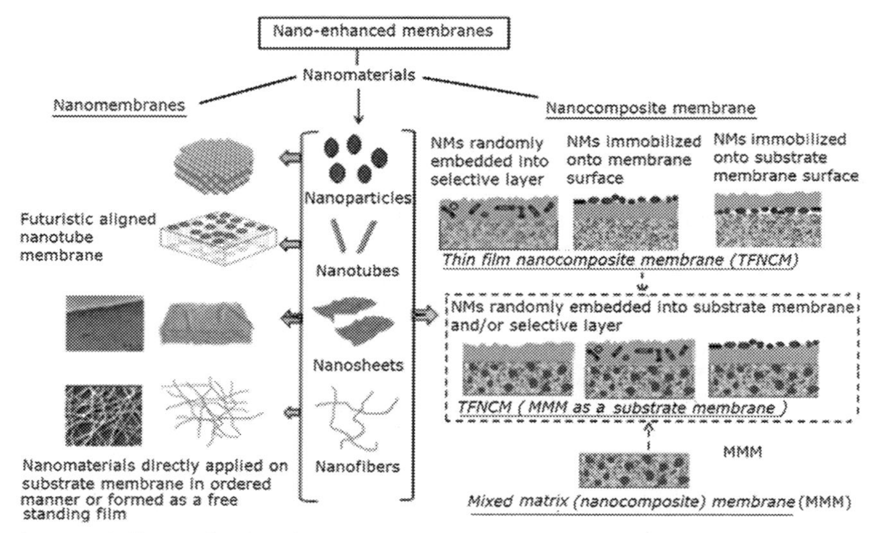

Fig. 1.3 Different kinds of nano-engineered membranes. *(From J.M. Gohil, R.R. Choudhury, S. Thomas, D. Pasquini, S.-Y. Leu, D.A. Gopakumar, Introduction to nanostructured and nano-enhanced polymeric membranes: preparation, function, and application for water purification, in: Micro and Nano Technologies, Elsevier, 2019, pp. 25–57. https://doi.org/10.1016/B978-0-12-813926-4.00038-0. (Chapter 2))*

Developing an easy procedure for the fabrication of TFN membranes with excellent antibiofouling and separation performance is very important. Simultaneous formation of the polymer matrix and nanoparticles, which is called in-situ hybridization, has many advantages and has been widely employed in various composites [14]. These advantages are the creation of an interpenetrating polymer network (IPN) between nanoparticles, and better dispersity/compatibility of the nanoparticles in hybrid networks. This procedure can also simplify the fabrication process of composite membranes (Fig. 1.4) [14]. Fig. 1.4 shows an easy in-situ hybridization procedure for the fabrication of membranes of Ag nanoparticle (AgNP)-incorporated TFN (Ag-TFN) RO with excellent antibiofouling and separation performance. AgNP has robust antibacterial activity and hydrophilic nature that can improve water permeance and biofouling resistance of the prepared membrane. This procedure is based on the possible multifunctional ability of the materials commonly used in interfacial polymerization (IP). MPD (*m*-phenylenediamine) monomer contains the main amine groups that are applied to create a main PA network, and can be a reducing agent to generate AgNPs from an Ag precursor (silver nitrate, $AgNO_3$). Sodium dodecyl

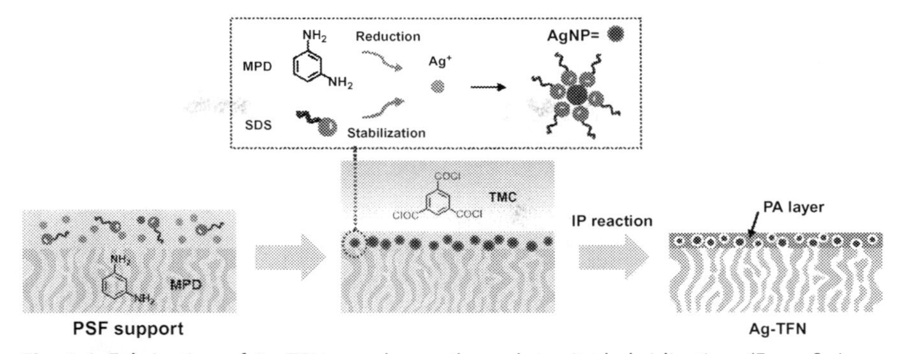

Fig. 1.4 Fabrication of Ag-TFN membrane through in-situ hybridization. *(From S. Jeon, J.-H. Lee, Rationally designed in-situ fabrication of thin film nanocomposite membranes with enhanced desalination and anti-biofouling performance, J. Membr. Sci. 615 (2020) 118542. https://doi.org/10.1016/j.memsci.2020.118542.)*

sulfate (SDS), which is widely applied as a surfactant to facilitate IP, can also be a stabilizer for AgNPs to improve their uniformity and dispersion [14].

1.4 Several aspects of nanocomposite membranes

The main purpose for preparing of the modified membrane is to improve the effectiveness of the physicochemical properties and separation performance. In this section, a summary of different nanoparticles used in the fabrication of composite membrane is provided [15].

1.4.1 TiO_2

TiO_2 is usually applied for the fabrication of nanocomposite membranes. This metal oxide always improves the mechanical strength, hydrophilicity, water permeability, and antifouling capability of a membrane. It also can improve the antibacterial property of the membrane surface [15]. Wei et al. [16] studied polyamide membranes of TFN nanofiltration prepared through interfacial polymerization of piperazine and trimesoyl chloride on a PES substrate using aminated titanium dioxide (APTES (3-aminopropyltriethoxysilane)-TiO_2). The results illustrated that a membrane containing 0.3 w/v% amine-functionalized TiO_2 seriously improved the pure water flux with a relatively high Na_2SO_4 rejection (>95%). This TFN membrane revealed a separation factor of 477.89 with the NaCl to Na_2SO_4 ratio of about 1:19.

1.4.2 Silver

Silver is a well-known antibacterial agent. Hence, this nanoparticle minimizes the biofouling of the membrane surface due to its antibacterial nature. This nanoparticle annihilates the microorganisms present in different effluents. Therefore, silver-incorporated composite membranes usually reveal better surface charge density, permselectivity, etc., in comparison with typical membranes [15]. Maiphetlho et al. [17] evaluated silver nanocomposite polymer inclusion membranes (PIMs) for trace metal transports. A polyvinyl chloride/di-(2-ethylhexyl) phosphoric acid (PVC-D2EHPA)-based PIM containing silver nanoparticles fabricated to remove and measure the trace metals in contaminated water. The contact angle results showed that the presence of AgNPs seriously modified the hydrophilic/hydrophobic characteristics of fabricated membranes. On the other hand, the membrane stability was reduced during the operations with repeated cycles despite an improvement of hydrophilicity by incorporating of AgNPs.

1.4.3 SiO_2

Silica is applied as an inherent semiconductor. Therefore, this nanoparticle improves hydrophilicity, permeability, and selectivity of the membrane. Thus, such a membrane can be mainly applied for the oil removal from water and gas separation. In addition, different kinds of silica-based composite membranes can be used in wastewater treatment [15]. Shakak et al. [18] have synthesized polysulfone/polyvinylpyrrolidone/SiO_2 nanocomposite ultrafiltration membranes and evaluated the amoxicillin removal from water. Ultrafiltration membranes were fabricated by the phase inversion method. Nanoparticles of SiO_2 were applied as a hydrophilic agent for modification of membrane. The results achieved by contact angle and Fourier-transform infrared spectroscopy (FTIR) analysis showed that the presence of hydroxyl groups of SiO_2 enhanced the membrane hydrophilicity. It was also found that by increasing the SiO_2 content, the spongiform and tear-like structure of the membranes shifted to canal-shaped, finger-shaped, and macro-void structures. Moreover, increasing the SiO_2 nanoparticle from 0 to 4 wt% caused an increase in the separation performance of amoxicillin from 66.52% to 89.81%.

1.4.4 Iron

FeO nanoparticle is a very suitable material for water purification due to its high surface area, accessibility, magnetic property, cheapness, and hydrophilic nature [13]. Copper removal was investigated by employing a polyethersulfone (PES)-iron oxide composite nanofiltration membrane [19]. The highest copper removal was in the presence of 0.1 wt% iron oxide. PES-iron oxide membrane treatment (reusability test) applying ethylene diamine tetraacetic acid (EDTA) had the best performance (in copper removal) at 0.1 wt% iron oxide [19]. FeO nanoparticles led to a considerable improvement in the removal of hydrophobic pollutants and antifouling for saline water treatment applications. At low FeO loading (0.05 wt%), peeling type structure was observed, while 0.2 wt% FeO led to the seed-like covered surface morphology. Excellent dispersion of FeO nanoparticles and finger-like structure were observed in the cross-section of membranes [20].

1.4.5 Zirconia

Zirconium oxide (zirconia or ZrO_2) has been used in the preparation of different NCMs. In the case of membrane of PVDF ultrafiltration, suitable changes in the membrane performance can be achieved via alterations to the solvent of PVDF or by varying the ZrO_2 content ratio of the PVDF/ZrO_2 mixture. ZrO_2 has also been used with sulfonated polyetherketone (PEK) to reduce the water and methanol permeabilities. This ability has made the PEK/ZrO_2 membrane a suitable candidate for utilization in applications related to fuel cells [13]. Noormohamadi et al. [21] studied the hydrophilic properties of nanoparticles of ZrO_2 and the role of nanoparticles of iron oxide in increasing the porosity of polymer membranes. To achieve this purpose, Fe_3O_4@ZrO_2/PAN (polyacrylonitrile) nanocomposite membrane was synthesized, and its performance was tested in decreasing biological fouling of the membrane. Analysis of FTIR showed the functional groups in nanoparticles of Fe_3O_4@ZrO_2 and confirmed their presence in the matrix of an Fe_3O_4@ZrO_2/PAN NCM. Analysis of contact angle verified the role of nanoparticles of ZrO_2 in increasing the hydrophilicity of membrane by 51%, and analysis of porosimetry verified the role of nanoparticles of iron oxide in improving porosity of the NCMs by 47%. Application of 1 wt% of nanoparticles of Fe_3O_4@ZrO_2 in the PAN membrane matrix resulted in a 40% increase in water flux.

1.4.6 Copper

Ben-Sasson et al. [22] employed a procedure for inserting copper nanoparticles (Cu-NPs) on the surface of a membrane of TFC polyamide reverse osmosis. Cu-NPs were synthesized, resulting in particles with an average radius of 34 nm. The positive charge of the Cu-NPs permitted an easy electrostatic functionalization of the RO membrane with a negative charge. It was found that functionalization of Cu-NPs had minimal effect on the transport parameters of the membrane. After functionalization, the surface charge of the membrane became positive. The functionalized membrane showed significant antibacterial activity.

1.4.7 Clay

Clay is relatively cheap, environmentally friendly, and available in nature. The most commonly embedded inorganic nanoparticles in the nanocomposites come from a group of 2:1 phyllosilicates that contain a sheet silicate structure. Hectorite, montmorillonite, and vermiculite are the most used fillers for polymer-clay nanocomposites. They significantly enhance the ability of filtration of membranes by adjusting the formation of macrovoids. In addition, they can increase hydrophilicity, surface pores, and porosity [13].

1.4.8 Other nanoparticles

Different NCMs can be fabricated using nonconventional nanoparticles such as Fe_3O_4, Al_2O_3, ZnO, Ag_2O, carbon nanotubes (CNTs), cellulose nanocrystals (CNCs), graphene oxide, zeolites, MOFs, etc. [15]. Functionalization of the NPs prohibits accumulating large NPs and develops other features of the selective layer. Considering the hydrophilic groups like carboxylic acid, amine, and sulfonic acid for NP functionalization, results in TFN membranes with superior surface hydrophilicity and water permeability. Functional groups of NPs can participate in interfacial polymerization, producing covalent bonds with the PA matrix [23].

1.5 Applications of NEMs

The advancement of membranes with multifunctional characteristics and minor consumption of energy and utilization of them for the elimination of contaminants have been significant in membrane separations. Unification of membrane and nanotechnologies has resulted in much progress in membrane configuration. This not only provides better control of the pore's shape of membranes but also improves the hydrophilicity and permeates flux of the nanocomposite membrane. Therefore, nano-architectured membranes are more "reactive" rather than just behaving as a physical barrier for transferring the solutes. Incorporating nanofillers in polymeric membranes leads to durable membranes, since it slows down the membrane fouling procedure that has resulted in a restriction in industrial usages [24].

1.5.1 Adsorption

Some NEMs, which are synthesized based on nano CNT, metal oxides, porous boron nitride, graphene, nanofibers, and zeolite, are applied for filtration and adsorption of heavy metal ions, etc. High porosity and active adsorption sites of these NMs facilitate the contaminant removal. For example, nanofiber-based membranes present suitable separation properties to remove the various trace amounts of pollutants from water by filtration and adsorption, owing to their porosity and high surface area. From water media, adsorbent could be captured through the chemical binding (owing to the presence of various functional groups in NMs) or physical adsorption (for example, due to the high surface area and porosity of NMs), or electrostatic attraction [12].

1.5.2 Water disinfection

Water disinfection technologies have been usefully combined according to the membrane and nanomaterials. NEMs are highly efficient for water disinfection via the killing/preventing of microorganisms and physical capture. Most of the NEMs fabricated for MF, NF, UF, and RO applications are intended to develop antifouling properties by imparting hydrophilic/antimicrobial properties to membranes. Molecular-level separation can be possible by NF membranes, and hence it produces water free from any toxic microorganisms [12].

1.5.3 Dissolved pollutants/salt ions removal by NF and RO

Removal of soluble salt ions from seawater and saline to make them drinkable is one of the significant applications of NF membrane technology. Commercial and laboratory NF membranes in hollow fiber and flat sheet configurations are TFCMs fabricated by interfacial polymerization (IP). NF membranes usually reject about 20%–80% of monovalent ions and >95% of divalent ions with the same charge on the membrane's surface.

NF is also used to remove the hardness in drinking water recycling as an alternative to RO or as a pre-RO process [12]. The TFN membrane, as a new generation of composite membrane, was explored in 2007 for the first time to enhance the separation performance and overcome the trade-off of selectivity-permeability of the TFC membrane in reverse osmosis applications. Incorporating a low content of zeolite into the polyamide (PA) layer of the TFC membrane improved the water permeability of the final membrane [13].

1.5.4 Gas separation

At an early step of the progress of NCMs, further studies have focused on the interactions between the polymer matrix and the nanofiller and the changes in the polymer matrix due to the presence of nanoparticles. The interfacial adhesion properties can be known better by investigating the findings achieved from experiments of gas permeation. The introduction of nanofiller leads to some significant changes in the polymer matrix, such as: (1) rigidification or tightening of polymer chains; (2) formation of tortuous paths; (3) creation of intermediate cavities or microvoids at polymer-filler interface as well as nonselective voids with distinctive properties; and (4) blockage of nanoparticles' pore openings [25]. Fig. 1.5 demonstrates the possible interactions of penetrant molecules with the nanofillers and polymer matrix. Various possible ways for the gas transports can be explained as: (1) solution-diffusion mechanism of gas transport in a conventional polymer membrane; (2) small gases can diffuse easier via the interfacial microvoids; (3) larger molecules should be transferred through a tortuous path in the filler's vicinity; (4) pore size restricts the larger gas molecules penetration while small gases can quickly utilize this diffusion path; and (5) the presence of functional moieties or adsorptive sites that attract target gas and increase overall solubility in the membrane [25].

It is significant to choose appropriate fillers related to the particle size, the pore diameter, functional groups of surface, and affinity of the fillers to the distinct gases in a gas mixture. Furthermore, the interaction of functional groups of the nanofiller surfaces with gases like CO_2 may increase permeability in the MMMs for gas separation membranes [26]. According to fillers' features in the membrane, gas transport can be increased via various routes. Filler particles can improve the features of the continuous polymeric phase, and hence change the general transport features. This is considerable for CO_2-inert particles that are appropriately dispersed in the polymer matrix,

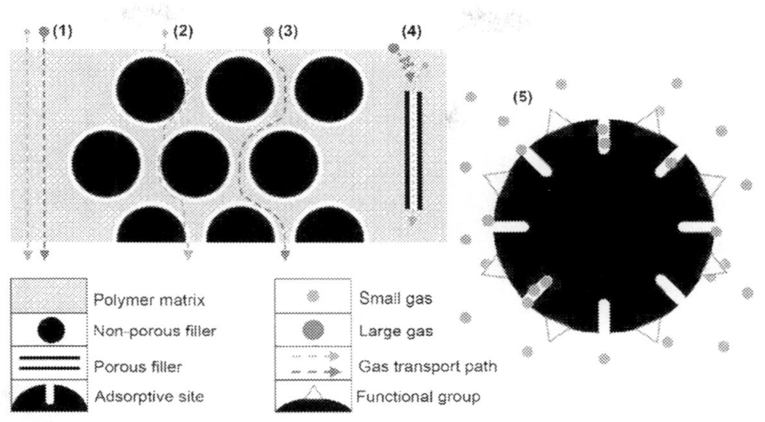

Fig. 1.5 A schematic view of the interaction of penetrant molecules with the polymer matrix and nanofillers in a nanocomposite membrane. *(From P.S. Goh, A.F. Ismail, Structure and gas transport of nanocomposite membranes, in: Micro and Nano Technologies, Elsevier, 2020, pp. 101–123. https://doi.org/10.1016/B978-0-12-819406-5.00003-4. (Chapter 3))*

where the fractional free volume (FFV) can be significantly increased, particularly for glassy polymers. Filler particles can change the dynamics or conformation of polymer chains near the particle surfaces, and consequently lead to an increase in diffusion. Chemical modification (or functionalization) of inorganic fillers has been applied to improve their scattering and separation features [27].

1.5.5 Applications of NEMs in modern engineering sciences

1.5.5.1 NEMs in biotechnology

Membranes are extensively applied in biotechnology, with applications ranging from bioproduct purification and laboratory filtration to use as a component of bioreactors, biosensors, and biomedical devices. Nano-engineered membranes, in addition to the function of the selective barrier, show advanced and tailorable properties, which are provided by functional groups of their nano-sized constituents [28]. High specific surface area and the possibility to combine separation with other advanced functions make membranes an optimal candidate for the development of hybrid systems. Biomolecule-membrane hybrid systems are examples of NEMs. A biomolecule is any molecule present in living organisms. Examples of biomolecules include large macromolecules such as proteins and nucleic acids, and small molecules like metabolites. Biomolecules are nanoscale materials with intrinsic structure and properties [28]. Nano-engineered membranes using biomolecules, i.e., the biofunctionalized membranes, have been employed in several biotechnological applications such as catalysis (membrane-based bioreactors), analysis (biosensors), separations (affinity membranes), and development of artificial organs. In bioreactors, the membranes can be applied just as an external separation unit downstream of a traditional reactor including the biocatalyst, as in membrane bioreactors (MBRs), or can be operated either as an internal site for the biocatalyst confinement or separation unit in biocatalytic membrane reactors (BMRs) [28].

1.5.5.2 NEMs in pharmaceutical applications

BMRs have been applied to produce amino acids, antibiotics, vitamins, anticancer drugs, and so on. These devices mainly use enzymes or cells immobilized on the membrane. In this field, BMR increases the yield and purity of the downstream concerning to the interest component, enabling cost reduction in downstream purification. BMRs represent higher conversion rates when a product inhibition or a reversible reaction happens [28].

1.5.5.3 NEMs in biomedical applications

Different types of synthetic paths have been applied for the surface modification of polymeric membranes. Functionalized biomedical polymeric membranes have shown excellent performance in the field of blood-contacting, which has led to the dramatic development of many critical medical applications such as blood purification, artificial limbs, and so on. In blood-contacting fields, polymeric functionalized biomedical membranes have shown outstanding performances and have fueled many diverse biomedical applications, such as blood purification, artificial organs, etc. Silk is a potential biopolymer for different biomedical applications, especially in the scaffold and regeneration membrane fields. Heparin-modified silk can yield a surface with high hemocompatibility. For example, the excellent anticoagulant activity of the heparin-coated graphene oxide (GO) makes it a suitable candidate for the fabrication of GO-based composite membranes with a blood compatible by the method of filtration. Many novel heparin-functionalized nanomaterials with specific functionalities will be developed in the future and used to prepare composite membranes with bioactive materials [29]. In another example, Ioniţă et al. [30] comparatively evaluated novel combinations of three kinds of carbon nanomaterial membranes (CNMs) with significant capability for biomedical usages. Biological performance and excellent cytocompatibility illustrated by (i) cellulose acetate/ammonia functionalized graphene oxide (CA/GO), (ii) cellulose acetate/ammonia functionalized carbon nanotubes (CA/CNT), and (iii) cellulose acetate/CNT-GO membranes suggested that these materials can be a novel type of composite membranes in different biomedical applications such as tissue regeneration and separation processes.

1.5.5.4 NEMs in food applications

Membrane technologies have several benefits in applications related to food industries [28]. For example, the electrospinning technique is used for the preparation of fibrous polymeric nonwoven membranes formed by polymer fibers with nanometric diameters for application in active food packaging systems and functional food. Bioactive compounds can be placed on the nanofiber's surface by adsorption to surface-functionalized nanofibers or by physical adsorption [31]. As another example, cellulose nanostructures have different applications in food packaging (including active packaging). These nanostructures are applied as a reinforcement phase in nanocomposites (as cellulose nanofibrils or cellulose nanocrystals). Cellulose nanostructures can be used as matrices for films—bacterial cellulose (BC), since it is fabricated naturally—nanostructured membranes that could grow in a medium including other biopolymers. They can be impregnated with other components or be disintegrated into nanocrystals or nanofibrils [32]. In addition, films based on synthetic polymers and biopolymers have been used in food packaging. Chitosan (CS) films are applied as materials in food packaging because they have nontoxic and antimicrobial properties. Moreover, polyvinyl alcohol (PVA) is a synthetic polymer and is nontoxic. In chitosan functionalized poly(vinyl alcohol) (CS/PVA) membranes, the combination of superior properties of PVA with chitosan is applied in food packaging due to the excellent chemical properties and generation costs [33].

1.5.5.5 NEMs for biosensor development

A biosensor is an analytical device that provides specific analytical information, applying a biological sensing element (bioreceptor) intimately related with or integrated within a physicochemical transducer [34]. Nanostructured membranes, which are composed of various materials with different porosity and geometry, have been widely applied in several medical and biological usages such as separation and sorting of biomolecules, analysis of single molecules, sensing of ions and proteins, and immunoisolation and drug delivery [34]. In biosensing applications, membrane technology offers multiple benefits. The high ratio of surface area to volume is an essential property of nanostructured membranes. The increased pore surface can interact with biomolecules that are trapped into the pores or flowing via the membrane, providing a significant and often highly reactive surface area that enables more effective detection of molecules on the sensing surface. Moreover, compared with other nanostructures like nanotubes and nanowires, membranes allow separating biomolecules according to their size, shape, charge, and interactions with nanopores that increase the selectivity of the sensor. They can be produced using different methods with high control in pore size, pore-size distribution, pore location, nanoarchitecture, nano-patterns, and the overall physical and chemical properties. In addition, surface modification methods can be applied to achieve better biocompatibility and antibiofouling capability to promote/control interaction with biomolecules [34].

The increased surface area, together with the selective barrier properties and the possibility of integration with both electrochemical and optical detection techniques, allowed the nanostructured membranes applications in many biosensing situations like glucose detection, nucleic acid detection, bacteria, and virus detection for either diagnostics or environmental and food analysis. The integration of nanostructured membranes in biosensor design has various benefits regarding to sensitivity, response time, and lower detection limits [34]. Microporous polyvinylidene fluoride (PVDF) membranes, which were examined between multiple layers of nanomaterials for continuous monitoring of glucose levels in vivo, were studied by Chen et al. [35]. These membranes, which were prepared using the layer-by-layer deposition method, were made by coating needle electrodes with polyaniline nanofiber, platinum nanoparticles, glucose oxidase, and porous layers. The purpose of the nanoparticles' presence in conductive polyaniline nanofillers was to create a superior surface to volume ratio and electrocatalytic activity for glucose enzyme. The glucose transport in the PVDF and nano-sphere Nafion was limited, and the lifetime of in vivo measurements was increased. In addition, the glucose biosensor showed a sensitivity of 0.23 μA/mM and a fast response time of less than 30 s. In addition, despite maintaining biocompatibility with surrounding tissues, implant tests showed an outstanding response to changes in blood glucose concentration. Furthermore, for rapid detection of pirimiphos-methyl in olive oil, an ultrasensitive electrochemical biosensor based on electrospun blended chitosan-poly(vinyl alcohol) nanofibrous enzymatically sensitized membranes was developed by El-Moghazy et al. [36]. This biosensor showed a good performance with a detection limit of 0.2 nm. Moreover, the biosensor was used in olive oil samples after a simple liquid-liquid extraction, and a 100% recovery rate was obtained.

1.5.5.6 Nanostructured membranes for engineering organs and tissues

Over the last decade, various kinds of nanomaterials have been designed, synthesized, and applied in the recovery of native tissues. In this scenario, nanostructured polymeric membranes have been successfully used to reconstruct tissue analogues due to the high control of the cell microenvironment at the molecular level. Nanostructured membrane systems are very suitable candidates for the development of novel tissue-engineered manufacturing [37]. For example, functionalized CNT (carbon nanotube) membranes were synthesized by Mata et al. [38] by combining a Diels-Alder cycloaddition reaction to produce cyclohexane ($—C_6H_{10}$) with gentle oxidation to produce carboxylic acid (—COOH). Functionalized CNT membranes showed maximum performance regarding in vitro proliferation and osteogenic differentiation of human osteoblastic cells. The in vivo subcutaneously implanted materials illustrated a higher biological reactivity, so inducing a slighter intense inflammatory response compared to nonfunctionalized CNT membranes but still showing a decreased cytotoxicity profile. Overall, the proposed application of Diels-Alder showed a better CNT biological response in terms of the biocompatibility and biodegradability profile. Therefore, in the application of bone tissue engineering without significant toxicological threats, the use of CNTs is recommended. Bacterial cellulose (BC), which is functionalized with silver nanoparticles (AgNPs), was also evaluated as an antimicrobial membrane for wound-healing treatment by Pal et al. [39]. UV light radiation was used green synthesis of silver nanoparticles inside a three-dimensional porous BC network. In this method, silver particles are photochemically deposited on the BC gel network and chemically bonded to cellulose fiber surfaces. X-ray diffraction measurements showed a highly crystalline nature of the BC membrane. The results showed that the synthesized composite could be useful in wound-healing due to its excellent properties, such as the fact that composite pellicles are maintained in a moist media that also prefers wound recovery.

1.5.5.7 Membrane systems with nonstructural features for neuroengineering applications

Recent advances in nanotechnology, biomaterials, and tissue engineering offer novel strategies developing novel in vitro devices that increase neuronal growth and differentiation by mimicking specific characteristics of the in vivo environment. In particular, polymeric membranes can promote the formation of membrane–neuron hybrid systems with selective structural, physicochemical, and mechanical features. They are capable of reconstructing neuronal tissue. Neuronal biohybrid membrane systems are extensively applied as in vitro brain tissue models as research platforms. Recently, applications of nanotechnology in the nervous system have mainly involved the investigation and application of new nanomaterials to improve diagnosis as well as therapy of neurological diseases [37].

1.5.5.8 Nanostructured functional membranes for self-cleaning separations

Generally, the addition of nanoparticles can be tailored to particular membrane applications with the choice of nanoparticle shape, size, and kind. For example, silver nanoparticles, owing to their antimicrobial property, could inhibit biofouling growth. Specific cases with different strategies have been applied to make self-cleaning nanostructured membranes. It should be reduced or minimized the membrane fouling by improving or optimizing operating process conditions. Membrane surface modification illustrates better membrane performance in reducing the fouling [40].

1.5.5.9 TiO_2-loaded self-cleaning membrane

Wu et al. [41] prepared an ultrafiltration membrane with simultaneous capabilities of self-protection and self-cleaning. A nanohybrid of TiO_2-polydopamine (PDA) was easily prepared and applied for functionalization of a polysulfone (PSf) membrane matrix. In a nanohybrid of TiO_2-PDA, spheres of PDA acted as an adhesive substrate to hold nanoparticles of covered photocatalytic TiO_2 densely, as well as serving as a free-radical scavenger to protect the PSf membrane against the damage by free radicals generated by TiO_2 during UV radiation. TiO_2-PDA with spherical shape could be easily doped into the PSf membrane through the method of phase inversion, which allowed the hydrophilic spheres of TiO_2-PDA to migrate to the surface of hydrophobic PSf and benefiting the ability of antifouling. Finally, the membrane of TiO_2-PDA/PSf hybrid showed excellent performances in separation and significant self-cleaning.

1.5.5.10 Antifouling membranes with silver nanoparticles

In an interesting work, a graphene oxide (GO)-silver-based metal organic framework (Ag-MOF) was incorporated in a selective layer of TFC, as investigated by Firouzjaei et al. [42]. The synthesized TFN membrane revealed superior antifouling and antibiofouling properties, higher hydrophilicity, more negative charge surface, and higher water permeability in comparison with the TFC membrane. Fluorescence imaging results displayed that the GO-Ag-MOF TFN membrane killed *Escherichia (E.) coli* more than the Ag-MOF TFN, GO TFN, and pristine TFC membranes. It was shown that the membrane of GO-Ag-MOF TFN had the lowest reduction of water flux through the other membranes, confirming the excellent properties of antifouling and antibiofouling of the membrane of GO-Ag-MOF TFN.

1.5.5.11 Carbon nanotubes-coated self-cleaning membranes

By spraying and crosslinking a selective ultra-thin layer of ultra-hydrophilic CNTs on a polyacrylonitrile (PAN) nanofiber matrix, Tian et al. [43] successfully created thin film nanocomposite (TFNC) membranes. By creating a superhydrophilic layer of ultra-thin CNTs, water is absorbed as a selective layer to remove oil droplets from surfactant-stabilized oil-in-water emulsions, which gives it an ultra-hydrophobic and underwater surface. The ultra-thin layer can maintain the support against contamination by oil droplets and surfactants.

1.6 Functionalization of nanoparticles in NEMs

1.6.1 Silica nanoparticles

Silica nanoparticles are embedded into the Nafion membrane matrix through the procedure of modification of nondestructive swelling-filling (SF) where silica is functionalized with O_2-including groups in the synthesis of sol-gel. In comparison with polymeric fillers, inorganic fillers are more chemically stable, and the possible damage of filler on oxidative consistency of the membrane of Nafion can be prevented. Functionalized silica (known as F-silica) may be firmly connected with polymer chains through hydrogen bonds between the oxygen functional groups on the surface of functionalized filler and —SO_3H groups of Nafion chains [44]. To prepare CH_4-selective MMMs for N_2 separation from natural gas, the copolymer of Pebax-1657 can be applied as a material of membrane, and silica nanoparticles (SNPs) can be used as the filler. Pebax-1657 is made of a soft polyethylene oxide segment and a hard polyamide segment. The soft-hard structure improves the adhesion between the filler nanoparticles and polymer chains. SNPs include hydroxyl groups on their surface and, accordingly, a powerful tendency to accumulate. Hence, the reduction of this functional group provides good dispersion of filler in the polymer matrix. Carboxylic groups can functionalize the surface of the fillers of SNPs. It prevents nano-fillers agglomeration in the polymer structure. Therefore, incorporating these particles enhances the performance of the ultimate membranes [45].

1.6.2 MOFs

Inorganic units, i.e., the clusters or metal ions and organic linkers, are conjugated to produce novel hybrid materials, known as metal organic frameworks or MOFs. MOFs are compounds with crystalline coordination. They have unique characteristics, high porosity, various pore sizes, large specific surface area, and multiple functional sites. Since the selective adsorption or molecular sieving of MOFs is significantly higher than the common particles, they have applications with great abilities [46]. Composite membranes contain polymers, and MOFs will overcome the trade-off between selectivity and permeability. MOFs are fillers with high potential over other porous materials due to excellent polymer chains/filler interfacial adhesion, owing to their partially organic nature [46]. The other benefit of MOFs is to compact multiple functional sites within a single structure. Varying the conditions of synthesis for initial reactants applied to produce the MOF will change the physical property and chemical structure of the MOFs that can be accounted as their benefits. With the presence of functional groups in a MOF structure will considerably affect its tendency toward adsorption of specific molecules from gas or liquid mixtures. Multifunctional sites (MFSs) are created through multiple-mono or multifunctional species of materials of MOF. Specified methods in the synthesis of MFS-MOFs are applying different functional sites. MOFs have an excellent ability for becoming functional [46]. Principal characteristics of MOFs, like their large surface areas and pore volumes, enable active guest species to be entered into the pores/channels and allow substrates to access the internal active sites. Thus, the combination of various kinds of active sites of MOF, functional organic linkers, like metal nodes, and guest species in the pores, establish MOFs as favorable multifunctional materials (Fig. 1.6) [47]. In addition, Fig. 1.7 shows a representation for a multifunctional MOF [48].

As an example, MOFs of UiO-66-type are widely investigated due to their significant resistance against chemical materials, stability against high temperatures, and good mechanical properties. By changing ligands with different functional groups, one can comfortably manipulate pore sizes and MOFs' chemical features of UiO-66-type. With the functionalization

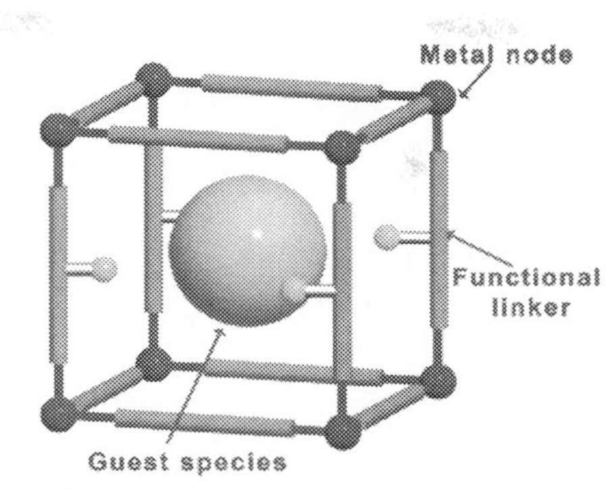

Fig. 1.6 Various kinds of active sites of MOF. *(From Y.-B. Huang, J. Liang, X.-S. Wang, R. Cao, Multifunctional metal–organic framework catalysts: synergistic catalysis and tandem reactions, Chem. Soc. Rev. 46(1) (2017) 126–157. https://doi.org/10.1039/C6CS00250A.)*

of UiO-66 with amine groups, UiO-66-NH_2 is extensively used in different applications [49]. For instance, Anjum et al. [50] observed that the gas permeability and selectivity were considerably increased by embedding UiO-66-NH_2 into Matrimid polyimide since the amine groups inside MOF pores simplified the transport of carbon dioxide. In addition, UiO-66-NH_2-based membranes illustrate a substantial ability for water treatment [49]. For example, Yin et al. [51] combined metal–organic frameworks (MOFs) functionalized with amine (UiO-66-NH_2) with ceramic membrane ultrafiltration in the removal of Pb (II) from wastewater. It was found that the MOFs could be completely retained by the membrane. At trans-membrane pressure (TMP) = 0.15 MPa,

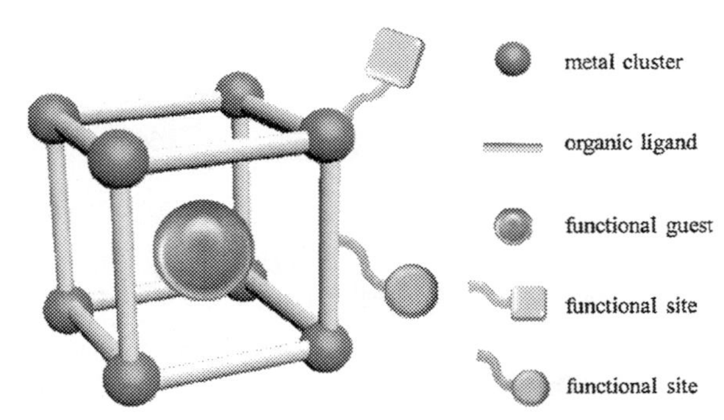

Fig. 1.7 A representation of a multifunctional MOF. *(From D. Ma, B. Li, Z. Shi, Multifunctional sites catalysts based on post-synthetic modification of metal-organic frameworks, Chin. Chem. Lett. 29(6) (2018) 827–830. https://doi.org/10.1016/j.cclet.2017.09.028.)*

cross-flow velocity (CFV) $= 4.0\,\mathrm{m\,s^{-1}}$ and temperature (T) $= 35°\mathrm{C}$, the highest removal (61.4%) of Pb (II) and lowest degree of membrane fouling were was obtained.

1.6.3 Fe_2O_3 nanoparticles

Hebbar et al. [52] proposed a novel approach applying excellent dispersion of nanoparticles of functionalized Fe_2O_3 to prepare a polyetherimide nanocomposite hollow fiber membrane with improved surface and properties of antibiofouling. The nanocomposite membrane showed a higher flux than the bare membrane during the filtration of natural organic materials with high risk with changing parameters like the pH of feed solution. The presence of nanoparticles of modified Fe_2O_3 (carboxylated Fe_2O_3 nanoparticles) in the membrane remarkably prevented the bacteria growth on the surface of membrane, leading to an improved antibiofouling property. The applied approach was fast and easy for Fe_2O_3 functionalization to improve the membrane properties and antibiofouling activity, which provided a considerable water treatment efficiently and stability. Mansourpanah et al. [53] studied variations in performance and structure of thin layer membranes of poly(piperazine amide) incorporated with nanoparticles of Fe_2O_3/SiO_2 to produce thin films that were modified. The incorporation of thin films with Fe_2O_3/SiO_2 offered excellent performance and properties of antifouling in comparison with unmodified thin layers. The results of AFM and SEM showed that the structure and morphology of the surface of thin layers changed in the presence of the nanoparticles. Furthermore, the ability of rejection of the thin films toward NaCl, Na_2SO_4, $PbNO_3$, and $CuSO_4$ illustrated desirable variations, leading to the excellent performance of the Fe_2O_3/SiO_2-modified poly(piperazine amide) thin-film composite membranes.

1.6.4 Nanodiamonds (NDs) nanoparticles

To functionalize the surface of nanodiamonds (NDs), hydrophilic cationic copolymers have been used to develop a highly selectively membrane-active nano-antibacterial agent [54]. Copolymers of poly(4-vinylpyridine-*co*-2-hydroxyethyl methacrylate) (P(VP-coHEMA)) were synthesized by copolymerizing HEMA (2-hydroxyethyl methacrylate) with 4-vinylpyridine (4-VP). Then, the quaternized P(VP-co-HEMA) functionalized NDs (QNDs) were obtained by conjugation with the silane-modified NDs (SiNDs) and quaternized with alkyl bromides. Finally, after the functionalization, the interaction with the solvent molecular network and the increased repulsive forces in the NDs result in the strongly aggregated pristine NDs breaking down into small nanoparticles with a particle size of 10–100 nm. In another example, Javadi et al. [55] studied the effect of nanoparticles of pure and functionalized nanodiamond (ND) on the performance and structure of membranes of polyvinylidene fluoride (PVDF). Firstly, ND was oxidized (O-ND) with heat, then silanized (S-ND), applying the reaction of esterification by vinyltrimethoxysilane, and eventually was functionalized by polyvinylpyrrolidone (PVP) (P-ND) through the procedure of surface-initiated free-radical polymerization. FTIR results showed that PVP was immobilized onto nanoparticles of ND. Various percentages of nanoparticles of ND and P-ND were then decorated into the PVDF membranes. It was illustrated that the water flux and hydrophilicity of membranes bettered with increasing the percentage of nanoparticles. Moreover, the presence of nanoparticles of ND and P-ND in the PVDF membranes resulted in membranes with good resistance against fouling.

1.6.5 Carbon nanotubes (CNTs)

Carbon nanotubes (CNTs) are sheets of graphitic carbon material twisted into an empty and seamless tube cylinder. The diameter of this cylinder is usually less than 10 nm, and its length is about a few millimeters. CNTs are in two types. The first one is single-walled carbon nanotubes (SWCNTs), and the second one is multiwalled carbon nanotubes (MWCNTs), which are made by a series of graphitic layers that are twisted around an axis. MWCNTs are composed of a tube restricted by graphitic layers with a distance from each other of 0.34 nm. CNTs have a smooth interior surface, a considerable ratio of length to diameter, and therefore precise diameter, and supreme stability against high temperatures. Consequently, they are very appropriate for advancements of MMMs [46]. Major usages of MMMs pertain to the CNTs dispersion in a polymer matrix effectively and appropriately. Decorating the CNTs vertically into the membrane matrix results in fast diffusion of low molecular weight molecules, like the permanent gases, via nanotubes' channels, leading to high (gas) permeability. For achieving the mentioned purposes, noncovalent functionalization of CNTs by attachment of molecules in noncovalent form, like beta-cyclodextrins (β-CD) and gamma-cyclodextrins (γ-CD), is a very effective method that significantly helps to improve the CNT distribution in MMMs [46]. Lee et al. [56] fabricated a novel MMM by embedding the functionalized multiwalled carbon nanotubes (MWCNTs-F) into the cellulose acetate butyrate (CAB) matrix to investigate the separation of CO_2/N_2. Fig. 1.8 illustrates the possible interactions of MWCNTs with β-CD and CAB, respectively. According to Fig. 1.8A, MWCNT with C=O and —OH groups was functionalized with β-CD and contained —OH and C—H groups. In this approach, the C—O—C group on the surface of MWCNTs-F is formed via a reaction of oxidation-reduction. In addition, due to the presence of ethanol, the C—H group forms. Furthermore, the C—O—H stretching vibration of α-pyranose could also be observed on the surface of MWCNTs-F. The synergistic effect of functional groups of CAB such as C—H, C=O, C—O, C—O—C, and —OH with the similar functional groups in MWCNTs-F results in a significant interaction with CO_2. Therefore, this MMM provides high performance for CO_2/N_2 separation.

Commonly, CNTs are vertically aligned in a polymer matrix. For a CNT membrane with vertical alignment, the CNT is arranged vertically on the support layer (Fig. 1.9A). In CNT-based MMMs, CNT is dispersed via the casting solution of the membrane to form a flat sheet membrane (Fig. 1.9B). For this case, the membrane does not have a regular alignment of filler. MMMs, including CNTs, can be prepared via both phase inversion and solution casting procedures. Compatibility between a polymer matrix and CNT is a significant factor for acquiring membranes with high mechanical consistency and well performance [57]. The functionalized CNT can be incorporated on the membrane surface [57]. For instance, Zeng et al. [58] attached the MWCNT-APTS (3-aminopropyltriethoxysilane) on the surface of a polyvinylidene fluoride (PVDF) membrane. For this reason, first, the functionalized CNTs were dispersed in deionized water to form suspensions containing 0.01–0.1 wt% of filler. Afterward, the suspensions were moved into a filtration cell under N_2 (0.2 MPa pressure) to permeate through the support membrane (average pore size 0.2 μm). Fabricated membranes revealed that the functionalized CNT was well embedded on the membrane surface (Fig. 1.9C). The membrane, including CNT, had an ability to reject of Cu^{2+} in the solution of feed. Wu et al. [59] prepared NF/MWCNT membranes with CNT deposition on support of a microfiltration (MF) membrane that was applied as an intermediate layer between the polyamide skin layer and support (Fig. 1.9D). The CNT was placed on the surface of the MF membrane by using vacuum filtration of a suspension of CNT before the interfacial polymerization. It was specified that the active layer thickness was enhanced with an increment of the CNT layer due to the monomer absorption on the coated membrane.

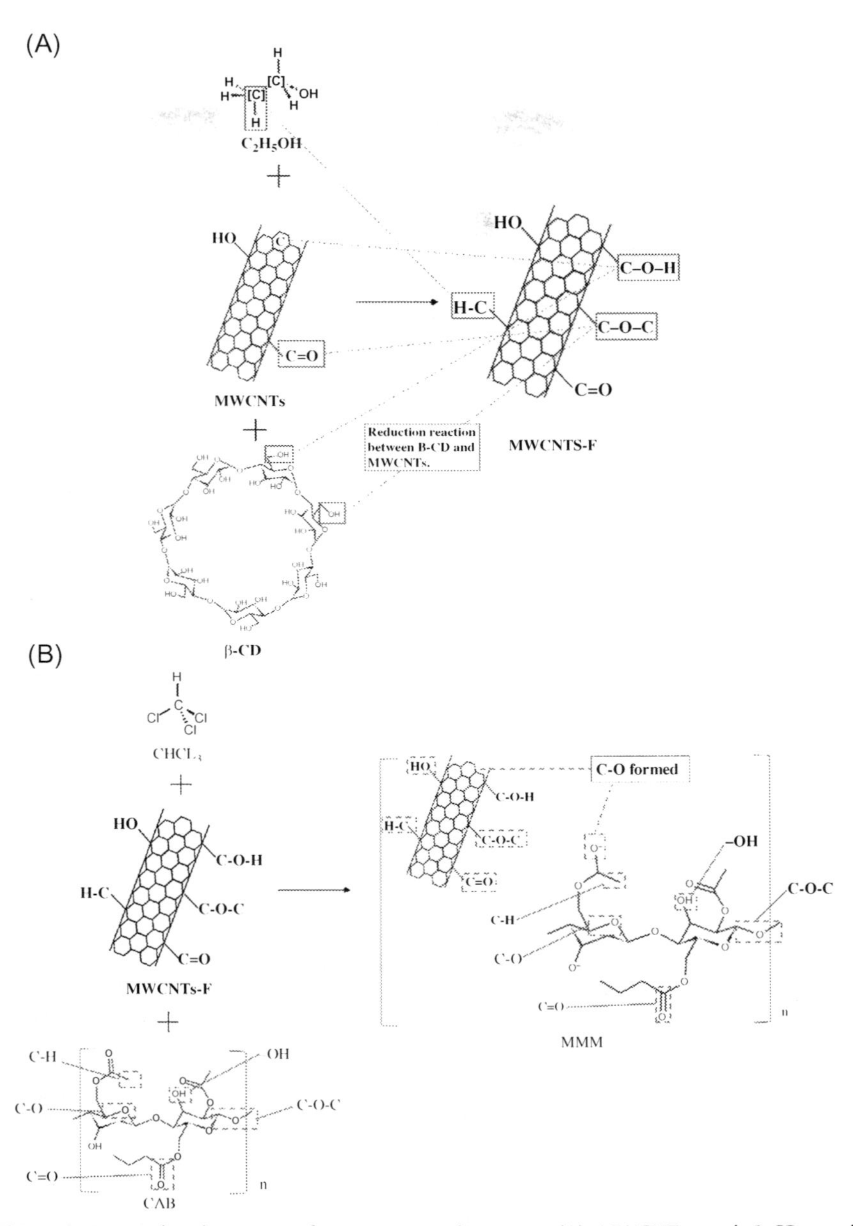

Fig. 1.8 A graphical image of interactions between (A) MWCNTs and β-CD, and (B) MWCNTs-F and CAB. *(From R.J. Lee, Z.A. Jawad, A.L. Ahmad, H.B. Chua, Incorporation of functionalized multi-walled carbon nanotubes (MWCNTs) into cellulose acetate butyrate (CAB) polymeric matrix to improve the CO2/N2 separation, Process. Saf. Environ. Prot. 117 (2018) 159–167. https://doi.org/10.1016/j.psep.2018.04.021.)*

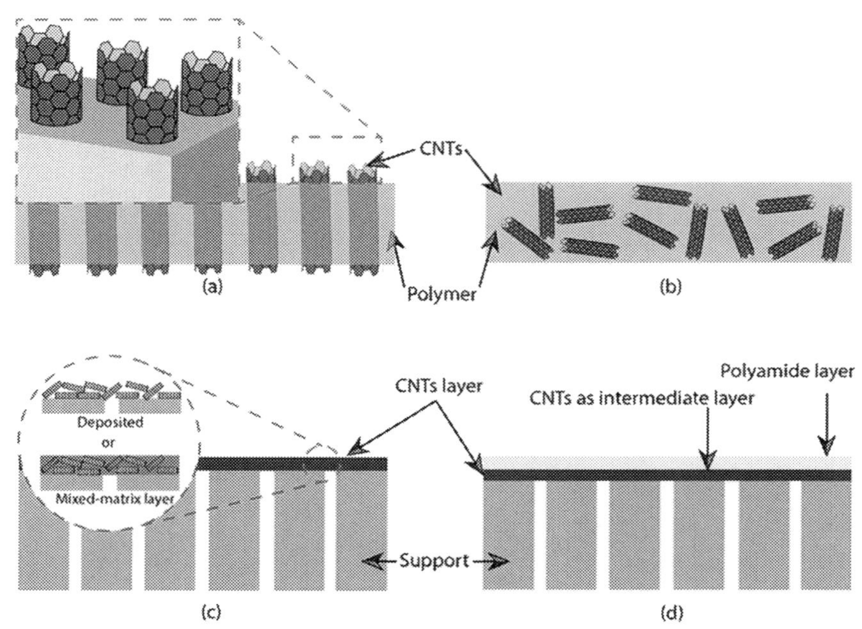

Fig. 1.9 CNT membranes with different structures. (A) CNT membrane with vertical alignment, (B) CNT membrane in the form of mixed-matrix, (C) coating of CNTs on the surface or support of the membrane, and (D) coating of CNTs on the surface (support) of membrane as intermediate layer. *(From M. Sianipar, S.H. Kim, Khoiruddin, F. Iskandar, I.G. Wenten, Functionalized carbon nanotube (CNT) membrane: progress and challenges, RSC Adv. 7(81) (2017) 51175–51198. https://doi.org/10.1039/C7RA08570B.)*

1.6.6 Graphene

Graphene is a single layer of sp^2 carbon atoms which are connected safely into the structure with a hexagonal network and provide a two-dimensional nanomaterial. Graphene is the basic structural unit for all forms of graphitic carbon, such as carbon nanotubes, graphite, and fullerenes. Monolayer graphene reveals excellent mechanical properties, but it is impermeable against gas. However, continuous films made from several to hundreds of graphene and GO flakes stuck on top of each other provide selective properties for different separations due to the space between the graphene sheets and/or holes produced in the oxidation step [46]. With good compatibility between polymer matrix and filler, permeability may be decreased. On the other hand, weak compatibility between the polymer and graphene may cause the chains of polymer to not stick firmly to the graphene. Hence, gas molecules move among the narrow gap or via the matrix of polymer instead of passing through the graphene. The graphene surface modification and functionalization might enhance the graphene/polymer matrix interfacial compatibility [46]. Membrane-based postcombustion carbon capture can reduce some of the problems of absorption separation. For this reason, membranes with excellent performance are required with a carbon dioxide permeance greater than 1000 GPU (gas permeation unit) and a separation factor greater than 20 for mixture of CO_2/N_2. He et al. fabricated a new type of graphene-based membrane in which an impermeable graphene lattice was exposed to oxygen plasma, creating a very good porosity of 18.5% and applied with a CO_2-philic polymer [60]. The preparation and structure of SPONG membranes are depicted in Fig. 1.10. Firstly, by exposing single-layer graphene to oxygen plasma and ozone, oxygen-functionalized nanoporous graphene (ONG) was attained. Afterward, to achieve a polymer functionalized ONG (PONG), CO_2-philic polymeric chains were grafted on the lattice of porous graphene. Finally, for preparing a film with a thickness less than 20 nm, a poly(ethylene glycol)-dimethyl-ether (PEGDE) oligomer was applied for the swelling of the CO_2-philic polymer layer in PONG, leading to a swollen PONG (SPONG) film. This type of membrane had excellent permeance of 6180 GPU with a CO_2/N_2 selectivity of 22.5. The results also showed that with the use of optimized graphene porosity, pore size, and functional groups, CO permeance increased to 11,790 GPU [60].

1.7 Sustainability of membranes

Cheap energy requirements for membrane processes make them good alternatives for separation processes with high intensity. They are strongly related

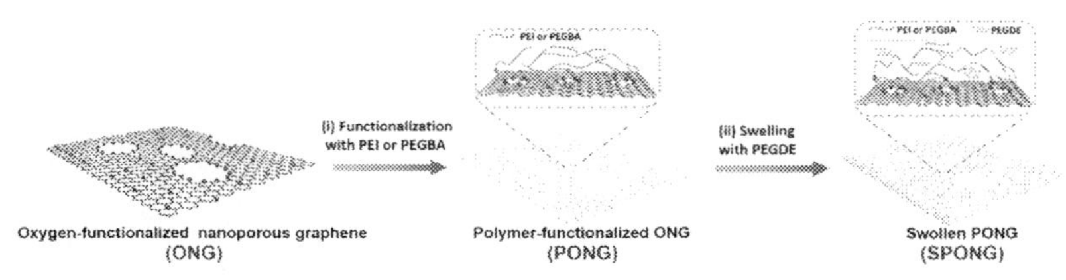

Fig. 1.10 Schematics of the fabrication and the structure of the membrane of SPONG. *(From G. He, S. Huang, L.F. Villalobos, J. Zhao, M. Mensi, E. Oveisi, M. Rezaei, K.V. Agrawal, High-permeance polymer-functionalized single-layer graphene membranes that surpass the postcombustion carbon capture target, Energy Environ. Sci. 12(11) (2019) 3305–3312. https://doi.org/10.1039/C9EE01238A.)*

in every strategy for sustainable development. For instance, for commercial/industrial implementation of membrane technology for gas separation, membranes must have sustainable performance in a long-term operation. Moreover, the effect of penetrants on gas separation performance through chemical and physical aging must be considered [61]. Membrane technology that is an energy-efficient and sustainable process must control its environmental effect as required by decreasing application of high-risk solvents and using "green" solvents and nano-sized materials. More comprehensive approaches like life-cycle assessment (LCA) are needed, where the entire process is assessed according to environmental effect [61]. LCA is a specific procedure to characterize the environmental effects on the life cycle of services and products. The LCA contains four stages:

(i) definitions of purpose, functional unit, scope, and boundaries of the system;
(ii) inventory life cycle (LCI);
(iii) life cycle impact assessment (LCIA); and
(iv) results' interpretation [62].

The methodology of LCA is now the most enough frameworks to evaluate the sustainability of a process or product. Environmental indicators are defined in an LCA investigation may be applied as indicators of sustainability. In addition, the process of inventory analysis and the procedures of effect assessment are considered. Thus, LCA is used to evaluate the sustainability of membrane technology, to recognize hotspots and improve their performance of sustainability [63]. LCA and its extensions are a fundamental section of frameworks according to life cycle thinking (LCT). This approach endeavors to decrease the environmental, social, and economic effects of human activities by taking into account all the steps of the associated life cycle. Developed strategies are near the optimum conditions and decrease the total effects of generation, applying, and disposing of a process, service, or product [63]. LCT evaluates the whole supply chain for a product, either downstream or upstream, and the economic, environmental, and social effects. Both quantitative and qualitative approaches can be applied. LCT can help recognize opportunities for betterment and decision-making in all sustainability dimensions [63]. As one example, Szekely et al. [64] assessed the sustainability of organic solvent nanofiltration according to an LCT perspective. They analyzed all stages of the membrane process, beginning with the fabrication of the membranes, operation of process, and options of end-of-life for membranes and other related units. Consumption of energy, carbon effect, and parameters of operation were the principal indicators applied in the assessment. The different options and process features in every step of the life cycle were compared with each other according to an extensive analysis of the literature to specify which ones were better and which conditions of operation were favorable.

1.8 Future perspective

Membrane technology is promising due to its credibility, facility of application, easy process scale-up, flexibility, and compatibility. To certify sustainable advancement of membrane technology, different membrane processes have been evaluated with various considerations such as economic, environmental, and technological effects. Procedures associated with process escalation and suitable material selection are essential for sustainable advancement. Research and development (R&D) approaches in membranes are growing. The utilization of nanomaterials as additives in manufacturing membranes results in excellent benefits in better efficiency. Many developments have been verified and validated in innovational and sustainable membrane technologies that have a possibility for decreasing the energy utilization of membrane processes by implementing the innovational nanostructured membranes. To certify sustainable advancement, a significant situation is developing of bio and green synthesis of nanomaterials to decrease the noxious by-products to the environment. The unique features of nanomaterials and final nanocomposite membranes have permitted significant progress in membrane efficiencies [65]. Depending on the types of the inserted nanomaterials in membranes, different characteristics such as antifouling features, antibacterial resistance, and absorbent capacities are illustrated. Many problems such as the long-term consistency of the nanomaterials inserted in the membrane matrices, production costs, and their mechanical characteristics should be solved before the membranes can be commercialized [66]. Numerous studies clarify that membrane structures can be adequately organized on a nanoscale to improve the macro-scale properties. Therefore, many interests are oriented toward the development of methods to manipulate and tune nanostructured membranes and visualize them [11].

Membranes with functionalized nanomaterials (FNMs) can be applied in pharmaceutical, biomedical, food, chromatography, biotechnology, etc. [28]. In terms of enhancing the synthetic membranes' efficiencies and realize their excellent potential in special areas like bioengineering, one of the primary problems is selectivity. The prevalent tendency is to advance membranes with nano-accuracy pore size. In addition, permeability must be acceptable for the actual application; however, selectivity appears to be the real bottleneck. Nanostructures with antifouling features have also received much attention [11]. Nanocapsule membranes are capable of releasing payloads wherever they are needed, thus reducing the number of consumed chemicals and making the treatment more efficient, which is promising for the controlled drug delivery. Nano-patterned biocompatible, bioactive, and biodegradable membranes in tissue engineering are capable of increasing cell adhesion, direct cell growth, and increase the transfer of nutrients to and catabolites from the cell media. Autologous structures for personalized medicine are also growing in the literature [11]. The rigidity of polymer is a fundamental factor to improve gas transfer efficiencies of the membrane. It can be further intensified with inherent microporosity and significant free volume of the membranes. Forthcoming advancements of nanoporous membranes for gas separation will need increased sorption of objective gas. Adjusting the sizes of the cavity is quite necessary for the objective gas to enhance the membrane performance [67].

Nanomaterials have occupied an extensive area of membrane development. Their scope is from molecular biology and tissue engineering to photonics, gas separation, and water treatment with a significant objective to attain superior efficiency and selectivity. Important features must be considered, such as production costs, operating time, ease of operation, and scaling up, f or nanomaterials to be commercialized [68]. Nanostructured membranes are predicted to be implemented in textiles markets, as there is a fascinating opportunity to make nano-assembled membranes for intelligent textiles. In this regard, adaptability and auto-adjustment are two imperatives for achieving criteria such as self-maintenance and long-distance interaction with the body and/or the surrounding environment. However, some worries remain, such as long-term constancy, persistence, and recyclability. These objectives are attained via the making of "an industrial research path" to produce novel science and fill the gap between investigations and manufacture [69].

1.9 Conclusions

In this chapter, we have discussed and summarized the knowledge and significance of functionalized nanomaterial (FNMs) in membranes, their effectiveness, and their advantages. The scale-up of membranes with FNMs at the pilot and industrial scales is an essential issue in developing stable membranes. Membranes with FNMs have a supreme ability in separation and purification processes by removing particles, microorganisms, organic compounds, and other new components. Choosing an appropriate membrane and its optimization with FNMs is vital since the FNMs directly result in lowering the costs of the separation process. More novel procedures for modifying of membranes with FNMs must be created and applied to improve separation efficiency and sustainability, especially in very long-term operations. The choice of a suitable procedure depends on the adaptability of materials of the membrane with FNMs, the persistence of materials of membrane for modifications, the aim of separation, economic issues, investment costs, kinds of operation, and the feasibility for industrial applications. It is predicted that the development of FNM-based membranes will continue in various separation and purification operations. Continuous efforts are required to solve challenges through a great deal of attention to research and development.

References

[1] M.M. Moftakhari Sharifzadeh, M. Zamani Pedram, A. Ebadi Amooghin, A new permeation model in porous filler based mixed matrix membranes for CO2 separation, Greenhouse Gases Sci. Technol. 9 (4) (2019) 719–742, https://doi.org/10.1002/ghg.1891.

[2] W. Hirunpinyopas, E. Prestat, S.D. Worrall, S.J. Haigh, R.A.W. Dryfe, M.A. Bissett, Desalination and nanofiltration through functionalized laminar MoS2 membranes, ACS Nano 11 (11) (2017) 11082–11090, https://doi.org/10.1021/acsnano.7b05124.

[3] A. Ebadi Amooghin, S. Mirrezaei, H. Sanaeepur, M.M. Moftakhari Sharifzadeh, Gas permeation modeling through a multilayer hollow Fiber composite membrane, J. Membr. Sci. Res. 6 (1) (2020) 125–134, https://doi.org/10.22079/jmsr.2019.112328.1281.

[4] H. Sanaeepur, R. Ahmadi, A. Ebadi Amooghin, D. Ghanbari, A novel ternary mixed matrix membrane containing glycerol-modified poly(ether-block-amide) (Pebax 1657)/copper nanoparticles for CO2 separation, J. Membr. Sci. 573 (2019) 234–246, https://doi.org/10.1016/j.memsci.2018.12.012.

[5] R.R. Gonzales, Y. Yang, M.J. Park, T.-H. Bae, A. Abdel-Wahab, S. Phuntsho, H.K. Shon, Enhanced water permeability and osmotic power generation with sulfonate-functionalized porous polymer-incorporated thin film nanocomposite membranes, Desalination 496 (2020) 114756, https://doi.org/10.1016/j.desal.2020.114756.

[6] K. Kalash, M. Kadhom, M. Al-Furaiji, Thin film nanocomposite membranes filled with MCM-41 and SBA-15 nanoparticles for brackish water desalination via reverse osmosis, Environ. Technol. Innov. 20 (2020) 101101, https://doi.org/10.1016/j.eti.2020.101101.

[7] L. Hao, Z. Chi, Q. Chen, H. Zhang, J. Wang, Constructing large loadings of dual pathways with Ti3C2Tx-CDs in thin film nanocomposite membrane for enhanced organic permeation, J. Membr. Sci. 620 (2021) 118872, https://doi.org/10.1016/j.memsci.2020.118872.

[8] M.R.A. Hamid, H.-K. Jeong, Recent advances on mixed-matrix membranes for gas separation: opportunities and engineering challenges, Korean J. Chem. Eng. 35 (8) (2018) 1577–1600, https://doi.org/10.1007/s11814-018-0081-1.
[9] S. Mashhadikhan, A. Moghadassi, A. Ebadi Amooghin, H. Sanaeepur, Interlocking a synthesized polymer and bifunctional filler containing the same polymer's monomer for conformable hybrid membrane systems, J. Mater. Chem. A 8 (7) (2020) 3942–3955, https://doi.org/10.1039/C9TA13375E.
[10] A. Ebadi Amooghin, H. Sanaeepur, M. Omidkhah, A. Kargari, "Ship-in-a-bottle", a new synthesis strategy for preparing novel hybrid host–guest nanocomposites for highly selective membrane gas separation, J. Mater. Chem. A 6 (4) (2018) 1751–1771, https://doi.org/10.1039/C7TA08081F.
[11] E. Drioli, L. Giorno, A. Gugliuzza, An introduction to nanostructured membranes for advanced applications in strategic fields, in: E. Drioli, L. Giorno, A. Gugliuzza (Eds.), Functional Nanostructured Membranes, first ed., Jenny Stanford Publishing, 2019, p. 620. https://doi.org/10.1201/9781351135115.
[12] J.M. Gohil, R.R. Choudhury, S. Thomas, D. Pasquini, S.-Y. Leu, D.A. Gopakumar, Introduction to nanostructured and nano-enhanced polymeric membranes: preparation, function, and application for water purification, in: Micro and Nano Technologies, Elsevier, 2019, pp. 25–57, https://doi.org/10.1016/B978-0-12-813926-4.00038-0 (Chapter 2).
[13] M. Bassyouni, M.H. Abdel-Aziz, M.S. Zoromba, S.M.S. Abdel-Hamid, E. Drioli, A review of polymeric nanocomposite membranes for water purification, J. Ind. Eng. Chem. 73 (2019) 19–46, https://doi.org/10.1016/j.jiec.2019.01.045.
[14] S. Jeon, J.-H. Lee, Rationally designed in-situ fabrication of thin film nanocomposite membranes with enhanced desalination and anti-biofouling performance, J. Membr. Sci. 615 (2020) 118542, https://doi.org/10.1016/j.memsci.2020.118542.
[15] S. Sarkar, S. Chakraborty, Nanocomposite polymeric membrane a new trend of water and wastewater treatment: a short review, Groundw. Sustain. Dev. 12 (2021) 100533, https://doi.org/10.1016/j.gsd.2020.100533.
[16] S. Wei, Y. Chen, X. Hu, C. Wang, X. Huang, D. Liu, Y. Zhang, Monovalent/divalent salts separation via thin film nanocomposite nanofiltration membrane containing aminated TiO2 nanoparticles, J. Taiwan Inst. Chem. Eng. 112 (2020) 169–179, https://doi.org/10.1016/j.jtice.2020.06.014.
[17] K. Maiphetlho, N. Shumbula, N. Motsoane, L. Chimuka, H. Richards, Evaluation of silver nanocomposite polymer inclusion membranes (PIMs) for trace metal transports: selectivity and stability studies, J. Water Process Eng. 37 (2020) 101527, https://doi.org/10.1016/j.jwpe.2020.101527.
[18] M. Shakak, R. Rezaee, A. Maleki, A. Jafari, M. Safari, B. Shahmoradi, H. Daraei, S.-M. Lee, Synthesis and characterization of nanocomposite ultrafiltration membrane (PSF/PVP/SiO2) and performance evaluation for the removal of amoxicillin from aqueous solutions, Environ. Technol. Innov. 17 (2020) 100529, https://doi.org/10.1016/j.eti.2019.100529.
[19] N. Ghaemi, S.S. Madaeni, P. Daraei, H. Rajabi, S. Zinadini, A. Alizadeh, R. Heydari, M. Beygzadeh, S. Ghouzivand, Polyethersulfone membrane enhanced with iron oxide nanoparticles for copper removal from water: application of new functionalized Fe3O4 nanoparticles, Chem. Eng. J. 263 (2015) 101–112, https://doi.org/10.1016/j.cej.2014.10.103.
[20] S.R. Lakhotia, M. Mukhopadhyay, P. Kumari, Iron oxide (FeO) nanoparticles embedded thin-film nanocomposite nanofiltration (NF) membrane for water treatment, Sep. Purif. Technol. 211 (2019) 98–107, https://doi.org/10.1016/j.seppur.2018.09.034.
[21] A. Noormohamadi, M. Homayoonfal, M.R. Mehrnia, F. Davar, Synergistic effect of concurrent presence of zirconium oxide and iron oxide in the form of core-shell nanoparticles on the performance of Fe3O4@ZrO2/PAN nanocomposite membrane,

Ceram. Int. 43 (18) (2017) 17174–17185, https://doi.org/10.1016/j.ceramint.2017.09.142.
[22] M. Ben-Sasson, K.R. Zodrow, Q. Genggeng, Y. Kang, E.P. Giannelis, M. Elimelech, Surface functionalization of thin-film composite membranes with copper nanoparticles for antimicrobial surface properties, Environ. Sci. Technol. 48 (1) (2014) 384–393, https://doi.org/10.1021/es404232s.
[23] F. Asempour, S. Akbari, D. Bai, D. Emadzadeh, T. Matsuura, B. Kruczek, Improvement of stability and performance of functionalized halloysite nano tubes-based thin film nanocomposite membranes, J. Membr. Sci. 563 (2018) 470–480, https://doi.org/10.1016/j.memsci.2018.05.070.
[24] N.P. Khumalo, G.D. Vilakati, S.D. Mhlanga, A.T. Kuvarega, B.B. Mamba, J. Li, D.S. Dlamini, Dual-functional ultrafiltration nano-enabled PSf/PVA membrane for the removal of Congo red dye, J. Water Process Eng. 31 (2019) 100878, https://doi.org/10.1016/j.jwpe.2019.100878.
[25] P.S. Goh, A.F. Ismail, Structure and gas transport of nanocomposite membranes, Nanocomposite Membranes for Gas Separation, Elsevier, 2020, pp. 101–123 (Chapter 3).
[26] A.D. Kiadehi, M. Jahanshahi, A. Rahimpour, S.A.A. Ghoreyshi, The effect of functionalized carbon nano-fiber (CNF) on gas separation performance of polysulfone (PSf) membranes, Chem. Eng. Process. Process Intensif. 90 (2015) 41–48, https://doi.org/10.1016/j.cep.2015.02.005.
[27] G. Guerrero, D. Venturi, T. Peters, N. Rival, C. Denonville, C. Simon, P.P. Henriksen, M.-B. Hägg, Influence of functionalized nanoparticles on the CO2/N2 separation properties of PVA-based gas separation membranes, in: 13th International Conference on Greenhouse Gas Control Technologies, GHGT-13, 14–18 November 2016, Lausanne, Switzerland, 114, 2017, pp. 627–635, https://doi.org/10.1016/j.egypro.2017.03.1205.
[28] G. Vitola, R. Mazzei, L. Giorno, Nanoengineered membranes in biotechnology, in: E. Drioli, L. Giorno, A. Gugliuzza (Eds.), Functional Nanostructured Membranes, first ed., Pan Stanford Publishing Pte. Ltd., 2018, pp. 473–523. https://www.taylorfrancis.com/chapters/mono/10.1201/9781351135115-19/nanoengineered-membranes-biotechnology-giuseppe-vitola-rosalinda-mazzei-lidietta-giorno-enrico-drioli-lidietta-giorno-annarosa-gugliuzza.
[29] C. Cheng, S. Sun, C. Zhao, Progress in heparin and heparin-like/mimicking polymer-functionalized biomedical membranes, J. Mater. Chem. B 2 (44) (2014) 7649–7672, https://doi.org/10.1039/C4TB01390E.
[30] M. Ioniță, L.E. Crică, S.I. Voicu, S. Dinescu, F. Miculescu, M. Costache, H. Iovu, Synergistic effect of carbon nanotubes and graphene for high performance cellulose acetate membranes in biomedical applications, Carbohydr. Polym. 183 (2018) 50–61, https://doi.org/10.1016/j.carbpol.2017.10.095.
[31] A. Brandelli, L.F.W. Brum, J.H.Z. dos Santos, Nanostructured bioactive compounds for ecological food packaging, Environ. Chem. Lett. 15 (2) (2017) 193–204, https://doi.org/10.1007/s10311-017-0621-7.
[32] H.M.C. Azeredo, M.F. Rosa, L.H.C. Mattoso, Nanocellulose in bio-based food packaging applications, Ind. Crops Prod. 97 (2017) 664–671, https://doi.org/10.1016/j.indcrop.2016.03.013.
[33] A. Rafique, K. Mahmood Zia, M. Zuber, S. Tabasum, S. Rehman, Chitosan functionalized poly(vinyl alcohol) for prospects biomedical and industrial applications: a review, Int. J. Biol. Macromol. 87 (2016) 141–154, https://doi.org/10.1016/j.ijbiomac.2016.02.035.
[34] F. Militano, T. Poerio, R. Mazzei, L. Giorno, Nano structured membranes for biosensor development, in: E. Drioli, L. Giorno, A. Gugliuzza (Eds.), Functional Nanostructured Membranes, first ed., Pan Stanford Publishing Pte. Ltd., 2018. https://www.taylorfrancis.com/chapters/mono/10.1201/9781351135115-18/nanostructured-

membranes-biosensor-development-francesca-militano-teresa-poerio-rosalinda-mazzei-lidietta-giorno-enrico-drioli-lidietta-giorno-annarosa-gugliuzza.
[35] D. Chen, C. Wang, W. Chen, Y. Chen, J.X.J. Zhang, PVDF-Nafion nanomembranes coated microneedles for in vivo transcutaneous implantable glucose sensing, Biosens. Bioelectron. 74 (2015) 1047–1052, https://doi.org/10.1016/j.bios.2015.07.036.
[36] A.Y. El-Moghazy, E.A. Soliman, H.Z. Ibrahim, J.-L. Marty, G. Istamboulie, T.Noguer, Biosensor based on electrospun blended chitosan-poly (vinyl alcohol) nanofibrous enzymatically sensitized membranes for pirimiphos-methyl detection in olive oil, Talanta 155 (2016) 258–264, https://doi.org/10.1016/j.talanta.2016.04.018.
[37] S. Morelli, S. Salerno, A. Piscioneri, L. De Bartolo, in: E. Drioli, L. Giorno, A. Gugliuzza (Eds.), Nanostructured Membranes for Engineering Organs and Tissues, first ed., Pan Stanford Publishing Pte. Ltd., 2018. https://www.taylorfrancis.com/chapters/mono/10.1201/9781351135115-21/nanostructured-membranes-engineering-organs-tissues-sabrina-morelli-simona-salerno-antonella-piscioneri-loredana-de-bartolo-enrico-drioli-lidietta-giorno-annarosa-gugliuzza?context=ubx&refId=7fe8eafe-d975-41f1-a52b-18a7ff651c81.
[38] D. Mata, M. Amaral, A.J.S. Fernandes, B. Colaço, A. Gama, M.C. Paiva, P.S. Gomes, R.F. Silva, M.H. Fernandes, Diels–Alder functionalized carbon nanotubes for bone tissue engineering: in vitro/in vivo biocompatibility and biodegradability, Nanoscale 7 (20) (2015) 9238–9251, https://doi.org/10.1039/C5NR01829C.
[39] S. Pal, R. Nisi, M. Stoppa, A. Licciulli, Silver-functionalized bacterial cellulose as antibacterial membrane for wound-healing applications, ACS Omega 2 (7) (2017) 3632–3639, https://doi.org/10.1021/acsomega.7b00442.
[40] A.Y. Gebreyohannes, A. Gugliuzza, L. Giorno, in: E. Drioli, L. Giorno, A. Gugliuzza (Eds.), Nanostructured Functional Membranes for Self-Cleaning Separations, first ed., Pan Stanford Publishing Pte. Ltd., 2018. https://www.taylorfrancis.com/chapters/mono/10.1201/9781351135115-14/nanostructured-functional-membranes-self-cleaning-separations-abaynesh-yihdego-gebreyohannes-annarosa-gugliuzza-lidietta-giorno-enrico-drioli-lidietta-giorno-annarosa-gugliuzza?context=ubx&refId=cd360bb3-eef3-4c3a-a39f-a75a667b3591.
[41] H. Wu, Y. Liu, L. Mao, C. Jiang, J. Ang, X. Lu, Doping polysulfone ultrafiltration membrane with TiO2-PDA nanohybrid for simultaneous self-cleaning and self-protection, J. Membr. Sci. 532 (2017) 20–29, https://doi.org/10.1016/j.memsci.2017.03.010.
[42] M.D. Firouzjaei, A.A. Shamsabadi, S.A. Aktij, S.F. Seyedpour, M. Sharifian Gh., A. Rahimpour, M.R. Esfahani, M. Ulbricht, M. Soroush, Exploiting synergetic effects of graphene oxide and a silver-based metal–organic framework to enhance antifouling and anti-biofouling properties of thin-film nanocomposite membranes, ACS Appl. Mater. Interfaces 10 (49) (2018) 42967–42978, https://doi.org/10.1021/acsami.8b12714.
[43] M. Tian, Y. Liao, R. Wang, Engineering a superwetting thin film nanofibrous composite membrane with excellent antifouling and self-cleaning properties to separate surfactant-stabilized oil-in-water emulsions, J. Membr. Sci. 596 (2020) 117721, https://doi.org/10.1016/j.memsci.2019.117721.
[44] J. Li, G. Xu, X. Luo, J. Xiong, Z. Liu, W. Cai, Effect of nano-size of functionalized silica on overall performance of swelling-filling modified Nafion membrane for direct methanol fuel cell application, Appl. Energy 213 (2018) 408–414, https://doi.org/10.1016/j.apenergy.2018.01.052.
[45] S. Mohammdhadi Mousavi, A. Raisi, H. Hashemi Moghaddam, M. Salehi Maleh, CH4-selective mixed-matrix membranes containing functionalized silica for natural gas purification, Chem. Eng. Technol. 43 (11) (2020) 2167–2180, https://doi.org/10.1002/ceat.202000105.

[46] A. Ebadi Amooghin, S. Mashhadikhan, H. Sanaeepur, A. Moghadassi, T. Matsuura, S. Ramakrishna, Substantial breakthroughs on function-led design of advanced materials used in mixed matrix membranes (MMMs): a new horizon for efficient CO2 separation, Prog. Mater. Sci. 102 (2019) 222–295, https://doi.org/10.1016/j.pmatsci.2018.11.002.
[47] Y.-B. Huang, J. Liang, X.-S. Wang, R. Cao, Multifunctional metal–organic framework catalysts: synergistic catalysis and tandem reactions, Chem. Soc. Rev. 46 (1) (2017) 126–157, https://doi.org/10.1039/C6CS00250A.
[48] D. Ma, B. Li, Z. Shi, Multi-functional sites catalysts based on post-synthetic modification of metal-organic frameworks, Chin. Chem. Lett. 29 (6) (2018) 827–830, https://doi.org/10.1016/j.cclet.2017.09.028.
[49] Y.M. Xu, S. Japip, T.-S. Chung, Mixed matrix membranes with nano-sized functional UiO-66-type MOFs embedded in 6FDA-HAB/DABA polyimide for dehydration of C1-C3 alcohols via pervaporation, J. Membr. Sci. 549 (2018) 217–226, https://doi.org/10.1016/j.memsci.2017.12.001.
[50] M.W. Anjum, F. Vermoortele, A.L. Khan, B. Bueken, D.E. De Vos, I.F.J. Vankelecom, Modulated UiO-66-based mixed-matrix membranes for CO2 separation, ACS Appl. Mater. Interfaces 7 (45) (2015) 25193–25201, https://doi.org/10.1021/acsami.5b08964.
[51] N. Yin, K. Wang, L. Wang, Z. Li, Amino-functionalized MOFs combining ceramic membrane ultrafiltration for Pb (II) removal, Chem. Eng. J. 306 (2016) 619–628, https://doi.org/10.1016/j.cej.2016.07.064.
[52] R.S. Hebbar, A.M. Isloor, K. Ananda, M.S. Abdullah, A.F. Ismail, Fabrication of a novel hollow fiber membrane decorated with functionalized Fe2O3 nanoparticles: towards sustainable water treatment and biofouling control, New J. Chem. 41 (10) (2017) 4197–4211, https://doi.org/10.1039/C7NJ00221A.
[53] Y. Mansourpanah, A. Rahimpour, M. Tabatabaei, L. Bennett, Self-antifouling properties of magnetic Fe2O3/SiO2-modified poly (piperazine amide) active layer for desalting of water: characterization and performance, Desalination 419 (2017) 79–87, https://doi.org/10.1016/j.desal.2017.06.006.
[54] W. Cao, X. Wang, Q. Li, X. Peng, L. Wang, P. Li, Z. Ye, X. Xing, Designing of membrane-active nano-antimicrobials based on cationic copolymer functionalized nanodiamond: influence of hydrophilic segment on antimicrobial activity and selectivity, Mater. Sci. Eng. C 92 (2018) 307–316, https://doi.org/10.1016/j.msec.2018.06.067.
[55] M. Javadi, Y. Jafarzadeh, R. Yegani, S. Kazemi, PVDF membranes embedded with PVP functionalized nanodiamond for pharmaceutical wastewater treatment, Chem. Eng. Res. Des. 140 (2018) 241–250, https://doi.org/10.1016/j.cherd.2018.10.029.
[56] R.J. Lee, Z.A. Jawad, A.L. Ahmad, H.B. Chua, Incorporation of functionalized multi-walled carbon nanotubes (MWCNTs) into cellulose acetate butyrate (CAB) polymeric matrix to improve the CO2/N2 separation, Process Saf. Environ. Prot. 117 (2018) 159–167, https://doi.org/10.1016/j.psep.2018.04.021.
[57] M. Sianipar, S.H. Kim, Khoiruddin, F. Iskandar, I.G. Wenten, Functionalized carbon nanotube (CNT) membrane: progress and challenges, RSC Adv. 7 (81) (2017) 51175–51198, https://doi.org/10.1039/C7RA08570B.
[58] G. Zeng, Y. He, Z. Yu, X. Yang, R. Yang, L. Zhang, Preparation of novel high copper ions removal membranes by embedding organosilane-functionalized multi-walled carbon nanotube, J. Chem. Technol. Biotechnol. 91 (8) (2016) 2322–2330, https://doi.org/10.1002/jctb.4820.
[59] M.B. Wu, Y. Lv, H.-C. Yang, L.F. Liu, X. Zhang, Z.K. Xu, Thin film composite membranes combining carbon nanotube intermediate layer and microfiltration support for high nanofiltration performances, J. Membr. Sci. 515 (2016) 238–244, https://doi.org/10.1016/j.memsci.2016.05.056.

[60] G. He, S. Huang, L.F. Villalobos, J. Zhao, M. Mensi, E. Oveisi, M. Rezaei, K.V. Agrawal, High-permeance polymer-functionalized single-layer graphene membranes that surpass the postcombustion carbon capture target, Energy Environ. Sci. 12 (11) (2019) 3305–3312, https://doi.org/10.1039/C9EE01238A.
[61] E. Lasseuguette, M.-C. Ferrari, G. Szekely, A. Livingston, Polymer membranes for sustainable gas separation, Elsevier, 2020, pp. 265–296, https://doi.org/10.1016/B978-0-12-814681-1.00010-2 (Chapter 10).
[62] K.C. Ho, Y.X. Teoh, Y.H. Teow, A.W. Mohammad, Life cycle assessment (LCA) of electrically-enhanced POME filtration: environmental impacts of conductive-membrane formulation and process operating parameters, J. Environ. Manag. 277 (2021) 111434, https://doi.org/10.1016/j.jenvman.2020.111434.
[63] A.A. Martins, N.S. Caetano, T.M. Mata, LCA for Membrane Processes, Springer, Singapore, 2017, pp. 23–66, https://doi.org/10.1007/978-981-10-5623-9_2.
[64] G. Szekely, M.F. Jimenez-Solomon, P. Marchetti, J.F. Kim, A.G. Livingston, Sustainability assessment of organic solvent nanofiltration: from fabrication to application, Green Chem. 16 (10) (2014) 4440–4473, https://doi.org/10.1039/C4GC00701H.
[65] P.S. Goh, T. Whye Wong, J. Wei Lim, A.F. Ismail, N. Hilal, Innovative and sustainable membrane technology for wastewater treatment and desalination application, in: Innovation Strategies in Environmental Science, Elsevier, 2020, pp. 291–319, https://doi.org/10.1016/B978-0-12-817382-4.00009-5 (Chapter 9).
[66] C.H. Koo, W.J. Lau, Mixed-matrix membranes incorporated with functionalized nanomaterials for water applications, Handbook of Functionalized Nanomaterials for Industrial Applications, Elsevier, 2020, pp. 15–21 (Chapter 2).
[67] E. Tocci, C. Rizzuto, J.C. Jansen, E. Drioli, in: E. Drioli, L. Giorno, A. Gugliuzza (Eds.), Nanostructured Membranes for selective Separations: Modeling and Simulations, first ed., Pan Stanford Publishing Pte. Ltd., 2018. https://www.taylorfrancis.com/chapters/mono/10.1201/9781351135115-7/nanostructured-membranes-selective-separations-modeling-simulations-elena-tocci-carmen-rizzuto-johannes-carolus-jansen-enrico-drioli-enrico-drioli-lidietta-giorno-annarosa-gugliuzza?context=ubx&refId=7dff5299-838a-4744-8df8-3446e4c79528.
[68] M.L. Perrotta, A. Gugliuzza, in: E. Drioli, L. Giorno, A. Gugliuzza (Eds.), Fabrication of Ordered Micro- and Nanoporous Membranes, first ed., Pan Stanford Publishing Pte. Ltd., 2018. https://www.taylorfrancis.com/chapters/mono/10.1201/9781351135115-8/fabrication-ordered-micro-nanoporous-membranes-maria-luisa-perrotta-annarosa-gugliuzza-enrico-drioli-lidietta-giorno-annarosa-gugliuzza?context=ubx&refId=c13e6e2f-ec15-4644-a462-4b9d8d871acc.
[69] A. Gugliuzza, in: E. Drioli, L. Giorno, A. Gugliuzza (Eds.), Nano-Assembled Membranes for Smart Textiles Solutions, first ed., Pan Stanford Publishing Pte. Ltd., 2018. https://www.taylorfrancis.com/chapters/mono/10.1201/9781351135115-16/nano-assembled-membranes-smart-textiles-solutions-annarosa-gugliuzza-enrico-drioli-lidietta-giorno-annarosa-gugliuzza?context=ubx&refId=e2f51fdc-031f-48d3-8a30-07fec2109bb2.

CHAPTER 2

Theoretical concepts of membrane-nanomaterial composites

Deepshikha Datta[a], K.S. Deepak[b], Krishna Priyadarshini Das[c], and Bimal Das[b]

[a]Department of Chemical Engineering, GMR Institute of Technology, Rajam, India
[b]Department of Chemical Engineering, National Institute of Technology Durgapur, Durgapur, India
[c]Department of Material Science and Engineering, Indian Institute of Technology, Delhi, India

2.1 Nanomaterials

2.1.1 Introduction

Nanotechnology is a buzzword in modern research and is dedicated to the synthesis, strategy, and manipulation of particles whose size approximately ranges from 1 to 100 nm (Fig. 2.1) [2]. It is an ever-exploring field that never fails to surprise researchers as all the properties of these nanoparticles differ from those of the bulk materials in fundamental ways. The links of the word "nano" can be traced back to the word "nanos," which is an Indo-European (Greek) word meaning "dwarf" [3]. The International Organization for Standardization (ISO) in 2008 defined nanoparticle (NP) as an object whose all three Cartesian dimensions do not exceed 100 nm. In the early 2000s, the European Union commission endorsed a much broader and technical definition [4]. The notion of nanotechnology started with Feynman's groundbreaking speech in 1959, titled "There is plenty of room at the bottom" [5].

This love for nanomaterials can be credited to their inherited peculiar properties due to their nanosize [6]. Their ability to get their properties tailored to meet our varied requirements has made them invaluable [7]. Their high surface area-to-volume ratio makes them ideal for catalytic activities [8], whereas a phenomenon such as localized surface plasmon resonance opens up their application in optics and sensors [9]. In addition to these characteristics, they possess some physical properties, making their applicability even more significant [10]. Today, with the advancement of research, nanotechnology has found its worthiness in the field of biosensors [11], labeling [12,13], drug delivery [12], food packaging [13], cosmetics [14], clothing

Membranes with Functionalized Nanomaterials
https://doi.org/10.1016/B978-0-323-85946-2.00007-2

Fig. 2.1 Schematic representation of nanosize [1].

[15], etc. From the toothpaste to the drugs we intake, there has not been any field that has been bereft of its potential [16,17].

2.1.2 Properties of nanomaterials

Researchers are in a continuous race to exploit the nanomaterials to their full extent. This craze for the nanomaterials can be attributed to their scale as it causes a drastic change in their features compared to their corresponding bulk materials. These drastic changes in their properties can be credited to the quantization of electronic states and high surface area-to-volume ratio [18]. Further, their 1D, 2D, and 3D arranged lattices have well-defined properties, thus opening up many applications. Among the numerous ways of classifying NPs, the most prominent one groups it into five broad categories: semiconductor quantum dots, magnetic NPs, polymeric NPs, carbon-based nanostructures (CNSs), and metallic NPs [16]. Being prominently human-made, they are sometimes referred to as engineered nanoparticles. Each of the categories mentioned above of NPs has been made to fit a particular problem by exploiting any of their enhanced properties. Whether it is the ease of synthesis or control over size, shape, structure, or composition, metallic nanomaterials are the most flexible and owing to this reason, they are in the spotlight in contrast to other engineered nanoparticles. This flexible nature of metallic nanoparticles makes it possible for researchers to finely tune their various properties, increasing their applicability [17] (Table 2.1). Surface phenomenon resonance (SPR) is a phenomenon attributed to gold's metallic nature, which occurs when light interacts with the surface of gold nanoparticles. Surface plasmon resonance is a phenomenon specific to AuNPs. It takes place when the e_s^- in the conduction band oscillates with a frequency equal to that of the incoming light, thus forming a plasmon band [36]. The potential of AuNPs for biomedical applications can be credited to their plasmonic properties [37]. Metallic nanoparticles, unlike their bulk counterparts, exhibit a broader absorption band. Due to this presence of the band in the visible spectrum, these NPs in colloidal solutions exhibit bright colors. One of the conditions to achieve SPR is that the NPs should be less than the wavelength of the incoming light [38].

2.1.2.1 Physical properties

Physical properties such as color, density, melting point, mechanical strength, elasticity, and conductivity of nanoparticles show a dependency on morphology and dimensions of nanoparticles. Another exciting feature that has been seen is that the hardness and mechanical strength of NPs

Table 2.1 Overview of applications derived by the peculiar properties of nanomaterials.

Nanomaterials	Properties and characteristics	Potential applications	Reference
Oligonucleotide modified AuNP	Surface plasmon resonance	Colorimetric detection of polynucleotide	[19]
Unmodified AuNP	Surface plasmon resonance	Colorimetric detection of DNA	[20]
	Broad excitation spectra	Fluorescent analysis of DNA	[21]
CNTs	Broad excitation spectra and superior photochemical stability	Analysis of DNA hybridization	[21]
Single walled/ multiwalled CNTs	Structure	Adsorption of zinc from water	[22]
Graphene oxide	Broad excitation spectra, superior photochemical stability, and quenching ability	Analysis of DNA hybridization	[21]
3D-honeycomb structure like graphene sheets	Catalytic activity	Solar cells	[23]
Oligonucleotide modified AgNP	Surface plasmon resonance and higher extinction coefficient	Colorimetric detection of DNA with enhanced sensitivity	[24]
Reduced graphene oxide	Specific surface area and water dispersibility	Detection of DNA mutations	[25]
Quantum dots	Peculiar photophysical properties and superior Photochemical Stability	Analysis of DNA hybridization	[26]

Table 2.1 Overview of applications derived by the peculiar properties of nanomaterials—cont'd

Nanomaterials	Properties and characteristics	Potential applications	Reference
Modified quantum dots	Peculiar photophysical properties and superior photochemical stability	Targeting cancer cell	[27]
Functionalized iron oxide NP	Magnetic properties	Catalyst in the hydrogenation of DMEC	[28]
Iron oxide nanosheets	Electrical properties	Solar energy conversion	[29]
Iron oxide	Superior photothermal stability and magnetization	MRI	[30]
Growth hormone conjugated fluorescent nanodiamond	Fluorescence	Raman mapping of cancer cell	[31]
Au conjugated fluorescent nanodiamonds coated with PEI800	Enhanced photoacoustic signals	Gene delivery	[32]
Multilayered-mesoporous titanium dioxide	Power conversion efficiency	Solar cells	[33]
Silica nanoparticles	Bio-efficacy	Reducing infectivity of *Bombyx mori* nucleopolyhedrovirus	[34]
Polymer stabilized iron sulfide	Catalysis	Removal of lindane from water	[35]

increase with the decrease in their size considerably. This dependency of properties helps researchers mold the particles according to their needs [39]. The surface area-to-volume ratio plays a huge role in determining the physical properties of nanoparticles. For a spherical particle, it increases with a decrease in radius. Thus, making several physical properties dependent on attributes such as size and shape [40]. The physical properties of nanomaterials differ significantly from their bulk counterparts. This difference can be attributed to the large surface to inner atoms ratio. This causes weak attractive forces in the core atoms, allowing the surface atoms to have considerably large surface energy. The nanomaterials also have electromagnetic properties, which vary substantially from that of their bulk counterparts. This variation in electrical properties is due to increased surface scattering, which is, in turn, a consequence of the reduction in dimensions.

2.1.2.2 Chemical properties

Nanomaterials have shown massive potential in catalytic activities thanks to their large surface-to-volume ratio. The high ratio of surface to inner atoms also favors reactivity as a more significant number of atoms are now subjected to reactions. These large numbers of surface atoms have low coordination numbers, thus becoming sites for facilitating reactions [41]. With advancements in nanotechnology, researchers have successfully utilized their catalytic ability as a viable alternative for Pt or Pd [20,21].

2.1.2.3 Optical properties

The potential of nanomaterials in optics has been exploited long back. From theLycurgus Cup to ruby-color-stained glasses, people had shown the impressive things nanomaterials can help achieve. The conducting electrons that oscillate collectively under the subjection of incoming electromagnetic radiation undergo a phenomenon called surface phenomenon resonance. The high conductivity and impressive optical features can be attributed to these free electrons. This phenomenon in gold nanoparticles is responsible for the bright colors it exhibits. Like gold, copper, and silver, Noble metals have the most strong plasmon resonance [42]. The appearance of this resonance in our visible spectrum makes their nanomaterials applicable in many exciting fields. The fact that the conglomeration of nanoparticles leads to a redshift in the absorption spectrum and the fact that the dielectric constant of the environment is highly correlated to their intensity of absorption and wavelength maximum makes them an ideal candidate for adsorption sensors [43]. Photoluminescence is another exciting property that has found immense potential. It has been found that the luminescent properties are near correlated to the size of the particle and their environment. This added with the fluorescence quenching ability and the ability to act as electron acceptors can be used in various sensor fabrication techniques [44].

2.1.3 Historical and critical milestones related to nanomaterials

In December 1959, Richard Feynman gave the benchmark speech of nanotechnology titled "There is Plenty of Room at the Bottom" at an American Physical Society conference. This speech is regarded as the first lecture on technology at an atomic scale [45]. Norio Taniguchi, in 1974 came up with the term "nanotechnology" while explaining precision machining at an atomic scale [17,36]. In 1937 Manfred V. Ardenne invented the scanning tunneling microscope and a revolution in microelectronics [46]. This revolution helped researchers get a boost in exploiting the potential of nanomaterials [47]. Soon after realizing the potential, the European Commission became the first body to adopt a strategy in 2004 to accelerate R&D in this field [48]. A snip of critical milestones in the field of nanotechnology is shown in Fig. 2.2. Several other research bodies followed the path and started preparing engineered nanomaterials [40,41]. The advantage of engineered nanomaterials was that several properties associated with the nanosize could be maneuvered according to our needs [54,55]. This flexibility in properties opened up various areas of application for the nanomaterials.

2.1.4 Functionalized nanoparticles

Making nanoparticles perform a specific task by introducing a functional group, thus altering the chemical behavior is termed functionalization. Since the emergence of nanotechnology, researchers have been working on synthesizing various functionalized nanoparticles to meet their specific demands. The functionalized NPs can be grouped based on the fabricated material and a gist represented in Fig. 2.3. The root cause of functionalizing NPs can be attributed to several reasons such as agglomeration, toxicity, and inertness, which strips down potential applications of nanoparticles [56]. By functionalizing them with eligible entities, not only can we mitigate their limitations but can also tailor and tweak a particular property of the nanoparticle such as selectivity, thermal stability, and permeability for making them the ideal candidate for specific applications such as biosensing [54], clinical clothing, drug delivery [55], and agricultural engineering [57].

2.2 Membranes

2.2.1 Introduction

Membranes have been used for several decades, and studies related to membrane-based processes can be traced back to the 18th century. The word "osmosis" was coined by Abee Nolet in 1748 while describing the permeation of water through the membrane [58]. The recent surge of research in this field can be attributed to the changing dynamics of world nature and climate change, leading to depleting water resources. This entwined with the clean track record of reverse osmosis, and ultra-filtration processes have helped keep the membrane-based process in the spotlight [59]. Today membrane technology has been exploited in a plethora of fields [60]. The European membrane society defines a

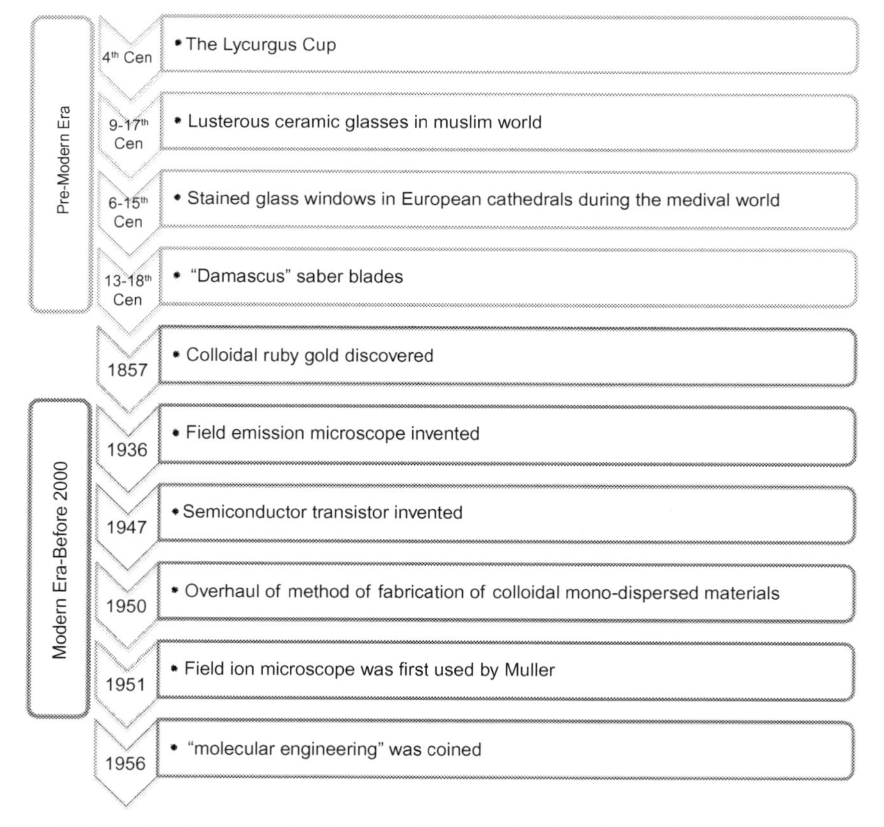

Fig. 2.2 Key developments in the area of nanotechnology [49–53].

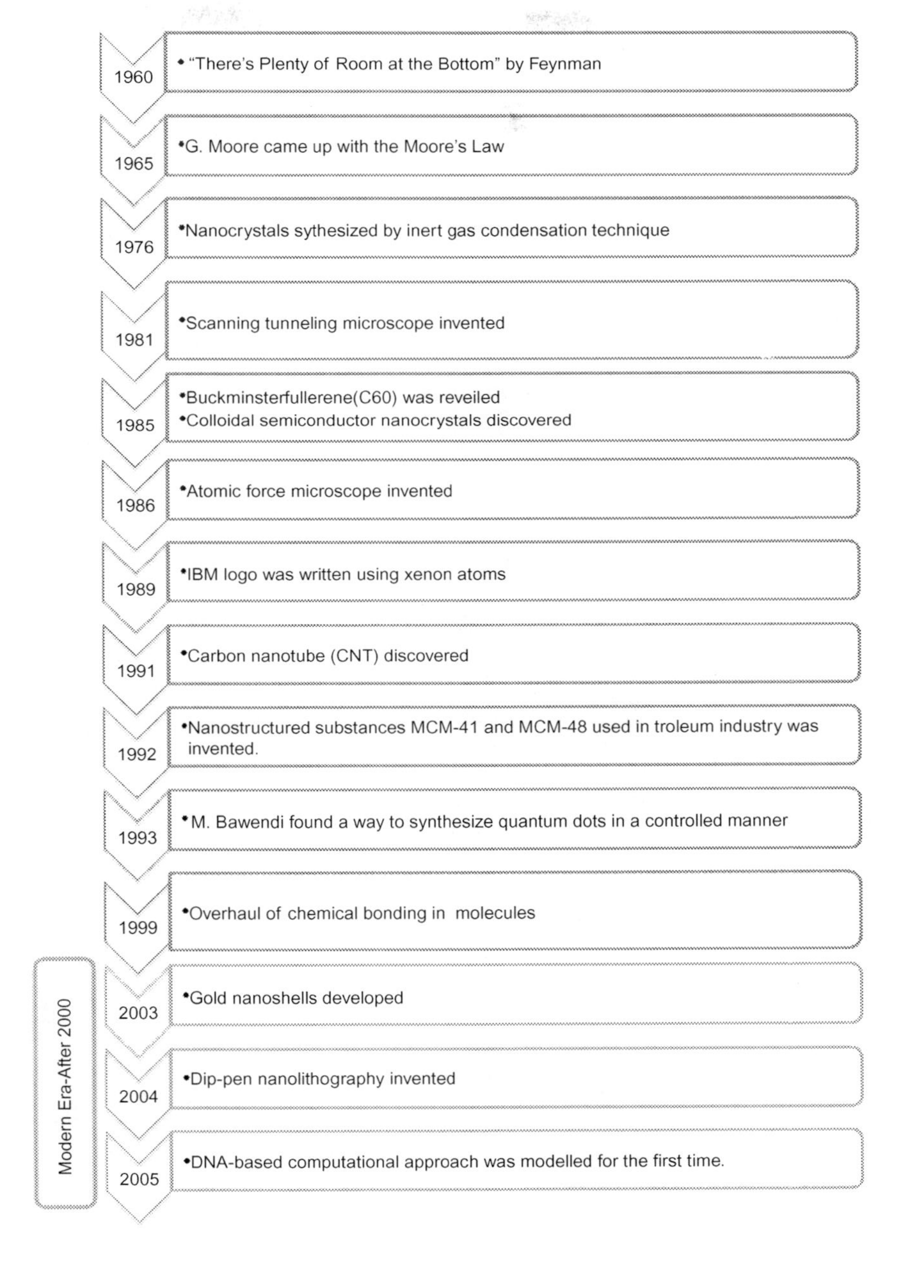

Fig. 2.2, Cont'd

(Continued)

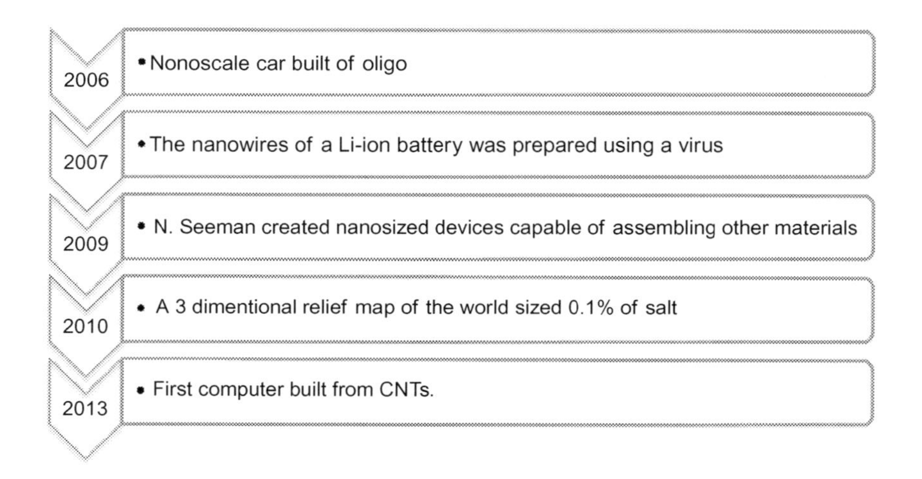

Fig. 2.2, Cont'd

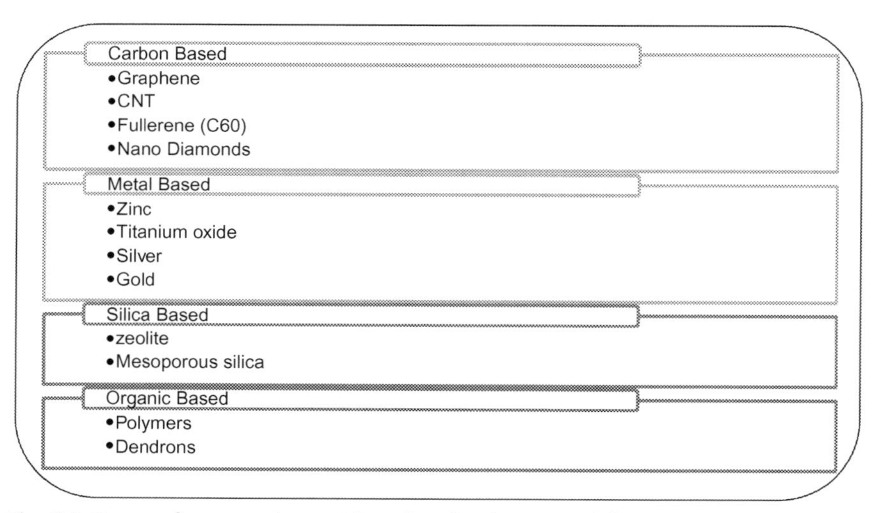

Fig. 2.3 Types of commonly used functionalized nanoparticles.

membrane as an intervening phase separating two phases and acting as an active or a passive barrier to the transport of matter between the phases adjacent to it (Fig. 2.4) [61].

Other terms that find numerous mentions are membrane science and membrane technology. The former deals with designing the materials for membranes, their characterization, and evaluating membrane processes

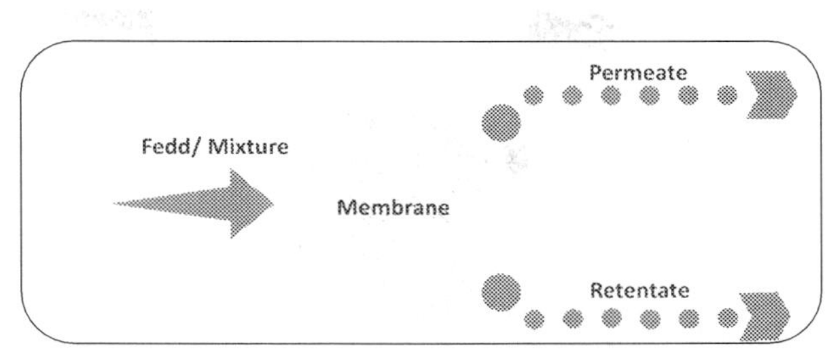

Fig. 2.4 Schematic representation showing basic membrane process.

[58–61], whereas the latter is a much broader field encompassing the scientific and engineering aspects for membrane-based processes such as diffusion, capillary, dialysis, and osmosis [62].

2.2.2 Introduction to membrane processes

2.2.2.1 Types of membranes

With due course of time and advancement in technology, researchers have developed various problem-specific membranes. Various parameters of the membranes can be used as a criterion for their segregation. Segregating membranes using the cross-sectional area and the constituting material is the most common one. Fundamentally, a membrane is nothing more than a thin film of homo/heterogeneous physical and chemical properties such as pore structures and compositions. The microporous membranes have a varied pore size distribution making them suitable for solute separation. The solute separability of such membranes is a function of their dimensions and pore size of the membranes. Thus, it ensures that the solutes having a considerable difference in the pore and molecular size are separated. Such functionalities make such membranes suitable for ultrafiltration and microfiltration. Due to high permeability, selectivity, chemical stability, thermal tolerance, and resistivity toward corrosion, membranes have been used in various blooming applications [63,64].

The nonporous/dense membranes are much denser, and the separation is facilitated by the rate of transportation, which is a function of the diffusion and other parameters of the solutes incorporated in the membrane. The mixture is subjected to various kinds of gradients to separate the permeate from the retentate. Owing to these properties, they are used for gas separation, RO, and pervaporation membranes. One disadvantage of using this membrane comes from its low permeability, which limits its application in industries. The third category of isotropic membranes is the electrically charged membranes, consisting of cation or anionically charged ions. The separation of solutes in such membranes depends on the charge and concentration of ions in the solution. Their electrical property is exploited for processes such as electrodialysis, where electrolytic solutions are used. Another main classification criterion is the material used. The inorganic membranes are the most common ones used for gas separation, microfiltration, and nanofiltration [67]. This wide usability can be attributed to the variability of parameters such as pore size, materials, and module configurations [68]. These membranes can then be segregated into porous and nonporous membranes based on the cross sections, as shown in Fig. 2.5. Owing to the excellent track record and simplicity in design, the organic polymers also constitute a large fraction of membranes used in industries. Over time researchers have also utilized several organic polymers for varied property-specific applications.

2.2.2.2 Advantages and limitations of various membranes

These membranes shown in Table 2.2 have their pros and cons, which must be considered before finalizing the type of membrane to be used. The mixed matrix membranes are an ice breaker in the field of membrane research. Its success and fame can be attributed to its ability to be tailor-made. The use of nanomaterials in these membranes establishes properties such as antibacterial [64,67,68,70,71], antifouling [50–53], and photocatalytic behavior [72–74]. This ability of mixed matrix membranes to be specifically made for a purpose has opened up its potential in processes photocatalytic gas separation [75,76], desalination [75,77,78], integrated wastewater treatment [79], energy production [80], and recovering required materials [81].

2.2.2.3 Membrane processes

The filtration processes can be segregated based on the pore size of the membranes. The pore dimensions and materials can be varied depending on their primary applicability of materials to separate, thus providing several membrane processes (Fig. 2.6). For the water treatment process, ultrafiltration and microfiltration are used, whereas reverse osmosis has found its potential in water purification and desalination. Membrane distillation is a relatively new process with its water desalination expertise with high salinity (10,000–35,000 ppm). The pore size and its distribution over the membrane are two characteristics that decide the membrane's commerciality. Having an appropriate pore size to facilitate required separation is as important as having a narrow pore size distribution. Table 2.3 showcases the applicability

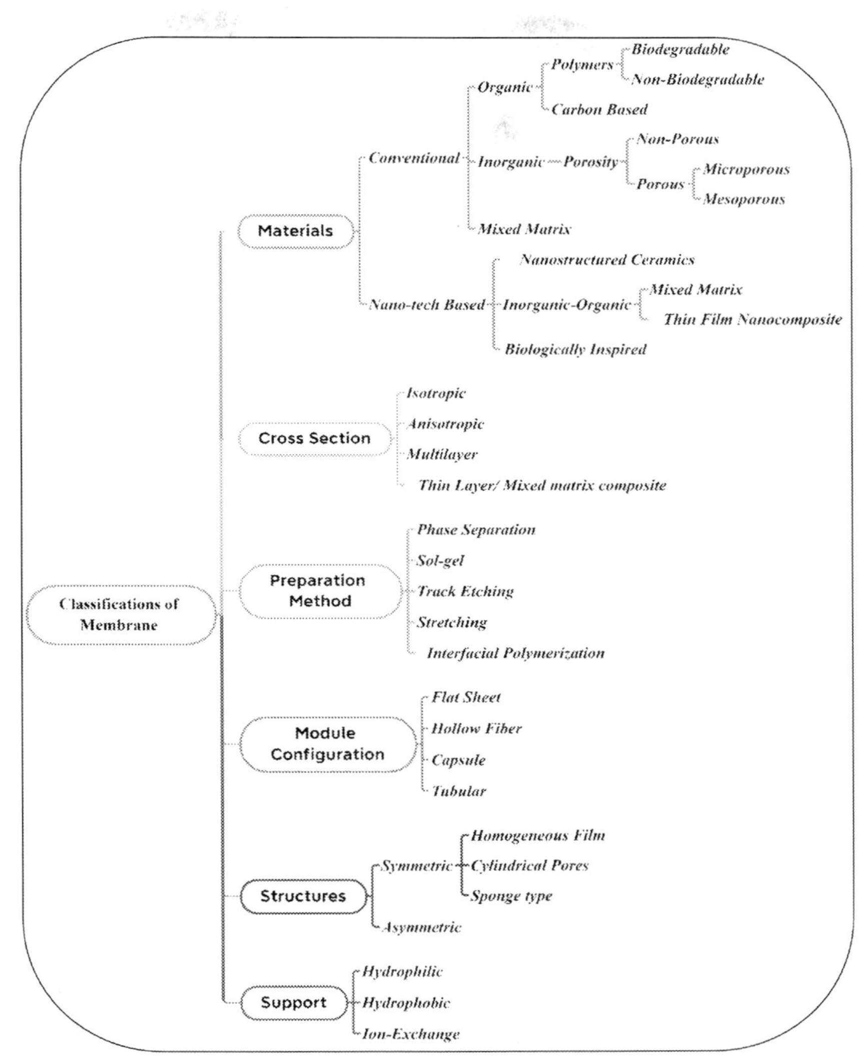

Fig. 2.5 Various criteria for the classification of the membranes [23,48,65,66].

of membrane processes on which the pore size plays a pivotal role. Membranes with much smaller pore size would not allow necessary solutes to pass by and would require enormous pressure to drive the process. On the other hand, having a larger pore size than required would allow unwanted components to pass by. Thus, strict control over the pore size is a sine qua non to the membrane's economic efficiency.

Table 2.2 Strengths and weaknesses of various membranes [69].

Organic membranes	*Pros*	Ease of fabrication and synthesizability Minimal manufacturing cost Superior mechanical strength Ease of tailoring parameters Separation mechanism: Solution diffusion Weaker thermal stability
	Cons	Deformation of its structure No controllability of the pore size Bargain between permeability and selectivity
Inorganic membranes	*Pros*	Superior physical and chemical strength Controllability of pore size Weaker bargain between permeability and selectivity Reliable operability Separation mechanism: molecular sieving, capillary condensation, surface diffusion, Knudsen diffusion
	Cons	Brittle Costly Tough to enlarge
Mixed matrix membranes	*Pros*	Enhanced physical strength Relatively less deformation Less energy intensive Compaction at extreme pressures Free of bargain between permeability and selectivity Better separability Separation mechanism: combination of principles governing polymeric and inorganic membrane
	Cons	Dependency on chemical and thermal stabilities Addition of fillers makes it fragile

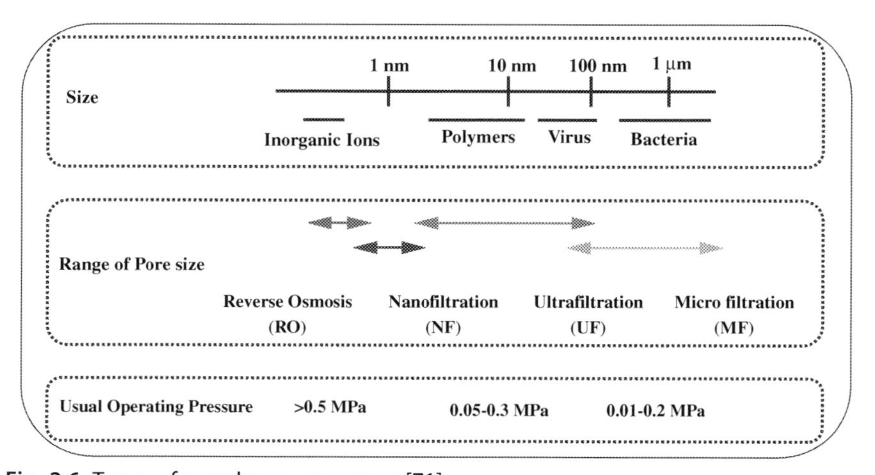

Fig. 2.6 Types of membrane processes [71].

Table 2.3 Membrane processes with their character and applications.

Membrane process	Application	Character
Microfiltration	Removal of suspended solids and colloids	Nonequilibrium pressure driven
Nanofiltration	Removal of organic compounds	
Ultrafiltration	Removal of viruses and macromolecules	
Reverse osmosis	Removal of salts	
Pervaporation	Separation of liquid and vapor mixture	
Gas separation	Separation of one or more gas from a mixture of gases	
Dialysis	Removal of very small solutes from a solution	Nonequilibrium and nonpressure driven
Electrodialysis	Selective transportation of ions	
Forward osmosis	Separation of water from dissolved solutes	
Liquid membrane	Liquid separation barrier for facilitating transportation through solution-diffusion mechanism	
Membrane distillation	Separation of a vapor phase from the liquid phase through a hydrophobic membrane	

2.2.3 Membrane fabrication processes

The two aspects while fabricating a membrane is the selection of material and the preparation method. Transport properties such as permeability and selectivity have a considerable say in the commerciality of the membrane. Its effectiveness and the pore cross section depend on the type of method employed. For preparing synthetic membranes, several fabrication processes are employed (Fig. 2.16), among which the most common ones are listed below:

2.2.3.1 Fabrication methods for mixed matrix membranes

Phase inversion

In 1960 Loeb and Sourirajan invented the first cellulose acetate membrane via phase inversion, and this event started a chain of novel methods for membrane fabrication [82]. Nevertheless, even after numerous methods, the phase inversion method remains the most common one. In this process, a

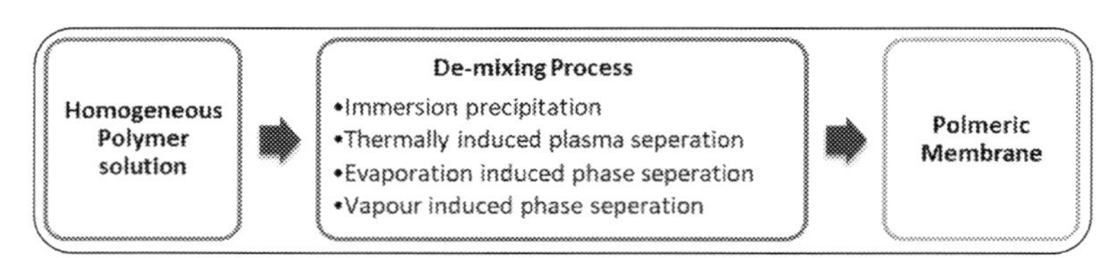

Fig. 2.7 Flow diagram showing phase inversion method of membrane fabrication.

homogeneous polymer solution is converted to solid state by a demixing process. This transformation can be achieved in several ways (Fig. 2.7).

The most prevalent one among these is the immersion precipitation, where the polymeric solution is made to exchange with a nonsolvent by immersing it in a coagulation bath having the nonsolvent (Fig. 2.8).

The exchange between the immiscible solvent and nonsolvent occurs until the process becomes thermodynamically unstable and the demixing starts to occur. At the end of the demixing process, an asymmetric solid polymeric membrane precipitates whose pore size can vary by varying the solvent's diffusion fluxes and nonsolvent[84]. By controlling the fluxes, microfiltration ($J_1 = J_2$) and ultrafiltration ($J_1 \ll J_2$) membranes can be produced. Where J_1 = Flux of nonsolvent and J_2 = Flux of solvent.

Stretching

Zhu W. et al. in the 1970s developed a novel method for preparing microporous membranes called stretching, and the polypropylene microporous membranes produced via this method were commercially available by the name of Celgard [85]. Being a solvent-free process, it is economically and

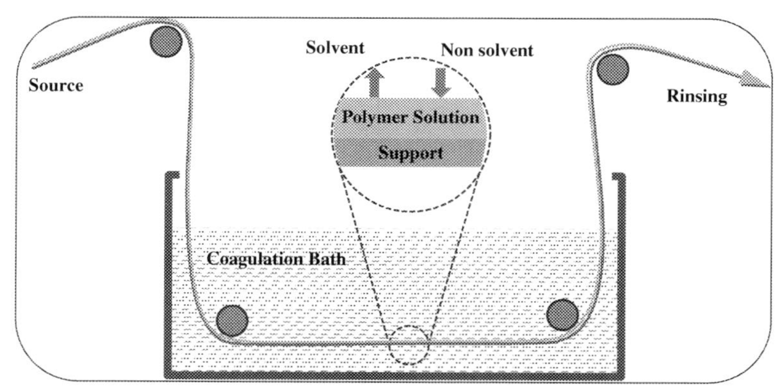

Fig. 2.8 Schematic representation showing phase inversion via immersion precipitation [83].

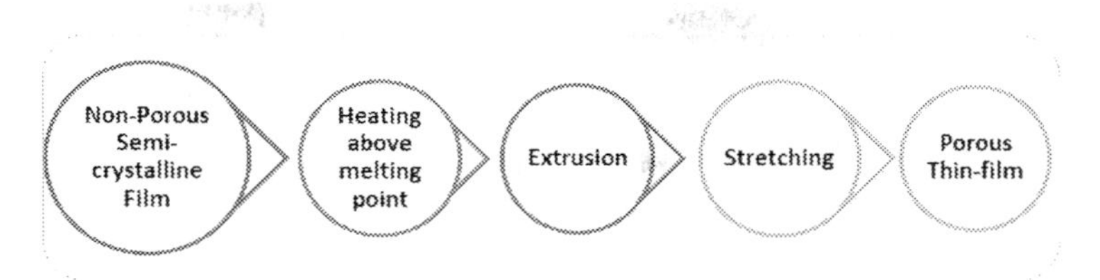

Fig. 2.9 Sequence of steps in the stretching process.

technologically convenient. In this process, the membrane is first heated above the melting point followed by extrusion and stretching, which results in the formation of pores in the thin sheets (Fig. 2.9).

The stretching process is done in two steps, namely cold and hot stretching. The cold stretching nucleates the pores in the membranes. The resulting pores size can then be controlled via hot stretching. Being a mechanical process, the physical properties and processing parameters define the resulting membranes [86].

The engineering parameters related to uniaxial and biaxial stretching are defined as follows:

$$\text{The stretching ratio of the } i^{\text{th}}\text{stretching operation } (R_s(i)) = \left[\frac{l_{stretched\ membrane}}{l_{original}} - 1\right] \times 100 \tag{2.1}$$

where $l_{stretched\ membrane}$ is length of the stretched membrane in the direction of the ith stretching operation and $l_{original}$ is original length of the rolled sheet before being stretched.

$$\text{Stretching rate } (r_s(i)) = \text{time derivative of } (R_s(i))\ (\%/s) \tag{2.2}$$

Track etching

Fleischer et al. in 1964 developed thin sheets by nuclear tracks for the production of membranes. Soon after this discovery, polycarbonate track membranes synthesized via this method was made available in the market [87]. The track etched membranes have precise pore structures than membranes produced via conventional methods. In this process, the duration of irradiation determines the porosity of the membrane and the etching time and temperature determine the size of the pores [88]. Following are the three types of track etching method employed in the industry [89]:

Fission fragment tracking In this method, a heavy nucleus such as uranium undergoes fission when subjected to neutron flux from a nuclear

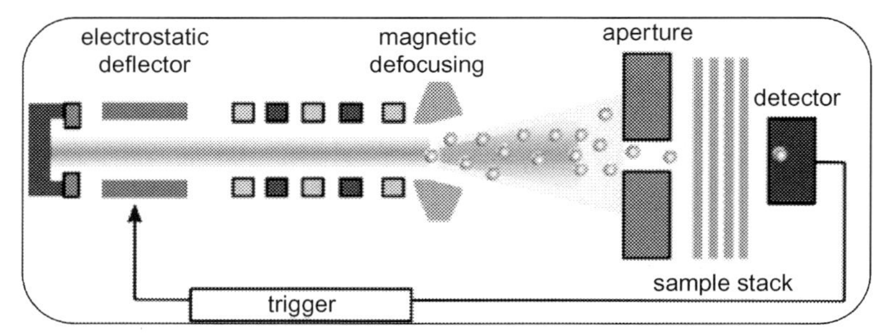

Fig. 2.10 Schematic representation showing single-step ion irradiation setup [90].

reactor (Fig. 2.10) [91]. The fragments then arising from fission are made to fall on the foil using a collimator. This induces linear damage to the tracks, thus producing pores in the film.

Accelerator tracking method This method facilitates porous membranes' production by subjecting the membranes to accelerated ions having energies of few MeV per nucleon. One major drawback of this method is the flux's lack of controllability from an accelerator and the hefty cost associated with this method [89].

Chemical etching method The chemical etching method pertains to a procedure consisting of ions' bombardment, followed by separating the damaged portions of the irradiated track via chemical etching. The mechanism of this additional step determines the pore size and shape. Parameters such as etching temperature and time can be tweaked to tailor the formation of favorable pore sizes [92].

Electrospinning

The word "electrospinning" came from the word "electrostatic spinning." Taylor's work on electrically driven jets can be credited as the progenitor of electrospinning [93]. It uses electrostatic force to produce fibers from the polymer solutions [94]. The fibers produced through this method usually have a smaller diameter and a larger surface area than those produced via conventional methods. Another plus point of this process is that by manipulating the viscosity, environmental conditions, electric potential, and the flow rate, the aspect ratio, and the fibers' morphology can be tailored as per requirements. This process involves subjecting the polymeric solution placed in a syringe to electrostatic potential until it overcomes the solution's surface tension. At this point, droplets erupt from the aperture forming a Taylor cone, which then undergoes stretching due to the electrostatic

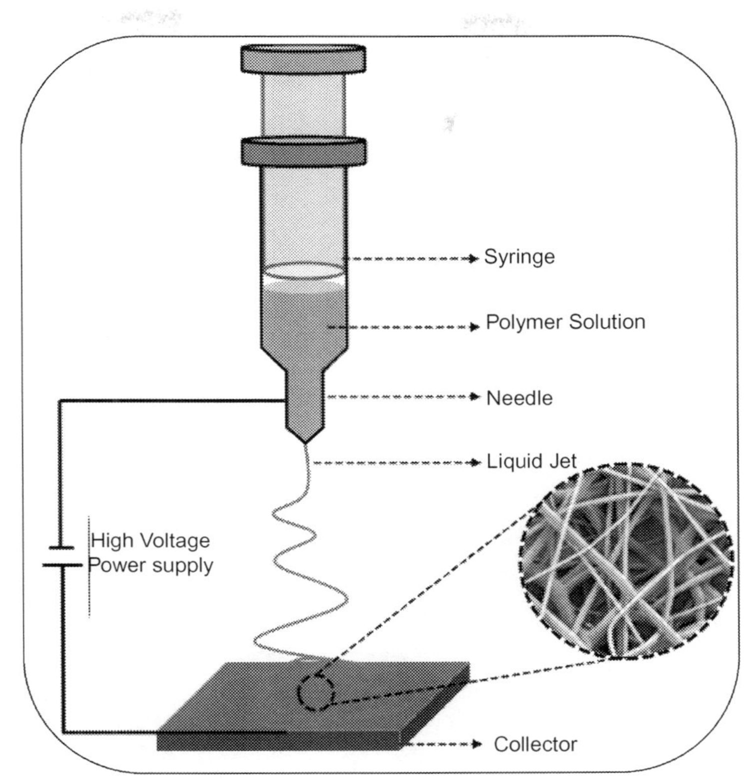

Fig. 2.11 Schematic representation showing the electrospinning procedure [95].

repulsion applied across the syringe and the collector (Fig. 2.11). This stretching process leads to the evaporation of solvent from the solution; thus, forming the fibrous polymer [96].

2.2.3.2 Fabrication methods for thin-film nanocomposite membranes

Interfacial polymerization

Cadotte et al. in 1960 came up with the first interfacially polymerized thin-film composite for reverse osmosis [97]. This acted as an ice breaker and then was extended to nanofiltration. In this process, microporous polysulfone support is dipped in an aqueous solution, leading to the formation of an amine impregnated membrane. This membrane is then submerged in a diisocyanate solution in the presence of hexane, as depicted in Fig. 2.12 [83,98]. This process resulted in a TFC polyamide membrane with enhanced salt rejection capabilities [99].

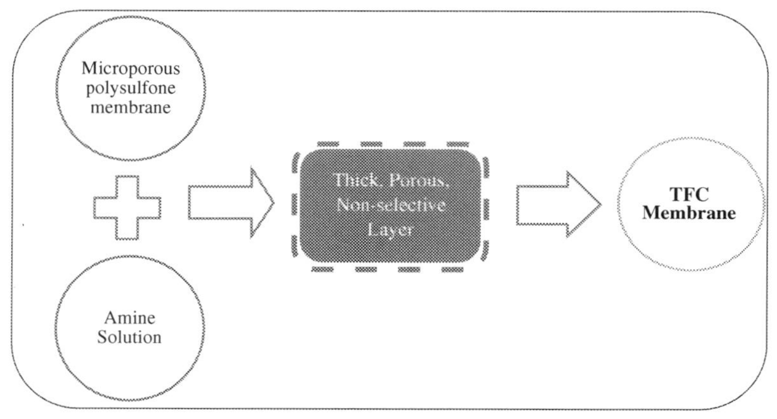

Fig. 2.12 Schematic representation showing the Interfacial Polymerization Process.

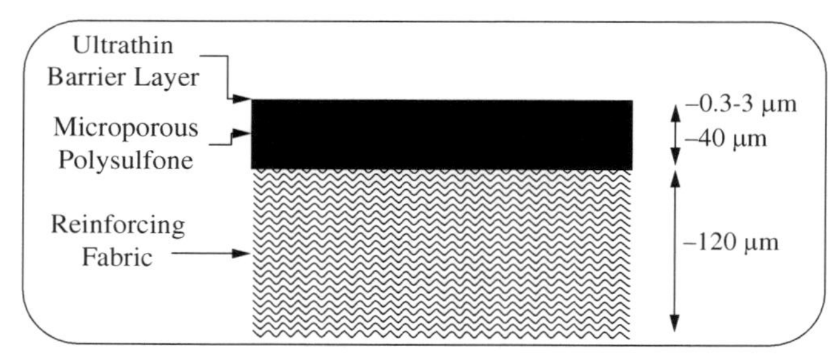

Fig. 2.13 Schematic representation of the Thin-Film Composite membrane [100].

The membranes produced by this technique have an ultrathin layer on top, thus acting as a support for the membrane (Fig. 2.13). The SEM and TEM images of thus synthesized membranes are displayed in Fig. 2.14. In addition to this, the ability to tailor properties of the top layer and the substrate has opened up new dimensions for researchers to explore. For the Interfacial polymerization process, it has been found that parameters such as concentration of monomers, solvent type, reaction time, and posttreatment conditions have a substantial effect on the shape and morphology of the synthesized membranes.

Dip coating

It is a widely used technique for the fabrication of TFN and TFC membranes [102]. It is a two-step process in which the matrix is first dipped in a coating solution followed by cross-linking and heating, thus forming a thin film of

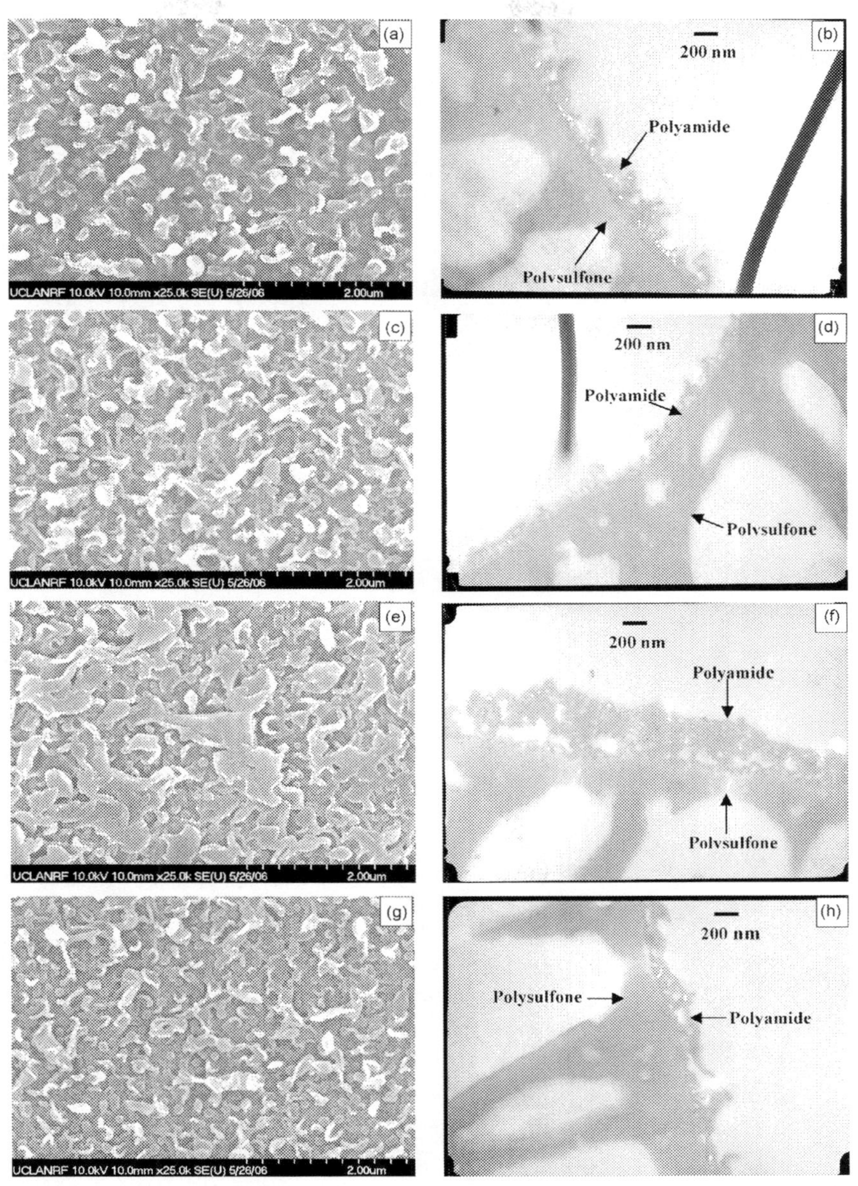

Fig. 2.14 SEM and TEM images of Reverse Osmosis membrane prepared in (1,2) C_6H_{14}, (3,4) C_7H_{16}, (5,6) C_6H_{12}, and (7,8) isopar [101].

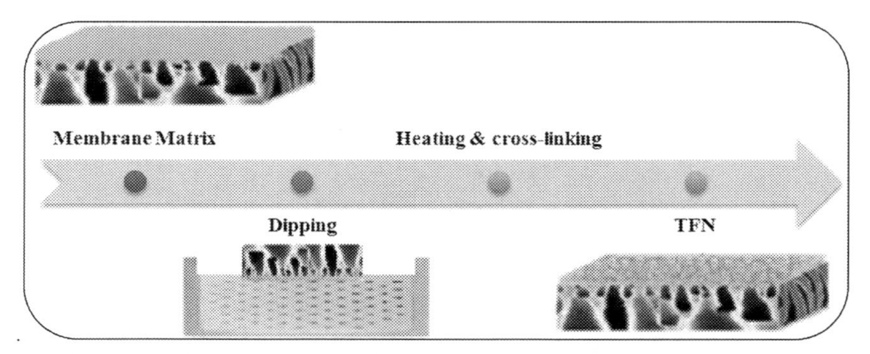

Fig. 2.15 Schematic representation showing steps involved in dip coating [104].

layer above the membrane matrix [103]. A schematic diagram illuminating the various steps involved in the fabrication of TFN membranes via dip coating is shown in Fig. 2.15.

The film thickness is dependent on the physical properties of the coating solution. Usually, membranes for processes such as gas separation, pervaporation, nanofiltration, and RO are fabricated by this method [105] (Fig. 2.16).

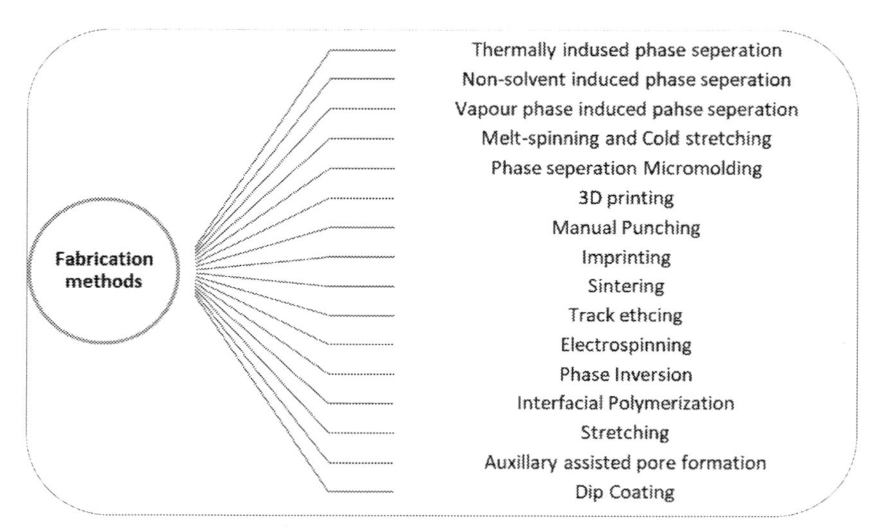

Fig. 2.16 Various types of fabrication methods of membranes.

2.2.4 Historical and key developments in the field of membranes

See Fig 2.17.

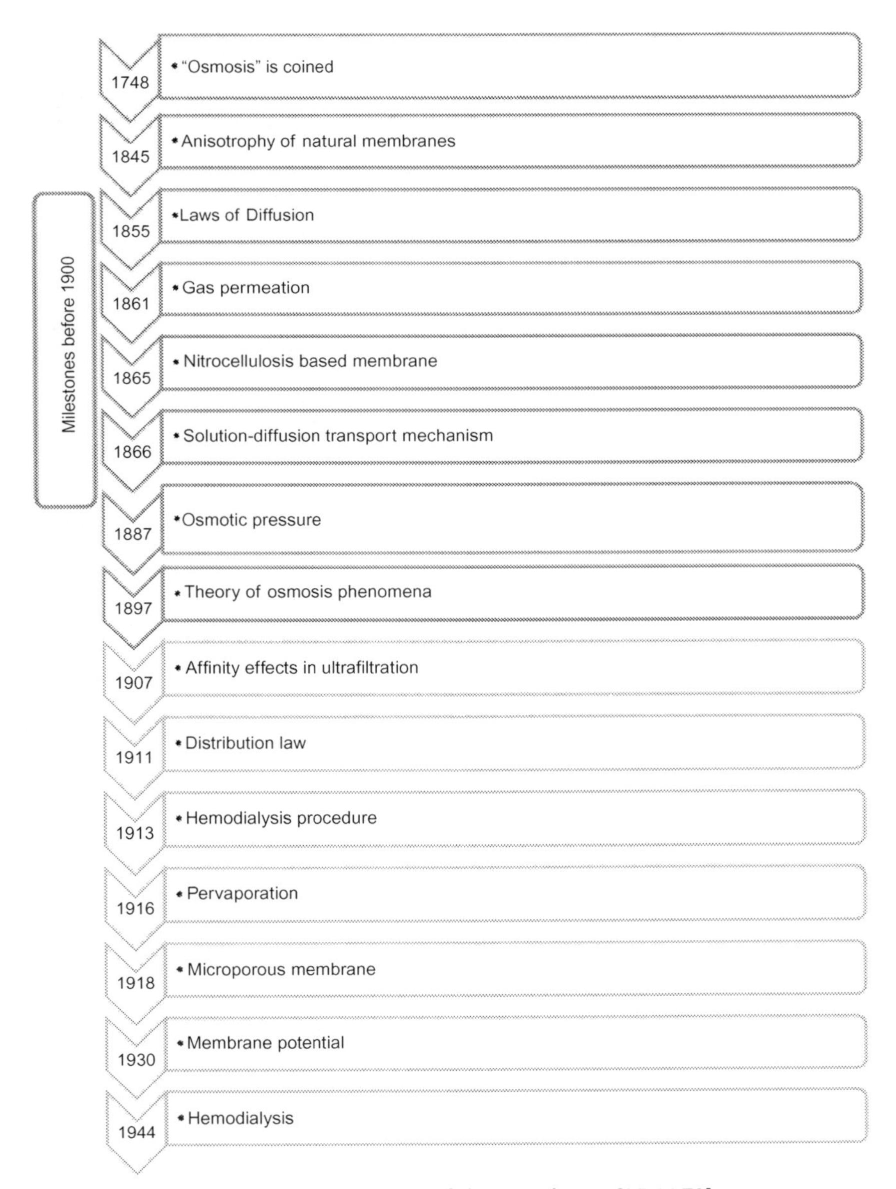

Fig. 2.17 Key developments in the area of the membrane [25,46,72].

(Continued)

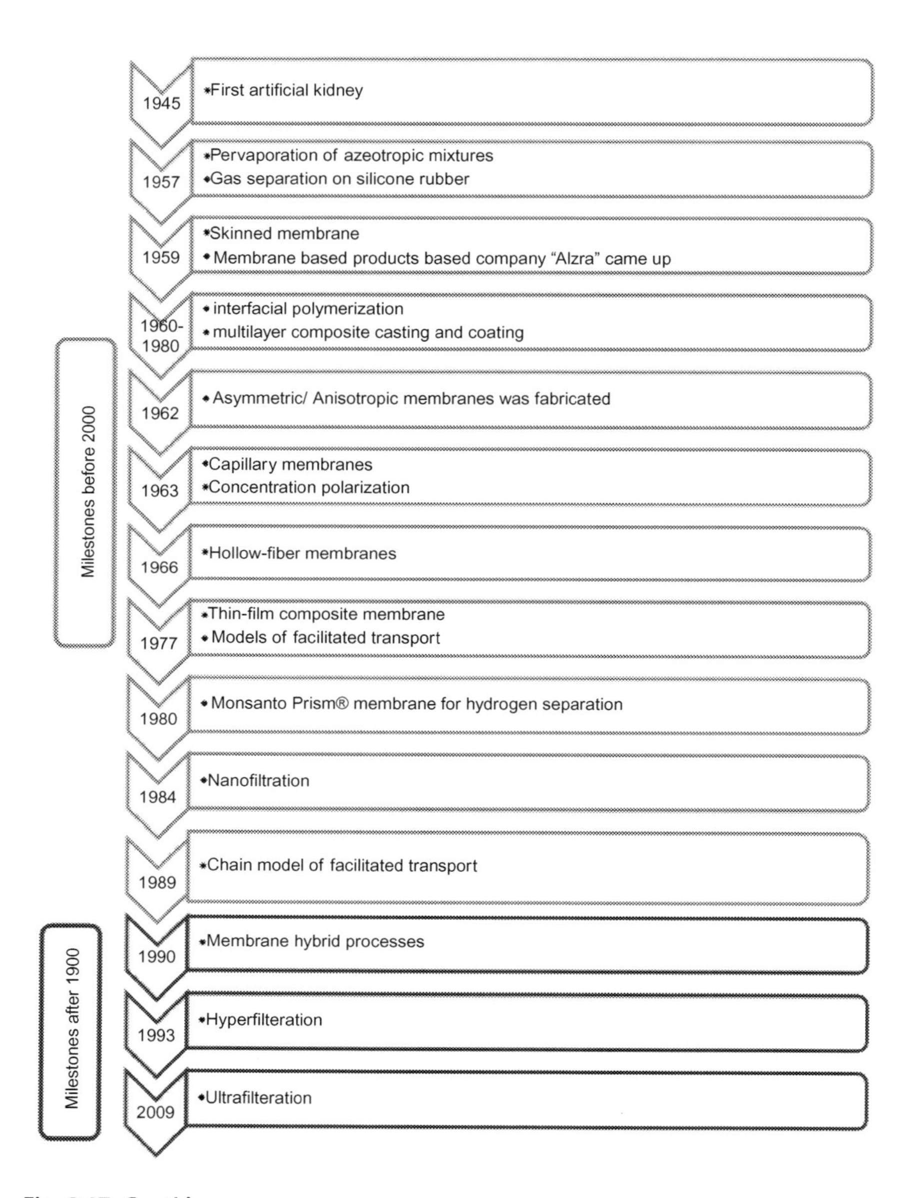

Fig. 2.17, Cont'd

2.3 Membrane-nanomaterial composites

2.3.1 Introduction

Nanocomposites are defined as two-phase systems of matrix and dispersed particles of nanoscale. This entwinement allows exploiting the functionalities of both matrixes and that of nanoparticles. Owing to quantum and macro-quantum tunneling effects, nanomaterials possess some unique and tailorable chemical and physical properties. The paradigm shift toward membrane-nanomaterial composites can be attributed to the increased awareness of sustainability (Fig. 2.18). Depleting resources and increasing economic constraints have long craved solutions that are economical and sustainable.

2.3.1.1 Need for membrane-nanomaterial composites

Polymeric membranes are widely used conventional membranes due to their simple pore-forming mechanism, superior flexibility, and lesser footprints while installation and relatively cheaper than their inorganic counterparts. However, resistance toward chemicals, better mechanical strength, better conversions, and better selectivity have made inorganic membranes a fierce commercial competitor to polymeric membranes [108,109]. However, these conventional membranes also have some limitations. One such limitation is the bargain between permeability and selectivity. Altering the

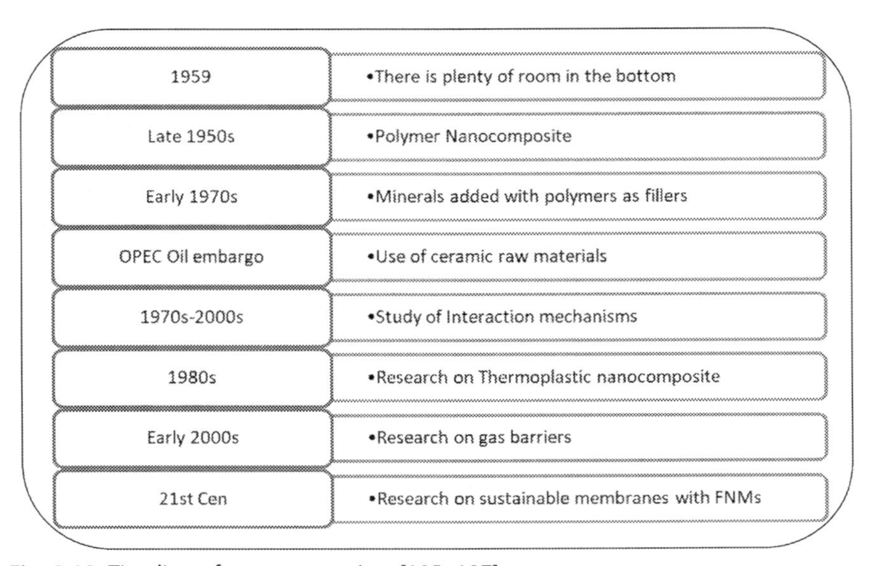

Fig. 2.18 Timeline of nanocomposites [105–107].

chemical structure of a material to improve permeability is often achieved at the cost of selectivity and vice versa [110]. Another limitation is the considerable energy consumption, fouling, and total cost of operating the process, limiting its use commercially. Thus, on the one hand, membrane processes driven by pressure fouling limit their commercialization potential by making them hostile to permeate quality deterioration and premature membrane replacement [111]. On the other hand, membrane processes that are not driven by pressure are limited by reduced permeation and fouling propensity [112]. Such limitations have led to research of superior membrane-nanomaterial composites.

2.3.2 Types of membrane-nanomaterial composites

The nanocomposite membranes can be segregated into various groups based on different criteria. The most common among these criteria is the cross section and fabrication method and structure. The fabrication method and structure can be grouped into four categories: conventional nanocomposite, TFN, TFC with nanocomposite substrate, and surface located nanocomposite (Fig. 2.19).

The conventional ones can be further segregated based on the materials used, namely organic, inorganic, biomaterial, and hybrid (Table 2.4). Another general classification is based on their cross sections and can be grouped as MMMs and TFCs. The mixed matrix membranes (MMMs) are synthesized by dispersing nanoparticles on casting solution followed by membrane casting. In contrast, the latter class is synthesized by subjecting the nanoparticles to assemble onto the membrane surface, followed by the creation of pores via dip-coating or pressurized deposition [140,141].

2.3.3 Theoretical approaches of binding functional nanomaterial on the membrane matrix

The type, amount, size, and way of incorporating nanomaterials affect the synthesized nanocomposite membranes properties (Fig. 2.20). Broadly the approaches of binding functionalized nanoparticles into the membrane matrix can be classified into blending sol-gel, and infiltration/in situ [111,137]. The former method involves blending the nanoparticles into a casting solution, followed by the phase inversion method of membrane fabrication. The last class involves depositing the NPs onto synthesized membranes and is much more efficient [143]. The profoundness of this later class of approaches for binding FNPs is because it allows for tweaking the

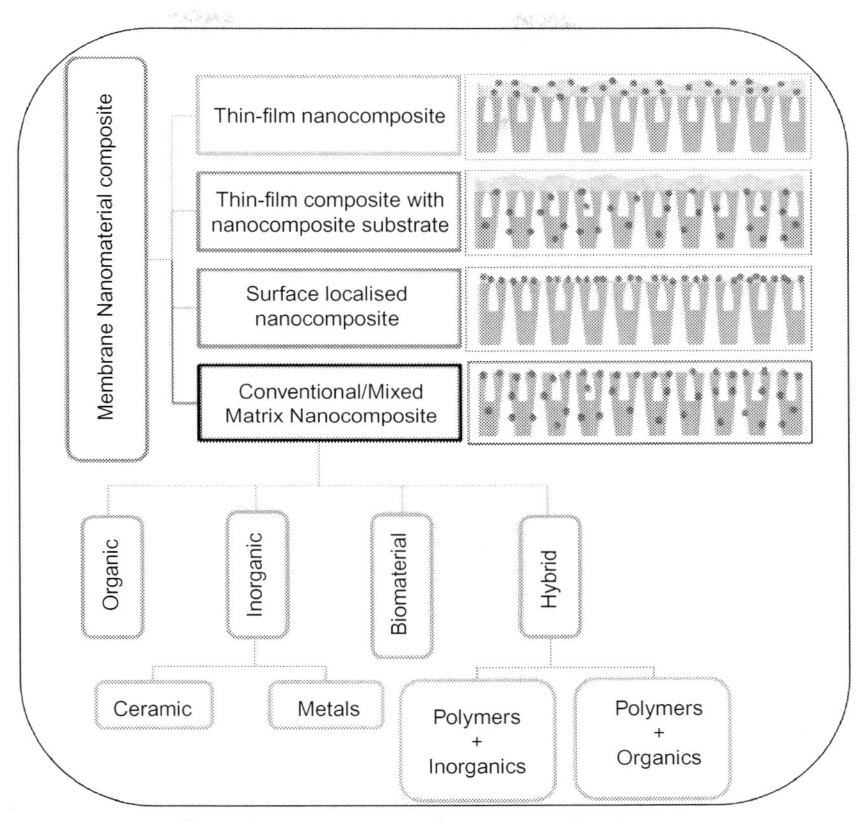

Fig. 2.19 Types of membrane-nanomaterial composites [113].

concentration and density of deposits with relatively improved controllability [144]. Another prevalent classification of approaches consists of four types: phased inversion mixed matrix, thin-film nanocomposite, surface layer coating, and infiltration (Fig. 2.21).

The phase inversion mixed matrix involves blending the nanomaterials into dope solution followed by membrane fabrication. This method is relatively more straightforward and compatible with an extensive range of solvents but faces backlash in controlling the distribution of NPs through the membrane [146,147]. The thin-film nanocomposite approach is relatively new and involves placing the NPs in between the support matrix and a thin film on the top. Unlike the previous approach, the incorporation of NPs onto the matrix can be controlled [148]. However, since the thin film is highly dependent on the IP conditions, it must be altered for its formation, thus adding manufacturing complexities [149]. Surface layer coating is a

Table 2.4 Types of membrane-nanomaterial composites and the list of properties altered.

Nanoparticle	Polymer	Properties altered	Reference
Mixed matrix			
Organic material			
PANI	PSf	Higher hydrophilicity	[114]
		Fouling propensity	
		Bigger pore size and porosity	
		Mechanical stability	
		Higher pure water flux	
		Mitigated elongation at break	[115]
PRh	PSf	Hydrophilicity	
		Porosity and Pore interconnection	
		Antifouling properties	
		Antibacterial properties	
POSS	CA	Membrane compaction	[116]
		Mechanical strength	
		Higher pure water flux	
Inorganic material			
Alumina	PES	Antifouling	[117]
		Higher flux	
	PVDF	Higher hydrophilicity	[118]
		Mechanical stability	
		Higher pure water flux	
CNT	PSf	Increased surface roughness	[119]
		Higher hydrophilicity	
		Mechanical stability	[120]
		Improved particle rejection	
Fe-Pd	PVDF	Required dechlorination	[121]
		Higher pure water flux	
		In situ formation of NPs at pore walls	
Silica	PSf	Mechanical stability	[122]
		Compaction	
		Higher pure water flux	
Ag	PSf	Superior mechanical strength	[123]
		Weakened hydraulic resistance	
		Antibacterial	

Table 2.4 Types of membrane-nanomaterial composites and the list of properties altered—cont'd

Nanoparticle	Polymer	Properties altered	Reference
Titania	PES	Hydrophilicity Higher pure water flux Superior rejection of solutes	[124]
		Thermal stability Mechanical stability	[125]
	PVDF	Thermal stability Reduced compaction under pressure Higher pure water flux	[126]
Hybrid material			
Cu NPs@ HNTs	PES	Antibacterial properties Increased membrane roughness Increased hydrophilicity Higher pure water flux	[127]
Ag NPs@ HNTs	PES	Hydrophilicity Higher pure water flux Antibacterial properties Superior rejection of solutes	[128]
Thin-film nanocomposite			
Ag-zeolite	PA-PSf	Hydrophilicity Water permeability Biofouling propensity	[129]
Titania	PA-PES	Improved permeability Salt rejection	[130]
Zeolite	PA-PSf	Hydrophilicity Surface charge Increased surface roughness Water permeability	[131]
		Salt rejection	[132]
NaA Zeolite	PSf	Improved permeability	[133]
Ag NPs	PSf	Increased water flux	[134]
		Permeability	[135]
Modified GO nanosheets	PSf	Antibacterial	[136]
		Permeability	[137]
SiO_2	PSf	Water permeability	[138]
		Salt rejection	[138]
GO QDs	PAN	Antifouling	[139]
TiO_2	PES	Increased pure water flux	[140]
		Salt rejection	[140]

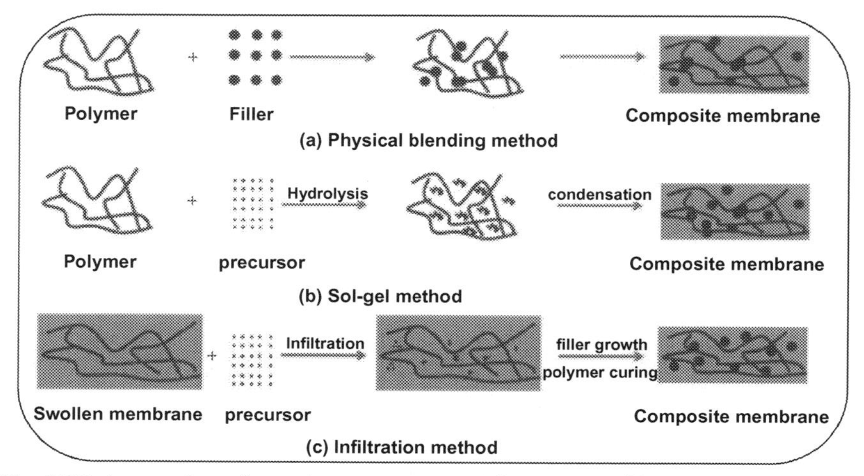

Fig. 2.20 Approaches of membrane-nanomaterial composite fabrication [142].

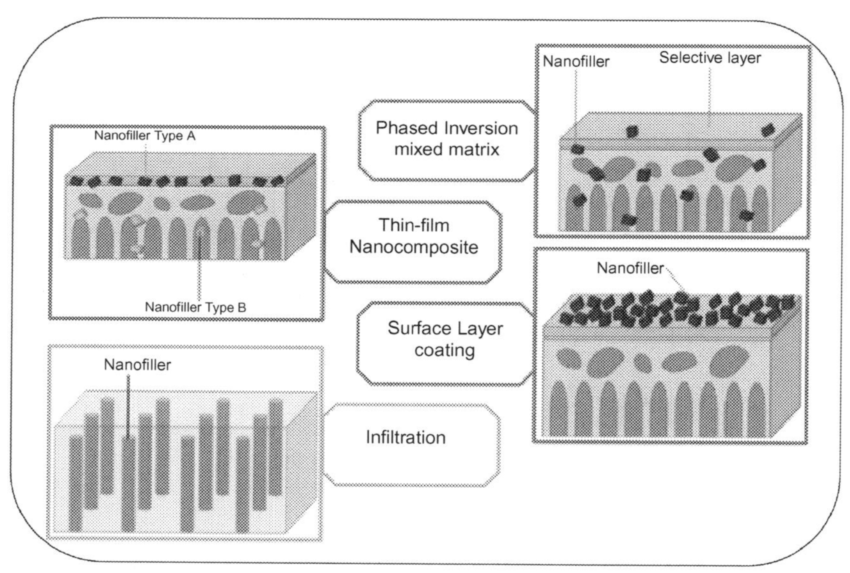

Fig. 2.21 Approaches used to incorporate NPs in membrane matrix [145].

deposition method where NPs are deposited on the already synthesized membranes. This opens up active sites of NPs, thus allowing more possibilities to tune the membrane properties [150]. Simultaneously, the lack of interfacial interactions between the support matrix and NPs leaching occurs, limiting its usability [151]. Infiltration is an entirely different approach and involves stuffing the support matrix with NP tubes, thus controlling the density and concentration. However, this ability to control the distribution comes with a considerable cost and is thus feasible for small membrane areas [144].

2.3.4 Applications of membrane-nanomaterial composites

Nanomaterials have opened up new possibilities and capabilities in membrane processes. Ever since the awareness of fast depleting clean water resources, researchers have come up with tons of modifications and functionalization in the membrane-nanomaterial composites, making it available for many applications whose absolute description is not within the scope of this chapter. However, a sneaky peek at some of the significant applications is shown in this chapter (Fig. 2.22).

2.3.4.1 Desalination

With the world toiling to cope with depleting freshwater resources, desalination has become a viable alternative occupying over 53% of clean water

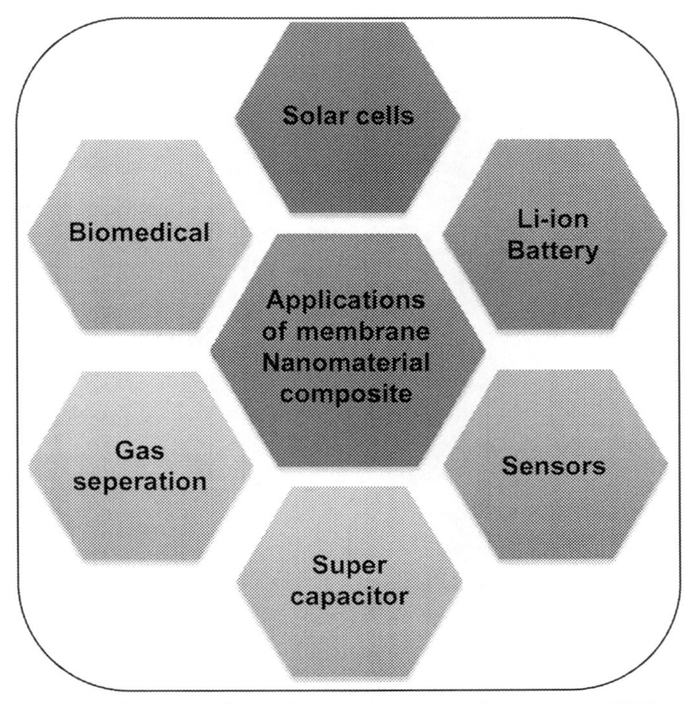

Fig. 2.22 Some applications of membrane-nanomaterial composite [152].

processes [153]. One major hurdle faced during the commercialization of membrane-based desalination is fouling. Fouling is described as the formation of an unwanted layer on the membrane or at the pores, which leads to the depletion of permeation flux, selectivity, and membrane life [154]. The addition of nanomaterials in the support matrix allows exploiting the antibacterial, antifouling, and other relevant properties of the NPs in the membrane, thus increasing its life span.

2.3.4.2 Photocatalysis

For decades chemicals such as chlorine and other disinfectants have been taken for granted in water treatment. These chemicals are used due to their residuals, which can produce carcinogenic products when they react with water. This process facilitates decontamination by using high-energy low-wavelength UV protons, which is detrimental to the cell walls and DNA [155].

Thus, removing pollutants via semiconductor-based photocatalysis with superior photoactivity, chemical stability, nontoxicity, and low cost was seen as a better alternative [156].

2.3.4.3 Fuel cells

Another field where NPs can have huge potential is in fuel cells. Irrespective of the field, researchers have shown that the electrochemical properties of NPs can be altered. Fuel cells are devices that by electrocatalysis convert chemical to electrical energy. The proton exchange membrane fuel cell is a subcategory of the fuel cell in which the fuel splits into protons and electrons, which then moves through a polymer electrolyte membrane placed in between the electrodes [157].

Owing to the ability of PEM to allow protons by Vehicle mechanism and Grotthuss mechanism and reject electrons to reach cathode avoiding short circuiting [158]. It is in this membrane where the NPs incorporated membranes find colossal potential.

2.3.4.4 Sensors

Sensors are one of the great beneficiaries of membrane-nanomaterial composites. With the advent of biomedical research, the sensors' biocompatibility has been in the arena of focus for several decades. The molecularly imprinted technology has been the answer to several such limitations and has proven its potential in garnishing biosensors with less toxicity, superior mechanical strength [159], low cost, and high selectivity [160]. Also, electrochemical sensors' ability to identify simultaneous molecules has exponentially increased its fame [161].

Several researchers have shown that using molecular imprinting technology has shown that synthesizing nanoparticles incorporated polymer films with the ability to recognize specific molecules [162,163]. The scientific diaspora is now focusing on exploiting computational technologies for synthesizing highly selective sensors due to the clean track record and potential of molecular imprinting technology, thanks to membrane-nanomaterial composites [164].

2.4 Conclusions

The present work offers a portrayal of the various concepts associated with membrane-nanomaterial composites. Being a booming field with its immense potential in exploring the plethora of other sustainable and cost-effective applications such as biosensors, desalination, and solar cells, the fundamentals of this field must be polished to build new applications on it. For years, researchers have eliminated each of the limitations encountered with conventional membranes such as fouling, lower mechanical strength, less selectivity, and membrane-nanomaterial composites have been the panacea for most hurdles.

The chapter has employed a locus standi with a piecemeal approach with an overhaul of nanomaterials and membranes, followed by the need to amalgamate these two engineered materials to synthesize membrane-nanomaterial composites.

With the kick start in nanotechnology from the 1960s, researchers have tweaked peculiar preferential properties of nanoparticles such as thermal stability, antibacterial properties, and localized surface plasmon resonance to their advantage, hence improvising them for specific applications such as biosensors, fuel cells, and agricultural engineering. This surge in nanotechnology research also brought several limitations such as agglomeration, inertness, and toxicity, which was mitigated by introducing functionalized nanomaterials, where specific entities such as dendrons and polymers were combined with the conventional NPs to synthesize superior and specialized NPs.

On the other hand, membranes have been the blue-eyed material, being exploited as a panacea for our toil to a sustainable future. The membrane processes such as reverse osmosis, nanofiltration, ultrafiltration, and microfiltration have been the pillars of membrane's applicability. However, conventional membranes came with a handful of limitations and for mitigation of which membrane-nanomaterial composites were produced. Unlike their ancestors, this engineering marvel was blessed with superior features such as lesser fouling, improved antibacterial properties, higher thermal stability, and better selectivity. These superiorities opened up newer doors of applications in drug delivery, water filtration, gas separation, and sensors.

In this book's later course, we would build upon these fundamental concepts and explore the fine details and potential offered by membrane-nanomaterial composites.

References

[1] Nano, Size of the Nanoscale, Nano, 2020. https://www.nano.gov/nanotech-101/what/nano-size. (Accessed 12 September 2020).

[2] K.N. Thakkar, S.S. Mhatre, R.Y. Parikh, Biological synthesis of metallic nanoparticles, Nanomed. Nanotechnol. Biol. Med. 6 (2) (2010) 257–262, https://doi.org/10.1016/j.nano.2009.07.002.

[3] M. Boholm, Å. Boholm, The many faces of nano in newspaper reporting, J. Nanopart. Res. 14 (2) (2012) 1–18, https://doi.org/10.1007/s11051-012-0722-y.

[4] P. Dobson, H. Jarvie, S. King, Nanoparticle, in: Definition, Size Range, & Applications, Britannica, 2019. https://www.britannica.com/science/nanoparticle. (accessed Sep. 12, 2020).

[5] R.P. Feynman, Plenty of Room at the Bottom, 1959.

[6] R. Geethalakshmi, D.V.L. Sarada, Gold and silver nanoparticles from Trianthema decandra: synthesis, characterization, and antimicrobial properties, Int. J. Nanomed. 7 (2012) 5375–5384, https://doi.org/10.2147/IJN.S36516.

[7] A. Matei, I. Cernica, O. Cadar, C. Roman, V. Schiopu, Synthesis and characterization of ZnO—polymer nanocomposites, Int. J. Mater. Form. 1 (Suppl. 1) (2008) 767–770, https://doi.org/10.1007/s12289-008-0288-5.

[8] R.M. Rioux, H. Song, J.D. Hoefelmeyer, P. Yang, G.A. Somorjai, High-surface-area catalyst design: synthesis, characterization, and reaction studies of platinum nanoparticles in mesoporous SBA-15 silica, J. Phys. Chem. B 109 (6) (2005) 2192–2202, https://doi.org/10.1021/jp048867x.

[9] S. Szunerits, R. Boukherroub, Sensing using localised surface plasmon resonance sensors, Chem. Commun. 48 (72) (2012) 8999–9010, https://doi.org/10.1039/c2cc33266c.

[10] Y. Guo, K. Xu, C. Wu, J. Zhao, Y. Xie, Surface chemical-modification for engineering the intrinsic physical properties of inorganic two-dimensional nanomaterials, Chem. Soc. Rev. 44 (3) (2015) 637–646, https://doi.org/10.1039/c4cs00302k. Royal Society of Chemistry.

[11] J.M. Pingarrón, P. Yáñez-Sedeño, A. González-Cortés, Gold nanoparticle-based electrochemical biosensors, Electrochim. Acta 53 (19) (2008) 5848–5866, https://doi.org/10.1016/j.electacta.2008.03.005.

[12] W.H. De Jong, P.J.A. Borm, Drug delivery and nanoparticles: applications and hazards, Int. J. Nanomed. 3 (2) (2008) 133–149, https://doi.org/10.2147/ijn.s596. Dove Press.

[13] N. Bumbudsanpharoke, J. Choi, S. Ko, Applications of nanomaterials in food packaging, J. Nanosci. Nanotechnol. 15 (9) (2015) 6357–6372, https://doi.org/10.1166/jnn.2015.10847. American Scientific Publishers.

[14] S. Kokura, O. Handa, T. Takagi, T. Ishikawa, Y. Naito, T. Yoshikawa, Silver nanoparticles as a safe preservative for use in cosmetics, Nanomed. Nanotechnol. Biol. Med. 6 (4) (2010) 570–574, https://doi.org/10.1016/j.nano.2009.12.002.
[15] L. Vinches, N. Testori, P. Dolez, G. Perron, K.J. Wilkinson, S. Hallé, Experimental evaluation of the penetration of TiO2 nanoparticles through protective clothing and gloves under conditions simulating occupational use, Nanosci. Methods 2 (1) (2013) 1–15, https://doi.org/10.1080/21642311.2013.771840.
[16] X. Huang, P.K. Jain, I.H. El-Sayed, M.A. El-Sayed, Gold nanoparticles: Interesting optical properties and recent applications in cancer diagnostics and therapy, Nanomedicine 2 (5) (2007) 681–693, https://doi.org/10.2217/17435889.2.5.681. Future Medicine Ltd London, UK.
[17] M.A. El-Sayed, Some interesting properties of metals confined in time and nanometer space of different shapes, Acc. Chem. Res. 34 (4) (2001) 257–264, https://doi.org/10.1021/ar960016n.
[18] M. Homberger, U. Simon, On the application potential of gold nanoparticles in nanoelectronics and biomedicine, Philos. Trans. R. Soc. A Math. Phys. Eng. Sci. 368 (1915) (2010) 1405–1453, https://doi.org/10.1098/rsta.2009.0275. Royal Society.
[19] R. Elghanian, J.J. Storhoff, R.C. Mucic, R.L. Letsinger, C.A. Mirkin, Selective colorimetric detection of polynucleotides based on the distance-dependent optical properties of gold nanoparticles, Science 277 (5329) (1997) 1078–1081, https://doi.org/10.1126/science.277.5329.1078.
[20] R. Kanjanawarut, X. Su, Colorimetric detection of DNA using unmodified metallic nanoparticles and peptide nucleic acid probes, Anal. Chem. 81 (15) (2009) 6122–6129, https://doi.org/10.1021/ac900525k.
[21] F. Li, et al., Nanomaterial-based fluorescent DNA analysis: a comparative study of the quenching effects of graphene oxide, carbon nanotubes, and gold nanoparticles, Adv. Funct. Mater. 23 (33) (2013) 4140–4148, https://doi.org/10.1002/adfm.201203816.
[22] C. Lu, H. Chiu, Adsorption of zinc(II) from water with purified carbon nanotubes, Chem. Eng. Sci. 61 (4) (2006) 1138–1145, https://doi.org/10.1016/j.ces.2005.08.007.
[23] H. Wang, K. Sun, F. Tao, D.J. Stacchiola, Y.H. Hu, 3D honeycomb-like structured graphene and its high efficiency as a counter-electrode catalyst for dye-sensitized solar cells, Angew. Chem. Int. Ed. 52 (35) (2013) 9210–9214, https://doi.org/10.1002/anie.201303497.
[24] D.G. Thompson, A. Enright, K. Faulds, W.E. Smith, D. Graham, Ultrasensitive DNA detection using oligonucleotide-silver nanoparticle conjugates, Anal. Chem. 80 (8) (2008) 2805–2810, https://doi.org/10.1021/ac702403w.
[25] T. Kang, H. Choi, S.W. Joo, S.Y. Lee, K.A. Yoon, K. Lee, Peptide nucleic acid-mediated aggregation of reduced graphene oxides and label-free detection of DNA mutation, J. Phys. Chem. B 118 (23) (2014) 6297–6301, https://doi.org/10.1021/jp501820j.
[26] H. Peng, L. Zhang, T.H.M. Kjällman, C. Soeller, J. Travas-Sejdic, DNA hybridization detection with blue luminescent quantum dots and dye-labeled single-stranded DNA, J. Am. Chem. Soc. 129 (11) (2007) 3048–3049, https://doi.org/10.1021/ja0685452.
[27] Z. Li, et al., Aptamer-conjugated dendrimer-modified quantum dots for cancer cell targeting and imaging, Mater. Lett. 64 (3) (2010) 375–378, https://doi.org/10.1016/j.matlet.2009.11.022.
[28] S.H. Gage, et al., Functionalization of monodisperse iron oxide NPs and their properties as magnetically recoverable catalysts, Langmuir 29 (1) (2013) 466–473, https://doi.org/10.1021/la304410z.
[29] A. Queraltó, et al., Reduced graphene oxide/iron oxide nanohybrid flexible electrodes grown by laser-based technique for energy storage applications, Ceram. Int. 44 (16) (2018) 20409–20416, https://doi.org/10.1016/j.ceramint.2018.08.034.

[30] Z. Zhou, et al., Iron/iron oxide core/shell nanoparticles for magnetic targeting MRI and near-infrared photothermal therapy, Biomaterials 35 (26) (2014) 7470–7478, https://doi.org/10.1016/j.biomaterials.2014.04.063.
[31] C.Y. Cheng, et al., Direct and in vitro observation of growth hormone receptor molecules in A549 human lung epithelial cells by nanodiamond labeling, Appl. Phys. Lett. 90 (16) (2007), https://doi.org/10.1063/1.2727557, 163903.
[32] X.Q. Zhang, M. Chen, R. Lam, X. Xu, E. Osawa, D. Ho, Polymer-functionalized nanodiamond platforms as vehicles for gene delivery, ACS Nano 3 (9) (2009) 2609–2616, https://doi.org/10.1021/nn900865g.
[33] Z. Sun, J.H. Kim, Y. Zhao, F. Bijarbooneh, V. Malgras, S.X. Dou, Improved photovoltaic performance of dye-sensitized solar cells with modified self-assembling highly ordered mesoporous TiO 2 photoanodes, J. Mater. Chem. 22 (23) (2012) 11711–11719, https://doi.org/10.1039/c2jm30660c.
[34] S. Das, A. Bhattacharya, N. Debnath, A. Datta, A. Goswami, Nanoparticle-induced morphological transition of Bombyx mori nucleopolyhedrovirus: a novel method to treat silkworm grasserie disease, Appl. Microbiol. Biotechnol. 97 (13) (2013) 6019–6030, https://doi.org/10.1007/s00253-013-4868-z.
[35] K.M. Paknikar, V. Nagpal, A.V. Pethkar, J.M. Rajwade, Degradation of lindane from aqueous solutions using iron sulfide nanoparticles stabilized by biopolymers, Sci. Technol. Adv. Mater. 6 (3–4) (2005) 370–374, https://doi.org/10.1016/j.stam.2005.02.016.
[36] K.A. Willets, R.P. Van Duyne, Localized surface Plasmon resonance spectroscopy and sensing, Annu. Rev. Phys. Chem. 58 (1) (2007) 267–297, https://doi.org/10.1146/annurev.physchem.58.032806.104607.
[37] L. Dykman, N. Khlebtsov, Gold nanoparticles in biomedical applications: recent advances and perspectives, Chem. Soc. Rev. 41 (6) (2012) 2256–2282, https://doi.org/10.1039/c1cs15166e.
[38] J.A. Creighton, D.G. Eadon, Ultraviolet-visible absorption spectra of the colloidal metallic elements, J. Chem. Soc. Faraday Trans. 87 (24) (1991) 3881–3891, https://doi.org/10.1039/FT9918703881.
[39] M. Brust, C.J. Kiely, Some recent advances in nanostructure preparation from gold and silver particles: a short topical review, Colloids Surf. A Physicochem. Eng. Asp. 202 (2–3) (2002) 175–186. Accessed: Sep. 12, 2020. [Online]. Available https://www.academia.edu/1045119/Some_recent_advances_in_nanostructure_preparation_from_gold_and_silver_particles_a_short_topical_review.
[40] D.M. Schaefer, A. Patil, R.P. Andres, R. Reifenberger, Elastic properties of individual nanometer-size supported gold clusters, Phys. Rev. B 51 (8) (1995) 5322–5332, https://doi.org/10.1103/PhysRevB.51.5322.
[41] K.J. Klabunde, R.M. Richards, Nanoscale Materials in Chemistry, 2nd Edition | Wiley, second ed., Wiley, 2009.
[42] C. Burda, X. Chen, R. Narayanan, M.A. El-Sayed, Chemistry and properties of nanocrystals of different shapes, Chem. Rev. 105 (4) (2005) 1025–1102, https://doi.org/10.1021/cr030063a. Chem Rev.
[43] R. Wilson, The use of gold nanoparticles in diagnostics and detection, Chem. Soc. Rev. 37 (9) (2008) 2028–2045, https://doi.org/10.1039/b712179m.
[44] M.A. Habeeb Muhammed, et al., Luminescent quantum clusters of gold in bulk by albumin-induced core etching of nanoparticles: metal ion sensing, metal-enhanced luminescence, and biolabeling, Chem. A Eur. J. 16 (33) (2010) 10103–10112, https://doi.org/10.1002/chem.201000841.
[45] R.P. Feynman, There's Plenty of Room at the Bottom, 1960.
[46] G. Binnig, H. Rohrer, Scanning tunneling microscopy from birth to adolescence, Rev. Mod. Phys. 59 (3) (1987) 615–625, https://doi.org/10.1103/RevModPhys.59.615.

[47] E.W. Müller, Work function of tungsten single crystal planes measured by the field emission microscope, J. Appl. Phys. 26 (6) (1955) 732–737, https://doi.org/10.1063/1.1722081.
[48] P.O. of the E. Union, COM/2004/0338 final, CELEX1, Communication from the Commission—Towards a European Strategy for Nanotechnology, 2004.
[49] P. Sreeramana Aithal, S. Aithal, Nanotechnology Innovations and Commercialization-Opportunities, Challenges & Reasons for Delay, 2016, Accessed: Sep. 12, 2020. [Online]. Available https://ssrn.com/abstract=2866363.
[50] M.C. Roco, From vision to the implementation of the U.S. National Nanotechnology Initiative, J. Nanopart. Res. 3 (1) (2001) 5–11, https://doi.org/10.1023/A:1011429917892.
[51] Social Studies, Grand Challenges: Nanotechnology and the Social Studies, Social Studies, 2020. https://www.socialstudies.org/social-education/77/2/grand-challenges-nanotechnology-and-social-studies. (Accessed 12 September 2020).
[52] J.F. Sargent, Nanotechnology: A Policy Primer, 2016, Accessed: Sep. 12, 2020. [Online]. Available www.crs.gov.
[53] Nano, Nanotechnology Timeline, Nano, 2020. https://www.nano.gov/timeline. (Accessed 12 September 2020).
[54] S. Jiang, K.Y. Win, S. Liu, C.P. Teng, Y. Zheng, M.Y. Han, Surface-functionalized nanoparticles for biosensing and imaging-guided therapeutics, Nanoscale 5 (8) (2013) 3127–3148, https://doi.org/10.1039/c3nr34005h. Royal Society of Chemistry.
[55] R. Mout, D.F. Moyano, S. Rana, V.M. Rotello, Surface functionalization of nanoparticles for nanomedicine, Chem. Soc. Rev. 41 (7) (2012) 2539–2544, https://doi.org/10.1039/c2cs15294k.
[56] A. Ravindran, P. Chandran, S.S. Khan, Biofunctionalized silver nanoparticles: Advances and prospects, Colloids Surf. B Biointerfaces 105 (2013) 342–352, https://doi.org/10.1016/j.colsurfb.2012.07.036. Elsevier.
[57] T. Wang, et al., Adsorption of agricultural wastewater contaminated with antibiotics, pesticides and toxic metals by functionalized magnetic nanoparticles, J. Environ. Chem. Eng. 6 (5) (2018) 6468–6478, https://doi.org/10.1016/j.jece.2018.10.014. Elsevier Ltd.
[58] R.W. Baker, Membrane Technology and Applications—Richard W. Baker—Google Books, John Wiley and Sons, 2012.
[59] R. Rautenbach, R. Albrecht, Membrane Separation Processes, John Wiley and Sons Inc, New York, NY, 1989.
[60] S. Alzahrani, A.W. Mohammad, Challenges and trends in membrane technology implementation for produced water treatment: a review, J. Water Process Eng. 4 (C) (2014) 107–133, https://doi.org/10.1016/j.jwpe.2014.09.007. Elsevier Ltd.
[61] European Membrane Society. https://www.emsoc.eu/. (accessed Sep. 12, 2020).
[62] S. Nunes, Membrane Technology in the Chemical Industry, Wiley-VCH, Weinheim ; New York, 2001.
[63] A.B. Fuertes, T.A. Centeno, Preparation of supported asymmetric carbon molecular sieve membranes, J. Membr. Sci. 144 (1–2) (1998) 105–111, https://doi.org/10.1016/S0376-7388(98)00037-4.
[64] Y.D. Chen, R.T. Yang, Preparation of carbon molecular sieve membrane and diffusion of binary mixtures in the membrane, Ind. Eng. Chem. Res. 33 (12) (1994) 3146–3153, https://doi.org/10.1021/ie00036a033.
[65] I.C. Pereira, A.S. Duarte, A.S. Neto, J.M.F. Ferreira, Chitosan and polyethylene glycol based membranes with antibacterial properties for tissue regeneration, Mater. Sci. Eng. C 96 (2019) 606–615, https://doi.org/10.1016/j.msec.2018.11.029.
[66] A. Mollahosseini, A. Rahimpour, A new concept in polymeric thin-film composite nanofiltration membranes with antibacterial properties, Biofouling 29 (5) (2013) 537–548, https://doi.org/10.1080/08927014.2013.777953.

[67] T.A. Saleh, V.K. Gupta, Nanomaterial and Polymer Membranes: Synthesis, Characterization, and Applications, Elsevier Inc, 2016.

[68] R.M. De Vos, H. Verweij, High-selectivity, high-flux silica membranes for gas separation, Science 279 (5357) (1998) 1710–1711, https://doi.org/10.1126/science.279.5357.1710.

[69] M. Vinoba, M. Bhagiyalakshmi, Y. Alqaheem, A.A. Alomair, A. Pérez, M.S. Rana, Recent progress of fillers in mixed matrix membranes for CO2 separation: a review, Sep. Purif. Technol. 188 (2017) 431–450, https://doi.org/10.1016/j.seppur.2017.07.051. Elsevier B.V.

[70] D.R. Lloyd, K.E. Kinzer, H.S. Tseng, Microporous membrane formation via thermally induced phase separation. I. Solid-liquid phase separation, J. Membr. Sci. 52 (3) (1990) 239–261, https://doi.org/10.1016/S0376-7388(00)85130-3.

[71] L.Y. Ng, A.W. Mohammad, C.P. Leo, N. Hilal, Polymeric membranes incorporated with metal/metal oxide nanoparticles: a comprehensive review, Desalination 308 (2013) 15–33, https://doi.org/10.1016/j.desal.2010.11.033.

[72] D. Emadzadeh, et al., A novel thin film nanocomposite reverse osmosis membrane with superior anti-organic fouling affinity for water desalination, Desalination 368 (2015) 106–113, https://doi.org/10.1016/j.desal.2014.11.019.

[73] A. Alpatova, E.S. Kim, X. Sun, G. Hwang, Y. Liu, M. Gamal El-Din, Fabrication of porous polymeric nanocomposite membranes with enhanced anti-fouling properties: effect of casting composition, J. Membr. Sci. 444 (2013) 449–460, https://doi.org/10.1016/j.memsci.2013.05.034.

[74] L.Q. Shen, Z.K. Xu, Z.M. Liu, Y.Y. Xu, Ultrafiltration hollow fiber membranes of sulfonated polyetherimide/polyetherimide blends: preparation, morphologies and anti-fouling properties, J. Membr. Sci. 218 (1–2) (2003) 279–293, https://doi.org/10.1016/S0376-7388(03)00186-8.

[75] M.A. Aroon, A.F. Ismail, T. Matsuura, M.M. Montazer-Rahmati, Performance studies of mixed matrix membranes for gas separation: a review, Sep. Purif. Technol. 75 (3) (2010) 229–242, https://doi.org/10.1016/j.seppur.2010.08.023. Elsevier B.V.

[76] G. Dong, H. Li, V. Chen, Challenges and opportunities for mixed-matrix membranes for gas separation, J. Mater. Chem. A 1 (15) (2013) 4610–4630, https://doi.org/10.1039/c3ta00927k.

[77] O.K. Dalrymple, W. Isaacs, E. Stefanakos, M.A. Trotz, D.Y. Goswami, Lipid vesicles as model membranes in photocatalytic disinfection studies, J. Photochem. Photobiol. A Chem. 221 (1) (2011) 64–70, https://doi.org/10.1016/j.jphotochem.2011.04.025.

[78] L.M. Pastrana-Martínez, S. Morales-Torres, J.L. Figueiredo, J.L. Faria, A.M.T. Silva, Graphene oxide based ultrafiltration membranes for photocatalytic degradation of organic pollutants in salty water, Water Res. 77 (2015) 179–190, https://doi.org/10.1016/j.watres.2015.03.014.

[79] H.G. Choi, M. Son, H. Choi, Integrating seawater desalination and wastewater reclamation forward osmosis process using thin-film composite mixed matrix membrane with functionalized carbon nanotube blended polyethersulfone support layer, Chemosphere 185 (2017) 1181–1188, https://doi.org/10.1016/j.chemosphere.2017.06.136.

[80] E. Bakangura, L. Wu, L. Ge, Z. Yang, T. Xu, Mixed matrix proton exchange membranes for fuel cells: state of the art and perspectives, Prog. Polym. Sci. 57 (2016) 103–152, https://doi.org/10.1016/j.progpolymsci.2015.11.004. Elsevier Ltd.

[81] H.S. Samanta, S.K. Ray, Separation of ethanol from water by pervaporation using mixed matrix copolymer membranes, Sep. Purif. Technol. 146 (2015) 176–186, https://doi.org/10.1016/j.seppur.2015.03.006.

[82] S. Loeb, S. Sourirajan, Sea water demineralization by means of an osmotic membrane, Adv. Chem. 38 (1963) 117–132.

[83] P. Radovanovic, S.W. Thiel, S.T. Hwang, Formation of asymmetric polysulfone membranes by immersion precipitation. Part I. Modelling mass transport during gelation, J. Membr. Sci. 65 (3) (1992) 213–229, https://doi.org/10.1016/0376-7388(92)87024-R.

[84] K. Boussu, B. Van der Bruggen, C. Vandecasteele, Evaluation of self-made nanoporous polyethersulfone membranes, relative to commercial nanofiltration membranes, Desalination 200 (1–3) (2006) 416–418, https://doi.org/10.1016/j.desal.2006.03.353.

[85] W. Zhu, X. Zhang, C. Zhao, W. Wu, J. Hou, M. Xu, A novel polypropylene microporous film, Polym. Adv. Technol. 7 (9) (1996) 743–748, https://doi.org/10.1002/(SICI)1099-1581(199609)7:9<743::AID-PAT548>3.0.CO;2-W.

[86] T. Sarada, L.C. Sawyer, M.I. Ostler, Three dimensional structure of celgard® microporous membranes, J. Membr. Sci. 15 (1) (1983) 97–113, https://doi.org/10.1016/S0376-7388(00)81364-2.

[87] R.L. Fleischer, P.B. Price, R.M. Walker, R.M. Walker, Nuclear Tracks in Solids: Principles and Applications, Google Books, 1975. https://books.google.co.in/books?hl=en&lr=&id=yfTBvben3GoC&oi=fnd&pg=PR15&ots=F5utc0zlDD&sig=6SLKvCqbgOIwSh8huzuH_BQYQ0I&redir_esc=y#v=onepage&q&f=false. (accessed Sep. 17, 2020).

[88] Y. Komaki, S. Tsujimura, Growth of fine holes in polyethylenenaphthalate film irradiated by fission fragments, J. Appl. Phys. 47 (4) (1976) 1355–1358, https://doi.org/10.1063/1.322840.

[89] P. Apel, Track etching technique in membrane technology, Radiat. Meas. 34 (1–6) (2001) 559–566, https://doi.org/10.1016/S1350-4487(01)00228-1.

[90] M.E. Toimil-Molares, Characterization and properties of micro- and nanowires of controlled size, composition, and geometry fabricated by electrodeposition and ion-track technology, Beilstein J. Nanotechnol. 3 (1) (2012) 860–883, https://doi.org/10.3762/bjnano.3.97.

[91] R.L. Fleischer, P.B. Price, E.M. Symes, Novel filter for biological materials, Science 143 (3603) (1964) 249–250, https://doi.org/10.1126/science.143.3603.249.

[92] M. Grasselli, N. Betz, Making porous membranes by chemical etching of heavy-ion tracks in β-PVDF films, Nucl. Inst. Methods Phys. Res. B 236 (1–4) (2005) 501–507, https://doi.org/10.1016/j.nimb.2005.04.027.

[93] G.I. Taylor, Electrically driven jets, Proc. R. Soc. London. A. Math. Phys. Sci. 313 (1515) (1969) 453–475, https://doi.org/10.1098/rspa.1969.0205.

[94] J. Zeleny, The electrical discharge from liquid points, and a hydrostatic method of measuring the electric intensity at their surfaces, Phys. Rev. 3 (2) (1914) 69–91, https://doi.org/10.1103/PhysRev.3.69.

[95] B.S. Lalia, V. Kochkodan, R. Hashaikeh, N. Hilal, A review on membrane fabrication: structure, properties and performance relationship, Desalination 326 (2013) 77–95, https://doi.org/10.1016/j.desal.2013.06.016.

[96] N. Bhardwaj, S.C. Kundu, Electrospinning: a fascinating fiber fabrication technique, Biotechnol. Adv. 28 (3) (2010) 325–347, https://doi.org/10.1016/j.biotechadv.2010.01.004. Elsevier.

[97] J.E. Cadotte, R.J. Petersen, R.E. Larson, E.E. Erickson, A new thin-film composite seawater reverse osmosis membrane, Desalination 32 (C) (1980) 25–31, https://doi.org/10.1016/S0011-9164(00)86003-8.

[98] P. Radovanovic, S.W. Thiel, S.T. Hwang, Formation of asymmetric polysulfone membranes by immersion precipitation. Part II. The effects of casting solution and gelation bath compositions on membrane structure and skin formation, J. Membr. Sci. 65 (3) (1992) 231–246, https://doi.org/10.1016/0376-7388(92)87025-S.

[99] J. Cadotte, R. Forester, M. Kim, R. Petersen, T. Stocker, Nanofiltration membranes broaden the use of membrane separation technology, Desalination 70 (1–3) (1988) 77–88, https://doi.org/10.1016/0011-9164(88)85045-8.

[100] R.J. Petersen, Composite reverse osmosis and nanofiltration membranes, J. Membr. Sci. 83 (1) (1993) 81–150, https://doi.org/10.1016/0376-7388(93)80014-O. Elsevier.
[101] A.K. Ghosh, B.H. Jeong, X. Huang, E.M.V. Hoek, Impacts of reaction and curing conditions on polyamide composite reverse osmosis membrane properties, J. Membr. Sci. 311 (1–2) (2008) 34–45, https://doi.org/10.1016/j.memsci.2007.11.038.
[102] Y. Lin, H. Li, C. Liu, W. Xing, X. Ji, Surface-modified Nafion membranes with mesoporous SiO2 layers via a facile dip-coating approach for direct methanol fuel cells, J. Power Sources 185 (2) (2008) 904–908, https://doi.org/10.1016/j.jpowsour.2008.08.067.
[103] B.A. McCool, N. Hill, J. DiCarlo, W.J. DeSisto, Synthesis and characterization of mesoporous silica membranes via dip-coating and hydrothermal deposition techniques, J. Membr. Sci. 218 (1–2) (2003) 55–67, https://doi.org/10.1016/S0376-7388(03)00136-4.
[104] M.R. Esfahani, et al., Nanocomposite membranes for water separation and purification: fabrication, modification, and applications, Sep. Purif. Technol. 213 (2019) 465–499, https://doi.org/10.1016/j.seppur.2018.12.050. Elsevier B.V.
[105] T.S. Chung, L.Y. Jiang, Y. Li, S. Kulprathipanja, Mixed matrix membranes (MMMs) comprising organic polymers with dispersed inorganic fillers for gas separation, Prog. Polym. Sci. 32 (4) (2007) 483–507, https://doi.org/10.1016/j.progpolymsci.2007.01.008.
[106] L. Li, C. Visvanathan, Membrane technology for surface water treatment: advancement from microfiltration to membrane bioreactor, Rev. Environ. Sci. Biotechnol. 16 (4) (2017) 737–760, https://doi.org/10.1007/s11157-017-9442-1. Springer Netherlands.
[107] D. Zioui, Z. Tigrine, H. Aburideh, S. Hout, M. Abbas, N. Merzouk, Membrane Technology for Water Treatment Applications Solar Refrigeration View Project Solid Drying Solar Air Conditionning System View Project Membrane Technology for Water Treatment Applications, 2020, Accessed: Sep. 13, 2020. [Online]. Available https://www.researchgate.net/publication/301647575.
[108] J.M. Skluzacek, M.I. Tejedor, M.A. Anderson, NaCl rejection by an inorganic nanofiltration membrane in relation to its central pore potential, J. Membr. Sci. 289 (1–2) (2007) 32–39, https://doi.org/10.1016/j.memsci.2006.11.034.
[109] J. Zaman, A. Chakma, Inorganic membrane reactors, J. Membr. Sci. 92 (1) (1994) 1–28, https://doi.org/10.1016/0376-7388(94)80010-3. Elsevier.
[110] L.M. Robeson, Correlation of separation factor versus permeability for polymeric membranes, J. Membr. Sci. 62 (2) (1991) 165–185, https://doi.org/10.1016/0376-7388(91)80060-J.
[111] A.H.M.A. Sadmani, The Rejection of Pharmaceutically Active and Endocrine Disrupting Compounds via Nanofiltration as a Function of Natural Water Components, 2009.
[112] D.L. Shaffer, J.R. Werber, H. Jaramillo, S. Lin, M. Elimelech, Forward osmosis: where are we now? Desalination 356 (2015) 271–284, https://doi.org/10.1016/j.desal.2014.10.031. Elsevier.
[113] J. Yin, B. Deng, Polymer-matrix nanocomposite membranes for water treatment, J. Membr. Sci. 479 (2015) 256–275, https://doi.org/10.1016/j.memsci.2014.11.019.
[114] S. Zhao, et al., Performance improvement of polysulfone ultrafiltration membrane using well-dispersed polyaniline-poly(vinylpyrrolidone) nanocomposite as the additive, Ind. Eng. Chem. Res. 51 (12) (2012) 4661–4672, https://doi.org/10.1021/ie202503p.
[115] Z. Fan, Z. Wang, N. Sun, J. Wang, S. Wang, Performance improvement of polysulfone ultrafiltration membrane by blending with polyaniline nanofibers, J. Membr. Sci. 320 (1–2) (2008) 363–371, https://doi.org/10.1016/j.memsci.2008.04.019.

[116] C.H. Worthley, K.T. Constantopoulos, M. Ginic-Markovic, E. Markovic, S. Clarke, A study into the effect of POSS nanoparticles on cellulose acetate membranes, J. Membr. Sci. 431 (2013) 62–71, https://doi.org/10.1016/j.memsci.2012.12.025.

[117] N. Maximous, G. Nakhla, W. Wan, K. Wong, Preparation, characterization and performance of Al2O3/PES membrane for wastewater filtration, J. Membr. Sci. 341 (1–2) (2009) 67–75, https://doi.org/10.1016/j.memsci.2009.05.040.

[118] M. Safarpour, A. Khataee, V. Vatanpour, Preparation of a novel polyvinylidene fluoride (PVDF) ultrafiltration membrane modified with reduced graphene oxide/titanium dioxide (TiO2) nanocomposite with enhanced hydrophilicity and antifouling properties, Ind. Eng. Chem. Res. 53 (34) (2014) 13370–13382, https://doi.org/10.1021/ie502407g.

[119] L. Brunet, et al., Properties of membranes containing semi-dispersed carbon nanotubes, Environ. Eng. Sci. 25 (4) (2008) 565–575, https://doi.org/10.1089/ees.2007.0076.

[120] J.H. Choi, J. Jegal, W.N. Kim, Fabrication and characterization of multi-walled carbon nanotubes/polymer blend membranes, J. Membr. Sci. 284 (1–2) (2006) 406–415, https://doi.org/10.1016/j.memsci.2006.08.013.

[121] V. Smuleac, L. Bachas, D. Bhattacharyya, Aqueous-phase synthesis of PAA in PVDF membrane pores for nanoparticle synthesis and dichlorobiphenyl degradation, J. Membr. Sci. 346 (2) (2010) 310–317, https://doi.org/10.1016/j.memsci.2009.09.052.

[122] M.T.M. Pendergast, J.M. Nygaard, A.K. Ghosh, E.M.V. Hoek, Using nanocomposite materials technology to understand and control reverse osmosis membrane compaction, Desalination 261 (3) (2010) 255–263, https://doi.org/10.1016/j.desal.2010.06.008.

[123] J.S. Taurozzi, et al., Effect of filler incorporation route on the properties of polysulfone-silver nanocomposite membranes of different porosities, J. Membr. Sci. 325 (1) (2008) 58–68, https://doi.org/10.1016/j.memsci.2008.07.010.

[124] M. Luo, Q. Wen, J. Liu, H. Liu, Z. Jia, Fabrication of SPES/nano-TiO2 composite ultrafiltration membrane and its anti-fouling mechanism, Chin. J. Chem. Eng. 19 (1) (2011) 45–51, https://doi.org/10.1016/S1004-9541(09)60175-0.

[125] M.L. Luo, W. Tang, J.Q. Zhao, C.S. Pu, Hydrophilic modification of poly(ether sulfone) used TiO2 nanoparticles by a sol-gel process, J. Mater. Process. Technol. 172 (3) (2006) 431–436, https://doi.org/10.1016/j.jmatprotec.2005.11.004.

[126] K. Ebert, D. Fritsch, J. Koll, C. Tjahjawiguna, Influence of inorganic fillers on the compaction behaviour of porous polymer based membranes, J. Membr. Sci. 233 (1–2) (2004) 71–78, https://doi.org/10.1016/j.memsci.2003.12.012.

[127] Google, US8795565B2—Biaxially Oriented Microporous Membrane, Google Patents, 2020. https://patents.google.com/patent/US8795565B2/en. (Accessed 23 September 2020).

[128] A.S. Gozdz, C.N. SchmutzJean, M. Tarascon, P.C. Warren, Method of Making Polymeric Electrolytic Cell Separator Membrane, US5607485A, 1995.

[129] M.L. Lind, B.H. Jeong, A. Subramani, X. Huang, E.M.V. Hoek, Effect of mobile cation on zeolite-polyamide thin film nanocomposite membranes, J. Mater. Res. 24 (5) (2009) 1624–1631, https://doi.org/10.1557/jmr.2009.0189.

[130] H.S. Lee, S.J. Im, J.H. Kim, H.J. Kim, J.P. Kim, B.R. Min, Polyamide thin-film nanofiltration membranes containing TiO2 nanoparticles, Desalination 219 (1–3) (2008) 48–56, https://doi.org/10.1016/j.desal.2007.06.003.

[131] B.H. Jeong, et al., Interfacial polymerization of thin film nanocomposites: a new concept for reverse osmosis membranes, J. Membr. Sci. 294 (1–2) (2007) 1–7, https://doi.org/10.1016/j.memsci.2007.02.025.

[132] M.L. Lind, D.E. Suk, T.V. Nguyen, E.M.V. Hoek, Tailoring the structure of thin film nanocomposite membranes to achieve seawater RO membrane performance, Environ. Sci. Technol. 44 (21) (2010) 8230–8235, https://doi.org/10.1021/es101569p.

[133] S.F. Anis, B.S. Lalia, R. Hashaikeh, N. Hilal, Breaking through the selectivity-permeability tradeoff using nano zeolite-Y for micellar enhanced ultrafiltration dye rejection application, Sep. Purif. Technol. 242 (2020), https://doi.org/10.1016/j.seppur.2020.116824, 116824.
[134] M. Mofradi, H. Karimi, M. Ghaedi, Hydrophilic polymeric membrane supported on silver nanoparticle surface decorated polyester textile: toward enhancement of water flux and dye removal, Chin. J. Chem. Eng. 28 (3) (2020) 901–912, https://doi.org/10.1016/j.cjche.2019.09.011.
[135] Z. Yang, et al., A novel thin-film nano-templated composite membrane with in situ silver nanoparticles loading: separation performance enhancement and implications, J. Membr. Sci. 544 (2017) 351–358, https://doi.org/10.1016/j.memsci.2017.09.046.
[136] Y. Zhang, H. Ruan, C. Guo, J. Liao, J. Shen, C. Gao, Thin-film nanocomposite reverse osmosis membranes with enhanced antibacterial resistance by incorporating p-aminophenol-modified graphene oxide, Sep. Purif. Technol. 234 (2020), https://doi.org/10.1016/j.seppur.2019.116017, 116017.
[137] Y. Kang, M. Obaid, J. Jang, I.S. Kim, Sulfonated graphene oxide incorporated thin film nanocomposite nanofiltration membrane to enhance permeation and antifouling properties, Desalination 470 (2019), https://doi.org/10.1016/j.desal.2019.114125, 114125.
[138] N. Niksefat, M. Jahanshahi, A. Rahimpour, The effect of SiO2 nanoparticles on morphology and performance of thin film composite membranes for forward osmosis application, Desalination 343 (2014) 140–146, https://doi.org/10.1016/j.desal.2014.03.031.
[139] C. Zhang, K. Wei, W. Zhang, Y. Bai, Y. Sun, J. Gu, Graphene oxide quantum dots incorporated into a thin film nanocomposite membrane with high flux and antifouling properties for low-pressure nanofiltration, ACS Appl. Mater. Interfaces 9 (12) (2017) 11082–11094, https://doi.org/10.1021/acsami.6b12826.
[140] V. Vatanpour, S.S. Madaeni, A.R. Khataee, E. Salehi, S. Zinadini, H.A. Monfared, TiO_2 embedded mixed matrix PES nanocomposite membranes: influence of different sizes and types of nanoparticles on antifouling and performance, Desalination 292 (2012) 19–29, https://doi.org/10.1016/j.desal.2012.02.006.
[141] R.A. Damodar, S.J. You, H.H. Chou, Study the self cleaning, antibacterial and photocatalytic properties of TiO2 entrapped PVDF membranes, J. Hazard. Mater. 172 (2–3) (2009) 1321–1328, https://doi.org/10.1016/j.jhazmat.2009.07.139.
[142] S.L. Li, Q. Xu, Metal-organic frameworks as platforms for clean energy, Energy Environ. Sci. 6 (6) (2013) 1656–1683, https://doi.org/10.1039/c3ee40507a. The Royal Society of Chemistry.
[143] M.A. Alaei Shahmirzadi, S.S. Hosseini, G. Ruan, N.R. Tan, Tailoring PES nanofiltration membranes through systematic investigations of prominent design, fabrication and operational parameters, RSC Adv. 5 (61) (2015) 49080–49097, https://doi.org/10.1039/c5ra05985b.
[144] D. Mattia, K.P. Lee, F. Calabrò, Water permeation in carbon nanotube membranes, Curr. Opin. Chem. Eng. 4 (2014) 32–37, https://doi.org/10.1016/j.coche.2014.01.006. Elsevier Ltd.
[145] P.S. Goh, A.F. Ismail, Review: is interplay between nanomaterial and membrane technology the way forward for desalination? J. Chem. Technol. Biotechnol. 90 (6) (2015) 971–980, https://doi.org/10.1002/jctb.4531. John Wiley and Sons Ltd.
[146] R. Kumar, A.M. Isloor, A.F. Ismail, S.A. Rashid, A. Al Ahmed, Permeation, antifouling and desalination performance of TiO2 nanotube incorporated PSf/CS blend membranes, Desalination 316 (2013) 76–84, https://doi.org/10.1016/j.desal.2013.01.032.

[147] B.M. Ganesh, A.M. Isloor, A.F. Ismail, Enhanced hydrophilicity and salt rejection study of graphene oxide-polysulfone mixed matrix membrane, Desalination 313 (2013) 199–207, https://doi.org/10.1016/j.desal.2012.11.037.
[148] H. Wu, H. Sun, W. Hong, L. Mao, Y. Liu, Improvement of polyamide thin film nanocomposite membrane assisted by tannic acid-FeIII functionalized multiwall carbon nanotubes, ACS Appl. Mater. Interfaces 9 (37) (2017) 32255–32263, https://doi.org/10.1021/acsami.7b09680.
[149] M. Amini, M. Jahanshahi, A. Rahimpour, Synthesis of novel thin film nanocomposite (TFN) forward osmosis membranes using functionalized multi-walled carbon nanotubes, J. Membr. Sci. 435 (2013) 233–241, https://doi.org/10.1016/j.memsci.2013.01.041.
[150] X. Liu, S. Qi, Y. Li, L. Yang, B. Cao, C.Y. Tang, Synthesis and characterization of novel antibacterial silver nanocomposite nanofiltration and forward osmosis membranes based on layer-by-layer assembly, Water Res. 47 (9) (2013) 3081–3092, https://doi.org/10.1016/j.watres.2013.03.018.
[151] K. Goh, L. Setiawan, L. Wei, W. Jiang, R. Wang, Y. Chen, Fabrication of novel functionalized multi-walled carbon nanotube immobilized hollow fiber membranes for enhanced performance in forward osmosis process, J. Membr. Sci. 446 (2013) 244–254, https://doi.org/10.1016/j.memsci.2013.06.022.
[152] A. Kausar, Applications of polymer/graphene nanocomposite membranes: a review, Mater. Res. Innov. 23 (5) (2019) 276–287, https://doi.org/10.1080/14328917.2018.1456636. Taylor and Francis Ltd.
[153] V. Kochkodan, N. Hilal, A comprehensive review on surface modified polymer membranes for biofouling mitigation, Desalination 356 (2015) 187–207, https://doi.org/10.1016/j.desal.2014.09.015. Elsevier.
[154] M. Rezakazemi, A. Dashti, H. Riasat Harami, N. Hajilari, Inamuddin, Fouling-resistant membranes for water reuse, Environ. Chem. Lett. 16 (3) (2018) 715–763, https://doi.org/10.1007/s10311-018-0717-8. Springer Verlag.
[155] J. Blanco-Galvez, P. Fernández-Ibáñez, S. Malato-Rodríguez, Solar photo catalytic detoxification and disinfection of water: recent overview, J. Sol. Energy Eng. Trans. 129 (1) (2007) 4–15, https://doi.org/10.1115/1.2390948.
[156] M.M.A. Shirazi, A. Kargari, M. Tabatabaei, Evaluation of commercial PTFE membranes in desalination by direct contact membrane distillation, Chem. Eng. Process. Process Intensif. 76 (2014) 16–25, https://doi.org/10.1016/j.cep.2013.11.010.
[157] N. Cele, S.S. Ray, Recent progress on nafion-based nanocomposite membranes for fuel cell applications, Macromol. Mater. Eng. 294 (11) (2009) 719–738, https://doi.org/10.1002/mame.200900143.
[158] S. Chandra, Fast proton transport in solids, Mater. Sci. Forum 1 (1984) 153–169. Trans Tech Publications Ltd.
[159] F. Xi, L. Liu, Q. Wu, X. Lin, One-step construction of biosensor based on chitosan-ionic liquid-horseradish peroxidase biocomposite formed by electrodeposition, Biosens. Bioelectron. 24 (1) (2008) 29–34, https://doi.org/10.1016/j.bios.2008.03.023.
[160] Y. Zou, C. Xiang, L.X. Sun, F. Xu, Glucose biosensor based on electrodeposition of platinum nanoparticles onto carbon nanotubes and immobilizing enzyme with chitosan-SiO2 sol-gel, Biosens. Bioelectron. 23 (7) (2008) 1010–1016, https://doi.org/10.1016/j.bios.2007.10.009.
[161] C.L. Sun, H.H. Lee, J.M. Yang, C.C. Wu, The simultaneous electrochemical detection of ascorbic acid, dopamine, and uric acid using graphene/size-selected Pt nanocomposites, Biosens. Bioelectron. 26 (8) (2011) 3450–3455, https://doi.org/10.1016/j.bios.2011.01.023.

[162] Z. Wang, H. Li, J. Chen, Z. Xue, B. Wu, X. Lu, Acetylsalicylic acid electrochemical sensor based on PATP-AuNPs modified molecularly imprinted polymer film, Talanta 85 (3) (2011) 1672–1679, https://doi.org/10.1016/j.talanta.2011.06.067.
[163] W. Lian, et al., Electrochemical sensor based on gold nanoparticles fabricated molecularly imprinted polymer film at chitosan-platinum nanoparticles/graphene-gold nanoparticles double nanocomposites modified electrode for detection of erythromycin, Biosens. Bioelectron. 38 (1) (2012) 163–169, https://doi.org/10.1016/j.bios.2012.05.017.
[164] I.A. Nicholls, et al., Theoretical and computational strategies for rational molecularly imprinted polymer design, Biosens. Bioelectron. 25 (3) (2009) 543–552, https://doi.org/10.1016/j.bios.2009.03.038. Elsevier Ltd.

CHAPTER 3

Candidates of functionalized nanomaterial-based membranes

Deepshikha Datta[a], Krishna Priyadarshini Das[b], K.S. Deepak[c], and Bimal Das[c]
[a]Department of Chemical Engineering, GMR Institute of Technology, Rajam, India
[b]Department of Material Science and Engineering, Indian Institute of Technology, Delhi, India
[c]Department of Chemical Engineering, National Institute of Technology Durgapur, Durgapur, India

3.1 Introduction

The world is facing severe threats toward the rising demands of essential fundamental commodities (food-water-energy), finished products, and services, and to address these most formidable challenges; intensive research has been conducted while lessening and minimizing the impact of human activities on the global environment and climate. The merging of nanotechnology provides a versatile platform that can endow eco-friendly, cost-saving, and efficient solutions to meet global challenges, thereby achieving sustainable development goals [1]. The nub of nanotechnology lies in the application of nanomaterials in numerous areas of science and technology. Owing to their unique and distinct characteristics, nanomaterials have gained public interest and are extensively used in numerous fields such as electrical, mechanical, biomedical research, pharmaceutical applications, environmental fields, energy harvesting, etc. [2,3]. Indeed, the functionalization of material permits the addition of a myriad of functional entities that enhance the ever-existing characteristics of the nanomaterial along with imparting some desired property.

Recent progress in membrane technology research has remarkably engrossed the attention of many researchers to develop the next-generation membranes, which can deliver high performance and help to resolve the expected global crises [4]. The fundamental property of the membrane to control the permeation rate of a chemical species helps them to exploit their ability in chemical and industrial technology. The membrane can be mainly categorized into two types: organic and inorganic membranes. Inorganic membranes are manufactured from ceramics and metals (palladium, silver, and their alloys), while organic membranes are made from polymers.

Membranes with Functionalized Nanomaterials
https://doi.org/10.1016/B978-0-323-85946-2.00004-7

Ceramic membranes are extensively used in food, biotechnology, gas separation, and pharmaceutical application due to their inertness to chemicals, high-temperature stability, long durability, and high life [5]. The metal membrane has good heat insulation and is widely used for the separation of hydrogen from the gas mixtures with palladium and its alloy being the primary choice of material [6]. Although inorganic membranes possess many advantages such as temperature stability, resistance toward solvents, well-defined stable pore structure, and the possibility for sterilization, their high cost becomes a major drawback that limits their potential applications. Polymeric membranes were recognized as a promising tool in myriad applications involving transport processes due to their finely controlled structure and excessive separation potential, easy processability, and low cost [7]. These applications include gas membrane separation, water purification processes (reverse osmosis and nano-, ultra-, and microfiltration), membrane bioreactor to water purification, electrodialysis, dialysis, hemodialysis, cell and lithium battery, various optoelectronic applications, and conductive polymeric coatings and films [8,9].

Nanoscience and nanotechnology are the creative scientific advancements that have given vast opportunities to develop high-performance membranes for various emerging processes, especially in nanotechnology-based membrane processes. Nanocomposite membrane is a promising modified version of the traditionally used membrane with improved physicochemical properties such as porosity, hydrophilicity, charge density, membrane durability along with the overall stability (thermal, chemical, and mechanical) of the membrane [10]. The nanomaterial-based membranes such as nanofiber membranes, dense-packed nanoparticles, aligned nanotube membranes, two-dimensional material-based layer membranes, and the nanocomposites membranes exhibit some exceptional properties by opening a new avenue for a broad spectrum of applications [11].

Despite spectacular attainments, there are still some complications during the membrane fabrication that needs to be addressed for their large-scale practical applications. In order to overwhelm these challenges, various revolutionary approaches have been implemented, namely modification of nanomaterials surfaces by embedding moieties, the advancement of unique nanomaterials with distinct pore structure/charge properties, and escalation of high performance and durable nanocomposite membranes [12]. Among these methods, functionalized nanomaterial-based membranes have become the most radical candidate that offers a promising resolution for developing the next generation of desirable high-performance membranes.

This chapter encompasses the functionalized nanomaterials and their functionalization approach as well as the applicational prospect. First, the basic structures, properties of nanomaterials are briefly described. Subsequently, the functionalized nanomaterials and their different routes for functionalization are elucidated. Furthermore, the properties and application prospects of the functionalized nanomaterial-based membranes in the various emerging areas have been summarized. Lastly, this article offers insight into the future prospects and research direction in terms of functional modification for nanomaterials.

3.2 Nanomaterials

Nanomaterials are defined as materials that possess one or more peripheral nanoscale dimensions (range 1–100 nm). Nanomaterials have received considerable attention with unprecedented enthusiasm owing to the unique physicochemical properties that make them useful functional materials for sustainable technologies. More often, the distinct size-dependent property of nanomaterials allows their possible applications in many fields of human activities [13,14]. Different types of nanomaterials have been developed to modify and enhance the pharmacokinetics and pharmacodynamics properties of drug molecules [15]. Owing to the biocompatibility, controlled release property, and subcellular size with tissue and cells, polymeric nanomaterials have been broadly used as a particulate carrier in the medical and pharmaceutical fields [16]. The technology offers exceptional protection against the degradation of drugs and provides a reduction in drug frequency.

3.2.1 Types of nanomaterials

Nanomaterials can be broadly categorized into two main types based on the carbon content, i.e., organic and inorganic nanomaterials. Owing to the versatile applications and huge amount of studies, carbon-based nanomaterials are considered as a separate class of nanomaterial with a broad range of spectroscopy. The basic classification of nanomaterials is manifested in Fig. 3.1. The most recent nanomaterials can be classified into three material-based categories:

(a) *Carbon-based nanomaterials*: Due to the unique catenation property, carbon can form covalent bonds with other carbons in different hybridization states such as Sp, Sp^2, and Sp^3 in order to form a variety of structures of small molecules and longer chains. Carbon-based nanomaterials are found in morphological forms such as ellipsoids, hollow tubes, or spheres. Graphene (Gr), carbon nanotubes (CNTs), Fullerenes

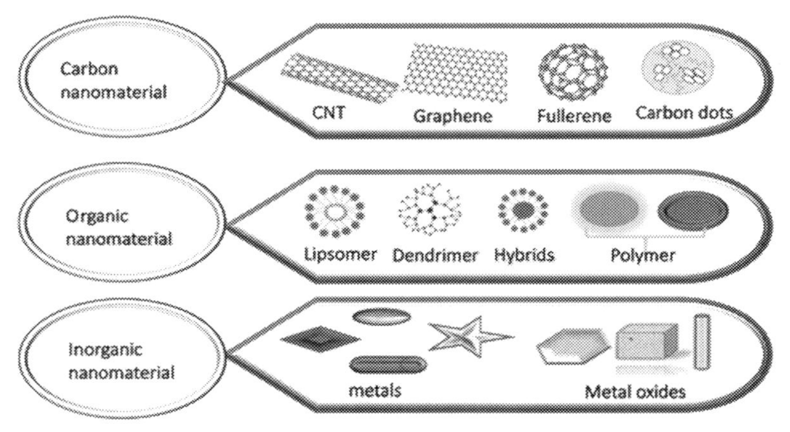

Fig. 3.1 Basic classification of nanomaterials.

(C60), carbon nanofibers, carbon onions, and carbon black are the different categories of carbon-based nanomaterials [17].

(b) *Inorganic-based nanomaterials*: These nanomaterials include metal-based nanoparticles, metal oxide/hydroxide nanoparticles, and transition metal chalcogenide (TMC) nanoparticles. These nanomaterials can be synthesized into metals like Ag, Au, Fe nanoparticles, and metal oxides such as ZnO, TiO_2, and Fe_3O_4, CeO_2.

(c) *Organic-based nanomaterials*: These nanoscale materials are made mostly from organic matter, aside from inorganic-based or carbon-based nanomaterials. The use of noncovalent interactions for self-assembling and molecular designing helps to transform the organic nanomaterials into coveted structures such as micelles, dendrimers, ferritin, micelles, compact polymers, and liposomes nanoparticles. These types of nanomaterials are usually biodegradable and nontoxic, and, therefore, considered environmentally friendly materials.

3.3 Potential of nanomaterials

Nanomaterials are at the forefront of advanced research owing to the genuinely revolutionary transformative capabilities for numerous emerging sustainable applications. Nanomaterials are materials having a size range between 1 and 100 nm in at least one dimension. It can be classified into various types, according to its material property, size, and shape. The different nanomaterials include carbon-based nanomaterials, metal nanoparticles, ceramics nanoparticles, semiconductor nanomaterials, and polymer nanoparticles [2]. Materials at nanoscale dimensions have novel and distinct physical and chemical properties owing to their high surface-to-volume ratio. The exceptional characteristics such as electrical, mechanical, optical, and thermal properties make nanomaterials rapid commercialization at the global market and looming as indispensable technologies. The potential applications of nanomaterials include medical and biomedical, catalysis, energy-based research, imaging, and environmental applications. The advancement of a variety of nanomaterials like functionalized nanomaterials, biological, polymeric, and polymer composite become the most attractive and extensively used materials in various applications, from biomedical to environmental [18,19]. Table 3.1 shows various properties and potential applications of emerging nanomaterials in multidisciplinary fields.

3.4 Limitations of nanomaterials

Despite the groundbreaking technologies in the nano-world, nanomaterials have some limitations that somehow limit their full-scale applications. The agglomeration of nanoparticles and poor materials surface interactions with the matrix are the two main factors that restrict its nanocomposites application. The toxicity to human beings and their impact on the environment has become one of the most crucial concerns that need to be addressed imperatively [20]. The potential of negative impacts on human health as well as ecosystem and safety considerations have emerged as a debatable topic for many researchers [21]. Numerous research has been devoted to mitigating the potentially adverse consequences of nanomaterials. Additionally, some further limitations that restrict the potential application in various fields are listed in Fig. 3.2.

The functionalization of nanomaterials has been proven as an adequate approach in order to resolve and extrapolate the current limitations. Such modification involves the introduction of various organic and inorganic moieties, thereby controlling the physiological and toxicological properties of nanomaterials.

3.5 Functionalized nanomaterials

The uniform dispersion of nanoparticles in bulk or matrices and their interactions with other molecules is needed to be explored for having a wide spectrum of potential applications. The nanomaterial's functionalization and modifications to their surfaces help in achieving good interactions of nanomaterials with the surrounding environments. Functionalized

Table 3.1 Various properties and potential applications of nanomaterials.

Sl. no	Nanomaterials	Properties	Applications	References
1	Carbon nanotube	High Young's modulus and tensile strength. Tunable surface chemistry, large surface area, high chemical stability electrical conductivity, bandgap fluorescence	Energy storage, gene and drug delivery, microwave absorption, Raman imaging and NIR imaging, magnetic force microscopy, nanocomposites membranes	[10,11]
2	Graphene	High electrical conductivity, strongest material with high tensile strength and Young's modulus, highly opaque	Gene and drug delivery, low-cost water desalination, biomedical technology, sensors, batteries, composites and coatings, electronics, catalyst supports	[12]
3	Fullerene	Free radical scavenging, hollow spherical shape, electron acceptor	Antioxidant therapy, cosmetics, biosensors, gene delivery	[10,13]
4	Mesoporous Silica	Ordered porous structure, high surface area, tunable particle size, functional surface, good biocompatibility	Cancer therapy, gene delivery	[14]
5	Gold (easily functionalize)	Relatively safe, nontoxic, easy to couple, biocompatibility	Drug-carriers, contrast agents, photothermal therapy, sensors	[15]
6	Silver	Antifungal, antibacterial	Antibacterial agents, water treatment, wound-dressing, coating surgical instruments, prostheses, cosmetics products	[16]
7	Titanium dioxide	Anti-microbial property, antibacterial property, good photostability	Pigment and coloring agents, solar cells, drug delivery, supercapacitors	[17]
8	Zinc oxide	Higher photostability, higher chemical stability, naturality, paramagnetism, a wide range of radiation absorption and higher electrochemical coupling coefficient	Solar cells, gas sensors, (Ultra-violet, UV) detectors, sunscreen, supercapacitors, Body-care products	[18]
10	Polymers (biodegradable)	Least toxicity and relatively safe, nonimmunologic, noninflammatory, biocompatibility	Drug delivery, biosensors, tracking, and therapeutic agent, organ implants, imaging, biotechnology, nanomedicines	[19,20]

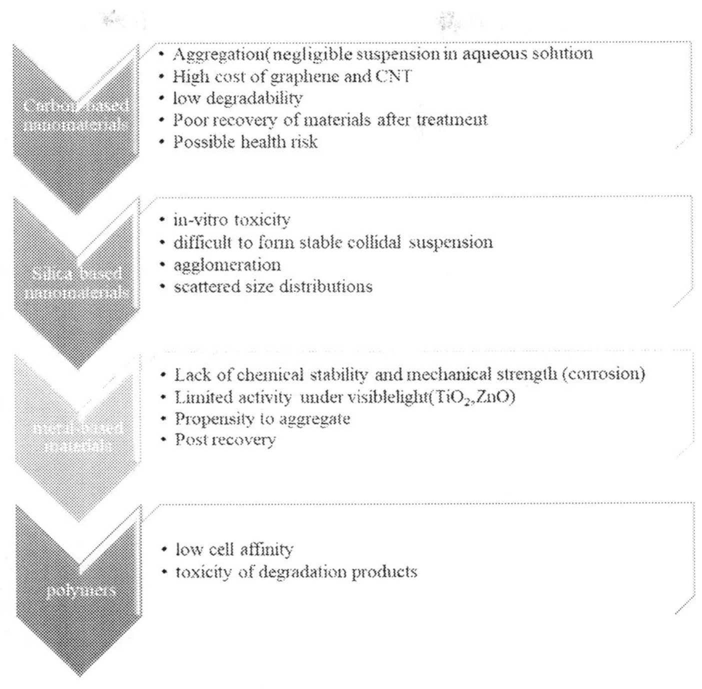

Fig. 3.2 Technical limitations of nanomaterials [2,22].

nanomaterials are deliberately designed to manifest the desired properties that we wished to incorporate in the nanomaterials through surface modifications [23]. The functionalization of nanomaterials involves the conjunction of certain functional entities and chemical groups that are physically and chemically decorated to the nanomaterial surface by changing the surface chemistry. The functionalization of nanomaterials has become a convenient strategy to overcome the drawbacks that limit the full-scale applications of nanomaterials in various looming research fields and industries. Functionalized nanomaterials not only enhanced existing characteristics but also help to impart some additional desired properties.

The proper functionalization dramatically influences the properties of nanomaterials, such as dispersibility, colloidal stability, as well as control the formation of clusters. Furthermore, the surface properties such as hydrophilicity, hydrophilicity, corrosivity, and conductivity can be easily tailored by the appropriate functionalization process [24]. As such, the unique characteristics of functionalized nanomaterials make them a striking candidate for numerous applications like water purifications, biomedical engineering science, cosmetics, pharmaceutical applications as well as drinking water and industrial wastewater treatment.

3.5.1 Types of functionalized nanomaterials

Nanomaterials have been functionalized with a variety of ligands by using different strategies (physical or chemical routes) to bestow novel properties, which are not manifested in their bulk counterparts. Functionalized nanomaterials can be categorized based on (a) functionalizing agent, (b) composition of nanomaterials [25], as depicted in Fig. 3.3.

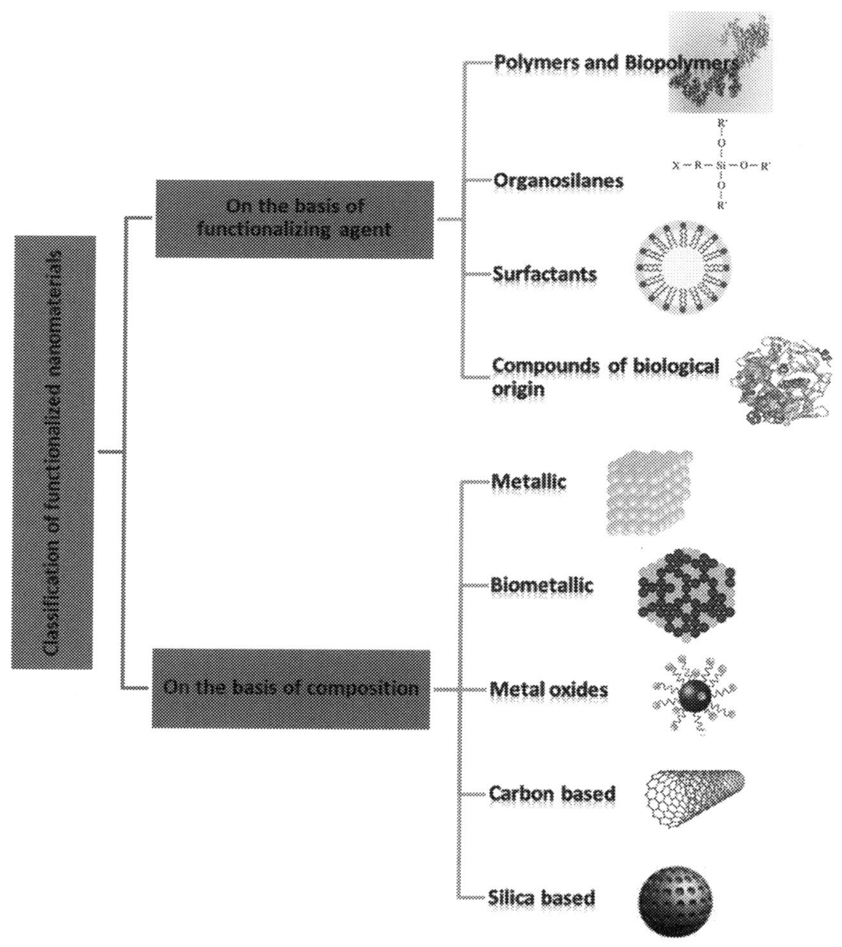

Fig. 3.3 Classification of functionalized nanomaterials based on (A) functionalizing agent, (B) composition of nanomaterials.

3.6 Methods of nanomaterials functionalization

Currently, researchers have adopted numerous methodologies for introducing surface functionalities, mainly via in situ synthesis and postsynthesis modifications. Various ligands like polymers, surfactants, inorganic materials, dendrimers, and biomolecules, are widely used to functionalize the nanomaterials [24,26]. An intensive research endeavor has been dedicated to the analysis of the nanomaterials interface modifications as well as functionalization. The functionalization of nanomaterials is carried out by using various synthetic strategies like covalent functionalization, noncovalent functionalization, intrinsic surface engineering, and amorphous nanoparticles coating, as described below.

3.6.1 Covalent functionalization

Covalent functionalization of the nanomaterials is the attachment of different chemical species to the nanomaterials through the formation of covalent bonds. This functionalization method is a promising approach to attach specific biomolecules, polymers, small organic molecules, and inorganic molecules to the nanomaterial surface in order to achieve better dispersion, high colloidal stability, and versatile properties. Covent functionalization is attractive as the covalent bond implies a very strong linkage and stable surface; thus, it is commonly used to modify common nanomaterials such as 2D nanomaterials (e.g., CNTs, GO, h-BN, and MoS2), nanoclays, and metal oxide nanoparticles. This type of functionalization significantly increases the utility of nanomaterials for numerous emerging fields like bioimaging, biosensors, packaging, environmental remediation, packaging, cosmetics, catalysis, and agriculture.

3.6.2 Noncovalent functionalization

In noncovalent functionalization, guest molecules bind to nanomaterials via the formation of fairly weak bonds. Noncovalent functionalization is mainly achieved via the p–p interactions, polymer wrapping hydrogen bonding, electron donor-acceptor ligand systems, and/or van der Waals interactions. This functionalization is very fruitful in close proximity and simply accessible in-room temperature as it is mainly based on weak forces like p–p interactions, van der Waals interactions, and hydrogen bonding. These weak forces are expected to influence the solvation of nanomaterials along with the surface interactions (hydrophobic and hydrophilic). Noncovalent functionalization can improve dispersibility, reactivity, binding capacity, catalytic activity, sensing behavior, and biocompatibility. Despite the fact that the energy released in the formation of noncovalent bonds is much less than the covalent interaction, the overall effects of the large interfaces are similar.

3.6.3 Intrinsic surface functionalization

Intrinsic surface functionalization refers to the atomic level modification in the crystal of nanomaterials that can be mainly achieved through defect engineering and heterogeneous incorporation. The introduction of electron-donating and electron-withdrawing elements (e.g., O, N, S, B, P, and other metals) alters the structural, electronic, and catalytic properties of the nanomaterials.

3.6.4 Amorphous nanoparticle coating

Inorganic nanomaterials, such as metallic nanomaterials coated with silica or other polymers introduce functional groups on their surface. Moreover, coating the surface of the nanomaterials with the polymers bestows additional functionality to the nanomaterials as it permits the anchoring of therapeutic ligands on their surface. Konduru et al. [27] coated silica (amorphous) on ZnO nanomaterials' surface in order to change their biological responses and biokinetic behavior in rats. Nanomaterials can be decorated with various entities by coating organic (monomer and polymer) and inorganic (metal and oxides) materials. Among metals, silica and gold are well explored to be coated on NPs that can spur high stability and give a conductive environment for successive functionalization.

3.7 Different types of nanomaterials and additives used for making the functionalized nanomaterial-based membrane

The advent of nanomaterial has explored immeasurable opportunities for developing high-performance membranes that can impart tailored properties such as high permeate flux, membrane durability, hydrophilicity, uniform porosity, antifouling resistance, photodegradation, and photocatalytic ability. There are numerous nanomaterials, which are extensively used in the development of various nanocomposite membranes, such as titanium dioxide (TiO_2), zinc oxide (ZnO), carbon nanotube (CNT), iron oxide (Fe_3O_4), silicon dioxide (SiO_2), graphene oxide (GO), silver (Ag), zirconium dioxide (ZrO_2), aluminum oxide (Al_2O_3), cobalt (Co), zeolite ($Na_2Al_2Si_2O_8$), copper oxide (CuO), and so on [28–30]. Despite the nanocomposite membrane's trailblazing applications, there are still some limitations that have become barriers to its full-scale applications.

Functionalizing nanomaterials emerge as a new family of high-performance materials that has offered a promising platform to modernize conventional membranes. The use of functionalized nanocomposites is a promising alternative for the membrane fabrication process to overcome the current limitations. The novel membranes composed of functionalized nanomaterials as additives, with desired state-of-the-art performance, demonstrate their great potential in various emerging sustainable technologies. These additional functions include antibiotic resistance, fouling resistance, and catalytical reactivity.

3.8 Carbon-based functionalized nanomaterials

Carbon, in its myriad forms, has been used in technical and artistic quarters from the primeval ages. Carbonous materials have their existence for several decades and are abundantly available in nature in the form of coal, natural graphite, and diamond. Human intellect has also engineered several other carbonous materials such as synthetic diamond, cokes, graphite fiber, lubricants, shoe polish, cutting wheel, printing inks. Carbonous material is also being used with synthetic polymers as a filler in tyres, tanks, aircraft, and spacecraft composites. Fullerenes (0D), nanotubes (1D), graphene sheets (2D), and diamond-like carbon (3D) are the different architectures of carbon [31]. Fig. 3.4 illustrated different architectural forms of carbon-based nanomaterials.

Carbon exists in several hybridization states (sp, sp^2, and sp^3) and creates a wide number of structures with other carbon atoms or nonmetallic elements ranging from small molecules to large polymeric ones. Carbon nanoallotropes could be generally sorted into two main groups by the predominant covalent bond in the structure which they have created; the first group is the build-up of mostly sp^2 hybridized carbon atoms that are positively packed in hexagonal (honeycomb) crystal lattices. Typical representatives of this group are graphene and graphene-like structures such as nanoribbons or quantum dots, CNTs, fullerenes, and the second group is the build-up of mostly sp^3 hybridized carbon atoms, and thus they cannot be considered as graphene derivatives. Fullerenes are spherical and hollow nanostructures that consist of sp^2 hybridized carbon atoms. C60 is the most bountiful and extensively used fullerene of all, even though there are also others with a lower

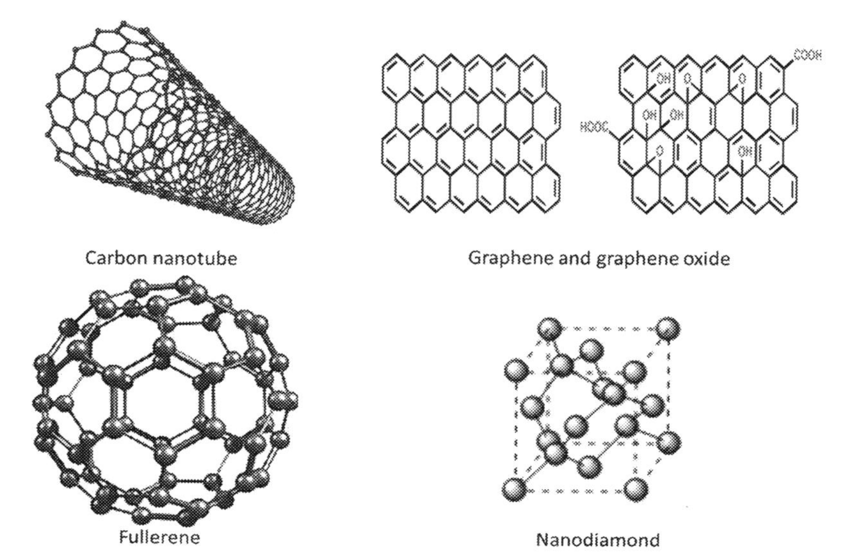

Fig. 3.4 Different architectural forms of carbon-based nanomaterials [32].

and higher amount of carbon atoms (e.g., C20, C36, C70, C96). A comparison of some properties of various carbon-based nanomaterials is illustrated in Table 3.2.

To fulfill the critical need for new material for nanocomposite application, graphene and CNT have the potential to alter the property of the material, which can be exploited in the broad spectrum of applications. Currently, peer-reviewed publications have perceived remarkable developments toward the use of carbon-based nanoparticles, mainly carbon nanotube (CNTs) and graphene oxide (GO) in nanocomposite membranes for a number of existing and emerging research fields [31]. CNTs and GO are favorable nanomaterials for the developing next generation of membranes, which exhibit high flux, high selectivity, and low fouling capabilities.

3.8.1 Carbon nanotube (CNT)

Carbon nanotubes can be visualized as concentrically rolled graphene sheets with a large number of potential helicities and chiralities [35]. In other words, it is a stack of multiple rolled graphene sheets that rolled up to form a long concentric cylinder. The chemical bonding in CNTs is basically sp^2 hybridized, but the circular curvature yields the quantum confinement and quantum conductance. Apart from this, circular curvature also induces

Table 3.2 Comparison of some properties of various carbon-based nanomaterials [33,34].

Carbon-based nanomaterials	Fullerene	Carbon nanotube	Graphene	Graphite
Dimensions	0	1	2	3
Hybridization	Mostly Sp^2	Mostly Sp^2	Sp^2	Sp^2
Availability	Industrial production, commercially available and laboratory-scale synthesis	Industrial production, commercially available	Laboratory-scale synthesis	Naturally occurring, commercially available, and laboratory production
Hardness	High	High	Uppermost (for single layer)	High
Tenacity	Elastic	Elastic and flexible	Elastic and flexible	Flexible and nonelastic
Special properties	Potentially high hardness in composite, semiconducting upon doping, optical, and electronic properties	Excellent electronic property and high strength	Strongest material and unique electronic properties	Lubricity and anisotropic electric conductivity

σ-π rehybridization, as three σ bonds are slightly out of the plane and for compensation, the π orbital is more delocalized outside the nanotube. This unique bonding makes CNT electrically and thermally more conductive, mechanically stronger, biologically and chemically more active than naturally occurring graphite [36]. The nanotubes' properties widely depend on the morphology, structure, length, and diameter of the tubes. According to the structure of the wall, nanotubes are classified into two types, namely single-wall carbon nanotube and multiwall carbon nanotube. SWCNT is long graphite with a length to diameter ratio of 1000 (approximately shows one-dimensional structure). MWCNT consists of several concentric layers of graphite that interlinked on themselves to form a tube shape; in other words, it is considered as a collection of concentric SWCNTs with different lengths, diameters, and properties [35,37].

Currently, carbon nanotube and their derivatives have engrossed significant attention of researchers due to their tremendous electrical property, high thermal resistance, superior mechanical strength as well as partial antibacterial properties. These novel and unique properties make carbon nanotube a favorable material in several existing and emerging applications such as biological applications, thermal conductors, catalyst supports, water and air filtration process, energy storage, and so on [38]. Carbon nanotube offers promising potential to alter the physicochemical properties of the membranes that prominently encouraged the broad applicational prospect of membrane across many fields [39]. Indeed, many researchers of different disciplines have prompted a considerable research effort toward producing CNT/polymer composite membranes in many structural and functional applications.

3.8.1.1 Functionalization of CNT

The properties and performance of CNT reinforced nanocomposites greatly depend upon the homogenous dispersion of individual carbon nanotube within the polymer matrix surface and interfacial interaction among the CNT and matrix phase. However, the smaller diameter, along with a higher aspect ratio (>1000) and remarkably larger surface area of make the CNT dispersibility more complicated within the matrix as compared to other conventional fillers [40]. The dispersion of nanotubes in the polymer phase is extremely tough as the commercial CNT exists in the cluster due to the Vander walls interactions. Another limitation encountered quite often is weak interfacial interaction with the polymer matrix. In general, carbon atoms situated on the CNT walls are chemically stable because of the

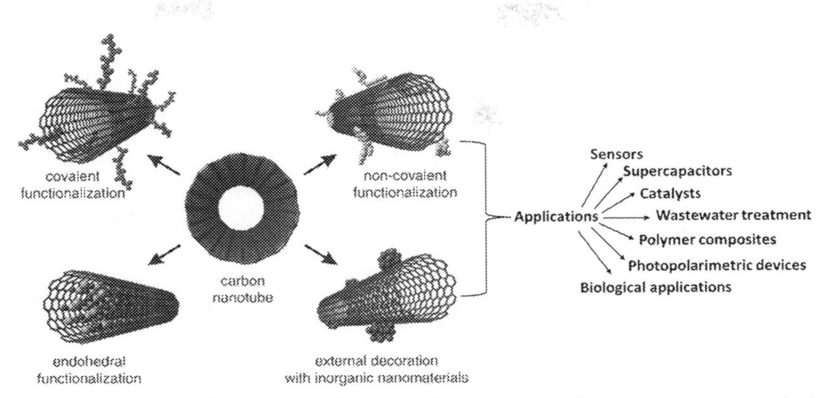

Fig. 3.5 Recent development in functionalization of carbon nanotubes and their emerging applications [42].

aromatic nature of the bond. As a reinforcing agent, CNT interacts with the surrounding matrix mainly via Vander walls' force of interactions, which is incapable of providing adequate load transfer across the nanotube/matrix interface [41]. In order to overcome the aforementioned problems, enormous efforts have been directed toward advanced methods to modify the surface property. The chemistry of functionalized CNT and the reaction mechanism of CNT and functional groups are described in many comprehensive review papers. Fig. 3.5 shows recent development in the functionalization of carbon nanotubes and their emerging applications. The functionalized methods can be roughly divided into covalent and noncovalent modifications in accordance with the interactions between the carbon atom on the CNT and active molecules. Fig. 3.6 manifested different types of bio-functionalizing agents used to modify CNT surface.

Covalent functionalization

Covalent functionalization of CNT is defined as the formation of covalent bonds among the chemical molecules and CNT by sharing one pair of electrons among them. In contrast with the noncovalent bonding interactions, the attachment of functional moieties to the tubular structure of CNT through covalent tethering provides a robust docking of functional entities that enable them to resist any desorption [44]. The covalent functionalization of nanotubes is often employed as it gives improved dispersibility and nucleation sites for further modifications. As such, the covalently functionalized materials possess improved stability that makes such methods more alluring for postmodification purposes, thereby extensively used to form a

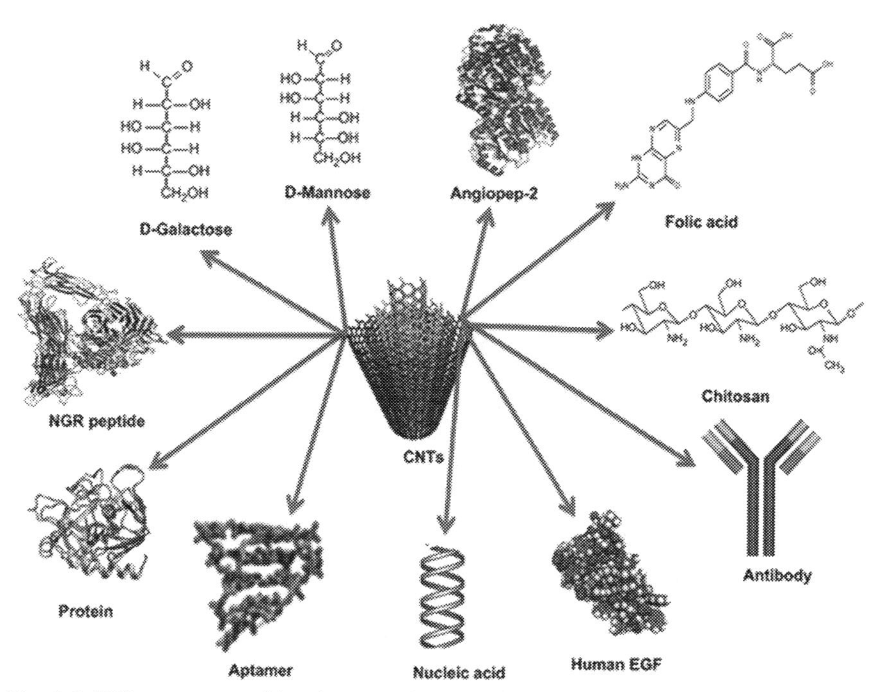

Fig. 3.6 Different types of bio-functionalizing agents used to modify CNT surface [43].

variety of nanostructures. Defect-group functionalization and direct sidewall functionalization are the two possible covalent functionalization methods for carbon nanotubes.

Noncovalent functionalization

In noncovalent functionalization of carbon nanotubes, the molecules are accumulated on the polyaromatic surface of the nanotube via Van der Waals force, Π-stacking, or electrostatic as well as hydrophobic interactions. Noncovalent functionalization has gained prominence owing to the higher adsorption capacity that facilitates higher surface energy. The noncovalent treatment involves the wrapping of nanotubes by DNA, polypeptides, and polycyclic aromatic hydrocarbons [45]. The interactions of aromatic moieties (e.g., pyrene) with the Π-orbital of nanotubes have been extensively employed for noncovalent functionalization. A larger Π-orbital system onto the aromatic molecules provides better attachment of the Π-orbital of CNT with moieties [46].

3.8.2 Graphene and its derivatives

3.8.2.1 Graphene

Graphene is a two-dimensional allotropic form of carbon that consists of a monoatomic hexagonal lattice structure. Each carbon atom in graphene is Sp^2 hybridized and bonded to the other three neighboring carbon atoms with an interatomic length of 1.4 Å. Graphene constitutes the basic structural elements of other carbon allotropes, namely, three-dimensional graphite, one-dimensional CNT, and zero-dimensional fullerenes [47]. This single-atom-thick sheet of carbon atom arranged in a honeycomb pattern is undoubtedly the most studied material owing to its exceptional thermal and electrical conductivity as well as excellent mechanical and electrochemical stability [47,48]. To now, a single layer of graphene is believed to be the world's strongest, thinnest, as well as stiffest material. The unique incredible and amazing properties of graphene originated from its unique structure include single-atom-thick, two-dimensional, and extensively conjugated; these structural elements endow graphene with a series of spectacular chemical and physical properties. The invention of graphene has opened a wide orifice of research and development in material science, its spectacular properties, and one atom thick structure has set tremendous results in composite applications.

3.8.2.2 Graphene oxide (GO)

Graphene oxide is the oxidized analog of graphene containing a single graphite monolayer. GO shows several domains like oxygenated aliphatic domains (Sp^3 carbon atoms) and randomly distributed aromatic regions (Sp^2 carbon atoms). Graphene oxide is showing excellent properties as it contains abundant oxygen functional groups like carboxyl, epoxy, hydroxyl, and carbonyl functional groups. In comparison to graphene, graphene oxide possesses the superiority of easy processing and mass production at less cost. Currently, GO is the most common precursor that has been widely used for the synthesis of reduced graphene oxide (RGO). GO contains oxygen-enriched functional groups or reduced doping elements that can be employed as an active catalytic center for covalent/noncovalent functionalization, thereby exploiting their application according to the requirements. Furthermore, the oxygen-enriched functional groups broaden the interlayer gap of GO, and it makes easy functionalization of GO with small moieties and polymer intercalations [49].

Functionalization of graphene and graphene oxide

All these miraculous properties expose graphene and its derivative's superiority over other nanostructured carbon allotropes. However, every model has its reverse. Although graphene and its derivatives have gained prominence in technological advancement, its easy agglomeration greatly limits its application in various composite and nanocomposites fields. The dispersion of graphene in water and organic molecules is very difficult to obtain due to its unique structure. In addition, graphene itself possesses zero band gap and is extremely inert for most of the chemical reactions that reduced the potential application of graphene in the field of sensors and semiconductors. Thence, the functionalization of graphene lessened their agglomeration as well as improved the graphene processability, thereby broadening its scope of applications in various sustainable looming fields. Besides, functionalization helps to modify the intrinsic features like electrical property that enables the control of bandgap and conductivity for the use of newfangled nanoelectronic devices [50]. The functionalization of graphene is generally carried out via three methods, i.e., covalent functionalization, noncovalent functionalization, and elemental doping [49].

Covalent functionalization Covalent functionalization of graphene involves the incorporation of newly introduced groups in graphene/GO via covalent bonds, thereby enhancing the processability and improving its performance. In order to develop organomodified graphene/graphene oxide, numerous studies and reviews followed the covalent functionalization method. As reported by Georgakilas, there are mainly two routes by which organic covalent functionalization of graphene is implemented: (1) through the formation of covalent bonds among free radicals or dienophiles and the C=C bond of pristine graphene (2) through the covalent bond formation between the organic functional groups and the oxygenated aliphatic domain of graphene oxide. Dienophiles and organic-free radicals are the most widely used organic species for covalent functionalization, and they can be attached to the sp^2 carbon atom of pure graphene by firstly decorating graphene with nitrophenyl groups. Additionally, graphene oxide is massively ornamented with oxygen-containing functional groups that make its covalent functionalization more convenient as compared to graphene. The surface of GO possesses abundant carboxyl groups, hydroxyl groups as well as epoxy groups, which can easily make common chemical reactions like epoxy ring-opening, carboxylic acylation, isocyanization, and diazotization. Although covalent functionalization improved the processability of graphene, it will cause defects and damage the initial graphene structure that could affect its chemical and physical properties.

Noncovalent functionalization Noncovalent functionalization of graphene oxide and graphene involves the introduction of functional moieties to the graphene structure through the poor interaction between target functional molecules and the graphene layer. Noncovalently functionalized graphene offers improved dispersibility along with excellent stability while maintaining the bulk structure and outstanding characteristics of graphene and graphene oxide. Numerous methods have been successfully implemented for the functional modification of surface via noncovalent bonds such as Π-Π stacking interaction, ionic bonding, hydrogen bonding, and electrostatic bonding interaction [51]. Among the ionic, metallic, or organometallic compounds employed in noncovalent functionalization synthesis, successful examples include the following: Mn^{2+}, Al^{3+}, CuO_2^{2-}, $MnFe_3O_4^{2+}$, Fe_3O_4, triphenylenes, polyvinyl imidazole, polyvinylpyrrolidone, and pyrene derivatives [52]. The main advantage of noncovalent functionalization is its simple operation with mild conditions without disturbing the initial structure and properties of the graphene. Nevertheless, the notable drawback of this method over covalent functionalization is because of the introduction of other compounds like surfactants onto the graphene surface. Table 3.3 described some advantages and disadvantages of the covalent and noncovalent functionalization of graphene and graphene oxide.

Elemental doping Elemental doping involves the incorporation of different elements into the graphene by adopting numerous processes such as ion bombardment, arc discharge, and annealing heat treatment. These modification results in the substitution of defects and valency defects in graphene while maintaining the intrinsic two-dimensional structure of graphene. In addition, doping enhances the surface property and adjusts the bond structure of graphene, thereby providing new performances. But the doping process is arduous to control quantitatively [49]. Table 3.4 presents brief details of the functionalization of graphene oxide and graphene.

3.8.3 Fullerene

A fullerene is a spherical carbon allotrope whose molecule is composed of at least 60 atoms of carbon connected by single or double bonds to form a discrete soccer-ball-shaped molecular structure. Fullerenes promise exciting new advances in nanotechnology and micro-electromechanical systems. Compounds of fullerene can be categorized into four subtypes namely hetero fullerene, alkali-doped fullerene, exohedral (inside the cage), and endohedral (outside the cage) [56]. The surface of fullerene consists of 20 six-member rings and 12 five-member rings depending on the basis of icosahedral symmetry closed cage structure. Presently, fullerene is used in numerous applications such as drug and gene delivery, photosensitizers, organic

Table 3.3 Advantage and disadvantage of graphene functionalization processes [53].

Functionalization methods	Interactions	Mechanism	Advantage	Disadvantage
Covalent functionalization	Polymers small organic molecules	Introduction of various active groups via chemical reaction on the edge or surface of the graphene	Make graphene more operable and workable	Cause defects and damage the initial graphene structure that could affect its chemical and physical properties
Noncovalent functionalization	Π-Π interaction hydrogen and ionic bonding chemical plating	Introduction of various active groups via chemical reaction on the edge or surface of the graphene.	Under a mild condition functionalizing the graphene take place. Graphene's initial structure and property remain unchanged	Easily Introduce other components like surfactant on the graphene's surface

Table 3.4 A brief details on the functionalization of graphene oxide and graphene [51,54,55].

Functionalization type	Modified group	Modification agent	Interaction type	Property
Covalent functionalization	–OH	2-Bromoisobutyryl bromide, NaN_3, HC=C–PS	Esterification	Good solubility
	–COOH	SOCl2	Esterification	Conductivity
	–OH	N2H4, DNA	Addition esterification	Good solubility
	–C=C–	4-Propargyloxydiazobenzenetetra-Forroborate	Diazotization	Water-soluble
Noncovalent functionalization	Carbon six-membered ring	Derivative of Tetrapyrene	Π-Π	Dispersed and stable conductive
	Carbon six-membered ring	Copolymer of butylene–styrene and Sulfonated styrene-ethylene	Copolymerization	Conductive
	–OH	DNA	Hydrogen bond interaction	Dispersed and stable, good solubility
	–COOH	SDBS	Ion interaction	Stably dispersed, conductive
	–COOH	Amine terminated polymers	Ion interaction	Dispersed and stable, good solubility
Elemental doping	–C–	P, N, and B		Band structure change

photovoltaics, endohedral fullerenes, antioxidants biopharmaceuticals, and diagnostic applications [56].

3.8.3.1 Functionalization of fullerene

Considerable research approaches have been devoted to the functionalization and application of fullerene and its derivatives due to their versatility since 1985. It is now well known that irrespective of their extreme conjugation, C_{60} generally behaves like an electron-deficient olefin, as a result, it can readily react with different nucleophilic reagents. These include active methylene compounds ($BrCH(COOEt)_2$, carbanions generated by silyl enol ethers, organometallics, the Bingel reaction), cyanide, and α-halo carbanions [57]. Indeed, by electrochemical reduction (constant voltage electrolysis or constant current electrolysis) or chemical reduction (alkali metals like K, Li, Cs, and Rb, alkaline earth metals such as Ba and Ca, and apart from that transition metals such as mercury), fullerene and its derivatives can be selectively reduced to fullerene anions (e.g., C_{60}^{n-}, $n=1$–6). Afterward, these species can act as nucleophiles for further derivatization and functionalization. In addition, fullerene is prone to undergo Diels–Alder reactions as a dienophile due to its unique lowest unoccupied molecular orbital (LUMO) level. Presently, functionalized fullerene is remarkable for numerous looming applications including self-assembly films, high-performance solar cells, gene delivery with polycationic fullerene, penta-haptofullerene metal complexes, liquid crystals, glycosidase inhibition with fullerene iminosugar balls, and bacterial antiadhesives [58].

3.8.4 Nanodiamond

Nanodiamonds are the resultant residue from a TNT or hexogen explosion in a contained space. They are now being studied in a variety of science, technology, and health applications due to their inexpensive large-scale synthesis based on the detonation of carbon-containing explosives, high biocompatibility, small size ($\leq$5 nm) with narrow size distribution, low-toxicity, easy surface functionalization, and bio-conjugation. As the diamond core is chemically inert with a highly tailorable and fully accessible surface, it facilitates easy binding of a number of functional groups, covalent or noncovalent attachment of biomolecules /drugs. Owing to these novel properties, nanodiamonds can be exploited for the development of therapeutic agents for diagnostic probes, antiviral and antibacterial treatments, gene therapy, delivery vehicles, novel medical devices, and tissue scaffolds [59].

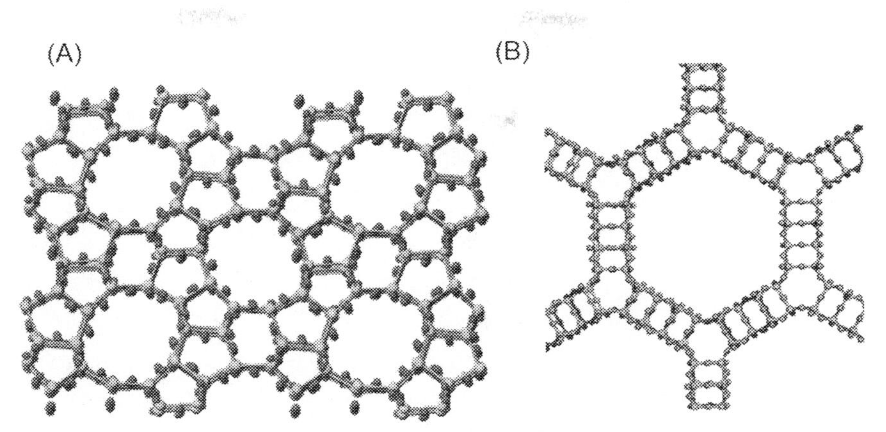

Fig. 3.7 Silica-based nanomaterials [61]. (A) Silicalite (MFI framework). (B) MCM-41.

3.9 Silica-based nanomaterials

Silicon is a tetravalent metalloid that belongs to group 14 of the periodic table and resembles neighborhood carbon in many aspects. Silicon possesses comparatively low reactivity, but its strong chemical affinity toward oxygen results in the formation of silica (SiO_2) and other silicate compounds [60]. Silica-based nanomaterials have received extensive research attention owing to their broad spectrum of applicability from structural materials to biomedical implants. Zeolite and mesoporous silica nanomaterials are the two well-known silica-based nanomaterials, as illustrated in Fig. 3.7.

Zeolites are the three-dimensional crystalline microporous aluminosilicates with a regular arrangement of micropore that has a large specific surface area. Besides, Zeolite possesses great ion exchange ability, excellent resistance to ionic radiations and temperature, and good compatibility with the environment. Zeolites offer an outstanding catalytic property by means of their crystalline aluminosilicate network [62]. Ordered mesoporous silica is the amorphous form of silica that possesses periodically arranged mesopores with tunable pore sizes and high surface area. Both zeolite and mesopore silica nanomaterials offer considerable properties such as high surface area, well-defined porous architecture, fast mass transport, and even more significantly, their unique properties coupled with the easiness of modification by various desirable functional groups.

However, despite the above spectacular properties, the potential applications of silica-based nanomaterials are restricted owing to their inorganic nature, aggregation property, and low thermal stability. Furthermore, the presence of fewer functional groups on the nanomaterial surface restricts its adsorption capacity and selectivity of adsorption [63]. Thankfully, Zeolite and mesoporous silica nanoparticles are readily modifiable as silanol functional groups cover their surfaces. The functionalized nanomaterials endow tunable properties such as hydrophilicity/hydrophobicity, surface charges.

3.9.1 Functionalization methods

The functionalization of silica-based nanomaterials enables their utility in numerous advanced applications by making them compatible with a wide variety of polymer molecules. In general, the same methods have been adopted to functionalize Zeolite and mesoporous nanosilica, i.e., silanol organ functionalization method. However, the functionalization of Zeolite slightly differs from mesoporous silica as it occurs solely on the external zeolite surface. The channel dimension of Zeolite is very narrow that restricts access to the internal surface for most of the reactants [64]. Two approaches commonly perform the functionalization of both nanomaterials, namely cocondensation and postsynthetic grafting [65], as manifested in Fig. 3.8.

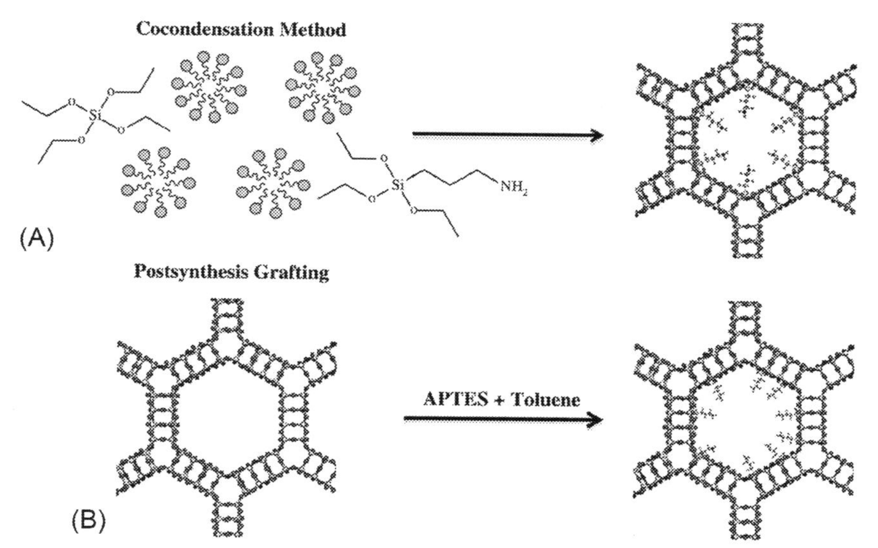

Fig. 3.8 Two functionalizing approach of silica-based nanomaterials: (A) cocondensation or one-pot synthesis, (B) postsynthetic grafting functionalization [22].

3.9.1.1 *Cocondensation (one-pot synthesis)*

Cocondensation is a favorable approach for functionalizing mesoporous silica nanomaterials as they have a propensity to crystallize more conveniently. This process involves the addition of one or more organosilane and silicon sources (more often use tetraethylorthosilicate) so that functionalization occurs concurrently with the framework's formation. The cocondensation method prompts a high degree of functionalization and uniform deposition of functional units on the entire inner pore surfaces with no pore blockage and shrinkage problem [66]. Besides, this process easily controls the particle morphology of the functionalized mesoporous silicate. However, the major drawback of the cocondensation functionalization method is that it affects the crystallinity of the final product [67].

3.9.1.2 *Postsynthetic*

The postsynthetic method refers to the subsequent modification of nanomaterials that have already been shaped prior to the functionalization. A well-known silylation process is most commonly carried out for surface functionalization of silica nanomaterials by grafting. The process is accomplished with the following three steps as shown below [Eqs. (i)–(iii)].

$$(\text{i}) \equiv \text{Si}-\text{OH}+\text{R}'\text{O}-\text{SiR}_3 \xrightarrow{100^\circ\text{C}} \equiv \text{Si}-\text{OSiR}_3+\text{HOR}'$$

$$(\text{ii}) \equiv \text{Si}-\text{OH}+\text{CL}-\text{SiR}_3 \xrightarrow{\text{base } 25^\circ\text{C}} \equiv \text{Si}-\text{OSiR}_3+\text{HCl}$$

$$(\text{iii})\, 2 \equiv \text{Si}-\text{OH}+\text{HN}(\text{SiR}_3) \xrightarrow{25^\circ\text{C}} 2\equiv \text{Si}-\text{OSiR}_3+\text{NH}_3$$

According to the above equations, silylation is carried out by replacing acidic hydrogen of silanol groups (Si-OH) on the material surface with an alkyl silyl group of organosilanes of the type $(R'O)_3SiR$, or less often with chlorosilanes $ClSiR_3$ or silazanes $HN(SiR_3)_3$ [66,68]. Such modification endows nanomaterials with enhanced surface reactivity, suitable adsorption property as well as excellent stability. Nearly all functional groups like amino, thiol, sulfonic acid, and carboxylic acid have been incorporated on the material surface by using this functionalization approach. Nevertheless, the heterogeneous distribution of the functional entities on the pore surface is the major drawback of this method. Usually, lots of functional groups are decorated around the outer surface of the nanomaterials and the entrance of the pores [69]. Table 3.5 summarizes fewer examples of cocondensation and postsynthetic functionalization methods of silica-based nanomaterials.

Table 3.5 Functionalization of silica-based nanomaterials by different reagents.

Sl no	Method of functionalization	Functionalizing agent	References
1	Co-condensation	3-Aminopropyltrimethoxysilane	[70]
		N-Dodecyltriethoxysilane	[71]
		N-(3-Trimethoxysilyl)-propyl) Diethylenetriamine	[72]
		3-Mercaptopropyl)trimethoxysilane (MPTS)	[73]
		3-Aminopropyl(diethoxy)methylsilane	[74]
2	Postsynthetic	Trimethylchlorosilane	[75]
		1-Benzoyl-3-propylthiourea	[76]
		3-Aminophenylboronic acid monohydrate and 3-Glycidyloxypropyltrimethoxysil ane	[77]
		1-Allyl-3-propylthiourea	[78]

3.10 Metallic inorganic nanomaterials

Metal and metal oxide-supported nanomaterials are in the limelight of modern nanotechnology, mostly Au, Ag, and Cu, owing to their excellent physicochemical properties along with their exceptional antimicrobial and antibacterial activity. The unique size-dependent property of metal makes the more promising materials in comparison to other large-scale materials. The basic forms of metal nanomaterials such as metal-based nanomaterials, metal oxide-based nanomaterials, doped metals, metal sulfides, and metal–organic frameworks [79] with their examples are depicted in Fig. 3.9.

3.10.1 Metal-based nanomaterials

The metal-based nanomaterials possess high robustness along with a high surface-to-volume ratio. Furthermore, the ease of functionalization enables them to target specific target bonding properties. Despite all these developments, there are still some drawbacks that need to be addressed for their more sweeping performance in modern challenges. During the synthesis, the interaction between the nanoparticles and the high surface area leads to agglomeration [22], which can alter some size-dependent material properties. To make them worthy for practical and commercial applications, it is necessary to functionalize and stabilize the metal nanomaterials [80].

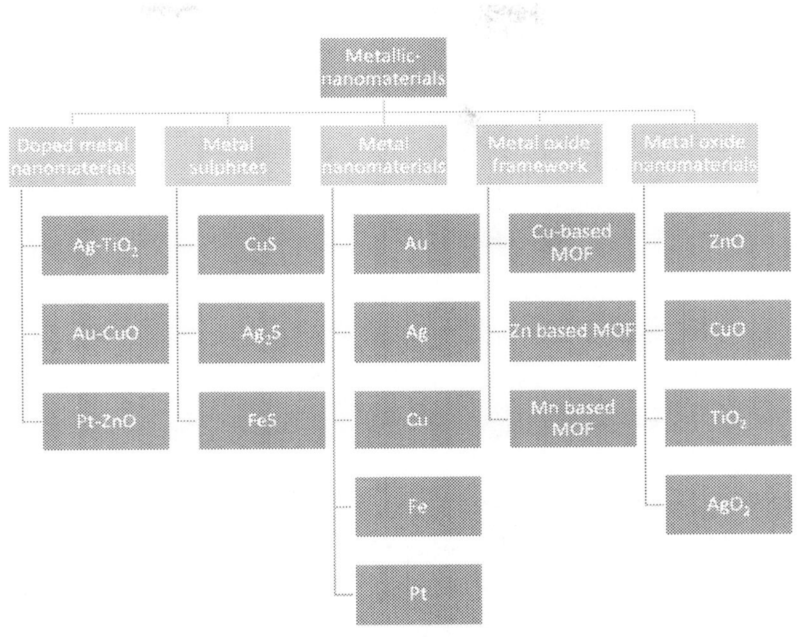

Fig. 3.9 Basic forms of metal-based nanomaterials [79].

Presently, metallic nanomaterials are synthesized and modified with various chemical functional groups. This process enables them to be assembled with ligands, antibodies, and drugs. As a result, it has a remarkable wide range of potential applications in magnetic separation, biotechnology, preconcentration of target analytes, targeted drug delivery, diagnostic imaging, and vehicles for gene and drug delivery. There are mainly two approaches by which we can modify the surface of the material to introduce functionalities and stabilization: (1) in situ functionalization and (2) postsynthetic process.

3.10.1.1 Gold nanomaterials

Gold nanoparticles possess a high X-ray absorption coefficient with ease of artificial manipulation that enables them for precise control over the particle's physico-chemical properties. The different forms of gold nanoparticles are displayed in Fig. 3.10. Additionally, it has a well-built binding affinity toward amines, disulfides, and thiols along with unique tunable optical and distinct electronic properties. These unique characteristics, along with the multiple surface functionalities, make them highly attractive and promising materials in numerous looming fields. Gold nanoparticles have sparked enormous interest in numerous potential applications, especially biomedical

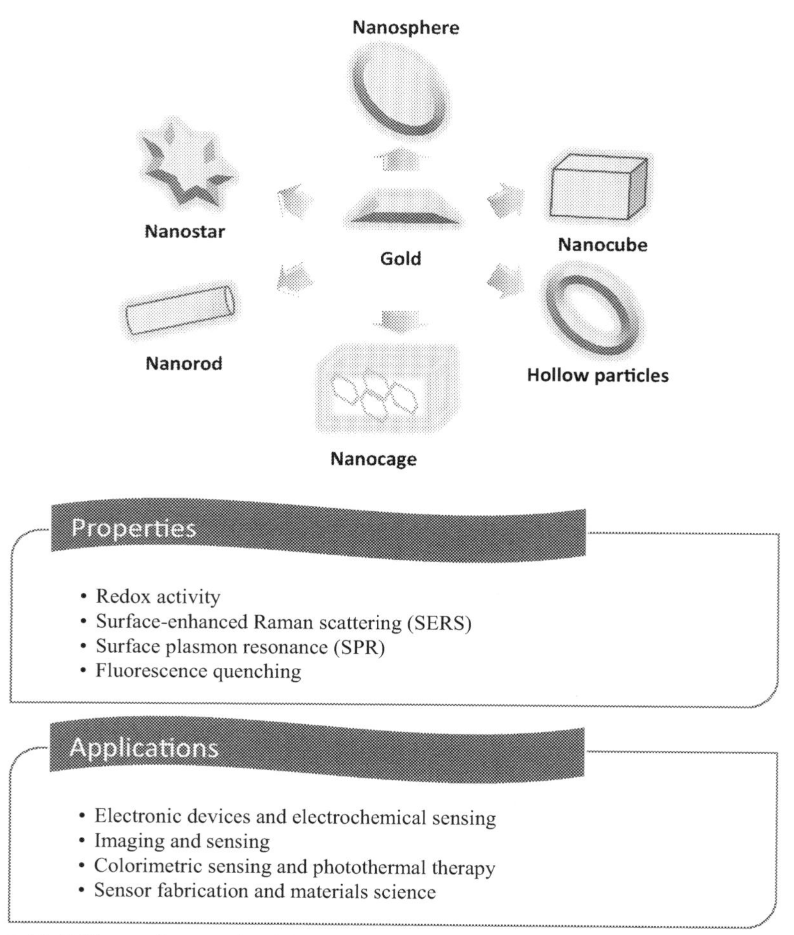

Fig. 3.10 Different forms of gold nanoparticles.

fields and environmental monitoring applications such as biosensors, theranostics, bioimaging applications, hazardous chemical sensing, and targeted drug delivery field [81]. The amenability of AuNP functionalization allows the assembly of nanobiological molecules such as proteins, antibodies, and oligonucleotides. The easy assembly of bio conjugates on the gold surface makes them the most favorable candidates in the design of novel biomaterials that could be exploited to investigate biological systems.

Despite these breakthroughs, many efforts have been devoted to designing gold nanoparticles in order to improve the bioavailability of these materials with less cytotoxicity and immunogenicity [82], thereby making it preferable to be used in vivo applications [83]. The most widely used modification is the functionalization of AuNPs with surfactants, polymers, ligands, drugs, peptides, proteins, and oligonucleotides that can modulate the physio-chemical properties for their numerous multidisciplinary applications [81]. Table 3.6 displays a brief description of the functionalization of gold nanomaterials and their potential applications.

3.10.1.2 Silver nanomaterials

Silver nanoparticles have achieved great success in a wide range of potential applications due to their superior physical, chemical, and optical properties [86]. Additionally, silver nanoparticles (AgNPs) display incomparable robust inhibitory and biocidal performance among various biocidal materials against a variety of microorganisms [87]. Substantial efforts have recently been fermented to assimilate AgNPs into polymer nanocomposite membrane surfaces by either in situ or ex situ processes. Besides the well-known antifouling properties of silver (Ag) nanoparticles, the membrane's hydrophilicity nature raises its importance. Ag is stable and exists in water without any interaction. Its electronic configuration shows that it is very much inert toward water molecules, so it exists in water without any chemical reaction. But Ag atom lacks one electron to fulfill its d-orbital. So, there is always a secondary interaction that makes Ag a hydrophilic species. The interaction between the Ag and membrane is purely secondary, but this interaction is sufficient enough to resist the force of water, which comes with pressure. The concentration of the Ag nanoparticles affects the membrane property a lot. Membranes containing different percentages of Ag nanoparticles were taken, and it was found that the flux rate goes up with an increase in the concentration of Ag nanoparticles.

Recent reviews evidenced that extensive usage of AgNPs leads to enhanced potential environmental contamination and human exposure. Silver nanoparticles show noticeable toxicity against tested cell eukaryotic lines and laboratory animals such as rodents or aquatic organisms [88]. The agglomeration and nanotoxicity are the two main downsides of silver nanoparticles. In this regard, numerous research has been devoted to mitigating the drawbacks.

The higher surface-to-volume ratio of silver nanomaterials offers the possibility of altering their surface to immobilize many ligands molecules. Such modification endows silver nanomaterials with district and novel functionalities, thus further extending their possibility of broad applications.

Table 3.6 A brief description of the functionalization of gold nanomaterials and their potential applications [81,84,85].

Sl no	Functional group	Ligands/carrier molecule	Key feature	Applications
1	Amino group	Cell-penetrating peptide (VG-21)	Adhere to the membrane of cell	Biodistribution studies and cellular as well as intercellular targeting
2	Amine Group	PEG	siRNA carrier	Mostly used in RNAi technology
3	Carboxyl group	Proteins		Useful in numerous fields depending in the proteins used
4	Peptide	Cell surface receptors, amyloid inhibitory peptide + sweet arrow peptide, antibody, octrotide peptide	Cytoplasmic and nuclear translocation, Adjuvant, targeting Carcinoma cells Analogue of somatostatin	Bioimaging (mostly imaging of cancer cells) cellular and intercellular targeting
5	DNA	Aptamer, antisense DNA Oligonucleotides Thiolated ssDNA of RNAi Gene, PEGylated gold, poly(β-amino-ester),	Targeting prostate Cancer cells, siRNA Carrier binds to Antisense RNA of p53	Drug and gene delivery, bioimaging, detection of specific genes such as microbial detection
6	RNA	Nanoconjugates of polyvalent RNA-gold		RNAi (RNA interference)
7	Antibodies	scFv Antibodies against numerous pathogens	Label, smaller size and fidelity	Diagnosis and immunoassay treatment, for example, antibiotic against aflatoxins

Besides, the unique properties fueled the development of functionalized materials in commercial applications like contraceptive devices, feminine hygiene products as well as biomedical applications, including drug delivery, medical device coating, diagnosis, and treatments. Table 3.7 exhibits different functionalizing agents of silver nanomaterials.

3.10.2 Metal oxides

Metal oxide nanoparticles are synthesized using pure metal targets, namely titanium, zinc, cerium, aluminum, nickel, tin, magnesium, and iron [94]. Metal oxides display a wide range of morphology, such as nanorods, nanowires, nanotubes, nanobelts, and nano rings. Metal oxide-based nanomaterials have engrossed their attention due to their fascinating optical, physical, and chemical properties that make them suitable for applications in sensors, electronics, solar cells, optoelectronics, batteries, optics, and photonics applications [95]. These exceptional properties arise from their limited size with a high density of edge and corner surface sites. In the looming field of nanotechnology, the current goal is to make nanoarrays or nanostructure materials that can endow desirable properties, while mitigating the significant drawbacks. Table 3.8 demonstrates the potential applications of various metal oxides.

3.10.2.1 Zinc oxide (ZnO) nanomaterials

Zinc is the first-row last transition element and is in +2 oxidation state in ZnO, so although its d-orbital is completely filled, its 4s-orbital remains vacant, so it has the capacity to accommodate electrons, and behaves a little bit like lewis acid. Zinc Oxide (ZnO) as a nanoparticle provides various promising properties which include:

- resistance to ultraviolet
- cheap as compared to TiO_2 and Al_2O_3 nanoparticles (one-fourth as compared to the latter)
- good antimicrobial and bactericidal properties [96,97]
- good adsorption capacity for gases such as H_2, CO, and CO_2 [97]
- has high catalytic activity
- can be easily implanted in a membrane which leads to
- increase in physical and chemical properties
- increase in hydrophilicity and mechanical properties

Due to the above-listed properties of ZnO nanoparticles, various studies have been steered to improve the properties of the membrane [98]. Zno in the nanometer-scale has been widely used for solar energy conversion,

Table 3.7 Functionalized silver nanoparticles with their various functional entities.

Sl no	Functionalizing agent	Properties	Applications	Reference
1	Tetracycline	Higher antimicrobial property and better antibacterial activity against multidrug resistant bacteria	Effective antibacterial agent against *S. aureus*, *P. aeruginosa*, and K. pneumonia	[89]
2	mPEGylated luteolin	Utilized to detect Hg^{2+} with high sensitivity and selectivity in the presence of other metal cations	Sensors to detect heavy metal like Hg^{2+} from tap water	[90]
3	Deltamelathrin	Enhance the ability to kill insects, specifically vectors of human pathogens	Widely used for vector insect control	[91]
4	Thiobarbituric acid or 11-mercaptoundecanoic acid residues	Slightly modulating its toxicity properties	Modulate AgNPs' cellular uptake and toxicity	[92]
5	Monosaccharides	Modulates their cellular uptake and toxicity		[93]

Table 3.8 Metal oxides with their potential applications [79,95].

Metal oxide	Potential applications
Silver oxide	Antimicrobial, gene therapies, drug delivery, imaging, and tissue developments,
Iron oxide	Environmental remediation applications and Magnetic imaging,
Zinc oxide	Skin protectant, gas sensors
Nickel (oxide)	Biomedical applications like anticancer therapy
Copper oxide CuO	Widely used in biomedical applications like antiviral, antimicrobial, antibiotic, antifouling, and antifungal treatment Other emerging applications like coating materials, inks, catalyst factor and lubricants
Gold oxide	Cellular imaging, cancer treatment, surgical devices, drug delivery and photodynamic therapies
Titanium dioxide	Coating material, painting, wastewater treatment, sterilization
Bismuth oxide	Drugs delivery systems

nonlinear optics, catalysis, varistors, pigments, gas sensors, cosmetics, etc. The surface of ZnO nanoparticles possesses abundant hydroxyl groups and their surface area is comparatively higher than other organic nanoparticles [99] that tend to absorb more hydroxyl groups.

3.10.2.2 Titanium oxide (TiO_2)

Unification of TiO_2 nanoparticles with membrane has been acknowledged more considerable attention due to its outstanding physicochemical performance, photo-induced hydrophilicity, and higher oxidation power. Tio_2 is currently widely used in a variety of applications such as disinfection agents, and some other industrial uses such as food color additive, white pigment, food color additive, decomposition of organic compounds, and flavor enhancer [100].

TiO_2 exists in three different naturally occurring phases, i.e., anatase, rutile (thermodynamically more stable bulk phase), and brookite, which could mainly be differentiated by their structural properties. While in nanoscale, the anatase and brookite phase of TiO_2 is considerably more stable as compared to the rutile phase due to its differences in surface reactivity. Anatase TiO_2 nanoparticles are smaller in size and form smaller pores membrane surface and more apertures as compared to larger size rutile TiO_2, which in turn results in higher permeability and antifouling performance. Efficient photoactivity, low cost, high stability, toxic-free property as well as easy availability with a nearly endless lifetime makes titanium potentially attractive in membrane technology [101]. The different functionalizing agents used for the functionalization of metal oxides are displayed in Table 3.9.

3.11 Organic nanomaterials

Common types of nanomaterials include liposomes, ferritin, micelles, dendrimers, compact polymers, and hybrid molecules. Most of the inorganic nanoparticles are considered environmentally friendly owing to their non-toxic and biodegradable property [2]. Liposomes and micelles pose hollow cores and are sensitive to electromagnetic and thermal radiation energies such as light and heat [112]. Most of the inorganic nanomaterials have gained significant attention in drug delivery applications due to their biocompatibility, high stability, surface morphology, delivery efficiency, and drug-carrying capacity [113]. Liposomes are spherical vesicles comprising a phospholipid bilayer that is only composed of lipidic compounds. Moreover, the sizes of liposomes are mostly ranged from 50 to 100 nm, while unilamellar liposomes vary from 100 to 800 nm [114]. Micelle nanomaterials are formed from amphiphilic molecules (viz., polymers or lipids) with a mean size of 10–100 nm. Dendrimer nanoparticles are star-shaped and hyperbranched structure (<10 nm) that is developed from one or more cores [115]. Dendrimer NPs comprise an approximately monodisperse polymer system synthesized by controlled polymerization with three structural parts: core, branch, and surface [116]. Compact polymeric NPs are formed as nanocapsules and nanospheres and consist only of synthetic or natural polymers with sizes ranging 10–1000 nm. Inorganic nanoparticles have been gaining considerable attention in various fields including water filtration and desalination, environmental application, energy storage, biomedical applications such as diagnostic and therapeutic systems in oncology, tumor imaging, and drug delivery.

3.11.1 Polymeric nanomaterials

The inertness of the commercial polymeric nanomaterials restricts their development for sustainable applications in various industries. Thence, modification of polymer surface must be implemented to enhance their printing, wetting, and adhesion properties. This can be achieved by

Table 3.9 Functionalization of metal oxides (TiO_2 and ZnO).

Sl no	Metal oxides	Functionalizing agent	Properties	Application	References
1	TiO_2	Chondroitin sulfate	Increase the dentine adhesiveness property	Desensitizing and remineralizing dentifrices	[102]
2	TiO_2	3-(2-aminoethylamino) propyltrimethoxy-silane (AAPTMS)	Best photocatalytic activity for NO oxidation and NOx removal	Air pollutants oxidation	[103]
3	TiO_2	Derivatives of Bidentate Benzene like 2-hydroxybenzoic acid, 3,4-dihydroxybenzoic acid, 2,3-dihydroxybenzoic acid and catechol 2,5-dihydroxybenzoic acid,	Electrical property and optical property		[104]
4	TiO_2	PEG chains	Improves the photocatalytic properties		[105]
5	TiO_2	3-(4-aminophenyl) propionic acid (APPA) or 3-mercaptopropionic acid (MPA) spacer with DMP1 peptides	Biocompatibility, corrosion and tribo-corrosion resistance		[106]
6	ZnO	3-Mercaptopropionic acid (MPA)	Higher cytotoxicity against the tumor cells	Drug delivery, cancer treatment such as chemotherapy, photodynamic therapy (PDT), and imaging	[107]

Continued

Table 3.9 Functionalization of metal oxides (TiO_2 and ZnO)—cont'd

Sl no	Metal oxides	Functionalizing agent	Properties	Application	References
7	ZnO	Aminopropyltriethoxysilane (APTS)	Remarkable biocidal activity	Sensors and antibacterial activity	[108]
8	ZnO	Monoethanolamine (MEA)	Enhances the activity for photocatalytic CO_2 reduction	Efficient CO_2 Capture and photoreduction, solar fuels	[109]
9	ZnO	Mercaptoundecanoic acid (MUA)	Water solubility and biocompatibility	Use for collisional quenching-based glucose sensing	[110]
10	ZnO	Mercaptoacetic acid (MAA)	Good fluorescence property, stability and suitable capability of resisting electrolyte	Bio-analysis markers	[111]

decorating a variety of polar and functional groups on polymeric surfaces and nanostructures. During the past decades, a number of surface functionalization methods have been performed which commonly pursue the following paths: firstly, primary reactive functional groups are bonded on the surface of the polymer chain. Secondly, the reactive surface is modified by active/bioactive agents, oligomers or polymers, hydrophobic and hydrophilic monomers, in order to attain specific surface characteristics that fulfill the needs of the end use [117]. The incorporation of active/bioactive entities on a polymeric nanomaterial surface is generally performed by covalent bonds, ligand-receptor pairing, and electrostatic interactions. Table 3.10 illustrates the functionalization of polymeric materials along with their potential applications.

3.12 Application of functionalized nanomaterials

As water pollution is growing at a rapid rate, the scarcity of drinking water is decreasing in a fueling way. To overcome this scenario and to provide people across the globe with fresh drinking water, membrane technology plays an indispensable role in producing potable drinking water from the ground, surface, and seawater sources in addition to the superior treatment of wastewater and desalination. Functionalized nanomaterial membrane has a tremendous application prospect to overcome a few faults in traditionally used wastewater treatment processes. Consequently, membrane technology becomes the optimized technology in brackish and seawater desalination processes and wastewater treatment processes. Nowadays, surface-functionalized magnetic nanomaterials have become a new kind of adsorbents and have shown great promise as the best feasible material for wastewater treatment. This occurred due to their large surface area, weak coagulation, and high stability, showing super-paramagnetic properties in a magnetic field [123]. The mixed matrix membrane incorporated with functionalized inorganic nanomaterials have earned its place in the water and wastewater treatment process as it has provided an opportunity for evolving a novel type of polymer-based membrane with tremendous physiochemical properties and excellent filtration performance [124]. Various functionalized nanomaterials such as carbon-based nanomaterial, titanium-based nanomaterial, binary metal oxides, and metal organic framework, are incorporated in MMMs to enhance their hydrophilicity, flux resistance, solute rejection property, antifouling, and antibacterial properties [124]. The functionalized nanomaterial embedded MMMs

Table 3.10 Functionalization of polymeric materials along with their potential applications.

Sl. no	Polymers	Functionalizing agent	Functional group incorporated	Properties	Applications	References
1	Polypropylene	Chitosen	Amine and amino group	Improve the polymer biocompatibity and higher cell metabolic activity enhance drug delivery properties	Enhance human fibroblast viability	[118]
3	PVDF	Beta-cyclodextrin	Hydroxyl group	Enhance drug delivery properties (excellent cytotoxicity and good adaption of L132)	Drug delivery guided tissue regeneration (GTR) device	[119]
4	PVDF	Biocidal oxine/TiO_2 nanocomposite	Hydroxyl group	Increase porosity and pore size along with enhanced antifouling properties	Auto-cleaning functionalized hybrid membrane with a long lasting antibiofilm effect	[120]
5	PMMA	L-3,4-Dihydroxyphenylalanine (L-DOPA)	Hydroxy, amine and carboxylic acid	To Enhance its biocompatibility and adhesion to corneal tissue	Implanted artificial cornea for patients with severe corneal diseases	[121]
6	Polyethylene terephthalate (PET)	Immobilized NTPDase and cysteine	Carboxyl group	Enhance the surface roughness and mechanical strength		[122]

are also showing their great potential in the heavy-metal removal application [125]. Additionally, advances in functionalized nanomaterials have created new opportunities for creating many powerful tools that can be used in clinical applications and biomedical research. Highly functionalized nanomaterials that contain well-controlled size and shape have become a new class of building blocks for monitoring molecular signals in biological systems and living organisms. The main potential application of functionalized nanomaterials is in the field of smart drug delivery systems. For instance, nanomaterials like carbon-based nanomaterials, nano formulations, polymeric nanostructure, and magnetic nanomaterials functionalized with bioactive moieties have given a potential site-target delivery of nutrients, drugs, and active organic entities to the plant and animal cells. In the 21st century, nanotechnology is considered an imminent technology and has contemplated a pivotal role in the cosmetic industry. The altered properties and characteristics like transparency, chemical reactivity, solubility, deeper skin penetration, excellent UV protection, long-lasting effects, ability to enhance the color, and fineness quality, make them an enticing material for personal care and cosmetics industry application [126]. Additionally, the incorporation of functionalized nanomaterials into synthetic fibers used in the textile industry allows to give a solution to traditional problems for textiles such as flammability, robustness against ultraviolet radiation, and many others [127].

3.13 Conclusions

The unexpected and unprecedented discoveries and the rapid progress in nanotechnology bear tremendous potential in innovative technological applications. In the present chapter, we have provided an overview of the development of nanotechnology in numerous emerging fields of membrane science and technology. Various types of functionalized nanomaterials such as carbonaceous, silica, metal, and polymer functionalized nanomaterials have been discussed along with their functionalization techniques. This work also elucidated numerous sustainable emerging applications of functionalized nanomaterial-based membrane. However, the research in functionalized nanomaterials in the membrane technology is still in the milestone and more focus is needed to employ what has been achieved so far to come up with feasible applications along with the search for novel technologies.

References

[1] M.S. Diallo, N.A. Fromer, M.S. Jhon, Nanotechnology for sustainable development: retrospective and outlook, J. Nanopart. Res. 15 (11) (2013), https://doi.org/10.1007/s11051-013-2044-0.

[2] I. Khan, K. Saeed, I. Khan, Nanoparticles: properties, applications and toxicities, Arab. J. Chem. 12 (7) (2019) 908–931, https://doi.org/10.1016/j.arabjc.2017.05.011.

[3] A. Bratovcic, Different applications of nanomaterials and their impact on the environment, Int. J. Mater. Sci. Eng. 5 (1) (2019) 1–7, https://doi.org/10.14445/23948884/ijmse-v5i1p101.

[4] E. Rodríguez et al., "We are IntechOpen, the world ' s leading publisher of Open Access books Built by scientists, for scientists TOP 1%" Intech, vol. 32, no. tourism, pp. 137–144, 1989, [Online]. Available: https://www.intechopen.com/books/advanced-biometric-technologies/liveness-detection-in-biometrics.

[5] R. Sondhi, R. Bhave, G. Jung, Applications and benefits of ceramic membranes, Membr. Technol. (11 November) (2003) 5–8, https://doi.org/10.1016/s0958-2118(03)11016-6.

[6] G.S. Burkhanov, N.B. Gorina, N.B. Kolchugina, N.R. Roshan, D.I. Slovetsky, E.M. Chistov, Palladium-based alloy membranes for separation of high purity hydrogen from hydrogen-containing gas mixtures, Platin. Met. Rev. 55 (1) (2011) 3–12, https://doi.org/10.1595/147106711X540346.

[7] S. Tonzani, Polymers in membrane science, J. Appl. Polym. Sci. 124 (Suppl. 1) (2012) 191–240, https://doi.org/10.1002/app.36932.

[8] H. Lin, Y. Ding, Polymeric membranes: chemistry, physics, and applications, J. Polym. Sci. A 58 (18) (2020) 2433–2434, https://doi.org/10.1002/pol.20200622.

[9] M. Dunleavy, Polymeric membranes. A review of applications, Med. Device Technol. 7 (4) (1996) 14–16. 18–21,.

[10] M.R. Esfahani, et al., Nanocomposite membranes for water separation and purification: fabrication, modification, and applications, Sep. Purif. Technol. 213 (December 2018) (2019) 465–499, https://doi.org/10.1016/j.seppur.2018.12.050.

[11] Y. Ying, et al., Recent advances of nanomaterial-based membrane for water purification, Appl. Mater. Today 7 (2017) 144–158, https://doi.org/10.1016/j.apmt.2017.02.010.

[12] M.A. Alaei Shahmirzadi, A. Kargari, 9—Nanocomposite membranes, in: Emerging Technologies for Sustainable Desalination Handbook, Butterworth-Heinemann, 2018, pp. 285–330.

[13] K. Miyazaki, N. Islam, Nanotechnology systems of innovation—An analysis of industry and academia research activities, Technovation 27 (11) (2007) 661–675, https://doi.org/10.1016/j.technovation.2007.05.009.

[14] O.V. Salata, Applications of nanoparticles in biology and medicine, J. Nanobiotechnol. 2 (2004) 1–6, https://doi.org/10.1186/1477-3155-2-3.

[15] Y. Peng, et al., Research and development of drug delivery systems based on drug transporter and nano-formulation, Asian J. Pharm. Sci. 15 (2) (2020) 220–236, https://doi.org/10.1016/j.ajps.2020.02.004.

[16] C. Pinto Reis, R.J. Neufeld, A.J. Ribeiro, F. Veiga, Nanoencapsulation I. Methods for preparation of drug-loaded polymeric nanoparticles, Nanomed. Nanotechnol. Biol. Med. 2 (1) (2006) 8–21, https://doi.org/10.1016/j.nano.2005.12.003.

[17] V. Georgakilas, J.A. Perman, J. Tucek, R. Zboril, Broad family of carbon nanoallotropes: classification, chemistry, and applications of fullerenes, carbon dots, nanotubes, graphene, nanodiamonds, and combined superstructures, Chem. Rev. 115 (11) (2015) 4744–4822, https://doi.org/10.1021/cr500304f.

[18] A.P. Ramos, Biomedical applications of nanotechnology, Biophys. Rev. (2017) 79–89, https://doi.org/10.1007/s12551-016-0246-2.

[19] F. Dong, R.T. Koodali, H. Wang, W. Ho, Nanomaterials for environmental applications, J. Nanomater. (2014). 2014.
[20] P.C. Ray, P.P. Fu, Toxicity and environmental risks of nanomaterials: challenges and future needs, J. Environ. Sci. Health C 27 (1) (2009) 1–35, https://doi.org/10.1080/10590500802708267.
[21] R.I.D.H. Andy, D.E.Y.L. Yon, S.H.M. Ahendra, M.I.J.M.C.L. Aughlin, J.A.R.L. Ead, Critical review and effects, Environ. Toxicol. Chem. 27 (9) (2008) 1825–1851, https://doi.org/10.1897/08-090.1.
[22] M. Darwish, A. Mohammadi, Functionalized nanomaterial for environmental techniques, Nanotechnology in Environmental Science, vol. 1–2, Wiley, 2018.
[23] R. Thiruppathi, S. Mishra, M. Ganapathy, P. Padmanabhan, B. Gulyás, Nanoparticle functionalization and its potentials for molecular imaging, Adv. Sci. 4 (3) (2017), https://doi.org/10.1002/advs.201600279.
[24] N. Kumar, S. Sinha Ray, Synthesis and Functionalization of Nanomaterials, vol. 277, Springer International Publishing, 2018.
[25] D. Rawtani, M. Tharmavaram, G. Pandey, C.M. Hussain, Functionalized nanomaterial for forensic sample analysis, TrAC Trends Anal. Chem. 120 (2019), https://doi.org/10.1016/j.trac.2019.115661, 115661.
[26] W. Wu, Q. He, C. Jiang, Magnetic iron oxide nanoparticles: synthesis and surface functionalization strategies, Nanoscale Res. Lett. 3 (11) (2008) 397–415, https://doi.org/10.1007/s11671-008-9174-9.
[27] N.V. Konduru, et al., Surface modification of zinc oxide nanoparticles with amorphous silica alters their fate in the circulation, Nanotoxicology 10 (6) (2016) 720–727, https://doi.org/10.3109/17435390.2015.1113322.
[28] S. Mondal, Carbon nanomaterials based membranes, J. Membr. Sci. Technol. 06 (04) (2016) 18–20, https://doi.org/10.4172/2155-9589.1000e122.
[29] C. Ursino, R. Castro-Muñoz, E. Drioli, L. Gzara, M.H. Albeirutty, A. Figoli, Progress of nanocomposite membranes for water treatment, Membranes (Basel) 8 (2) (2018) 1–40, https://doi.org/10.3390/membranes8020018.
[30] C.R. Martin, Membrane-based synthesis of nanomaterials, Chem. Mater. 8 (8) (1996) 1739–1746, https://doi.org/10.1021/cm960166s.
[31] A. Barhoum, et al., A broad family of carbon nanomaterials: classification, properties, synthesis, and emerging applications, Handbook of Nanofibers, Springer, Cham, 2019.
[32] A.C. Tripathi, S.A. Saraf, S.K. Saraf, Carbon Nanotropes: A Contemporary Paradigm in Drug Delivery, 2015, pp. 3068–3100, https://doi.org/10.3390/ma8063068.
[33] R. Raccichini, A. Varzi, S. Passerini, B. Scrosati, The role of graphene for electrochemical energy storage, Nat. Mater. 14 (3) (2015) 271–279, https://doi.org/10.1038/nmat4170.
[34] R. Shah, A. Kausar, B. Muhammad, S. Shah, Progression from graphene and graphene oxide to high performance polymer-based nanocomposite: a review, Polym.-Plast. Technol. Eng. 54 (2) (2015) 173–183, https://doi.org/10.1080/03602559.2014.955202.
[35] E. Dervishi, et al., Carbon nanotubes: synthesis, properties, and applications, Part. Sci. Technol. 27 (2) (2009) 107–125, https://doi.org/10.1080/02726350902775962.
[36] M. Jose, M. Baeza, R. Olive-Monllau, F. Cespedes, J. Bartroli, Development of tunable nanocomposites made from carbon nanotubes for electrochemical applications, Adv. Compos. Mater. Med. Nanotechnol (April) (2011) 2011, https://doi.org/10.5772/14697.
[37] N. Saifuddin, A.Z. Raziah, A.R. Junizah, Carbon nanotubes: a review on structure and their interaction with proteins, J. Chem. 2013 (2013), https://doi.org/10.1155/2013/676815.
[38] A. Venkataraman, E.V. Amadi, Y. Chen, C. Papadopoulos, Carbon nanotube assembly and integration for applications, Nanoscale Res. Lett. 14 (1) (2019), https://doi.org/10.1186/s11671-019-3046-3.

[39] M. Sianipar, S.H. Kim, F.I. Khoiruddin, I.G. Wenten, Functionalized carbon nanotube (CNT) membrane: progress and challenges, RSC Adv. 7 (81) (2017) 51175–51198, https://doi.org/10.1039/c7ra08570b.
[40] P.C. Ma, N.A. Siddiqui, G. Marom, J.K. Kim, Dispersion and functionalization of carbon nanotubes for polymer-based nanocomposites: a review, Compos. Part A Appl. Sci. Manuf. 41 (10) (2010) 1345–1367, https://doi.org/10.1016/j.compositesa.2010.07.003.
[41] J.M. Wernik, B.J. Cornwell-Mott, S.A. Meguid, Determination of the interfacial properties of carbon nanotube reinforced polymer composites using atomistic-based continuum model, Int. J. Solids Struct. 49 (13) (2012) 1852–1863, https://doi.org/10.1016/j.ijsolstr.2012.03.024.
[42] S. Mallakpour, S. Soltanian, Surface functionalization of carbon nanotubes: fabrication and applications, RSC Adv. 6 (111) (2016) 109916–109935, https://doi.org/10.1039/C6RA24522F.
[43] M. Karimi, A. Ghasemi, S. Mirkiani, S.M. Moosavi Basri, M.R. Hamblin, Carbon Nanotubes in Drug and Gene Delivery, Morgan & Claypool Publishers, 2017.
[44] Z. Syrgiannis, M. Melchionna, M. Prato, Covalent Carbon Nanotube Functionalization Defect-Group Functionalization, no. 1, 2014, pp. 1–8, https://doi.org/10.1007/978-3-642-36199-9.
[45] L. Santiago-Rodríguez, G. Sánchez-Pomales, C.R. Cabrera, DNA-functionalized carbon nanotubes: synthesis, self-assembly, and applications, Isr. J. Chem. 50 (3) (2010) 277–290, https://doi.org/10.1002/ijch.201000034.
[46] T. Ohashi, Carbon nanotubes, Carbon Nanomaterials for Advanced Energy Systems: Advances in Materials Synthesis and Device Applications, John Wiley & Sons, Ltd, 2015, pp. 47–84.
[47] M. Xu, T. Liang, M. Shi, H. Chen, Graphene-like two-dimensional materials, Chem. Rev. 113 (5) (2013) 3766–3798, https://doi.org/10.1021/cr300263a.
[48] M. Ionita, M.A. Pandele, H. Iovu, Sodium alginate/graphene oxide composite films with enhanced thermal and mechanical properties, Carbohydr. Polym. 94 (1) (2013) 339–344, https://doi.org/10.1016/j.carbpol.2013.01.065.
[49] W. Yu, L. Sisi, Y. Haiyan, L. Jie, Progress in the functional modification of graphene/graphene oxide: a review, RSC Adv. 10 (26) (2020) 15328–15345, https://doi.org/10.1039/d0ra01068e.
[50] V. Georgakilas, et al., Functionalization of graphene: covalent and non-covalent approaches, derivatives and applications, Chem. Rev. 112 (11) (2012) 6156–6214, https://doi.org/10.1021/cr3000412.
[51] M. Ioniţă, G.M. Vlăsceanu, A.A. Watzlawek, S.I. Voicu, J.S. Burns, H. Iovu, Graphene and functionalized graphene: extraordinary prospects for nanobiocomposite materials, Compos. Part B Eng. 121 (2017) 34–57, https://doi.org/10.1016/j.compositesb.2017.03.031.
[52] T.J.M. Fraga, M.N. Carvalho, M.G. Ghislandi, M.A. Da Motta Sobrinho, Functionalized graphene-based materials as innovative adsorbents of organic pollutants: a concise overview, Braz. J. Chem. Eng. 36 (1) (2019) 21–31, https://doi.org/10.1590/0104-6632.20190361s20180283.
[53] A. Liang, X. Jiang, X. Hong, Y. Jiang, Z. Shao, D. Zhu, Recent developments concerning the dispersion methods and mechanisms of graphene, Coatings 8 (1) (2018), https://doi.org/10.3390/coatings8010033.
[54] C.C. Liu, et al., Multifunctionalization of graphene and graphene oxide for controlled release and targeted delivery of anticancer drugs, Am. J. Transl. Res. 9 (12) (2017) 5197–5219.
[55] W. Yu, L. Sisi, Y. Haiyan, L. Jie, Progress in the functional modification of graphene/graphene oxide: a review, RSC Adv. 10 (26) (2020) 15328–15345, https://doi.org/10.1039/d0ra01068e.

[56] S. Thakral, R. Mehta, Fullerenes: an introduction and overview of their biological properties, Indian J. Pharm. Sci. 68 (1) (2006) 13–19, https://doi.org/10.4103/0250-474X.22957.

[57] H.S. Lin, Y. Matsuo, Functionalization of [60]fullerene through fullerene cation intermediates, Chem. Commun. 54 (80) (2018) 11244–11259, https://doi.org/10.1039/C8CC05965A.

[58] W. Yan, S.M. Seifermann, P. Pierrat, S. Bräse, Synthesis of highly functionalized C60 fullerene derivatives and their applications in material and life sciences, Org. Biomol. Chem. 13 (1) (2015) 25–54, https://doi.org/10.1039/c4ob01663g.

[59] A.M. Schrand, S.A.C. Hens, O.A. Shenderova, Nanodiamond particles: properties and perspectives for bioapplications, Crit. Rev. Solid State Mater. Sci. 34 (1–2) (2009) 18–74, https://doi.org/10.1080/10408430902831987.

[60] S. Barua, S. Gogoi, R. Khan, N. Karak, Silicon-Based Nanomaterials and Their Polymer Nanocomposites, no. January, Elsevier Inc., 2018.

[61] S.E. Lehman, S.C. Larsen, Zeolite and mesoporous silica nanomaterials: greener syntheses, environmental applications and biological toxicity, Environ. Sci. Nano 1 (3) (2014) 200–213, https://doi.org/10.1039/C4EN00031E.

[62] Y. Li, L. Li, J. Yu, Applications of zeolites in sustainable chemistry, Chem 3 (6) (2017) 928–949, https://doi.org/10.1016/j.chempr.2017.10.009.

[63] M. Manyangadze, N.H.M. Chikuruwo, T.B. Narsaiah, C.S. Chakra, M. Radhakumari, G. Danha, Enhancing adsorption capacity of nano-adsorbents via surface modification: a review, S. Afr. J. Chem. Eng 31 (September 2019) (2020) 25–32, https://doi.org/10.1016/j.sajce.2019.11.003.

[64] B.Z. Zhan, M.A. White, M. Lumsden, Bonding of organic amino, vinyl, and acryl groups to nanometer-sized NaX zeolite crystal surfaces, Langmuir 19 (10) (2003) 4205–4210, https://doi.org/10.1021/la026737e.

[65] A.M. Putz, L. Almásy, A. Len, C. Ianăşi, Functionalized silica materials synthesized via co-condensation and post-grafting methods, Fullerenes, Nanotubes, Carbon Nanostruct. 27 (4) (2019) 323–332, https://doi.org/10.1080/1536383X.2019.1593154.

[66] F. Hoffmann, M. Cornelius, J. Morell, M. Fröba, Silica-based mesoporous organic-inorganic hybrid materials, Angew. Chem. Int. Ed. 45 (20) (2006) 3216–3251, https://doi.org/10.1002/anie.200503075.

[67] Y. Huang, Functionalization of mesoporous silica nanoparticles and their applications in organo-, metallic and organometallic catalysis, in: Functionalization of Mesoporous Silica Nanoparticles and Their Applications in Organo-, Metallic and Organometallic Catalysis, 2009, pp. 1–12. 37–42, [Online]. Available: http://lib.dr.iastate.edu/cgi/viewcontent.cgi?article=1953&context=etd.

[68] B.A. Stein, B.J. Melde, R.C. Schroden, Hybrid inorganic–organic mesoporous silicates—nanoscopic reactors coming of age, Adv. Mater. 12 (19) (2000) 1403–1419, https://doi.org/10.1002/1521-4095(200010)12:19<1403::AID-ADMA1403>3.0.CO;2-X.

[69] C. Delacôte, F.O.M. Gaslain, B. Lebeau, A. Walcarius, Factors affecting the reactivity of thiol-functionalized mesoporous silica adsorbents toward mercury(II), Talanta 79 (3) (2009) 877–886, https://doi.org/10.1016/j.talanta.2009.05.020.

[70] S.H. Cheng, et al., Tri-functionalization of mesoporous silica nanoparticles for comprehensive cancer theranostics—the trio of imaging, targeting and therapy, J. Mater. Chem. 20 (29) (2010) 6149–6157, https://doi.org/10.1039/c0jm00645a.

[71] Z.F. Li, H. Zhang, Q. Liu, Y. Liu, L. Stanciu, J. Xie, Novel pyrolyzed polyaniline-grafted silicon nanoparticles encapsulated in graphene sheets as Li-ion battery anodes, ACS Appl. Mater. Interfaces 6 (8) (2014) 5996–6002, https://doi.org/10.1021/am501239r.

[72] G.P. Knowles, S.W. Delaney, A.L. Chaffee, Diethylenetriamine[propyl(silyl)]-functionalized (DT) mesoporous silicas as CO2 adsorbents, Ind. Eng. Chem. Res. 45 (8) (2006) 2626–2633, https://doi.org/10.1021/ie050589g.
[73] C. Von Baeckmann, R. Guillet-Nicolas, D. Renfer, H. Kählig, F. Kleitz, A toolbox for the synthesis of multifunctionalized mesoporous silica nanoparticles for biomedical applications, ACS Omega 3 (12) (2018) 17496–17510, https://doi.org/10.1021/acsomega.8b02784.
[74] M. Pálmai, et al., Preparation, purification, and characterization of aminopropyl-functionalized silica sol, J. Colloid Interface Sci. 390 (1) (2013) 34–40, https://doi.org/10.1016/j.jcis.2012.09.025.
[75] N.I. Taib, S. Endud, M.N. Katun, Functionalization of mesoporous Si-MCM-41 by grafting with trimethylchlorosilane, Int. J. Chem. 3 (3) (2011) 2–10, https://doi.org/10.5539/ijc.v3n3p2.
[76] V. Antochshuk, O. Olkhovyk, M. Jaroniec, I.S. Park, R. Ryoo, Benzoylthiourea-modified mesoporous silica for mercury(II) removal, Langmuir 19 (7) (2003) 3031–3034, https://doi.org/10.1021/la026739z.
[77] Y. Xu, et al., Highly specific enrichment of glycopeptides using boronic acid-functionalized mesoporous silica, Anal. Chem. 81 (1) (2009) 503–508, https://doi.org/10.1021/ac801912t.
[78] O. Olkhovyk, M. Jaroniec, Adsorption characterization of ordered mesoporous silicas with mercury-specific immobilized ligands, Adsorption 11 (suppl. 1) (2005) 685–690, https://doi.org/10.1007/s10450-005-6007-3.
[79] A.A. Yaqoob, et al., Recent advances in metal decorated nanomaterials and their various biological applications: a review, Front. Chem. 8 (May) (2020) 1–23, https://doi.org/10.3389/fchem.2020.00341.
[80] V.V. Vodnik, U. Bogdanović, Metal nanoparticles and their composites: a promising multifunctional nanomaterial for biomedical and related applications, in: Materials for Biomedical Engineering: Inorganic Micro- and Nanostructures, 2019, pp. 397–426, https://doi.org/10.1016/B978-0-08-102814-8.00014-7.
[81] K. Mahato, et al., Gold nanoparticle surface engineering strategies and their applications in biomedicine and diagnostics, 3 Biotech 9 (2) (2019) 1–19, https://doi.org/10.1007/s13205-019-1577-z.
[82] G. Vales, S. Suhonen, K.M. Siivola, K.M. Savolainen, J. Catalán, H. Norppa, Genotoxicity and cytotoxicity of gold nanoparticles in vitro: role of surface functionalization and particle size, Nanomaterials 10 (2) (2020) 271, https://doi.org/10.3390/nano10020271.
[83] P. Baptista, et al., Gold nanoparticle-based theranostics: disease diagnostics and treatment using a single nanomaterial, Nanobiosensors Dis. Diagn. (June) (2015) 11, https://doi.org/10.2147/ndd.s60285.
[84] P. Tiwari, K. Vig, V. Dennis, S. Singh, Functionalized gold nanoparticles and their biomedical applications, Nanomaterials 1 (1) (2011) 31–63, https://doi.org/10.3390/nano1010031.
[85] F. Dumur, E. Dumas, C.R. Mayer, Functionalization of gold nanoparticles by inorganic entities, Nanomaterials 10 (3) (2020), https://doi.org/10.3390/nano10030548.
[86] A.S. Salmiati, M.R. Salim, A.B.H. Kueh, T. Hadibarata, H. Nur, A review of silver nanoparticles: research trends, global consumption, synthesis, properties, and future challenges, J. Chin. Chem. Soc. 64 (7) (2017) 732–756, https://doi.org/10.1002/jccs.201700067.
[87] X. Zhang, Z. Liu, W. Shen, S. Gurunathan, Silver Nanoparticles : Synthesis, Characterization, Properties, Applications, and Therapeutic Approaches, 2016, https://doi.org/10.3390/ijms17091534.

[88] M.C. Stensberg, Q. Wei, E.S. McLamore, D.M. Porterfield, A. Wei, M.S. Sepúlveda, Toxicological studies on silver nanoparticles: challenges and opportunities in assessment, monitoring and imaging, Nanomedicine (Lond.) 6 (5) (2011) 879–898, https://doi.org/10.2217/nnm.11.78.
[89] B. Buszewski, K. Rafińska, P. Pomastowski, J. Walczak, A. Rogowska, Novel aspects of silver nanoparticles functionalization, Colloids Surf. A Physicochem. Eng. Asp. 506 (2016) 170–178, https://doi.org/10.1016/j.colsurfa.2016.05.058.
[90] W. Qing, M. Zhao, C. Kou, M. Lu, Y. Wang, Functionalization of silver nanoparticles with mPEGylated luteolin for selective visual detection of Hg2+ in water sample, RSC Adv. 8 (51) (2018) 28843–28846, https://doi.org/10.1039/c8ra05243c.
[91] A. Sooresh, H. Kwon, R. Taylor, P. Pietrantonio, M. Pine, C.M. Sayes, Surface functionalization of silver nanoparticles: novel applications for insect vector control, ACS Appl. Mater. Interfaces 3 (10) (2011) 3779–3787, https://doi.org/10.1021/am201167v.
[92] A. Borowik, et al., The impact of surface functionalization on the biophysical properties of silver nanoparticles, Nanomaterials 9 (7) (2019), https://doi.org/10.3390/nano9070973.
[93] D.C. Kennedy, et al., Carbohydrate functionalization of silver nanoparticles modulates cytotoxicity and cellular uptake, J. Nanobiotechnol. 12 (1) (2014) 1–8, https://doi.org/10.1186/s12951-014-0059-z.
[94] M. Fernndez-Garca, J.A. Rodriguez, Metal oxide nanoparticles, Encycl. Inorg. Chem. (October) (2009), https://doi.org/10.1002/0470862106.ia377.
[95] M.S. Chavali, M.P. Nikolova, Metal Oxide Nanoparticles and their Applications in Nanotechnology, vol. 1, Springer International Publishing, 2019. no. 6.
[96] K.S. Siddiqi, A.u. Rahman, Tajuddin, A. Husen, Properties of zinc oxide nanoparticles and their activity against microbes, Nanoscale Res. Lett. 13 (2018), https://doi.org/10.1186/s11671-018-2532-3.
[97] V. Abbasi-Chianeh, B. Bostani, Z. Noroozi, M.R. Akbarpour, F. Yahyavi, Enhanced structural, adsorption, and antibacterial properties of ZnO nanoparticles, J. Aust. Ceram. Soc. 55 (3) (2019) 639–644, https://doi.org/10.1007/s41779-018-0273-5.
[98] M. Sheikh, et al., Application of ZnO nanostructures in ceramic and polymeric membranes for water and wastewater technologies: a review, Chem. Eng. J. 391 (August 2019) (2020) 123475, https://doi.org/10.1016/j.cej.2019.123475.
[99] C.P. Leo, W.P. Cathie Lee, A.L. Ahmad, A.W. Mohammad, Polysulfone membranes blended with ZnO nanoparticles for reducing fouling by oleic acid, Sep. Purif. Technol. 89 (2012) 51–56, https://doi.org/10.1016/j.seppur.2012.01.002.
[100] H. Mosaddeghi, B. Rezaei, H. Mosaddeghi, Applications of titanium dioxide nanoparticles, Appl. Titan. Dioxide Nanocoatings (January) (2009) 1–4, https://doi.org/10.13140/RG.2.1.2184.6006.
[101] E. Livari, A. Aroujalian, A. Raisi, M. Fathizadeh, The effect of Ti02 nanoparticles on PES UF membrane fouling in water-oil sepration, Procedia Eng. 44 (6–8) (2012) 1783–1785, https://doi.org/10.1016/j.proeng.2012.08.949.
[102] G. Sereda, K. Rashwan, B. Karels, and A. Fritza, "Novel materials for desensitizing and remineralizing dentifrices," Adv. Mater.—TechConnect Briefs 2016, vol. 1, pp. 135–138, 2016.
[103] S. Karapati, T. Giannakopoulou, N. Todorova, N. Boukos, D. Dimotikali, C. Trapalis, Eco-efficient TiO2 modification for air pollutants oxidation, Appl. Catal. Environ. 176–177 (2015) 578–585, https://doi.org/10.1016/j.apcatb.2015.04.012.
[104] I.A. Janković, Z.V. Šaponjić, M.I. Čomor, J.M. Nedeljkovic, Surface modification of colloidal tiO2 nanoparticles with bidentate benzene derivatives, J. Phys. Chem. C 113 (29) (2009) 12645–12652, https://doi.org/10.1021/jp9013338.
[105] S. Rahim, M. Sasani Ghamsari, S. Radiman, Surface modification of titanium oxide nanocrystals with PEG, Sci. Iran. 19 (3) (2012) 948–953, https://doi.org/10.1016/j.scient.2012.03.009.

[106] P.E. Bunney, A.N. Zink, A.A. Holm, C.J. Billington, C.M. Kotz, Orexin activation counteracts decreases in nonexercise activity thermogenesis (NEAT) caused by high-fat diet, Physiol. Behav. 176 (3) (2017) 139–148, https://doi.org/10.1016/j.physbeh.2017.03.040.
[107] S.B. Ghaffari, M.H. Sarrafzadeh, Z. Fakhroueian, S. Shahriari, M.R. Khorramizadeh, Functionalization of ZnO nanoparticles by 3-mercaptopropionic acid for aqueous curcumin delivery: synthesis, characterization, and anticancer assessment, Mater. Sci. Eng. C 79 (2017) 465–472, https://doi.org/10.1016/j.msec.2017.05.065.
[108] P. Saravanan, K. Jayamoorthy, S. Ananda Kumar, Switch-on fluorescence and photo-induced electron transfer of 3-aminopropyltriethoxysilane to ZnO: dual applications in sensors and antibacterial activity, Sens. Actuators B 221 (2015) 784–791, https://doi.org/10.1016/j.snb.2015.05.069.
[109] Y. Liao, Z. Hu, Q. Gu, C. Xue, Amine-functionalized ZnO nanosheets for efficient CO2 capture and photoreduction, Molecules 20 (10) (2015) 18847–18855, https://doi.org/10.3390/molecules201018847.
[110] K.-E. Kim, T.G. Kim, Y.-M. Sung, Enzyme-conjugated ZnO nanocrystals for collisional quenching-based glucose sensing, CrstEngComm 14 (8) (2012) 2859–2865, https://doi.org/10.1039/C2CE06410C.
[111] J. Zhuang, M. Liu, H. Liu, MAA-modified and luminescence properties of ZnO quantum dots, Sci. China, Ser. B: Chem. 52 (12) (2009) 2125–2133, https://doi.org/10.1007/s11426-009-0198-5.
[112] A. Puri, et al., Lipid-based nanoparticles as pharmaceutical drug carriers: from concepts to clinic, Crit. Rev. Ther. Drug Carrier Syst. 26 (6) (2009) 523–580, https://doi.org/10.1615/CritRevTherDrugCarrierSyst.v26.i6.10.
[113] J.K. Patra, et al., Nano based drug delivery systems: recent developments and future prospects 10 technology 1007 nanotechnology 03 chemical sciences 0306 physical chemistry (incl. Structural) 03 chemical sciences 0303 macromolecular and materials chemistry 11 medical and He, J. Nanobiotechnol. 16 (1) (2018) 1–33, https://doi.org/10.1186/s12951-018-0392-8.
[114] A. Akbarzadeh, et al., Liposome: classification, preparation, and applications, Nanoscale Res. Lett. 8 (1) (2013) 1, https://doi.org/10.1186/1556-276X-8-102.
[115] E. Abbasi, et al., Dendrimers: synthesis, applications, and properties, Nanoscale Res. Lett. 9 (1) (2014) 1–10, https://doi.org/10.1186/1556-276X-9-247.
[116] G.M. Dykes, Dendrimers: a review of their appeal and applications, J. Chem. Technol. Biotechnol. 76 (9) (2001) 903–918, https://doi.org/10.1002/jctb.464.
[117] J.M. Goddard, J.H. Hotchkiss, Polymer surface modification for the attachment of bioactive compounds, Prog. Polym. Sci. 32 (7) (2007) 698–725, https://doi.org/10.1016/j.progpolymsci.2007.04.002.
[118] D.S. Morais, et al., Surface functionalization of polypropylene (PP) by chitosan immobilization to enhance human fibroblasts viability, Polym. Test 86 (January) (2020), https://doi.org/10.1016/j.polymertesting.2020.106507.
[119] M. Li, M.J. Mondrinos, X. Chen, M.R. Gandhi, F.K. Ko, P.I. Lelkes, Elastin blends for tissue engineering scaffolds, J. Biomed. Mater. Res. A 79 (4) (2006) 963–973, https://doi.org/10.1002/jbm.a.
[120] R.K. Manoharan, S. Ayyaru, Y.H. Ahn, Auto-cleaning functionalization of the polyvinylidene fluoride membrane by the biocidal oxine/TiO2 nanocomposite for anti-biofouling properties, New J. Chem. 44 (3) (2020) 807–816, https://doi.org/10.1039/c9nj05300j.
[121] R. Sharifi, et al., Poly(methyl methacrylate): covalent functionalization of PMMA surface with L-3,4-dihydroxyphenylalanine (L-DOPA) to enhance its biocompatibility and adhesion to corneal tissue (Adv. Mater. Interfaces 1/2020), Adv. Mater. Interfaces 7 (1) (2020) 2070001, https://doi.org/10.1002/admi.202070001.

[122] V. Muthuvijayan, J. Gu, R.S. Lewis, Analysis of functionalized polyethylene terephthalate with immobilized NTPDase and cysteine, Acta Biomater. 5 (9) (2009) 3382–3393, https://doi.org/10.1016/j.actbio.2009.05.020.
[123] S.M. Abdelbasir, A.E. Shalan, An overview of nanomaterials for industrial wastewater treatment, Korean J. Chem. Eng. 36 (8) (2019) 1209–1225, https://doi.org/10.1007/s11814-019-0306-y.
[124] D. Qadir, H. Mukhtar, L.K. Keong, Mixed matrix membranes for water purification applications, Sep. Purif. Rev. 46 (1) (2017) 62–80, https://doi.org/10.1080/15422119.2016.1196460.
[125] S. Singh, K.C. Barick, D. Bahadur, Functional oxide nanomaterials and nanocomposites for the removal of heavy metals and dyes, Nanomater. Nanotechnol. 3 (1) (2013) 1–19, https://doi.org/10.5772/57237.
[126] S. Kaul, N. Gulati, D. Verma, S. Mukherjee, U. Nagaich, Role of nanotechnology in cosmeceuticals: a review of recent advances, J. Pharm. 2018 (2018) 1–19, https://doi.org/10.1155/2018/3420204.
[127] P.J. Rivero, A. Urrutia, J. Goicoechea, F.J. Arregui, Nanomaterials for functional textiles and fibers, Nanoscale Res. Lett. 10 (1) (2015) 1–22, https://doi.org/10.1186/s11671-015-1195-6.

CHAPTER 4

Fabrication of sustainable membranes with functionalized nanomaterials (FNMs)

Jasir Jawad and Syed Javaid Zaidi
Centre for Advanced Materials, Qatar University, Doha, Qatar

4.1 Introduction

Potable water supply is essential for human life and the ecosystem. Due to population growth, contamination of freshwater, and global warming, there is a lack of clean water and a huge demand to be fulfilled [1]. According to World Health Organization (WHO), water shortages will affect 4 billion people by 2050 [2] and 98% of the world's water comprises brackish and seawater. As the freshwater sources are depleting, there is a need for efficient and low-cost treatment methods to meet the global requirement and achieve water sustainability. Although several techniques are available for water treatment applications, membrane-based technology has gained special attention in the past few years due to its high stability, low cost, ease of operation, high performance, and low energy consumption [3,4].

Membranes can be porous and dense that provide a barrier between two phases and allow selective transport from one side to another. The membrane-based technologies for water treatment include microfiltration (MF), ultrafiltration (UF), nanofiltration (NF), reverse osmosis (RO), forward osmosis (FO), and membrane distillation (MD). The membrane's separation performance is limited to the transport properties and selectivity of the membrane, which depends heavily on the pore structure within the membrane [5]. Furthermore, traditional membranes have a trade-off between permeability and rejection and high fouling propensity [6]. Conventional membranes can be classified into organic polymeric membranes and inorganic ceramic-based membranes. Both types of membranes have been extensively used for desalination and water treatment applications on a laboratory and industrial scale. However, due to their limitations, a recent trend is toward developing a sustainable membrane with a longer lifetime and enhanced properties, such as high permeability, antifouling, antibiofouling, and high stability.

Membranes with Functionalized Nanomaterials
https://doi.org/10.1016/B978-0-323-85946-2.00001-1

Current water treatment systems, distribution, and discharge are unsustainable due to their centralized system and reliance excessively on conveyance [7]. Nanomaterials have given rise to many innovations and significant knowledge enhancements in environmental applications, both technologically and in industrial development [8]. The advances in nanotechnology can help overcome present-day problems associated with potable water supplies using nanostructured and nanoengineered materials [9]. It is the latest approach to upgrade membrane surface properties for enhanced efficiency for water treatment applications. Several types of nanomaterials exist with various physical and chemical properties. Typically, nanomaterials are defined as materials less than 100 nm, at least in one dimension. The zero-dimensional nanomaterials are nanoparticles, and one-dimensional nanomaterials consist of nanofibers, nanotubes, nanorods, and nanowires, and two-dimensional nanomaterials include nanosheets [10].

Some of the nanomaterials used extensively for water treatment applications include, but are not limited to, carbon nanotubes, nanoscale metal oxides, zeolites, nanofibers, graphene, magnetic nanoparticles, and aquaporin. Using these nanomaterials, various types of nanocomposite membranes can be fabricated. The thin-film nanocomposite (TFN) membranes have nanoparticles integrated into the surfaces and the substrate structure [11]. Nanofibrous membranes are nonwoven filter media prepared using nanofibers with remarkable properties such as high porosity and pore size distribution [9]. Nanocomposite ceramic membranes have high chemical and thermal stabilities and high performance in water treatment application due to the existence of nanoparticles [12]. Polymeric and ceramic carbon nanotubes (CNT)-based composite membranes provide excellent water transport properties, high surface area, high chemical inertness, and mechanical strength [4]. These are fabricated using single-walled and multiwalled carbon nanotubes.

The current trend in the ongoing research aims to decrease capital investments, operational costs, including energy consumption, pretreatment, and maintenance [8]. Consequently, nanomaterials present themselves as the best opportunity to achieve these goals by enhancing the water treatment membranes to improve process performance. With much development in the laboratory scale, functionalized nanomaterials in the membrane show a promising future in the desalination industry. This chapter focuses on the fabrication methods employed for the synthesis of nanocomposite membranes with functionalized nanomaterials. The fabrication methods have been divided into different sections based on membranes produced, such as TFN, nanofibrous, nanocomposite ceramic, and CNT-based membranes. The fabrication techniques described in this paper are limited to the most popular and advantageous methods in the literature. Moreover, a discussion over the fabricated membrane performances and their advantages and drawbacks for each method have been included.

4.2 Fabrication methods

4.2.1 Thin-film nanocomposite membranes

The thin-film composite (TFC) membranes are dominantly used in the nanofiltration and reverse osmosis processes [13]. Although TFC membrane operates without major technical issues, there is a trade-off between salt rejection and water permeability [14,15]. Furthermore, it is vulnerable to organic or inorganic fouling along with chlorine attack. Several efforts have been made to improve the membrane characteristics and performance by incorporating inorganic nanomaterials into the polyamide (PA) layer or the microporous substrate. The resulting membrane is known as thin-film nanocomposite, which has become popular due to its enhanced TFC membrane properties. Jeong et al. first reported the use of interfacial polymerization to fabricate the TFN membranes [16]. Another method widely used for the fabrication of TFN membranes is the dip-coating method. During the fabrication of the TFN membrane, nanoparticles ranging from 20 to 200 nm are integrated into the ultrathin active layer or the support layer to enhance the properties of the membrane [11]. The PA layer modification has greater significance in water treatment applications than the nanoparticles' modification of the substrate, as the PA layer is directly in contact with the feed.

4.2.1.1 Interfacial polymerization

Interfacial polymerization is the most commonly used method for the fabrication of TFC/TFN membranes. The IP method is primarily used to manufacture polyamide membranes that are commercially used in water treatment and purification processes. The highly cross-linked polyamide active layer is formed on the surface of the porous support layer copolymerization between two immiscible reactive monomers in organic and aqueous phases [11]. The porous substrate encounters m-phenyldiamine (MPD), an amine monomer, in the aqueous phase. The percentage concentration for the MPD is reported to be about 2 wt% [17]. In the next step, the membrane is immersed in an organic solution with a second reactive monomer called trimesoyl chloride (TMC), which ranges from 0.1 to 0.2 wt%. The monomers react at the interface of the two immiscible solvents forming a highly

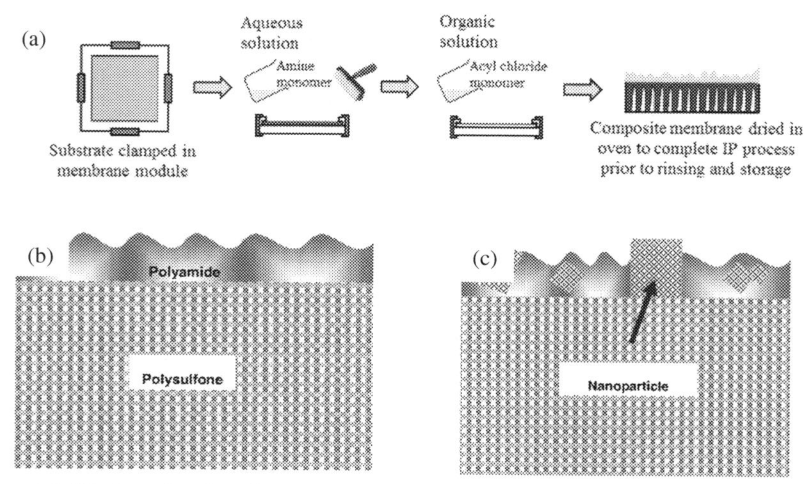

Fig. 4.1 (A) Interfacial polymerization technique for the fabrication of TFC/TFN membranes [13]; illustration of (B) pure TFC membrane, and (C) TFN membrane with nanoparticles [16].

cross-linked thin film layer on the support layer of the membrane. The diffusion of MPD monomer into the TMC phase is limited due to the polymerization process, which creates a self-limited film thickness [18]. The TFC membrane is then heated to 60–80°C to improve the adhesion between the film and the support layer. The selection of monomer and its properties aid in the control of film density, thickness, chemical resistance, roughness, and hydrophilicity [19]. Fig. 4.1A shows the interfacial polymerization technique for the fabrication of TFC and TFN membranes. Fig. 4.1B and C show the illustration of the TFC membrane with and without nanoparticles.

Jeoung et al. developed zeolite-polyamide nanocomposite membrane by dispersing 0.004%–0.4% (*w*/*v*) of synthesized zeolite A nanoparticles in hexane-TMC solution [16]. The TFN membrane showed dramatic advancements in terms of water permeability and interfacial properties. The nanoparticle molecular sieve pore provides a preferential flow path to the water molecules while keeping the salt rejection similar to the TFC membrane. The zwitterion-based polymers enhance the antifouling properties of the conventional TFC membranes due to their durability, super hydrophilicity, and environmental stability [20]. Similarly, TFN membranes with zwitterionic nanogels fabricated using the IP technique showed improved performance for separating salts and organics from water [21]. Nanofillers such as single and multicarbon nanotube, graphene oxide, titanium oxide (TiO_2), and silica

(SiO_2) nanoparticles have been used to modify the TFC membranes. TiO_2 is popular due to its small size and hydrophilicity. TiO based modifications have yielded highly improved flux and demonstrated a reduction in internal concentration polarization in FO membranes [22,23]. The unique properties of SiO_2 nanoparticles enable its use in the fabrication of TFN membranes, such as its ability to be thermally resistant, environmentally inert, high surface energy, and high surface area. TFN membrane consisting of MCM-41 SiO_2 nanoparticles developed using IP technique resulted in enhanced hydrophilicity, roughness, and zeta potential [24]. An increase of 60% in the permeate flux was observed with the TFN and 98% salt rejection.

Although TFN membranes provide enhanced performance, these are also met with a few challenges. Such laboratory-scale studies have yielded significant results. However, developing large-scale defect-free TFN membranes for long-term operations is a challenge. This may result from poor compatibility between polymer and nanomaterial and uneven distribution of nanoparticles on the surface of the PA layer [24,25]. Another significant issue is the loss of nanomaterials from the membrane due to leaching during the filtration process [26]. The rubber rolling process is employed to remove the excess aqueous solution, which also causes the expulsion of the amine monomer. This is necessary to prevent tiny water droplets on the substrate before interaction with the second monomer. Otherwise, the acyl chloride monomer reacts with water droplets causing a lower degree of cross-linking and defects. In recent years, the newly developed IP processes and modifications to the IP process have been introduced to overcome these challenges, such as the spin-based IP [27] and filtration-based IP [28] techniques.

4.2.1.2 Dip coating

Dip coating is another simple technique to fabricate TFC and TFN membranes with a dense active layer. It is a widely used inexpensive method to fabricate membranes utilized in gas separation, RO and NF processes. The process involves coating the substrate membrane by dipping it into the dilute solution of polymer (coating solution), where the cross-linking occurs. The targeted nanomaterials are dissolved in the solutions, which are directly coated on the substrate surface. It is followed by the application of heat in order to dry the coated layer on the membrane. When the membrane is removed from the solution, a thin layer of coating forms on the surface followed by removing the solvent through evaporation [29,30].

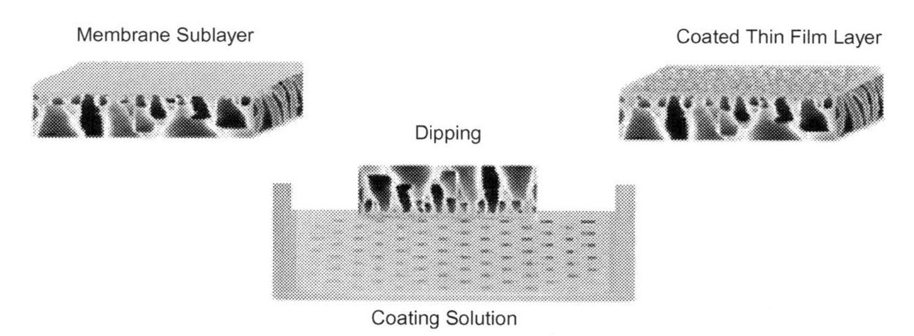

Fig. 4.2 Schematic representation for membrane fabrication using the dip-coating method [11].

Fig. 4.2 shows a general schematic representation of the membrane fabrication using the dip-coating method.

Although the process seems simple, it involves complex chemical and physical changes [31]. The thickness and morphology of the deposited thin film depend on several factors such as immersion time, dip-coating cycles, and withdrawal speed. It also depends on some of the coating solution properties such as density, viscosity and surface tension, and the condition of substrate surface and evaporation. At first, solution dipping and coating were implemented to enhance the surface properties of the fibrous material. Later, modifications were added to the simple dipping method in order to improve the properties further. One such modification is the sol-gel-based dip coating method, which can control the deposition of a film on the substrate by regulating the sol and gelation process [32,33]. The advantages of sol-gel dip coating include its efficiency, low cost, high performance, and less chemical consumption [34]. Vacuum-assisted and spin-assisted dip-coating help improve infiltration between coating solution and substrate [31]. Photo-assisted dip-coating assists the coating solution evaporation process, and the irradiation effect is advantageous toward film disposition.

The dip-coating method has been used to prepare an omniphobic nanofiber membrane for membrane distillation for desalination of highly saline solution. Lee et al. used electrospinning (ES) to fabricate a nanofibrous substrate. Then, negatively charged silica nanoparticles were grafted on the substrate using dip coating to attain multilevel reentrant structures [35]. The resulting membrane resists wetting to several low surface tension liquids, including organic solvents. Glutaraldehyde (GA) has been used as a cross-linker due to its enhancement capabilities regarding the chemical, thermal, and mechanical stability of the composite membrane [36,37]. Hollow fiber NF membrane dip-coated with sodium carboxymethyl cellulose (CMCNa) on a polypropylene microporous support using $FeCl_3$ as cross-linker exhibited long term stability and anti-fouling properties [38]. The dip-coating method shows a promising technique for fabricating organic solvent nanofiltration (OSN) membranes with defect-free and minimal thickness of the layer [39]. FO polyamide membranes have also been enhanced using functionalized nanoparticles to optimize the surface [40,41]. Silver nanoparticle (AgNP)-decorated graphene oxide (GO) nanosheets functionalized membranes showed super hydrophilicity and anti-bacterial properties. Commercial polyamide membranes are prone to chlorine attacks. Kang et al. presented the dip coating of the membrane with poly (*N*, *N*-dimethylaminoethyl methacrylate) (PDMAEMA)-ethanol and a cross-linking agent, i.e., p-xylylene dichloride-ethanol solution, leading to an excellent chlorine oxidation resistance [42].

4.2.2 Nanofiber membranes

The nanofibrous membrane (NFM) is a relatively new approach in desalination and water treatment applications. Nanofibers have been employed as a strategy to improve the membrane surface properties. Nanofibers belong to the one-dimensional class of nanomaterials where one of the dimensions is in the nanometer scale [10,43]. These also include nanorods, nanotubes, and nanowires. The remarkable characteristics of nanofiber, i.e., their significantly high surface area to volume ratio and high porosity, distinguish it from other nanomaterials [44]. The electrospinning method is the most widely used technique for the fabrication of nanofiber membranes. The electrospun nanofiber membranes have outstanding multifunctional properties that can be utilized for water purification materials. Due to their high porosity, these are limited to the separation of microsized particles such as in microfiltration separation. However, the electrospun membranes can be further enhanced by adding a layer of thin selective coating to extend their separation to complex colloidal solutions [9]. These modifications can be achieved via interfacial polymerization and other physical and chemical modification. IP modification transforms the membrane into a nonporous composite membrane that can be utilized in an RO and NF application. In general, the modifications with nanoparticles have resulted in the enhancement of both salt rejection and water flux [45,46]. The subsequent sections present the electrospinning method, phase inversion, and melt-blown techniques for membrane fabrication.

4.2.2.1 Electrospinning

Electrospinning is an economical and straightforward technique for fabricating cost-effective nanofibers in different shapes and sizes. It is also suitable for a wide range of inorganic and organic systems with a promising control over the nanomaterial size distribution [47]. Well-controlled operating conditions and parameters can lead to a highly porous and defect-free nonwoven nanofiber membrane [48]. In the ES method, a solution jet is generated using a high voltage electric field on a dope polymer solution [49]. The solvent evaporates as the jet thins, resulting in polymer fibers ranging from several microns to nanometer-scale, obtained in a grounded collector [3,50]. A schematic representation of the ES setup is shown in Fig. 4.3. A basic ES system consists of three major parts: a feeding area, a collector area, and a high voltage supply. The feeding system consists of feed storage for the polymer solution, a thin metallic needle known as a spinneret, and an injection pump for the polymer solution to be injected at a constant flow rate.

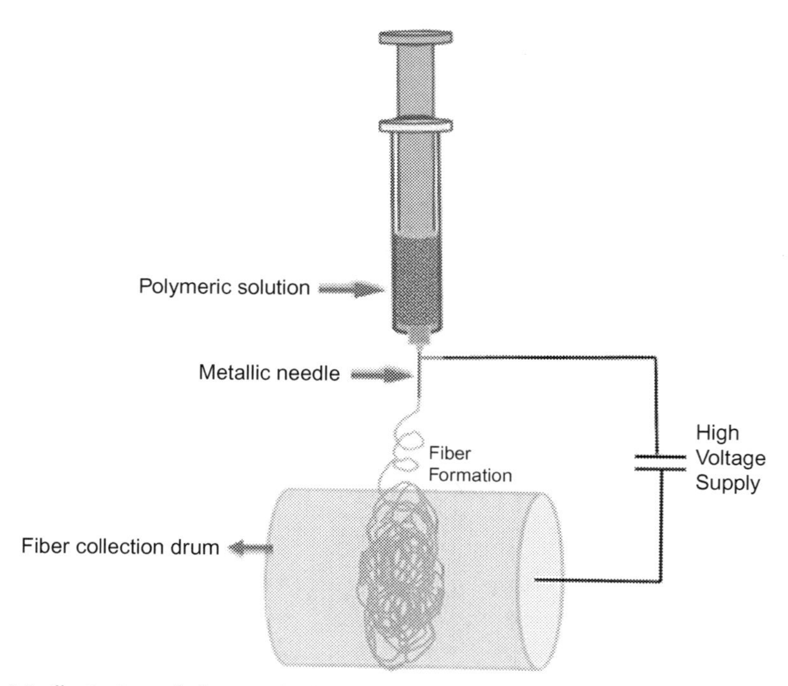

Fig. 4.3 Illustration of electrospinning process for the production of nanofibers [3].

The fiber properties are affected by operating conditions such as voltage, flow rate, and collector distance [51]. Furthermore, the morphology and topography are also altered by the ambient conditions such as temperature and humidity of the electrospinning chamber. Variation of the applied voltage to the electrospinning process has shown the effective fabrication of close looped columnar networks of polymer fiber [52]. However, high voltages cause the increased formation of beads [53]. The conductivity of the polymer solution also causes to change in the properties of the nanofibers. Therefore, some researchers have attempted using several salts to increase the conductivity that resulted in favorable outcomes [51].

Typical membranes have a 2D framework, whereas the electrospun nanofibrous membranes are beneficial due to their porous and interconnected 3D fibrous system. This helps achieve high internal surface area and separation performance from the membrane. Most instabilities during the ES process may cause breakage and stop the continuous formation of fibers [3]. However, bending instability is essential to the fabrication of fiber as the polymer jet bends in expanding loops collected over the substrate. In the case of low viscosity polymer, instabilities occur when polymer entanglement is insufficient for fiber formation. A phenomenon called "electrospraying" occurs where particles are formed rather than fiber [48].

4.2.2.2 Phase inversion

Phase inversion is another widely used technique to prepare nanofibers due to its simplicity and cost effectiveness. In this method, the phases separate due to physical incompatibility removing the solvent phase while retaining the other phases. The fabrication of nanofiber is achieved in four steps [9]. The first step requires the preparation of the solvent. In the second step, a homogeneous polymer solution is made by the dissolution of the polymer in the solvent at elevated temperatures. The third step is the gelation step, where the solution is retained at the gelation temperature resulting in phase separation. Finally, the solvent is extracted, and the matrix is dried, which prompts the generation of a nanofibrous framework. This process of nanofiber production is shown in Fig. 4.4. The concentration of the polymer has a significant impact on the properties of the produced nanofiber. For example, at a high polymer concentration, the porosity of fiber decreases, whereas its mechanical properties are improved [55,56]. Several factors such as solvent type, polymer type, gelation temperature, thermal treatment, and gelation duration affect the morphology of the nanofibers. The size of nanofiber produced ranges from 50 to 500 nm. One limitation of phase inversion is that

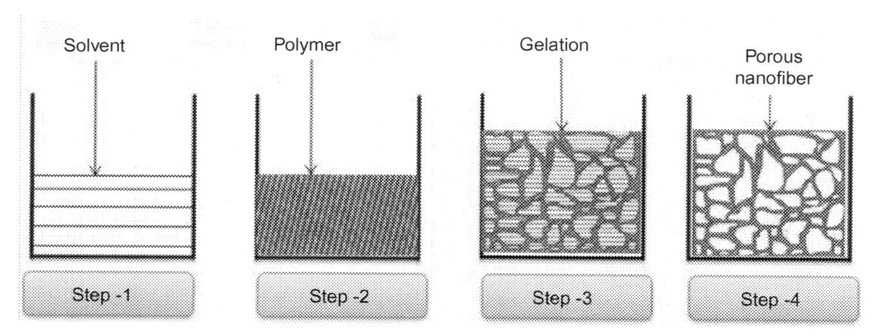

Fig. 4.4 Schematic representation of nanofiber fabrication process using phase inversion method [54].

only certain polymers like polylactide (PLA) and polyglycolide can be fabricated into nanofibers using this technique [57]. This may be because not all polymers can experience phase separation due to limited gelation ability. Other disadvantages include high time consumption, hard to retain porosity, and structural instability [58].

4.2.2.3 Melt blowing

Melt blowing is a popular option for fabricating superfine fibers in the range of 2–5 μm. It is used in the fabrication of almost all nonwoven fiber barrier membranes [59]. Fig. 4.5 represents the schematic representation of the melt blowing process. A thermoplastic polymer undergoes extrusion, and as it passes the MB die, high-velocity hot air attenuates the fiber rapidly, reducing its diameter by hundred times that of the nozzle diameter [59,60] thereby creating fine fibers in the process, which are located on a drum collector

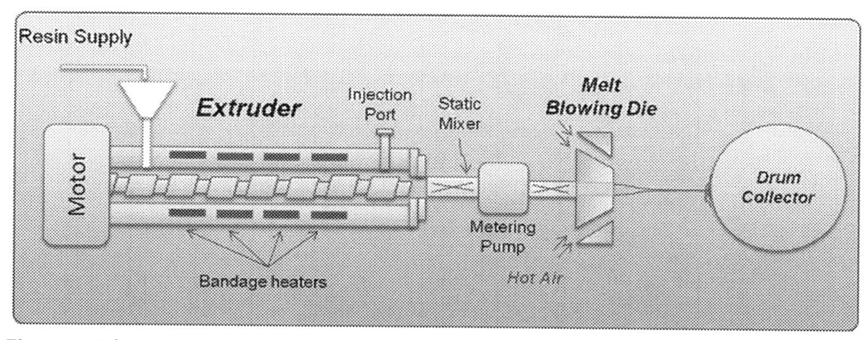

Fig. 4.5 Schematic representation for the fabrication of nanofibers using melt blowing technique [59].

to generate a self-bonded and fine-filtered web. The fiber-to-fiber bonding and entanglement result in proper web cohesion that can be used directly without additional bonding. The diameter size of the fiber produced has been reduced to as much as 500 nm [61]. Recent studies have shown that nanofibers as fine as 75 nm can be produced using the melt blowing technique [62]. Research has shown that a decrease in fiber size (less than 1 μm) has led to superior filtration properties due to less weight of the fiber web than conventional melt-blown webs [59]. Nanofibers can be achieved using special dies with smaller orifices, decreasing the viscosity of the polymer melt and reasonable modification to the melt blowing setup. This would provide an alternative to a quicker and inexpensive strategy to fabricate nanofibers than the ES method.

Melt blowing is an emerging technology, and much research is being conducted on the technique from a different perspective. Several studies have been conducted on the influence of slot die geometry on the fiber size and airflow field below the melt blowing die [63,64]. The influence of the nose-piece shape on the flow field below the die was evaluated both experimentally and theoretically using CFD simulations [65]. The study showed that higher maximum centerline air velocity led to an increased attenuation rate, generating finer fiber at a specific airflow rate. The effect of air velocity and air temperature has also been investigated on the melt-blown fibers [66,67]. The air permeability and pore size decrease with an increase in the air velocity and temperature. With the application of high extrusion pressure of 1500 psi with 64 orifices per inch, nanofiber of an average diameter of 400 nm was made possible [59]. However, it was noted that a lot of "fly" was observed, which is broken fiber debris resulting from high air velocity. Nanofibers have been fabricated using several types of dies, including stacked plates comprising a row of orifices [68] and a single hole die [69].

4.2.3 Functionalized ceramic membranes

The use of ceramic membranes for desalination and water treatment is relatively new. Ceramic membranes have been used in microfiltration [70,71], ultrafiltration [72], nanofiltration [73], membrane bioreactors [74], and reverse osmosis [75]. Ceramic membranes are advantageous over polymeric membranes due to several characteristics such as low operational cost, low fouling, long lifetime, stronger mechanical strength, high chemical, and thermal stability [12]. Due to this reason, they are a preferable choice for applications in water and wastewater treatment and oil/water separation.

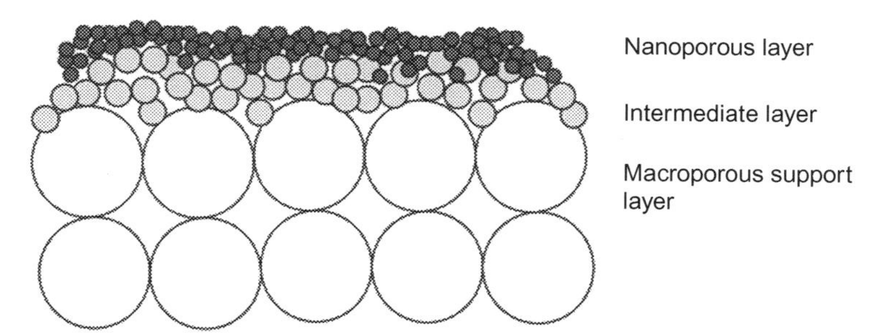

Fig. 4.6 Illustration of a ceramic membrane functionalized with nanocomposite [79].

Typically, the ceramic membrane comprises an active layer, a support layer, and an optional intermediate layer. The ceramic materials consisting of alumina, silica, zirconia, and other metal oxide mixtures, combine with binders and peptizing agents to fabricate the support layer [76]. The support layer provides mechanical strength, whereas the porous intermediate layer provides increased permeability, smooth surface, and reduces the pore size of the membrane [77]. The active layer plays a vital role in the separation process as it determines the porosity and thermal and mechanical properties [78]. Fig. 4.6 shows the schematic representation of nanoparticle functionalized ceramic membrane structure.

Recently, similar to polymeric membranes, nanoparticles have been used to enhance the active layer of the ceramic membrane for their physicochemical properties [80–82]. These modifications have led to an improvement in the performance of the ceramic membranes. Aluminum oxide (Al_2O_3) and titanium dioxide (TiO_2) are the most widely used nanoparticles for nanostructured ceramic membranes in desalination and water treatment [79]. Iron oxide (Fe_2O_3) is also a popular option, and silver nanoparticles are used to prevent biofouling. Modifications with metal oxide nanoparticles also add antifouling capabilities to the membrane by catalyzing reactions to degrade foulants. Moreover, casting nanoparticles increase the nanoscale pores compared to the traditional sintering of ceramics [83]. Several methods of fabrication of nanoparticle functionalized ceramic membranes are given in the subsequent sections.

4.2.3.1 Slip casting method

The slip casting method has proven to be a simple, effective, and inexpensive technique for fabricating ceramic membranes [84]. It is one of the favorable options for large-scale production due to the controllability over the degree

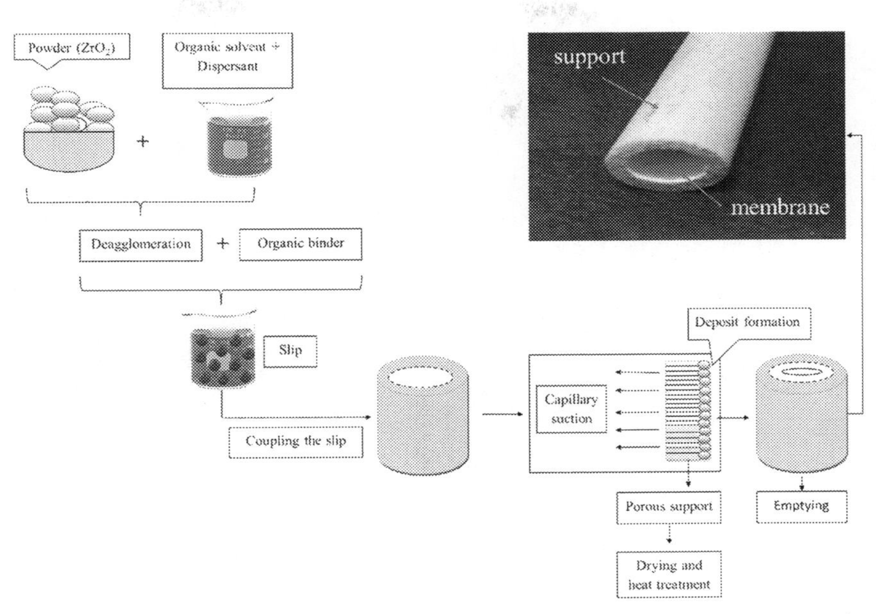

Fig. 4.7 Schematic representation of the fabrication of ceramic membranes using slip casting method [87].

of dispersion in the ceramic phase and porosity of the material [85]. Moreover, another advantage of the slip-cast method is that it can produce complex shapes of membranes. The first step in the slip casting method is the preparation of the slurry. Ceramic powder is mixed with an organic solvent/demineralized water, chemical deagglomeration, and an organic binder to produce the slurry. The slurry is then transferred to porous mold in which the solvents diffuse through the pores due to capillary suction pressure [86]. Particle precipitation causes the formation of a gel layer on the inner surface of the mold. Lastly, to avoid particles penetrating the pores, consolidation of the internal layer is carried out. Fig. 4.7 shows the steps involved in the slip casting method for the production of ceramic membranes.

Several studies have been published to improve the slip casting method for membrane fabrication. Tubular oxygen separation membranes were fabricated using a slip casting method with a mixer of $SrCO_3$, $BaCO_3$, F_2CO_3, and CO_3O_4 powder, and stabilized zirconia balls [88]. The study used the combination of Cerasperse 5468 and ammonium salt of polycarboxylic acid as the dispersant and polyethylene glycol (PEG 400) as the binder. Kaoline and fly ash have been used as low-cost materials to prepare cost-effective ceramic membranes using the slip casting method [89,90]. The parameters involved in the fabrication of the membrane have a significant effect on its quality and performance. The capillary suction pressure is inversely proportional to the radius of the pore in the mold [91]. Moreover, the percentage of solids in the slurry and the casting time directly impact the consolidation rate.

The increase in sintering temperature results in the reduction of grain size and membrane density, leading to a decrease in the mechanical strength of the membrane [84]. The slurry density and viscosity are dependent on particle size, morphology, thickness, suspension pH, and casting time [92]. One of the issues associated with the slip casting method is the weak strength and low toughness of the ceramic membrane due to the high porosity of the membrane [12]. Furthermore, the casting times are usually long, and the wall thickness is difficult to control during the casting process, which typically results in thick walls [86].

4.2.3.2 Tape casting

The tape casting technique is used for the fabrication of smooth and thin flat sheet ceramic membranes. It is also popularly used in the production of thin piezoelectric materials. The main advantage of this method is the production of flat and porous ceramic components with customized thicknesses [93], as the thickness of the membrane plays a vital role in the membrane separation processes. The thickness of membranes fabricated with tape casting is typically a few millimeters thick. The process involves transferring the powder suspension in the reservoir, which is then leveled to the desired thickness using an adjustable casting knife. This is done using the gap between the blade and the moving carrier. The membrane is then dried up through solvent evaporation in the drying zone [94]. Fig. 4.8 shows the tape casting method setup to produce thin ceramic membranes. The produced membrane is affected by several parameters, such as powder suspension viscosity, the gap between blade and carrier, and the reservoir depth.

Various membranes have been developed using the tape casting method so far [96–98]. Typically, the slurry consists of additives, plasticizers, and binders that yield the dried tape. A disc-type aluminum oxide ceramic membrane was prepared from a slurry made by mixing alumina powder with an azeotropic mixture (methyl ethyl ketone $\pm$ ethyl alcohol) [99]. The study used polyvinyl butyral as the binder, benzyl butyl phthalate (BBP) and PEG as plasticizers. Moreover, it was found that a high amount of binder in the mixer leads to the agglomeration of particles, which also decreases the pore size in the membrane. The tape casting has been combined with freeze casting and the phase inversion method to alter the morphology

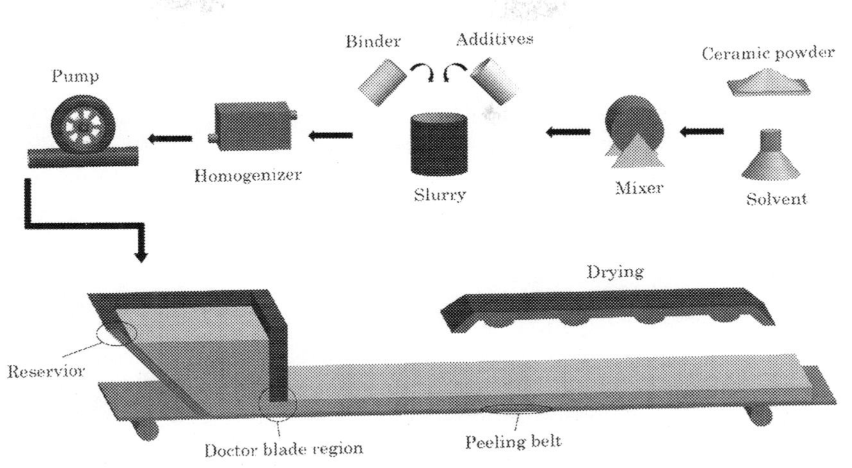

Fig. 4.8 Schematic representation of the slip casting setup for ceramic membrane fabrication [95].

[100,101]. One of the disadvantages of tape casting is the irregular shape of the produced membrane, resulting from the plaster mold corrosion [102]. This is common for inexpensive materials such as kaolin in the slurry mixture. Moreover, cracks are formed during the de-binding and de-sintering process. It is also observed that the processing time for slurries with tiny powders is long.

4.2.3.3 Pressing method

The pressing method is widely known and is the simplest technique for the fabrication of disc-shaped ceramic membranes. In this method, there is no need for the preparation of slurry [103]. This method requires mixing a dry powder with a pore-forming agent, which is then pressed uniaxially with a hydraulic pressing machine. Typically, a pressure greater than 100 MPa is applied to produce a sintered dense membrane [86]. The pressing method is only limited to the fabrication of ceramic membranes that are only permeable to oxygen and hydrogen. Binders can be added to enhance the mechanical strength of the membrane [104]. After the pressing, heat treatment is provided to the membranes to produce a crack-free surface followed by the sintering process.

Porous zirconia (thin top layer) and alumina (support) membranes were fabricated using the pressing method where calcined alumina and tetragonal zirconia were used [105]. Isopropyl alcohol and polyvinyl butyral resin (PVB) were used as binders, whereas sucrose was used as a porogenic agent.

The pressing method has produced membranes with homogenous physical properties and uniform porosity [106,107]. Sintering of ceramic material before fabrication results in an increase in membrane strength [108]. However, the sintering temperature above 850°C leads to the weakening of the membrane. Various studies have shown the fabrication of ceramic membranes using low-cost materials with uniform distribution of pores and high porosity [109–111].

4.2.3.4 Extrusion

Extrusion is a preferred method for the mass production of ceramic membranes [86]. It is specifically used to fabricate the tubular configuration of porous ceramic membranes. In extrusion, a die exhaust is used instead of the mold. The extruder may be of two types, i.e., screw and plunger. A plunger extruder is advantageous as it decreases the chances of contamination; however, it results in pressure loss at the nozzle [103]. The screw extruder maintains the pressure at the ram and the nozzle. Furthermore, the slurry preparation requires the moisture content to be more than 15%.

The membrane fabrication process consists of five steps: blending, pugging, extrusion, cutting and drying, and sintering. During the blending process, several chemicals such as surfactant, coagulant, lubricant, plasticizer, binder, and additives are added to control the moisture content of the slurry. The mixture is kept for several hours under room temperature and high humidity conditions in the pugging step. The homogenous mixture is then transferred to the extruder to be forced through a nozzle. The ceramic membrane properties such as shape, porosity, and pore distribution depend on the die and the force in the extruder [112]. The product is cut into the desired length and dried to evaporate any remaining solvent, binder, and plasticizer to keep its final shape and form. The last step of the process is the sintering of the membrane, which helps avoid cracks and defects in the membrane [113].

Several researchers have published studies concerning the investigation of favorable extrusion conditions such as temperature and viscosity and the impact on the geometry of the membrane. Sintering of tubular ceramic membrane at the temperature of 950°C helps prevent crack, bends, and other defects [114]. Membranes made from clay can be sintered to a 1250°C temperature resulting in the uniform distribution of pore size [115,116]. The geometry of the product depends on the rheological properties of the slurry and the die dimension, and the sample length cut [117]. Typically, the extrusion process is conducted at high temperatures and low speeds to avoid fracture. The effect of additives, particle size, and binder content has been studied on the alumina support for ceramic membranes using the extrusion process [118]. Several low-cost materials such as kaolin have been used to produce ceramic membranes using the extrusion process [119,120].

4.2.4 Carbon nanotube (CNT)-based composite membranes

Carbon nanotubes are made of one or more cylindrical graphene sheets rolled to form a tube-like structure [1,121]. The two types of CNTs are single-walled carbon nanotubes (SWCNTs) and multiwalled carbon nanotubes (MWCNTs). The SWCNTs consist of a single graphene sheet in a seamless cylindrical tube. In contrast, MWCNTs are made up of multiple layers of the concentric graphene sheet arranged coaxially around the hollow core. Van der Waals forces exist between consecutive layers of the MWCNTs. Both CNTs have been utilized in desalination and water treatment applications [122–124].

Furthermore, CNTs provide low-energy solutions for the challenges encountered in membrane technology for water treatment and desalination. CNT membranes allow water to pass through it frictionlessly, at the same time retaining several different pollutants. This is due to their inner pore diameter, high aspect ratios, and smooth hydrophobic walls. CNTs have excellent properties such as high specific surface area, chemical inertness, high mechanical strength, and exceptional water transport properties that make them an attractive option in the fabrication of composite membranes for water treatment [1,125,126]. CNTs are limited by their tendency to aggregate and form clusters due to high van der Waals forces between the tubes. Due to this phenomenon, their interaction with other compounds and solubility in conventional solvents gets affected [121]. However, functionalization of the CNTs surface may be able to help overcome these challenges [127]. Popular CNT-based composite membrane fabrication methods include chemical vapor deposition, in situ polymerization, and direct coating. These fabrication methods are detailed in the subsequent sections.

4.2.4.1 Chemical vapor deposition

Chemical vapor deposition (CVD) is one of the most popular methods for the fabrication of CNT-based composite membranes. In addition to being a low-cost process, it is also easy to operate and flexible process [4]. In CVD, thermal dehydrogenation reaction occurs in tubular reactor decomposing hydrocarbons under metal catalysts such as nickel, cobalt or iron at 600–1200°C temperature

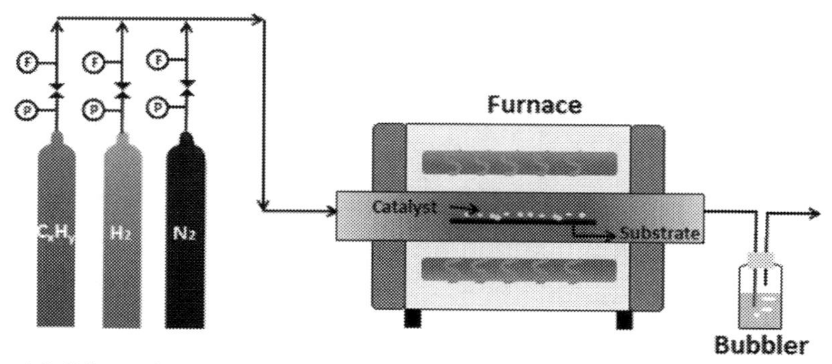

Fig. 4.9 Schematic representation of the chemical vapor deposition setup for production of CNTs [4].

[128]. Fig. 4.9 shows the schematic representation of the experimental setup of the CVD for the generation of CNTs. In general, CVD is used for the fabrication of inorganic CNT-based composite membranes. Like the fabrication of CNTs, the CNT is directly grown on the substrate to obtain the composite membrane. Therefore, the stability of the membrane at high temperatures is an important aspect of the CVD process. The various substrate, including ceramic membranes made from alumina (Al_2O_3), yttri stabilized zirconia (YSZ), and mullite, have been utilized to grow CNT due to their outstanding stability.

Vertically aligned CNTs have been developed on the ceramic membrane (Al_2O_3) using CVD for water treatment [129] and gas transport [130] application. Compared to conventional ceramic membranes, the CNT-based composite membranes have shown high filtration efficiency and low energy consumption due to a significant decrease in the pressure drop [131]. One of the limitations of the CVD method is the challenge of fabricating a uniform distribution of size-controlled CNTs. For this reason, the CVD-template method is utilized for the high density and uniform distribution of CNT on the membrane. Anodic aluminum oxide (AAO) is the most commonly used template for CNT growth, fabricated using an electrochemical anodization process [132,133].

4.2.4.2 In situ polymerization method

Similar to the fabrication of thin-film composite membranes, interfacial polymerization or in situ polymerization can be used to develop polymeric CNT composite membranes. This technique involves the fabrication of CNT in a random arrangement state in the polymeric membrane. The

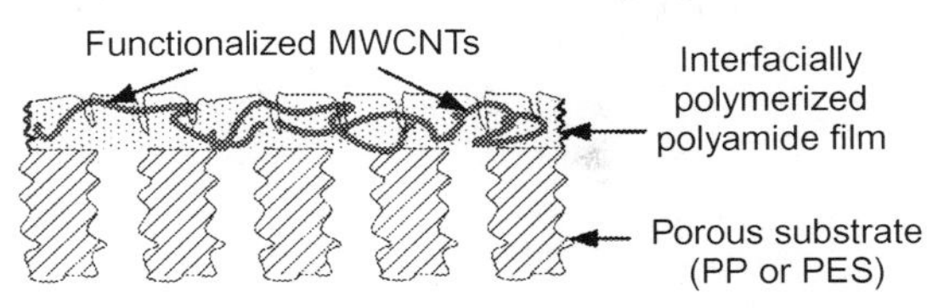

Fig. 4.10 Illustration of interfacial polymerized polyamide film with MWCNTs [136].

process is divided into two steps. The first step involves the blending of CNTs and a monomer. Secondly, the polymerization of the monomer occurs under specific conditions to produce CNT-based composite membranes [134,135]. In this way, a chemical bond strengthens the interfacial adhesion between the polymer and the CNTs. Fig. 4.10 shows functionalized MWCNTs on a polyamide film using the IP technique.

Kim et al. fabricated polyamide membranes with CNTs for the application in reverse osmosis [137]. IP was employed to fabricate the membrane using TMC solution in n-hexane and functionalized CNTs in the aqueous solution of MPD. The produced membrane demonstrated high stability and performance in terms of water flux and salt rejection. Similarly, Zhao et al. pretreated MWCNTs with mixed acids to increase their chemical activity and dispersity [135]. Later, the MWCNTs were embedded in the skin layer of the nanocomposite membrane. Moreover, the CNT membrane also developed antifouling and antioxidative properties compared to the polyamide membrane.

Chan et al. showed that zwitterion functionalized CNTs enhanced water flux and salt rejection in desalination membranes [134]. The CNTs were partially aligned on the polyamide layer using vacuum filtration during membrane fabrication. Studies on vertically Aligned CNTs on polymeric membranes are scarce. In most cases, the CNTs are dispersed as fillers in the polymer matrix for the water treatment process. Kim et al. presented novel vertically aligned CNTs with polymer monomers for the in situ polymerization [138]. The CNT membrane exhibited high gas and water flux due to high permeability, durability, and flexibility, showing potential future applications in water treatment and desalination.

4.2.4.3 Direct coating method

The CVD method is generally used to fabricate inorganic CNT-based membranes, whereas the IP method allows the fabrication of polymeric CNT-based membranes. The direct coating method is advantageous as it can be used to synthesize both polymeric and inorganic CNT-based membranes. It is a simple method that can be used to deposit CNTs on the surface of the membrane substrate. The CNTs are first suspended in the water and later dispersed using sonication. The suspension is then loaded onto the clean membrane substrates such as PA [139], polyvinylidene fluoride (PVDF) [140], or polyethersulfone (PES) [141]. Filtering, syringe-filtering, and vacuum filtering may be used to deposit the CNTs on the surface of the membrane uniformly.

PVDF membrane with CNTs prepared using the syringe-filtering coating method showed high antifouling properties with minimal water permeability reduction [140]. However, the membrane was unstable due to weak interaction between the CNT layer and the membrane substrate. Consequently, the CNTs were also released from the membrane layer during application. This limitation reduces the value of the product and presents a hazard as CNTs could be present in the treated water. The vacuum filtration-pyrolysis process can significantly enhance the interaction between CNTs and the substrate layer [142]. Using this process, CNTs on Al_2O_3 ceramic membrane showed high stability. Another novel CNT-based membrane was fabricated using a vacuum filtering coating method loaded with reduced graphene oxide and CNTs on an AAO membrane [143]. For water treatment applications, the membrane showed high permeability and performance.

4.3 Summary and future perspective

The use of functionalized nanomaterials in membrane technology has attracted significant attention due to their enhanced properties, such as their chemical and mechanical stabilities, antifouling properties, and high water and gas permeation rates. Thin-film nanocomposites, nanofibrous, ceramic, and CNT-based composite membranes are some of the most popular options for separation processes, especially in water treatment and desalination applications. In this chapter, several fabrication methods specific to each membrane type are presented with the resulting enhancement, their advantages and limitations. Methods such as IP and surface coating have been used to fabricate thin-film nanocomposite, polymeric, and ceramic membranes enhanced with CNTs. Electrospinning is a more economical option compared to melt blowing and extrusion for the fabrication of nanofibrous membranes. Direct coating methods have been used for the preparation of both polymeric and ceramic CNT-based membranes.

The addition of nanomaterials to the membrane has led to an increase in the performance of the membrane. Specifically, the improved hydrophilicity and pore size reduction in the FO and RO membranes result from using TiO_2-modified TFC membrane via IP technique. Functionalized silver nanoparticles fabricated membrane show enhanced antibiofouling properties. TFN membrane fabrication has shown defects and irregularities and monomer loss during the rubber rolling process on a large scale. This needs to be investigated and improved by introducing new TFN fabrication methods or modified IP methods to tackle these challenges. Coating methods have shown weak interaction between the nanomaterial layer and the substrate. Improvement in this process may result in a high performance and defect-free membrane using an inexpensive and economical coating process. Phase inversion technique is limited to only a particular type of polymer capable of gelation. The sustainable aspect of this technique depends on exploring more polymers compatible with this method of membrane fabrication. Research toward decreasing the time consumption for some of the processes such as phase inversion and slip-casting methods can make these techniques more applicable. Recent developments have been focused on the economic, social, environmental, and technological aspects to ensure sustainable membranes. The trend is shifting toward finding low-cost materials to fabricate advanced membranes with multiple properties and high performance. Future research should consider more cost-effective and environment-friendly fabrication methods instead of using toxic chemicals hazardous to health and the environment.

Acknowledgment

The authors thank Qatar University for the financial support through grant number IRCC-2019-04.

References

[1] R. Das, M.E. Ali, S.B.A. Hamid, S. Ramakrishna, Z.Z. Chowdhury, Carbon nanotube membranes for water purification: a bright future in water desalination, Desalination 336 (2014) 97–109, https://doi.org/10.1016/j.desal.2013.12.026.

[2] WHO, Progress on Drinking Water and Sanitation, 2012 Update, World Health Organization, 2012.

[3] F.E. Ahmed, B.S. Lalia, R. Hashaikeh, A review on electrospinning for membrane fabrication: challenges and applications, Desalination 356 (2015) 15–30, https://doi.org/10.1016/j.desal.2014.09.033.

[4] L. Ma, X. Dong, M. Chen, L. Zhu, C. Wang, F. Yang, Y. Dong, Fabrication and water treatment application of carbon nanotubes (CNTs)-based composite membranes: a review, Membranes (Basel) 7 (2017), https://doi.org/10.3390/membranes7010016.

[5] L. Li, M. Chen, Y. Dong, X. Dong, S. Cerneaux, S. Hampshire, J. Cao, L. Zhu, Z. Zhu, J. Liu, A low-cost alumina-mullite composite hollow fiber ceramic membrane fabricated via phase-inversion and sintering method, J. Eur. Ceram. Soc. 36 (2016) 2057–2066, https://doi.org/10.1016/j.jeurceramsoc.2016.02.020.

[6] A.F. Ismail, P.S. Goh, S.M. Sanip, M. Aziz, Transport and separation properties of carbon nanotube-mixed matrix membrane, Sep. Purif. Technol. 70 (2009) 12–26, https://doi.org/10.1016/j.seppur.2009.09.002.

[7] Y.H. Teow, A.W. Mohammad, New generation nanomaterials for water desalination: a review, Desalination 451 (2019) 2–17, https://doi.org/10.1016/j.desal.2017.11.041.

[8] P.S. Goh, A.F. Ismail, Review: is interplay between nanomaterial and membrane technology the way forward for desalination? J. Chem. Technol. Biotechnol. 90 (2015) 971–980, https://doi.org/10.1002/jctb.4531.

[9] H. Saleem, L. Trabzon, A. Kilic, S.J. Zaidi, Recent advances in nanofibrous membranes: production and applications in water treatment and desalination, Desalination 478 (2020), https://doi.org/10.1016/j.desal.2019.114178, 114178.

[10] J.N. Tiwari, R.N. Tiwari, K.S. Kim, Zero-dimensional, one-dimensional, two-dimensional and three-dimensional nanostructured materials for advanced electrochemical energy devices, Prog. Mater. Sci. 57 (2012), https://doi.org/10.1016/j.pmatsci.2011.08.003.

[11] M.R. Esfahani, S.A. Aktij, Z. Dabaghian, M.D. Firouzjaei, A. Rahimpour, J. Eke, I.C. Escobar, M. Abolhassani, L.F. Greenlee, A.R. Esfahani, A. Sadmani, N. Koutahzadeh, Nanocomposite membranes for water separation and purification: fabrication, modification, and applications, Sep. Purif. Technol. 213 (2019) 465–499, https://doi.org/10.1016/j.seppur.2018.12.050.

[12] D. Ewis, N.A. Ismail, M.A. Hafiz, A. Benamor, A.H. Hawari, Nanoparticles functionalized ceramic membranes: fabrication, surface modification, and performance, Environ. Sci. Pollut. Res. (2021), https://doi.org/10.1007/s11356-020-11847-0.

[13] M.Q. Seah, W.J. Lau, P.S. Goh, H.H. Tseng, R.A. Wahab, A.F. Ismail, Progress of interfacial polymerization techniques for polyamide thin film (Nano)composite membrane fabrication: a comprehensive review, Polymers (Basel) 12 (2020) 1–39, https://doi.org/10.3390/polym12122817.

[14] F. Zarei, R.M. Moattari, S. Rajabzadeh, M. Bagheri, A. Taghizadeh, T. Mohammadi, H. Matsuyama, Preparation of thin film composite nano-filtration membranes for brackish water softening based on the reaction between functionalized UF membranes and polyethyleneimine, J. Membr. Sci. 588 (2019), https://doi.org/10.1016/j.memsci.2019.117207.

[15] T.S. Jamil, E.S. Mansor, H. Abdallah, A.M. Shaban, Innovative high flux/low pressure blend thin film composite membranes for water softening, React. Funct. Polym. 131 (2018), https://doi.org/10.1016/j.reactfunctpolym.2018.08.007.

[16] B.H. Jeong, E.M.V. Hoek, Y. Yan, A. Subramani, X. Huang, G. Hurwitz, A.K. Ghosh, A. Jawor, Interfacial polymerization of thin film nanocomposites: a new concept for reverse osmosis membranes, J. Membr. Sci. 294 (2007) 1–7, https://doi.org/10.1016/j.memsci.2007.02.025.

[17] T.H. Lee, J.Y. Oh, S.P. Hong, J.M. Lee, S.M. Roh, S.H. Kim, H.B. Park, ZIF-8 particle size effects on reverse osmosis performance of polyamide thin-film nanocomposite membranes: importance of particle deposition, J. Membr. Sci. 570–571 (2019), https://doi.org/10.1016/j.memsci.2018.10.015.

[18] S.J. Park, W. Choi, S.E. Nam, S. Hong, J.S. Lee, J.H. Lee, Fabrication of polyamide thin film composite reverse osmosis membranes via support-free interfacial polymerization, J. Membr. Sci. 526 (2017), https://doi.org/10.1016/j.memsci.2016.12.027.

[19] A. Peyki, A. Rahimpour, M. Jahanshahi, Preparation and characterization of thin film composite reverse osmosis membranes incorporated with hydrophilic SiO2 nanoparticles, Desalination 368 (2015), https://doi.org/10.1016/j.desal.2014.05.025.
[20] C. Liu, J. Lee, J. Ma, M. Elimelech, Antifouling thin-film composite membranes by controlled architecture of zwitterionic polymer brush layer, Environ. Sci. Technol. 51 (2017), https://doi.org/10.1021/acs.est.6b05992.
[21] Y.L. Ji, Q.F. An, X.D. Weng, W.S. Hung, K.R. Lee, C.J. Gao, Microstructure and performance of zwitterionic polymeric nanoparticle/polyamide thin-film nanocomposite membranes for salts/organics separation, J. Membr. Sci. 548 (2018), https://doi.org/10.1016/j.memsci.2017.11.057.
[22] D. Emadzadeh, W.J. Lau, T. Matsuura, A.F. Ismail, M. Rahbari-Sisakht, Synthesis and characterization of thin film nanocomposite forward osmosis membrane with hydrophilic nanocomposite support to reduce internal concentration polarization, J. Membr. Sci. 449 (2014), https://doi.org/10.1016/j.memsci.2013.08.014.
[23] H.S. Lee, S.J. Im, J.H. Kim, H.J. Kim, J.P. Kim, B.R. Min, Polyamide thin-film nanofiltration membranes containing TiO2 nanoparticles, Desalination 219 (2008), https://doi.org/10.1016/j.desal.2007.06.003.
[24] J. Yin, E.S. Kim, J. Yang, B. Deng, Fabrication of a novel thin-film nanocomposite (TFN) membrane containing MCM-41 silica nanoparticles (NPs) for water purification, J. Membr. Sci. 423–424 (2012), https://doi.org/10.1016/j.memsci.2012.08.020.
[25] J. Wang, Y. Wang, Y. Zhang, A. Uliana, J. Zhu, J. Liu, B. Van Der Bruggen, Zeolitic Imidazolate framework/graphene oxide hybrid nanosheets functionalized thin film nanocomposite membrane for enhanced antimicrobial performance, ACS Appl. Mater. Interfaces 8 (2016), https://doi.org/10.1021/acsami.6b06992.
[26] E. Mahmoudi, L.Y. Ng, W.L. Ang, Y.T. Chung, R. Rohani, A.W. Mohammad, Enhancing morphology and separation performance of polyamide 6,6 membranes by minimal incorporation of silver decorated graphene oxide nanoparticles, Sci. Rep. 9 (2019), https://doi.org/10.1038/s41598-018-38060-x.
[27] X. Kang, X. Liu, J. Liu, Y. Wen, J. Qi, X. Li, Spin-assisted interfacial polymerization strategy for graphene oxide-polyamide composite nanofiltration membrane with high performance, Appl. Surf. Sci. 508 (2020), https://doi.org/10.1016/j.apsusc.2019.145198.
[28] S. Al Aani, A. Haroutounian, C.J. Wright, N. Hilal, Thin film nanocomposite (TFN) membranes modified with polydopamine coated metals/carbon-nanostructures for desalination applications, Desalination 427 (2018), https://doi.org/10.1016/j.desal.2017.10.011.
[29] Y. Lin, H. Li, C. Liu, W. Xing, X. Ji, Surface-modified Nafion membranes with mesoporous SiO2 layers via a facile dip-coating approach for direct methanol fuel cells, J. Power Sources 185 (2008), https://doi.org/10.1016/j.jpowsour.2008.08.067.
[30] B.A. McCool, N. Hill, J. DiCarlo, W.J. DeSisto, Synthesis and characterization of mesoporous silica membranes via dip-coating and hydrothermal deposition techniques, J. Membr. Sci. 218 (2003), https://doi.org/10.1016/S0376-7388(03)00136-4.
[31] X. Tang, X. Yan, Dip-coating for fibrous materials: mechanism, methods and applications, J. Sol-Gel Sci. Technol. 81 (2017) 378–404, https://doi.org/10.1007/s10971-016-4197-7.
[32] Y. Lu, R. Ganguli, C.A. Drewien, M.T. Anderson, C. Jeffrey Brinker, W. Gong, Y. Guo, H. Soyez, B. Dunn, M.H. Huang, J.I. Zink, Continuous formation of supported cubic and hexagonal mesoporous films by sol-gel dip-coating, Nature 389 (1997), https://doi.org/10.1038/38699.
[33] R.M. Almeida, M.C. Gonçalves, S. Portal, Sol-gel photonic bandgap materials and structures, J. Non Cryst. Solids (2004), https://doi.org/10.1016/j.jnoncrysol.2004.08.085.
[34] X. Wang, D. Pan, D. Weng, C.Y. Low, L. Rice, J. Han, Y. Lu, A general synthesis of Cu-In-S based multicomponent solid-solution nanocrystals with tunable band gap, size, and structure, J. Phys. Chem. C 114 (2010), https://doi.org/10.1021/jp103572g.

[35] J. Lee, C. Boo, W.H. Ryu, A.D. Taylor, M. Elimelech, Development of omniphobic desalination membranes using a charged electrospun nanofiber scaffold, ACS Appl. Mater. Interfaces 8 (2016), https://doi.org/10.1021/acsami.6b02419.
[36] M. Jahanshahi, A. Rahimpour, M. Peyravi, Developing thin film composite poly (piperazine-amide) and poly(vinyl-alcohol) nanofiltration membranes, Desalination 257 (2010), https://doi.org/10.1016/j.desal.2010.02.034.
[37] S. Pourjafar, A. Rahimpour, M. Jahanshahi, Synthesis and characterization of PVA/ PES thin film composite nanofiltration membrane modified with TiO 2 nanoparticles for better performance and surface properties, J. Ind. Eng. Chem. 18 (2012), https:// doi.org/10.1016/j.jiec.2012.01.041.
[38] S. Yu, Z. Chen, Q. Cheng, Z. Lü, M. Liu, C. Gao, Application of thin-film composite hollow fiber membrane to submerged nanofiltration of anionic dye aqueous solutions, Sep. Purif. Technol. 88 (2012), https://doi.org/10.1016/j.seppur.2011.12.024.
[39] L. Sarango, L. Paseta, M. Navarro, B. Zornoza, J. Coronas, Controlled deposition of MOFs by dip-coating in thin film nanocomposite membranes for organic solvent nanofiltration, J. Ind. Eng. Chem. 59 (2018), https://doi.org/10.1016/j.jiec.2017.09.053.
[40] A. Soroush, W. Ma, Y. Silvino, M.S. Rahaman, Surface modification of thin film composite forward osmosis membrane by silver-decorated graphene-oxide nanosheets, Environ. Sci. Nano 2 (2015), https://doi.org/10.1039/c5en00086f.
[41] A. Tiraferri, Y. Kang, E.P. Giannelis, M. Elimelech, Highly hydrophilic thin-film composite forward osmosis membranes functionalized with surface-tailored nanoparticles, ACS Appl. Mater. Interfaces 4 (2012), https://doi.org/10.1021/am301532g.
[42] G.D. Kang, C.J. Gao, W.D. Chen, X.M. Jie, Y.M. Cao, Q. Yuan, Study on hypochlorite degradation of aromatic polyamide reverse osmosis membrane, J. Membr. Sci. 300 (2007), https://doi.org/10.1016/j.memsci.2007.05.025.
[43] Kenry, C.T. Lim, Synthesis, optical properties, and chemical-biological sensing applications of one-dimensional inorganic semiconductor nanowires, Prog. Mater. Sci. 58 (2013), https://doi.org/10.1016/j.pmatsci.2013.01.001.
[44] Z.M. Huang, Y.Z. Zhang, M. Kotaki, S. Ramakrishna, A review on polymer nanofibers by electrospinning and their applications in nanocomposites, Compos. Sci. Technol. 63 (2003), https://doi.org/10.1016/S0266-3538(03)00178-7.
[45] S. Shokrollahzadeh, S. Tajik, Fabrication of thin film composite forward osmosis membrane using electrospun polysulfone/polyacrylonitrile blend nanofibers as porous substrate, Desalination 425 (2018), https://doi.org/10.1016/j.desal.2017.10.017.
[46] S.H. Park, J.H. Kim, S.J. Moon, E. Drioli, Y.M. Lee, Enhanced, hydrophobic, fluorine-containing, thermally rearranged (TR) nanofiber membranes for desalination via membrane distillation, J. Membr. Sci. 550 (2018), https://doi.org/10.1016/ j.memsci.2017.10.065.
[47] S. Subramanian, R. Seeram, New directions in nanofiltration applications—are nanofibers the right materials as membranes in desalination? Desalination 308 (2013), https://doi.org/10.1016/j.desal.2012.08.014.
[48] Z. Li, C. Wang, Effects of working parameters on electrospinning BT—one-dimensional nanostructures: electrospinning technique and unique nanofibers, in: Progress in Biomaterials, 2013.
[49] I. Sas, R.E. Gorga, J.A. Joines, K.A. Thoney, Literature review on superhydrophobic self-cleaning surfaces produced by electrospinning, J. Polym. Sci. B Polym. Phys. 50 (2012), https://doi.org/10.1002/polb.23070.
[50] J.H. He, H.Y. Kong, R.R. Yang, H. Dou, N. Faraz, L. Wang, C. Feng, Review on fiber morphology obtained by bubble electrospinning and blown bubble spinning, Therm. Sci. 16 (2012), https://doi.org/10.2298/TSCI1205263H.
[51] T. Jarusuwannapoom, W. Hongrojjanawiwat, S. Jitjaicham, L. Wannatong, M. Nithitanakul, C. Pattamaprom, P. Koombhongse, R. Rangkupan, P. Supaphol, Effect of solvents on electro-spinnability of polystyrene solutions and morphological

appearance of resulting electrospun polystyrene fibers, Eur. Polym. J. 41 (2005), https://doi.org/10.1016/j.eurpolymj.2004.10.010.
[52] Y. Xin, D.H. Reneker, Garland formation process in electrospinning, Polymer (Guildf) 53 (2012), https://doi.org/10.1016/j.polymer.2012.05.060.
[53] A.G. Sener, A.S. Altay, F. Altay, Effect of voltage on morphology of electrospun nanofibers, in: ELECO 2011—7th Int. Conf. Electr. Electron. Eng, 2011.
[54] T. Garg, G. Rath, A.K. Goyal, Biomaterials-based nanofiber scaffold: targeted and controlled carrier for cell and drug delivery, J. Drug Target. 23 (2014) 202–221, https://doi.org/10.3109/1061186X.2014.992899.
[55] P. Kumar, Effect of Collector on Electrospinning to Fabricate Aligned Nanofiber, Effect of Collector on Electrospinning to Fabricate Aligned Nanofiber, Natl. Inst. Technol., 2012.
[56] S. Ramakrishna, K. Fujihara, W.E. Teo, T.C. Lim, Z. Ma, An Introduction to Electrospinning and Nanofibers, 2005, https://doi.org/10.1142/5894.
[57] X. Zhang, Y. Lu, Centrifugal spinning: an alternative approach to fabricate nanofibers at high speed and low cost, Polym. Rev. 54 (2014), https://doi.org/10.1080/15583724.2014.935858.
[58] Y. Tsuboi, Y. Yoshida, K. Okada, N. Kitamura, Phase separation dynamics of aqueous solutions of thermoresponsive polymers studied by a laser T-jump technique, J. Phys. Chem. B 112 (2008), https://doi.org/10.1021/jp711128s.
[59] M.A. Hassan, B.Y. Yeom, A. Wilkie, B. Pourdeyhimi, S.A. Khan, Fabrication of nanofiber meltblown membranes and their filtration properties, J. Membr. Sci. 427 (2013) 336–344, https://doi.org/10.1016/j.memsci.2012.09.050.
[60] G.S. Bhat, S.R. Malkan, Extruded continuous filament nonwovens: advances in scientific aspects, J. Appl. Polym. Sci. 83 (2001), https://doi.org/10.1002/app.2259.
[61] A. Van Wente, Superfine thermoplastic fibers, Ind. Eng. Chem. 48 (1956), https://doi.org/10.1021/ie50560a034.
[62] R. Uppal, G. Bhat, C. Eash, K. Akato, Meltblown nanofiber media for enhanced quality factor, Fibers Polym. 14 (2013) 660–668, https://doi.org/10.1007/s12221-013-0660-z.
[63] H.M. Krutka, R.L. Shambaugh, D.V. Papavassiliou, Effects of temperature and geometry on the flow field of the melt blowing process, Ind. Eng. Chem. Res. 43 (2004), https://doi.org/10.1021/ie040043e.
[64] H.M. Krutka, R.L. Shambaugh, D.V. Papavassiliou, Effects of die geometry on the flow field of the melt-blowing process, Ind. Eng. Chem. Res. 42 (2003), https://doi.org/10.1021/ie030457s.
[65] H.M. Krutka, R.L. Shambaugh, D.V. Papavassiliou, Analysis of a melt-blowing die: comparison of CFD and experiments, Ind. Eng. Chem. Res. 41 (2002), https://doi.org/10.1021/ie020366f.
[66] E.M. Moore, D.V. Papavassiliou, R.L. Shambaugh, Air velocity, air temperature, fiber vibration and fiber diameter measurements on a practical melt blowing die, Int. Nonwovens J. 13 (2004), https://doi.org/10.1177/1558925004os-1300309.
[67] Y. Lee, L.C. Wadsworth, Structure and filtration properties of melt blown polypropylene webs, Polym. Eng. Sci. 30 (1990), https://doi.org/10.1002/pen.760302202.
[68] G.F. Ward, Meltblown nanofibres for nonwoven filtration applications, Filtr. Sep. 38 (2001), https://doi.org/10.1016/S0015-1882(01)80540-1.
[69] C.J. Ellison, A. Phatak, D.W. Giles, C.W. Macosko, F.S. Bates, Melt blown nanofibers: fiber diameter distributions and onset of fiber breakup, Polymer (Guildf) 48 (2007), https://doi.org/10.1016/j.polymer.2007.04.005.
[70] B.K. Nandi, R. Uppaluri, M.K. Purkait, Preparation and characterization of low cost ceramic membranes for micro-filtration applications, Appl. Clay Sci. 42 (2008), https://doi.org/10.1016/j.clay.2007.12.001.

[71] M.C. Almandoz, C.L. Pagliero, N.A. Ochoa, J. Marchese, Composite ceramic membranes from natural aluminosilicates for microfiltration applications, Ceram. Int. 41 (2015) 5621–5633, https://doi.org/10.1016/j.ceramint.2014.12.144.
[72] M. Ben Ali, N. Hamdi, M.A. Rodriguez, K. Mahmoudi, E. Srasra, Preparation and characterization of new ceramic membranes for ultrafiltration, Ceram. Int. 44 (2018) 2328–2335, https://doi.org/10.1016/j.ceramint.2017.10.199.
[73] H. Guo, S. Zhao, X. Wu, H. Qi, Fabrication and characterization of TiO2/ZrO2 ceramic membranes for nanofiltration, Microporous Mesoporous Mater. 260 (2018) 125–131, https://doi.org/10.1016/j.micromeso.2016.03.011.
[74] G. Skouteris, D. Hermosilla, P. López, C. Negro, Á. Blanco, Anaerobic membrane bioreactors for wastewater treatment: a review, Chem. Eng. J. 198–199 (2012) 138–148, https://doi.org/10.1016/j.cej.2012.05.070.
[75] M. Sheikh, M. Pazirofteh, M. Dehghani, M. Asghari, M. Rezakazemi, C. Valderrama, J.L. Cortina, Application of ZnO nanostructures in ceramic and polymeric membranes for water and wastewater technologies: a review, Chem. Eng. J. 391 (2020), https://doi.org/10.1016/j.cej.2019.123475, 123475.
[76] M.M. Cortalezzi, J. Rose, A.R. Barron, M.R. Wiesner, Characteristics of ultrafiltration ceramic membranes derived from alumoxane nanoparticles, J. Membr. Sci. 205 (2002) 33–43, https://doi.org/10.1016/S0376-7388(02)00049-2.
[77] S. Dong, Z. Wang, M. Sheng, Z. Qiao, J. Wang, Scaling up of defect-free flat membrane with ultra-high gas permeance used for intermediate layer of multi-layer composite membrane and oxygen enrichment, Sep. Purif. Technol. 239 (2020), https://doi.org/10.1016/j.seppur.2020.116580, 116580.
[78] Y. Gong, S. Gao, Y. Tian, Y. Zhu, W. Fang, Z. Wang, J. Jin, Thin-film nanocomposite nanofiltration membrane with an ultrathin polyamide/UIO-66-NH2 active layer for high-performance desalination, J. Membr. Sci. 600 (2020), https://doi.org/10.1016/j.memsci.2020.117874, 117874.
[79] J. Kim, B. Van Der Bruggen, The use of nanoparticles in polymeric and ceramic membrane structures: review of manufacturing procedures and performance improvement for water treatment, Environ. Pollut. 158 (2010) 2335–2349, https://doi.org/10.1016/j.envpol.2010.03.024.
[80] H. Park, Y. Kim, B. An, H. Choi, Characterization of natural organic matter treated by iron oxide nanoparticle incorporated ceramic membrane-ozonation process, Water Res. 46 (2012) 5861–5870, https://doi.org/10.1016/j.watres.2012.07.039.
[81] H. Liu, C. Li, X. Ren, K. Liu, J. Yang, Fine platinum nanoparticles supported on a porous ceramic membrane as efficient catalysts for the removal of benzene, Sci. Rep. 7 (2017) 1–8, https://doi.org/10.1038/s41598-017-16833-0.
[82] S. Peng, Y. Chen, X. Jin, W. Lu, M. Gou, X. Wei, J. Xie, Polyimide with half encapsulated silver nanoparticles grafted ceramic composite membrane: enhanced silver stability and lasting anti–biofouling performance, J. Membr. Sci. 611 (2020), https://doi.org/10.1016/j.memsci.2020.118340, 118340.
[83] F. DiGiano, In pursuit of innovative membrane technology, in: IWA Membr. Res. Conf, 2008.
[84] Y. Zhang, F. Zeng, C. Yu, C. Wu, W. Ding, X. Lu, Fabrication and characterization of dense BaCo0.7Fe0.2Nb0.1O3-δ tubular membrane by slip casting techniques, Ceram. Int. 41 (2015) 1401–1411, https://doi.org/10.1016/j.ceramint.2014.09.073.
[85] A.R. Studart, U.T. Gonzenbach, E. Tervoort, L.J. Gauckler, Processing routes to macroporous ceramics: a review, J. Am. Ceram. Soc. 89 (2006) 1771–1789, https://doi.org/10.1111/j.1551-2916.2006.01044.x.
[86] S.K. Amin, H.A.M. Abdallah, M.H. Roushdy, S.A. El-Sherbiny, An overview of production and development of ceramic membranes, Int. J. Appl. Eng. Res. 11 (2016) 7708–7721.

[87] M. Boussemghoune, M. Chikhi, F. Balaska, Y. Ozay, N. Dizge, B. Kebabi, Preparation of a zirconia-based ceramic membrane and its application for drinking water treatment, Symmetry (Basel) 12 (2020) 933, https://doi.org/10.3390/sym12060933.

[88] M.B. Choi, D.K. Lim, S.Y. Jeon, H.S. Kim, S.J. Song, Oxygen permeation properties of BSCF5582 tubular membrane fabricated by the slip casting method, Ceram. Int. 38 (2012) 1867–1872, https://doi.org/10.1016/j.ceramint.2011.10.012.

[89] M. Issaoui, L. Limousy, Low-cost ceramic membranes: synthesis, classifications, and applications, C. R. Chim. 22 (2019) 175–187, https://doi.org/10.1016/j.crci.2018.09.014.

[90] T. Mohammadi, A. Pak, Effect of calcination temperature of kaolin as a support for zeolite membranes, Sep. Purif. Technol. 30 (2003) 241–249, https://doi.org/10.1016/S1383-5866(02)00146-6.

[91] M.-B. Choi, S.-J. Song, T.-W. Lee, H.-I. Yoo, U.-D. Lee, B.-R. Bang, Preparation of asymmetric tubular oxygen separation membrane with oxygen permeable Pr2Ni0.75Cu0.25Ga0.05O4+δ, Int. J. Appl. Ceram. Technol. 8 (2011) 800–808, https://doi.org/10.1111/j.1744-7402.2010.02507.x.

[92] J.M. Kim, H.N. Kim, Y.J. Park, J.W. Ko, J.W. Lee, H.D. Kim, Fabrication of transparent MgAl2O4 spinel through homogenous green compaction by microfluidization and slip casting, Ceram. Int. 41 (2015) 13354–13360, https://doi.org/10.1016/j.ceramint.2015.07.121.

[93] R.K. Nishihora, P.L. Rachadel, M.G.N. Quadri, D. Hotza, Manufacturing porous ceramic materials by tape casting—a review, J. Eur. Ceram. Soc. 38 (2018) 988–1001, https://doi.org/10.1016/j.jeurceramsoc.2017.11.047.

[94] G. Etchegoyen, T. Chartier, P. Del-Gallo, An architectural approach to the oxygen permeability of a La0.6Sr0.4Fe0.9Ga0.1O3-δ perovskite membrane, J. Eur. Ceram. Soc. 26 (2006) 2807–2815, https://doi.org/10.1016/j.jeurceramsoc.2005.06.025.

[95] M. Jabbari, R. Bulatova, A.I.Y. Tok, C.R.H. Bahl, E. Mitsoulis, J.H. Hattel, Ceramic tape casting: a review of current methods and trends with emphasis on rheological behaviour and flow analysis, Mater. Sci. Eng. B 212 (2016) 39–61, https://doi.org/10.1016/j.mseb.2016.07.011.

[96] P.M. Geffroy, J.M. Bassat, A. Vivet, S. Fourcade, T. Chartier, P. Del Gallo, N. Richet, Oxygen semi-permeation, oxygen diffusion and surface exchange coefficient of La(1-x)SrxFe(1-y)GayO3- δ perovskite membranes, J. Membr. Sci. 354 (2010) 6–13, https://doi.org/10.1016/j.memsci.2010.03.001.

[97] M. Reichmann, P.M. Geffroy, J. Fouletier, N. Richet, T. Chartier, Effect of cation substitution in the A site on the oxygen semi-permeation flux in La0.5A0.5Fe0.7-Ga0.3O 3-δ and La0.5A0.5Fe0.7Co 0.3O3-δ dense perovskite membranes with A = Ca, Sr and Ba (part I), J. Power Sources 261 (2014) 175–183, https://doi.org/10.1016/j.jpowsour.2014.03.074.

[98] M. Reichmann, P.M. Geffroy, J. Fouletier, N. Richet, P. Del Gallo, T. Chartier, Effect of cation substitution at the B site on the oxygen semi-permeation flux in La0.5Ba0.5Fe0.7B0.3O3-δ dense perovskite membranes with B = Al, Co, Cu, Mg, Mn, Ni, Sn, Ti and Zn (part II), J. Power Sources 277 (2015) 17–25, https://doi.org/10.1016/j.jpowsour.2014.12.012.

[99] N. Das, H.S. Maiti, Formatation of pore structure in tape-cast alumina membranes—effects of binder content and firing temperature, J. Membr. Sci. 140 (1998) 205–212, https://doi.org/10.1016/S0376-7388(97)00282-2.

[100] H. Fang, C. Ren, Y. Liu, D. Lu, L. Winnubst, C. Chen, Phase-inversion tape casting and synchrotron-radiation computed tomography analysis of porous alumina, J. Eur. Ceram. Soc. 33 (2013) 2049–2051, https://doi.org/10.1016/j.jeurceramsoc.2013.02.032.

[101] S. Deville, Freeze-casting of porous ceramics: a review of current achievements and issues, Adv. Eng. Mater. 10 (2008) 155–169, https://doi.org/10.1002/adem.200700270.

[102] L. Hao, Y. Zhang, R. Kubomura, S. Ozeki, S. Liu, H. Yoshida, Y. Jin, Y. Lu, Preparation and thermoelectric properties of CuAlO2 compacts by tape casting followed by SPS, J. Alloys Compd. 853 (2021), https://doi.org/10.1016/j.jallcom.2020.157086.

[103] S.K. Hubadillah, M.H.D. Othman, T. Matsuura, A.F. Ismail, M.A. Rahman, Z. Harun, J. Jaafar, M. Nomura, Fabrications and applications of low cost ceramic membrane from kaolin: a comprehensive review, Ceram. Int. 44 (2018) 4538–4560, https://doi.org/10.1016/j.ceramint.2017.12.215.

[104] A. Manni, B. Achiou, A. Karim, A. Harrati, C. Sadik, M. Ouammou, S. Alami Younssi, A. El Bouari, New low-cost ceramic microfiltration membrane made from natural magnesite for industrial wastewater treatment, J. Environ. Chem. Eng. 8 (2020), https://doi.org/10.1016/j.jece.2020.103906, 103906.

[105] R. Del Colle, C.A. Fortulan, S.R. Fontes, Manufacture and characterization of ultra and microfiltration ceramic membranes by isostatic pressing, Ceram. Int. 37 (2011) 1161–1168, https://doi.org/10.1016/j.ceramint.2010.11.039.

[106] X. Li, J. Luo, Z. Feng, S. Zhou, J. Huang, In-situ hot pressing sintering behaviors of Y2O3-La2O3 co-doped AlON ceramic, Ceram. Int. 42 (2016) 17382–17386, https://doi.org/10.1016/j.ceramint.2016.08.037.

[107] V. Mosquim, B.M. Ferrairo, M. Vertuan, G. Magdalena, C.A. Fortulan, P. Noronha Lisboa-Filho, P.F. Cesar, E.A. Bonfante, H.M. Honório, A.F.S. Borges, Structural, chemical and optical characterizations of an experimental SiO 2-Y-TZP ceramic produced by the uniaxial/isostatic pressing technique, J. Mech. Behav. Biomed. Mater. (2020), https://doi.org/10.1016/j.jmbbm.2020.103749.

[108] I. Ivanets, V.E. Azarova, S. Agabekov, C. Azarov, D. Batsukh, V.G. Batsuren, A.A. Prozorovich, Rat'ko, effect of phase composition of natural quartz raw material on characterization of microfiltration ceramic membranes, Ceram. Int. 42 (2016) 16571–16578, https://doi.org/10.1016/j.ceramint.2016.07.077.

[109] S. Chakraborty, R. Uppaluri, C. Das, Optimal fabrication of carbonate free kaolin based low cost ceramic membranes using mixture model response surface methodology, Appl. Clay Sci. 162 (2018) 101–112, https://doi.org/10.1016/j.clay.2018.06.002.

[110] S. Mestre, A. Gozalbo, M.M. Lorente-Ayza, E. Sánchez, Low-cost ceramic membranes: a research opportunity for industrial application, J. Eur. Ceram. Soc. 39 (2019) 3392–3407, https://doi.org/10.1016/j.jeurceramsoc.2019.03.054.

[111] S. Saja, A. Bouazizi, B. Achiou, H. Ouaddari, A. Karim, M. Ouammou, A. Aaddane, J. Bennazha, S.A. Younssi, Fabrication of low-cost ceramic ultrafiltration membrane made from bentonite clay and its application for soluble dyes removal, J. Eur. Ceram. Soc. 40 (2020) 2453–2462, https://doi.org/10.1016/j.jeurceramsoc.2020.01.057.

[112] T. Isobe, Y. Kameshima, A. Nakajima, K. Okada, Y. Hotta, Extrusion method using nylon 66 fibers for the preparation of porous alumina ceramics with oriented pores, J. Eur. Ceram. Soc. 26 (2006) 2213–2217, https://doi.org/10.1016/j.jeurceramsoc.2005.04.014.

[113] A. Talidi, N. Saffaj, K. El Kacemi, S.A. Younssi, A. Albizane, A. Chakir, Processing and characterization of tubular ceramic support for microfiltration membrane prepared from pyrophyllite clay, Sci. Study Res. Chem. Chem. Eng. Biotechnol. Food Ind. 12 (2011) 263–268.

[114] R. Vinoth Kumar, A. Kumar Ghoshal, G. Pugazhenthi, Elaboration of novel tubular ceramic membrane from inexpensive raw materials by extrusion method and its performance in microfiltration of synthetic oily wastewater treatment, J. Membr. Sci. 490 (2015) 92–102, https://doi.org/10.1016/j.memsci.2015.04.066.

[115] N. Saffaj, M. Persin, S.A. Younsi, A. Albizane, M. Cretin, A. Larbot, Elaboration and characterization of microfiltration and ultrafiltration membranes deposited on raw support prepared from natural Moroccan clay: application to filtration of solution containing dyes and salts, Appl. Clay Sci. 31 (2006) 110–119, https://doi.org/10.1016/j.clay.2005.07.002.
[116] F. Bouzerara, A. Harabi, S. Achour, A. Larbot, Porous ceramic supports for membranes prepared from kaolin and doloma mixtures, J. Eur. Ceram. Soc. 26 (2006) 1663–1671, https://doi.org/10.1016/j.jeurceramsoc.2005.03.244.
[117] W.D. Callister, D.G. Rethwisch, Fundamentals of Materials Science and Engineering an Integrated Approach, Wiley, 2012.
[118] T. Isobe, T. Tomita, Y. Kameshima, A. Nakajima, K. Okada, Preparation and properties of porous alumina ceramics with oriented cylindrical pores produced by an extrusion method, J. Eur. Ceram. Soc. 26 (2006) 957–960, https://doi.org/10.1016/j.jeurceramsoc.2004.11.015.
[119] M.H. Abd Aziz, M.H.D. Othman, N.A. Hashim, M.R. Adam, A. Mustafa, Fabrication and characterization of mullite ceramic hollow fiber membrane from natural occurring ball clay, Appl. Clay Sci. 177 (2019) 51–62, https://doi.org/10.1016/j.clay.2019.05.003.
[120] M. Mouiya, A. Bouazizi, A. Abourriche, A. Benhammou, Y. El Hafiane, M. Ouammou, Y. Abouliatim, S.A. Younssi, A. Smith, H. Hannache, Fabrication and characterization of a ceramic membrane from clay and banana peel powder: application to industrial wastewater treatment, Mater. Chem. Phys. 227 (2019) 291–301, https://doi.org/10.1016/j.matchemphys.2019.02.011.
[121] S. Mallakpour, S. Soltanian, Surface functionalization of carbon nanotubes: fabrication and applications, RSC Adv. 6 (2016) 109916–109935, https://doi.org/10.1039/c6ra24522f.
[122] A.T. Nasrabadi, M. Foroutan, Ion-separation and water-purification using single-walled carbon nanotube electrodes, Desalination 277 (2011) 236–243, https://doi.org/10.1016/j.desal.2011.04.028.
[123] K. Dai, L. Shi, D. Zhang, J. Fang, NaCl adsorption in multi-walled carbon nanotube/active carbon combination electrode, Chem. Eng. Sci. 61 (2006) 428–433, https://doi.org/10.1016/j.ces.2005.07.030.
[124] L. Joseph, J. Heo, Y.G. Park, J.R.V. Flora, Y. Yoon, Adsorption of bisphenol A and 17α-ethinyl estradiol on single walled carbon nanotubes from seawater and brackish water, Desalination 281 (2011) 68–74, https://doi.org/10.1016/j.desal.2011.07.044.
[125] K. Goh, H.E. Karahan, L. Wei, T.H. Bae, A.G. Fane, R. Wang, Y. Chen, Carbon nanomaterials for advancing separation membranes: a strategic perspective, Carbon N. Y. 109 (2016) 694–710, https://doi.org/10.1016/j.carbon.2016.08.077.
[126] K.J. Lee, H.D. Park, Effect of transmembrane pressure, linear velocity, and temperature on permeate water flux of high-density vertically aligned carbon nanotube membranes, Desalin. Water Treat. 57 (2016) 26706–26717, https://doi.org/10.1080/19443994.2016.1189704.
[127] J. Liu, M.R.i. Zubiri, B. Vigolo, M. Dossot, Y. Fort, J.J. Ehrhardt, E. McRae, Efficient microwave-assisted radical functionalization of single-wall carbon nanotubes, Carbon N. Y. 45 (2007) 885–891, https://doi.org/10.1016/j.carbon.2006.11.006.
[128] N. Zhao, C. He, Z. Jiang, J. Li, Y. Li, Fabrication and growth mechanism of carbon nanotubes by catalytic chemical vapor deposition, Mater. Lett. 60 (2006) 159–163, https://doi.org/10.1016/j.matlet.2005.08.009.
[129] B. Lee, Y. Baek, M. Lee, D.H. Jeong, H.H. Lee, J. Yoon, Y.H. Kim, A carbon nanotube wall membrane for water treatment, Nat. Commun. 6 (2015) 1–7, https://doi.org/10.1038/ncomms8109.

[130] W. Mi, Y.S. Lin, Y. Li, Vertically aligned carbon nanotube membranes on macroporous alumina supports, J. Membr. Sci. 304 (2007) 1–7, https://doi.org/10.1016/j.memsci.2007.07.021.
[131] Y. Zhao, Z. Zhong, Z.X. Low, Z. Yao, A multifunctional multi-walled carbon nanotubes/ceramic membrane composite filter for air purification, RSC Adv. 5 (2015) 91951–91959, https://doi.org/10.1039/c5ra18200j.
[132] G. Pilatos, E.C. Vermisoglou, G.E. Romanos, G.N. Karanikolos, N. Boukos, V. Likodimos, N.K. Kanellopoulos, A closer look inside nanotubes: pore structure evaluation of anodized alumina templated carbon nanotube membranes through adsorption and permeability studies, Adv. Funct. Mater. 20 (2010) 2500–2510, https://doi.org/10.1002/adfm.200901429.
[133] M. Alsawat, T. Altalhi, T. Kumeria, A. Santos, D. Losic, Carbon nanotube-nanoporous anodic alumina composite membranes with controllable inner diameters and surface chemistry: influence on molecular transport and chemical selectivity, Carbon N. Y. 93 (2015) 681–692, https://doi.org/10.1016/j.carbon.2015.05.090.
[134] W.F. Chan, H.Y. Chen, A. Surapathi, M.G. Taylor, X. Shao, E. Marand, J.K. Johnson, Zwitterion functionalized carbon nanotube/polyamide nanocomposite membranes for water desalination, ACS Nano 7 (2013) 5308–5319, https://doi.org/10.1021/nn4011494.
[135] H. Zhao, S. Qiu, L. Wu, L. Zhang, H. Chen, C. Gao, Improving the performance of polyamide reverse osmosis membrane by incorporation of modified multi-walled carbon nanotubes, J. Membr. Sci. 450 (2014) 249–256, https://doi.org/10.1016/j.memsci.2013.09.014.
[136] S. Roy, S.A. Ntim, S. Mitra, K.K. Sirkar, Facile fabrication of superior nanofiltration membranes from interfacially polymerized CNT-polymer composites, J. Membr. Sci. 375 (2011) 81–87, https://doi.org/10.1016/j.memsci.2011.03.012.
[137] H.J. Kim, K. Choi, Y. Baek, D.G. Kim, J. Shim, J. Yoon, J.C. Lee, High-performance reverse osmosis CNT/polyamide nanocomposite membrane by controlled interfacial interactions, ACS Appl. Mater. Interfaces 6 (2014) 2819–2829, https://doi.org/10.1021/am405398f.
[138] S. Kim, F. Fornasiero, H.G. Park, J. Bin In, E. Meshot, G. Giraldo, M. Stadermann, M. Fireman, J. Shan, C.P. Grigoropoulos, O. Bakajin, Fabrication of flexible, aligned carbon nanotube/polymer composite membranes by in-situ polymerization, J. Membr. Sci. 460 (2014) 91–98, https://doi.org/10.1016/j.memsci.2014.02.016.
[139] H.J. Kim, Y. Baek, K. Choi, D.G. Kim, H. Kang, Y.S. Choi, J. Yoon, J.C. Lee, The improvement of antibiofouling properties of a reverse osmosis membrane by oxidized CNTs, RSC Adv. 4 (2014) 32802–32810, https://doi.org/10.1039/c4ra06489e.
[140] G.S. Ajmani, D. Goodwin, K. Marsh, D.H. Fairbrother, K.J. Schwab, J.G. Jacangelo, H. Huang, Modification of low pressure membranes with carbon nanotube layers for fouling control, Water Res. 46 (2012) 5645–5654, https://doi.org/10.1016/j.watres.2012.07.059.
[141] L. Bai, H. Liang, J. Crittenden, F. Qu, A. Ding, J. Ma, X. Du, S. Guo, G. Li, Surface modification of UF membranes with functionalized MWCNTs to control membrane fouling by NOM fractions, J. Membr. Sci. 492 (2015) 400–411, https://doi.org/10.1016/j.memsci.2015.06.006.
[142] X. Fan, H. Zhao, Y. Liu, X. Quan, H. Yu, S. Chen, Enhanced permeability, selectivity, and antifouling ability of CNTs/Al2O3 membrane under electrochemical assistance, Environ. Sci. Technol. 49 (2015) 2293–2300, https://doi.org/10.1021/es5039479.
[143] X. Chen, M. Qiu, H. Ding, K. Fu, Y. Fan, A reduced graphene oxide nanofiltration membrane intercalated by well-dispersed carbon nanotubes for drinking water purification, Nanoscale 8 (2016) 5696–5705, https://doi.org/10.1039/c5nr08697c.

CHAPTER 5

Sustainable membranes with FNs: Current and emerging research trends

P. Das[a] **and Suman Dutta**[b]
[a]CSIR-Central Institute of Mining and Fuel Research, Dhanbad, Jharkhand, India
[b]Department of Chemical Engineering, IIT(ISM), Dhanbad, India

5.1 Introduction

The domain of wastewater treatment and water resources conservation and management is receiving heightened research attention in recent years due to the shrinkage of water resources around the world. Newer water pollutants, coupled with more stringent environmental standards fuelled by an acute demand for clean water, are generating veritable challenges for water scientists, industries, water conservationists, and the general populace as a whole. Creative water treatment technologies are being devised to address all the challenges that look beyond the conventional wastewater treatment (WWT) techniques and also score high on the sustainability quotient. Areas such as value-added product recovery, zero liquid discharge, and the addition of renewable energy sources are being explored apart from viable cost-benefit analysis and complete contaminant degradation. Conventional bulky, energy-intensive multistage processes are being replaced by smaller, multifunctional process-intensified technologies that cater to environmental friendliness, minimize secondary contamination, and ensure high separation efficiency. A critical assessment integration of membrane separation technology and nanotechnology is one avenue that possesses the potential to assimilate all the aforesaid aspects and deliver clean water.

Membrane technology, a largely physical separation technique, has become a panacea in recent years for different forms of water-related problems. However, since it is largely a physical separation technique, the necessity of combining other treatment technologies for the passivation of membrane concentrates to completely degrade the contaminants. This gap can be effectively filled by nanotechnological solutions without mandating a major change in the existing infrastructure. Combining these two

Membranes with Functionalized Nanomaterials
https://doi.org/10.1016/B978-0-323-85946-2.00011-4

methods represents a potent integrated technique and has carved out a niche research area. Nanotechnological solutions can be integrated through various methods such as surface grafting of nanomaterials, blending or copolymerization, synthesis of functionalized membranes, and nanocomposite functionalized membrane fabrication [1]. The schematic shown in Fig. 5.1 provides a visual representation of the different classes of nanomaterial-functionalized membranes.

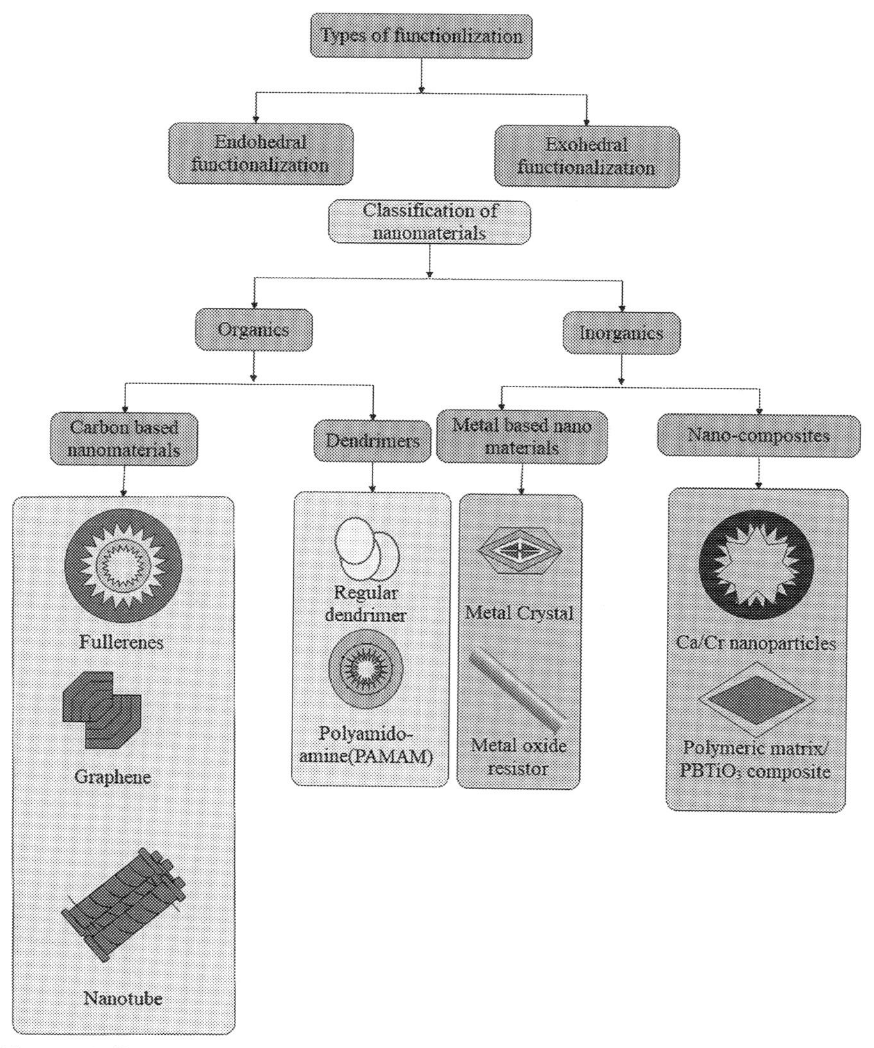

Fig. 5.1 Different classes of nanomaterial-functionalized membranes [2–4].

These open up new vistas of research and development as they provide the operational flexibility for the incorporation of desired properties. Furthermore, diverse analogous applications such as photocatalysis, electrocatalysis, magnetic force-driven separation, stimulus responsiveness, and a host of other methods can be coupled, which fortifies the remediation operations.

The combination of membranes with nanomaterials is a beneficial association where membrane properties such as porosity, permeability, and selectivity are improved. Simultaneously, the membrane provides a large contact surface in an efficient packing density of optimized surface area: volume ratio. This intensifies the charge distribution, electrostatic forces, adsorptive uptake capacity, and oxidation potential of the functionalized materials. Pollutant species properties such as antibacterial characteristics, rapid degradation of dissolved organics, and breakdown of heavy metals can be incorporated and coupled with the selective separation efficacy of the membranes. This process also allows the design flexibility of coupling renewable energy sources into the system increasing the energy efficiency and environment friendliness. As a result, highly process-intensified modular units can be developed, which are characterized by sustainability and high separation efficiency and offer complete wastewater treatment.

This chapter explores wastewater treatment using integrated functionalized-nanomaterial membranes. The fabrication techniques and their mode of action, and remediation mechanisms of such composite membranes have been studied in detail. The diverse applications in various domains ranging from osmotic gradient energy production to organic solvent recovery have been discussed along with coverage of specific problems in the domain of wastewater such as dissolved organics, heavy metals, or microbial content. Insights about comparative performance enhancements, pros and cons, and newer developments in the field were included in the discourse. The possibility of value-added product recovery as a means of incentivization was explored.

5.2 Fabrication techniques

5.2.1 Phase inversion

Phase inversion is a ubiquitous method of polymeric membrane fabrication. As the name suggests the polymer is converted from liquid state to solid state with this technique. The system consists of a membrane, polymer, and solvent in a distribution analogous to liquid-liquid extraction. The three components mix to produce a polymer-rich phase and a polymer-lean phase similar to the raffinate and extract phases. The membrane formation takes place in the polymer-rich phase, which is more thermodynamically stable compared to the polymer lean phase. The four steps involved in the membrane formation are polymer mix, polymer precipitation, polymer solidification, and finally membrane synthesis [5]. In-between the second and third stages there is a rapid demixing and casting of the target polymer with the dopants. Depending on the final pore structure of the membrane and its end usages, phase inversion possesses several variants such as captivate precipitation, immersion precipitation, and thermo-induced phase separation [6]. The most simple and widely used phase inversion technique is the dry wet technique invented by Loeb and Sourirajan [7]. This involves solution casting with controlled thickness followed by partial evaporation to produce a thin film. Subsequently, it is immersed in a nonsolvent filled coagulation bath where the solidification process of the membrane is completed.

5.2.2 Interfacial polymerization

This method of membrane synthesis is focused on building a layer-by-layer deposition of thin films following a step-growth polymerization pathway. Some of the frequently used monomers include water-soluble m-phenylenediamine, trimesoyl (benzene-1,3,5-tricarbonyl) chloride and isophthaloyl chloride [8]. Most of the thin film composite reverse osmosis and nanofiltration membranes are synthesized using the interfacial polymerization technique. The chief advantage is the ability of film thickness control by controlling the rate of impregnation in these polycondensation polymerization reactions. Since the process occurs at the phase boundaries, it ensures good control of the membrane porosity distribution and thereby the membrane performance [9,10]. There are several variants in this technique depending on the method, phase, and reaction mechanisms such as in situ techniques, vapor-phase interfacial polymerization, free radical propagations, etc.

5.2.3 Track-etching

The crux of the process focuses on chemically controlling the pore formation and the peripheral structural distribution following a controlled template through ion tracking and chemical etching. Here, a thin polymer film is irradiated with nuclear fission particles that produce a defined path. This film is then subjected to solution polymerization, which preferentially creates pores along the susceptible path produced by the bombardment of nuclear fission particles [11]. It allows the flexibility of synthesizing functional templates that can be repeated [12]. Owing to the precise control over the microstructure of the membrane offered by the track-etching technique, this is being used for the production of nanostructured membranes [13].

5.2.4 Electrospinning

This technique is specially used for producing nanofibrous membranes. Here, the polymeric solution is subjected to high voltage, which breaks the cohesive surface bonds and converts the polymer into fine drums. Depending on the target pore size of the final substrate, the voltage and the process conditions are controlled to yield microporous, ultraporous, or nanoporous membranes. Under the traction between the polymer and the externally applied electric field, the spherical shape of the droplets changes into a conical form called the Tyler cone [14].

Due to the control over the pore shape, membranes with accurately distributed cylindrical or conical pores can be created. This provides selectivity to the membranes in the separation process. The 3D structure of the membranes can be tweaked by introducing dopants or by polymer grafting to impart properties desirable for any particular applications such as desalination.

5.3 Classification of different forms of nanomaterial-functionalized membranes

5.3.1 Conventional asymmetric nanocomposite

These membranes adhere to the structure commonly found in all heterogeneous membranes where a thin selective layer on top is sandwiched with a thicker, porous support layer at the bottom. The thin permeable top layer is often synthesized by phase inversion. Integrally asymmetric nanocomposite membranes find widespread commercial use due to their comparatively lower production costs [15]. Owing to the selectivity, temperature and pH resistances, wide pore size distribution and the ability to sustain nanoparticle impregnations for specialized usages, these membranes are often used in gas separation and CO_2 enrichment [16]. Among the different classes, polyethersulfone blends are the most commonly used. This allows easy blending with nanowalled composites with high selectivities by coating with PDMS, which imparts high permeabilities in the range of 9 GPU [17,18].

Fig. 5.2 depicts various structural modifications in the hybrid membranes.

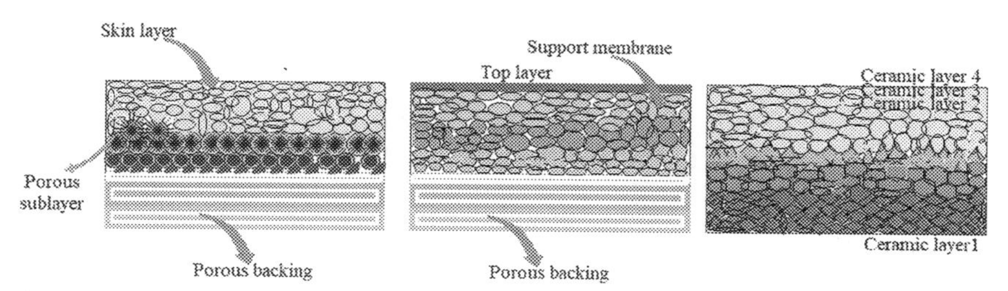

Fig. 5.2 Structural modifications in the hybrid membranes [19].

5.3.2 Thin film nanocomposite

The crux of this technology focuses on embedding nanoparticles to make the thin PA barrier layer more functional and to increase its degree of adhesion with the porous support layer. The thin film nanocomposite (TFN) membranes have been extensively studied for the removal of recalcitrant components present in wastewater such as dissolved organics, endocrine disruptors, and heavy metals [20–22]. Owing to the hydrophilicity and semipermeability, TFN membranes are also being researched upon for forward osmosis applications where the separation from the feed has taken place under the concentration gradient imposed by a draw solution [23].

Some newer variants (TFNi) focus on the fabrication of TFN membranes where nanoparticles are disposed in a separate layer between the barrier layer and the support layer. This layer called the interlayer comprises of single or more layers that aid in separation and have demonstrated superior separation flux values [24]. The TFNi membranes have also demonstrated reduced energy consumption and as well as decreased the values of the structural parameters by the introduction of a hydrophilic interlayer [25,26].

5.3.3 Thin Film Composite (TFC) with nanocomposite substrate

The third category comprises of thin-film membranes with nanocomposite substrates. Here the active surface of the membrane is synthesized in the conventional TFC fabrication. It is then dispersed on a support containing a nanocomposite substrate. Owing to the simplicity of fabrication, in recent years this has received widespread attention. The structural modification of support geometry with nanocomposites has shown marked improvement in terms of process performance such as reduction in internal concentration polarization (ICP) with a polyacetate TFC active surface and novel polysulfone nanocomposite substrate doped with a 0.05% zeolite loading [27]. Further modification of the structure of support has also been explored.

Dual casting to disperse graphene oxide in different polysulfone has imparted benefits of porosity, permeability, and permeability enhancement with a simultaneous reduction of ICP values by reduction in the mass transfer resistance [28]. Research in the field of modification of membrane support has highlighted the role of the support layer in the hydrodynamics and mass transfer of the separation process apart from providing a structural framework to the selective active layer. The incorporation of TiO_2 nanoparticles in the polysulfone support layer has significantly improved the permeate flux vis-à-vis TFC membranes sans the nanocomposite substrate [29].

5.3.4 Surface-located nanocomposite

This is another form of surface modification of the selective, active layer. Here, small amounts of nanocomposites are grafted onto the surface layer. In applications such as photocatalysis, this often proves to be economical as most of the photocatalytic nanomaterials located in the bulk matrix are often impervious to the light source. The incorporation of the nanocomposite on the surface greatly enhances the interfacial surface area and the sorption characteristics.

5.4 Diverse applications of functionalized nanomaterials

5.4.1 Pressure Retarded Osmosis (PRO)

The crux of the process follows the forward osmosis principles where a semipermeable membrane separates the feed and the draw solutions under the influence of the concentration gradient; the solvent moves from the feed side to the draw solution side. As a result, the volume of the DS increases. If the volume of the DS chamber is kept constant, then the fresh onslaught of water creates a force inside the fixed volume chamber. This force is harnessed to drive a turbine and generate electricity. This promising process of pressure-retarded osmosis can be harnessed to generate electricity from natural salinity gradients in estuaries [30]. The process can also serve the dual purpose of desalination with simultaneous energy production. However, despite the promising applications of this process, the main stumbling block in its practical application is the energy density and cost-effectiveness of the PRO membranes that are currently available [31]. Another drawback with the currently available TFC membranes used for osmotic energy generation is the high propensity of the internal concentration polarization, which causes flux decline. Herein, steps in the functionalized nanomaterials with a thinner and highly selective PA layer decreases the ICP and also ensures more structural cohesiveness leading to higher separation fluxes [32]. Surface-modified TFC membranes have also shown higher power densities of up to 3.9 w/m^2, whereas zeolite-impregnated TFC membranes have shown satisfactory performance while working with DS in the high concentration of 2M [33,34]. The use of TFN has shown up to 2.5 times increment in the flux values as compared to conventionally available TFC membranes [27]. Another problem faced by the conventional PRO membranes using TFC-RO is the problem of compaction and mechanical stability under high transmembrane pressure. Using the TFN membranes synthesized from tiered polyetherimide nanofiber and interspersed with CNTs helped to increase the pressure-bearing capacity of the PRO membranes up to 24 bars [35].

5.4.2 Application in oily wastewater treatment

Oily wastewater treatment is a prominent research challenge posed by the effluents generated from petroleum processing plants in particular and also includes the effluents produced with the oil industry in general. Several techniques have been used for mitigation of oily wastewater such as adsorption and nanomaterial-mediated chemical separations, among others Anjum et al. [36]. In this domain, mined matrix membrane, where nanofillers are embedded in the polymeric matrix increases the selective oil adsorption capacity of the membrane manifolds [37]. Incorporation of nanomaterials allows the flexibility of imparting desired properties such as fouling resistance, selectivity, hydrophilicity, etc., to the membranes. Surface-modified nanomembranes with improved internal networks and higher porosity can be specifically designed to address the problem of oily wastewater treatment. Newer research efforts in this direction have given rise to the fusion of photocatalytic TFN membranes for the remediation of oily wastewaters. Photocatalysis helps in the selective breakdown of the organics. The presence of the catalysts in the form of nanoparticles greatly enhances the effectiveness of the reaction. The entire porous structural framework reinforced by zeolites and alumina increases the flux and the separation of the membranes. Conventionally available organic-degrading photocatalysts such as Degussa P_{25} TiO_2 when impregnated in TFN membranes aids in oily WWT with a simultaneous photocatalyst recovery while improving the antifouling properties of the membrane leading to a win-win-situation [38]. Specially designed nanoparticles possessing magnetic properties can be synthesized for this purpose, which will impose attraction. This will facilitate coalescence and eventual precipitation of the oil droplets; other properties such as pH sensitivity, self-cleaning properties, and superoleophobicity can be imparted to the nanoparticles for aiding the process of separation [39,40].

In the existing nanocomposite membranes, special hydrophilicity can be introduced for the greater rejection of hydrophobic oily components. Similarly, additives such as ZnO and SiO_2 can be impregnated on the nanocomposite PES membranes for improvement of the permeate flux and throughput from the membrane [41].

5.4.3 Organic solvents recovery

Organic solvent recovery is a panoptic term encompassing diverse solvents that are generated during the course of chemical and organic compound synthesis. Many of the organic solvents that are used, if recovered viably, can be reused saving expenditure as well as pollution loadings. Furthermore, many organic solvents contain valuable compounds, active pharmaceutical ingredients and other components which, if recovered profitably, can be of great value addition [42]. Membrane technology has been extensively researched upon as a sustainable option for organic solvent recovery. Nanofiltration and RO membranes have also been studied for selective recovery of value-added products as well as recovery of organic solvents [43]. Notwithstanding all the advantages offered by membrane technology in this regard, one key disadvantage is the high fouling rate and concentration polarization that organic solvent separation entails [44]. This problem is ameliorated using nanocomposite membranes with fouling resistant properties [45]. One commonly used nanofiller substrate for polymeric solvent recovery membranes is graphene oxide due to its molecular sieving properties and structural cohesiveness [46]. Tweaking the molecular-level chemistry, graphene oxide-integrated membranes can be structurally modified to promote the size-selective separation of smaller molecules [47]. Another problem of long-term structural stability of polymeric membranes for prolonged exposure to organic solvents is also overcome using graphene oxide. Another class of dopants that are being researched upon as additives in the top selective layer of nanocomposite membranes are metal-organic frameworks (MOFs) [48]. The addition of MOFs greatly enhances the porosity and the surface area per unit volume, thereby increasing the overall permeate flux from the membranes. The predominant transport mechanism governing the mass transfer process is pore diffusion and interaction between the MOF and selective PA layers that regulates the solution diffusion mechanism [49]. However, the process also often mandates a modification process for better compatibility between the MOF additive and the base polymer matrix [50].

5.4.4 Development of photocatalytic membranes for removal of organic pollutants

Photocatalytic properties, when combined with highly selective thin film nanocomposite membranes, generated integrated, process-intensified units that can serve as complete effluent treatment technology solutions. When photocatalyst-embedded nanocomposite membranes are used for wastewater treatment, a combination of physical and chemical separation takes place. There is a selective physical exclusion of the organics accompanied by a simultaneous degradation of the organic compounds. This ensures that no toxic residues are generated during the process and, in a way, it provides a viable solution to deal with the problem of membrane concentrates from the process. Photocatalysts produce oxidizing radicals such as hydroxyl radicals, superoxides as well as they initiate complete mineralization pathways to break down the complex organic compounds into benign residues such as water and [51]. TiO_2 is the most ubiquitously used photocatalyst for this purpose along with several classes of TiO_2 dopants such as Ag-doped TiO_2, TiO_2 nanocomposites, La^{3+}-TiO_2 suspensions, and Fe-doped TiO_2 dioxide for the breakdown of different recalcitrant organic dyes such as Cibacron yellow, methyl red, basic violet 3, Reactive Orange 16, and Acid Blue 80 among others [52–55].

Photocatalysts have also proven to be effective in the breakdown of other organic pollutants such as endocrine-disrupting compounds, active pharmaceutical ingredients, and pharmaceutical and personal care products [56,57]. With the incorporation of nanoscopic photocatalysts in the active layer of the membranes, the 3D morphology of the membranes has confirmed marked improvements, such as the formation of a selective layer near the phase boundary or finger-like morphology through the membrane structure [58]. Several techniques such as blending of photocatalysts with membrane matrix dopants, immobilization of photocatalyst, and selective deposition of the photocatalyst onto the membrane surface have been used for the preparation of such composite membranes [59]. Research efforts directed toward increment of the photocatalytic reaction efficiency focused on concentrating more photocatalyst on the membrane surface by the development of dual-layer membranes. Here, the top layer acts as the active photocatalytic and separation layer, while the lower layer acts as a filter and structural support layer [60].

This allows the photocatalyst to be preferentially concentrated on the upper circumference rather than being uniformly distributed in the entire polymer blend. The inner circumference impermeable to incident light contains lower concentrations of photocatalysts, thus reducing loss to catalytic activity due to lack of light penetration. Apart from this, the overall separation efficiency and flux also greatly enhance due to the formation of the dual-layer mined matrix membranes. Photocatalysis-integrated nanocomposite membranes have shown great promise toward mitigation of organic pollutants as well as toward zero liquid discharge technology development. However, more focused research is needed for improving photocatalytic efficiency and for preventing of bandgap reduction and pilot level studies for this technology should to be commercially carried out.

5.5 Removal of different contaminants with functionalized nanomaterial membranes

5.5.1 Dissolved organics

Removal of nanomaterial-based membranes is emerging as one of the promising tools for the removal of organic compounds such as dyes, phenolics, and also naturally dissolved organic matter (NOM) [61]. The ability to incorporate charged nanomaterials ensures electrostatic interactive forces between the membrane surface and the polar functional groups associated with dyes and analogous compounds leading to their effective removal [62]. The colloids and organics are also adsorbed effectively onto the surface of the graphene oxide and CNT membranes. The multiwalled and hexagonal matrix increases the surface area, aiding in faster removals. This makes it possible for water-soluble dye removal [63,64]. The charged based and internal molecular size exclusion has been utilized in the synthesis of nanofabricated porous nanocrystalline silicon for selective protein and other selective macromolecular retention [65]. The incorporation of metallic organic frameworks also enlarges the structural matrix that helps in high dye rejections of 91% along with increasing the structural stability of the membranes [25,26]. TFN membranes with surface modification by MOF particles leads to few-defect-membrane syntheses and has been effectively used in organics separation from pharmaceutical application demonstrating a 91% tetracyline rejection [20,22]. Incorporation of nanocomposite grafts and Nafion-modified UF membranes displaying a mixed matrix template have also revealed 90% rejection of methyl orange [66]. The table lists some of the recalcitrant dye removals with FN membranes.

One of the new classes of membrane materials in this domain is organic-inorganic nanohybrids. Integration of flexible graphene oxide with polyelectrolyte complexes has led to the production of superior pervaporation nanohybrid membranes with FO-improved ethanol/water separation apart from dye and inorganic ion rejection [67]. The Donan separation mechanism combined with adsorptive removal in alumina nanoparticle-embedded mixed matrix membrane has demonstrated the broad-spectrum removal of different phenolic compounds such as catechol, paranitrophenol, phenol, orthochloro, and orthonitro phenols [68]. The addition of alumina nanoparticles even in small amounts has been proved to increase the selectivity, strength, and permeability in the cellulose acetate matrix as well as in hydrophobic polymers such as PVDF [69,70]. Surface modification of polymeric membranes with graphene oxide and TiO_2 nanoparticles imparts additional properties such as photocatalytic potential, which increases the removal of organic residues such as m-p nitrophenol [71].

These photocatalytic membranes can be reused and coupled with renewable sources such as solar photoreactors for the breakdown of complex organics such as tartrazine. This makes the entire operation highly sustainable while ensuring complete degradation of the contaminants. Similarly, immobilization of TiO_2 in zeolite and ceramic materials has also been studied for combining photocatalytic and membrane filtration [72]. Novel research developments in the field have also yielded ultrafast separation membranes for increasing the separation process [73].

5.5.2 Heavy metal removal

The electrostatic charge distribution and zero potential of the nanomembranes serves to be effective for size-exclusion and adsorptive removal of heavy metals. The occurrence of heavy metals from industrial wastewater as well as surface and groundwater poses an environmental challenge due to the high eco-toxicity associated with them. The presence of heavy metals in trace amounts also gives rise to issues of biomagnification, bioaccumulation, and metal poisoning [74,75]. Hence, heavy metal remediation is a pressing problem. Their dilute concentrations make their selective removal difficult. Conventional treatment techniques such as adsorption, coagulation, and ion exchange have demonstrated heavy metal removal efficiency. However, the aforesaid techniques often transfer the heavy metals from one phase to another. The sludge/concentrates necessitate passivation and often give rise to secondary contamination. The AOP's offer a viable solution by degradation and when combined with membrane separation, it possesses the potential to be developed into a complete treatment option.

Using functionalization, commonly available ultrafiltration membranes and nanocomposite membranes can be integrated into the PES, polymer backbone. Such mixed matrix membranes become capable of 99.6% lead removal. Interestingly PES membranes are embedded with Fe-Ni heavy metal nanoparticle metal-adsorbing properties [76].

The polymeric PVDF membranes on functionalization with ferric compounds such as ferric sulfide, ferroferric oxide develop improved hydrophilicity, uptake capacity, and salt rejection properties. This makes this capable of broad-spectrum heavy metal removal such as lead (Pb^{2+}), cadmium (Cd^{2+}) and chromium (Cr^{3+}), and arsenic (As) in a single-stage separation process from both industrial and groundwater (concentrated and diffuse sources) [77]. Nanocomposite membranes also facilitate practical implementations of parallel novel separation techniques, for instance, functionalization of forward osmosis membranes with graphene oxide and polyethylene glycol demonstrated improved hydrophilicity and reduction in reverse salt diffusion through the FO membranes. The resulting mixed matrix membranes demonstrated 99.9%, 99.7%, and 98.3% removal efficiencies of toxic heavy metals lead, cadmium, and chromium, respectively [78]. Similarly, surface functionalization of polyester-imide nanofiltration membranes with ZnO nanoparticles not only improved mechanical strength and antifouling resistance but also demonstrated more than 80% removal of monovalent and divalent salts such as Na_2SO_4 and $Pb(NO_3)_2$ [79].

The functionalization offers the flexibility of the addition of desirable properties to suit the process needs. The usage of nanocomposite membranes for heavy metal removal has proved to improve the organic and biofouling properties by grafting specific polymers such as piperazine amide in the membranes. The properties of conventionally used membranes can also be modified, which adds to the cost-effectiveness of the process. Microfiltration membrane geometry can be treated for synthesizing nanofibrous composite MF membranes.

Cross-linking polyvinyl alcohol and polyacrylonitrile electrospun fibers on a glutaraldehyde base yielded charge separation that enabled a highly permeable and selective microfiltration membrane for chromium and cadmium ions removal from water [80]. It is also possible to simultaneously remove heavy metals and other organic contaminants by the modification of a nanofiltration membrane with magnetic graphene particles [81].

5.5.3 Microbial content mitigation

Membrane technology, especially nanofiltration and RO membranes, have displayed the removal of microbial content by size exclusion [82,83]. However, that is not often sufficient to destroy all viruses, bacteria, and other pathogens [84]. Besides, it also leads to biofouling which destroys membrane performance by affecting selectivity and separation flux [85]. TiO_2 and Ag nanoparticles possess well-established bactericidal and antifungal characteristics and cause complete degradation of the microbial content. Hence, incorporation of such functionalized nanomaterials in the membranes creates a win-win situation for oxidation, degradation kinetics, and ultimate size- and charge-based exclusion of microbial and pathogenic content from the effluents and surface and groundwater. This parameter is of prime importance to ascertain the drinkability of surface/groundwaters. The presence of any fecal coliform/total coliforms, viruses, other bacteria, amoeba, and allied species of microorganisms leads to a host of water-borne diseases [86,87]. Hence, treatment of microbial content becomes absolutely necessary.

There are different techniques for the development of nonfunctionalized membrane synthesis membranes for this purpose. Nanoparticles can be surface-doped, blended in a casting solution, or copolymerized [88,89]. Copper nanoparticles synthesized with a polyethene-imine as a capping agent demonstrated antimicrobial activity by preventing microbial adhesion to the membrane surface in a cellulose acetate RO membrane [90]. Similar antibacterial resistance can also be generated by amide polymerization between graphene and carboxylic acid groups tweaking the active layer of the membrane [91]. These nanosheets with surface functionalization have also proved to be cost-effective by reducing the total nanomaterial requirement. The nanomaterials directly rupture the microbial cell membrane or generate active species that affects the growth of the microbial cells leading to its dehydration and degradation [92]. There are several mechanisms of functioning such as cleaning of DNA, changes in DNA patterns and nucleic acid sequencing, and active lipid-based peroxidases, which are responsible for the breakdown of the cells [93]. Nanomaterials of metals such as iron, silver, copper, gold, and their oxides generate reactive oxidation species (ROS) that disrupts the RNA and DNA and disintegrate the structural framework of the cells [94]. Another pathway of microbial degradation includes deactivation of microbial enzymes and deactivating them, disturbing the e-transport and thereby cutting off the energy supply to these cells, which effectively prevents their growth and proliferation [95].

5.6 Pilot studies and industrial applications

The promising genre of nanocomposite membranes is one of the most researched areas, with a host of high impact factor research publications made annually. However, a limited number of industrial trials have been undertaken to date. Still less are commercial applications. Nevertheless, the research wings of one prominent chemical and membrane company BASF have innovated upon the probable practical implementation of nanocomposite membranes [96]. Industrially, LG Chemicals have launched a range of water treatment filters where thin-film nanocomposite (TFN) RO membranes are being used for seawater desalination [97]. The presence of TFN in the active layer of the membranes has increased salt rejection in the range of 99% and high separation flux. This bears testimony to the efficacy and selectiveness of TFN membranes for the conversion of brackish water into drinking water. The American startup nano Inc has also successfully undertaken a real-life performance test at port Hueneme United States navy facility of a specially designed TFN element. Following satisfactory process performance, a commercial venture under the name Quantum Flux was launched by the company [98]. Before commercial level applications, 400 ft^2 TFN elements were also tested for recovering water from the membrane concentrates at the Lahat water facility. This was followed up by independent third-party testing of the efficacy of the membranes by Avista Technologies in San Mareos, California. This also established competitive desalination performance by the TFN RO membranes [99]. Pilot level comparative performance analysis with 0.01% TiO_2-integrated PA TFN membranes and PA TFC membranes revealed competitive total organic carbon removal from surface water accompanied by steadily increasing values of permeate flux with time [100].

5.7 Possibility of value-added product recovery

Recovery of value-added products from wastewaters represents a sustainable and economically attractive strategy of valorization. It simultaneously reduces effluent loadings, reducing the complexity of the treatment process. Different components considered as contaminants such as heavy metals, organic residues active pharmaceutical ingredients, brine, etc. serve as valuable feedstocks for a host of processes, if recovered profitably. This approach, apart from promoting eco-friendliness of the process, also promotes resource recovery. The trade-off's incurred in terms of CAPEX and OPEX for functionalized nanomaterial fabrication; the process of effluent treatment can be offset by the value addition from the recovered contaminants.

Membrane technology is already being researched upon for value-added product recovery from effluents and contaminated wastewaters. Novel technologies such as membrane crystallization are being tested on pilot and semi-commercial levels for Na_2SO_4 and Na_2CO_3 recovery from brines, struvite synthesis, and lithium chloride extraction from waste streams [42]. Wastewaters containing ammonium, nitrate, phosphate, brine, high BOD, and COD are being harnessed for slow-release fertilizer formation, isolation of valuable chemicals and energy production [101–103]. Specific hydrophobic membrane contactors enable phase separation and selectivity and have been used to recover valuables from liquid and gaseous streams [104].

The selectivity and yield of the target products can be enhanced by combining the high reactivity and specificity offered by the nanomaterials. Membranes with chelating ligands have been utilized for selective and isolation recovery of gold, uranium, and rare earths [105–107]. Incorporation of CNTs with membrane polymers and/or ion-specific ligands grafting into base membrane matrix can assist in selective extraction of the molecular sieving and ion exchange potential of nanocomposite membranes with polymer/ceramic/MOF integrations also offers the porosity, surfaces area, and selectivity for value-added product recovery [108].

Commonly used nanofiltration membranes can be modified to increase the MWCO values and nanoparticles such as graphene oxide nanocomposites grafted on the surface for greater selectivity and chemical resistance [109]. Table 5.1 describes valuable component recovery.

TFN membranes fabricated by embedding of amino-functionalized nanoparticles in a base of asymmetric polyethimide-modified SiO_2 substrate demonstrated recovery of dewaxing solvents at a low dosage of FN of 0.1 wt./vol.% of UZM in the polyamide active membrane layer of the organic solvent NF membranes demonstrated superior oil rejection and selectivity, which in turn allowed its recovery [116]. Enhancement of the resource recovery aspect of advanced separation processes such as pervaporation has also been established with the use of MOF nanocomposite hybrid membranes. An innovative casting of ZIF-based silicon rubber nanocomposite on an SS mesh frame developed into a highly efficient pervaporation membrane for the recovery of a valuable furfural molecule useful in biorefinery and biochemical applications [112]. In the domain of gas separation also, nanocomposite membranes blended with nanofillers, the carbon-zeolite assembly has demonstrated recovery of syngas and H_2 [117].

Table 5.1 Value added product recovery from wastewater streams.

Sr. no	Recovered component	Membrane type	Feed	Membrane separation process	References
1.	Ethanol, methanol	Polydimethylsiloxane-silica nanocomposite membranes	Alcohol-water mixture	Pervaporation	[110]
2.	Ethanol	Cellulose nanofiber nanocomposite	Ethanol-water mixture	Pervaporation	[111]
3.	Furfural	Metal-organic framework ZIF-8 nanocomposite membrane	Slurry	Pervaporation	[112]
4.	Iso-butanol	Capillary supported ultrathin silicalite-poly(dimethylsiloxane) nanocomposite	The aqueous solution of iso-butanol	Fermentation-pervaporation	[113]
5.	Phosphorus	Tetraamminezinc complex integrated nanocomposite network	Simulated wastewater solution: a mixture of zinc nitrate and ammonia solution	Nanofiltration	[114]
6.	Dye: sunset yellow	Tannic acid iron functionalized thin film composite	Brackish/salty water	Forward osmosis	[115]

These applications bear testimony to the efficacy in the recovery of commercially valuable components from wastewater. However, pilot-scale studies are required at a large scale to establish process scalability and techno-economic viability.

References

[1] M.A.A. Shahmirzadi, A. Kargari, Nanocomposite membranes, in: Emerging Technologies for Sustainable Desalination Handbook, 2018, pp. 285–330, https://doi.org/10.1016/b978-0-12-815818-0.00009-6.

[2] H. Al-Kayiem, S. Lin, A. Lukmon, Review on nanomaterials for thermal energy storage technologies, Nanosci. Nanotechnol. Asia 3 (2013) 60–71, https://doi.org/10.2174/2211352511313119990011.

[3] S. Raghav, R. Painuli, D. Kumar, Multifunctional nanomaterials for multifaceted applications in biomedical arena, Int. J. Pharmacol. 13 (2017) 890–906, https://doi.org/10.3923/ijp.2017.890.906.

[4] D. Teleanu, C. Chircov, A. Grumezescu, R. Teleanu, Neurotoxicity of nanomaterials: an up-to-date overview, Nano 9 (2019) 96, https://doi.org/10.3390/nano9010096.

[5] R.W. Baker, Membranes and Modules, Chapter 3: In Membrane Technology and Applications, Wiley, Chichester, England, 2004, pp. 89–155.

[6] X. Qu, P.J. Alvarez, Q. Li, Applications of nanotechnology in water and wastewater treatment, Water Res. 47 (2013) 3931–3946, https://doi.org/10.1016/j.watres.2012.09.058.

[7] S. Loeb, S. Sourirajan, High Flow Porous Membranes for Separation of Water from Saline Solutions, 1964. US patent 3,133,132.

[8] Z. Yong, Y. Sanchuan, L. Meihong, G. Congjie, Polyamide thin film composite membrane prepared from m-phenylenediamine and m-phenylenediamine-5-sulfonic acid, J. Membr. Sci. 270 (2006) 162–168, https://doi.org/10.1016/j.memsci.2005.06.053.

[9] M. Adamczak, G. Kamińska, J. Bohdziewicz, Preparation of polymer membranes by in situ interfacial polymerization, Int. J. Polym. Sci. 2019 (2019) 1–13, https://doi.org/10.1155/2019/6217924.

[10] Y. Li, Y. Su, Y. Dong, X. Zhao, Z. Jiang, R. Zhang, J. Zhao, Separation performance of thin-film composite nanofiltration membrane through interfacial polymerization using different amine monomers, Desalination 333 (2014) 59–65, https://doi.org/10.1016/j.desal.2013.11.035.

[11] P. Apel, Track Etching Technique in Membrane Technology, 2001, [WWW Document]. RadiationMeasurements.URL https://www.sciencedirect.com/science/article/pii/S1350448701002281. (Accessed 8 February 2021).

[12] S. Chakarvarti, J. Vetter, Template synthesis a membrane-based technology for generation of nano −/micro materials: a review, Radiat. Meas. 29 (1998) 149–159, https://doi.org/10.1016/s1350-4487(98)00009-2.

[13] L.D.-D. Pra, E. Ferain, R. Legras, S. Demoustier-Champagne, Fabrication of a new generation of track-etched templates and their use for the synthesis of metallic and organic nanostructures, Nucl. Instrum. Methods Phys. Res., Sect. B 196 (2002) 81–88, https://doi.org/10.1016/s0168-583x(02)01252-1.

[14] R.T. Collins, J.J. Jones, M.T. Harris, O.A. Basaran, Electrohydrodynamic tip streaming and emission of charged drops from liquidcones, Nat. Phys. 4 (2007) 149–154, https://doi.org/10.1038/nphys807.

[15] P. Vandezande, L.E.M. Gevers, I.F.J. Vankelecom, Solvent resistant nanofiltration: separating on a molecular level, Chem. Soc. Rev. 37 (2008) 365–405, https://doi.org/10.1039/b610848m.
[16] H. Adib, S. Hassanajili, D. Mowla, F. Esmaeilzadeh, Fabrication of integrally skinned asymmetric membranes based on nanocomposite polyethersulfone by supercritical CO2 for gas separation, J. Supercrit. Fluids 97 (2015) 6–15, https://doi.org/10.1016/j.supflu.2014.11.001.
[17] T.D. Kusworo, B. Budiyono, Enhanced biogas separation performance of nanocomposite polyethersulfone membranes using carbon nanotubes, Waste Technology 1 (2013), https://doi.org/10.12777/wastech.1.2.10-14.
[18] S. Saedi, S.S. Madaeni, K. Hassanzadeh, A.A. Shamsabadi, S. Laki, The effect of polyurethane on the structure and performance of PES membrane for separation of carbon dioxide from methane, J. Ind. Eng. Chem. 20 (2014) 1916–1929, https://doi.org/10.1016/j.jiec.2013.09.012.
[19] L. Cseri, T. Fodi, J. Kupai, G.T. Balogh, A. Garforth, G. Szekely, Membrane-assisted catalysis in organic media, Adv. Mater. Lett. 8 (2017) 1094–1124, https://doi.org/10.5185/amlett.2017.1541.
[20] H. Guo, Y. Deng, Z. Tao, Z. Yao, J. Wang, C. Lin, T. Zhang, B. Zhu, C.Y. Tang, Does hydrophilic polydopamine coating enhance membrane rejection of hydrophobic endocrine-disrupting compounds? Environ. Sci. Technol. Lett. 3 (2016) 332–338, https://doi.org/10.1021/acs.estlett.6b00263.
[21] H. Guo, Z. Yao, Z. Yang, X. Ma, J. Wang, C.Y. Tang, A one-step rapid assembly of thin film coating using green coordination complexes for enhanced removal of trace organic contaminants by membranes, Environ. Sci. Technol. 51 (2017) 12638–12643, https://doi.org/10.1021/acs.est.7b03478.
[22] X. Guo, D. Liu, T. Han, H. Huang, Q. Yang, C. Zhong, Preparation of thin film nanocomposite membranes with surface modified MOF for high flux organic solvent nanofiltration, AICHE J. 63 (2016) 1303–1312, https://doi.org/10.1002/aic.15508.
[23] W. Zhao, H. Liu, Y. Liu, M. Jian, L. Gao, H. Wang, X. Zhang, Thin-film nanocomposite forward-osmosis membranes on hydrophilic microfiltration support with an intermediate layer of graphene oxide and multiwall carbon nanotube, ACS Appl. Mater. Interfaces 10 (2018) 34464–34474, https://doi.org/10.1021/acsami.8b10550.
[24] S. Karan, Z. Jiang, A.G. Livingston, Sub-10 nm polyamide nanofilms with ultrafast solvent transport for molecular separation, Science 348 (2015) 1347–1351, https://doi.org/10.1126/science.aaa5058.
[25] Z. Yang, P.-F. Sun, X. Li, B. Gan, L. Wang, X. Song, H.-D. Park, C.Y. Tang, A critical review on thin-film nanocomposite membranes with interlayered structure: mechanisms, recent developments, and environmental applications, Environ. Sci. Technol. 54 (2020) 15563–15583, https://doi.org/10.1021/acs.est.0c05377.
[26] G. Yang, D. Zhang, G. Zhu, T. Zhou, M. Song, L. Qu, K. Xiong, H. Li, A Sm-MOF/GO nanocomposite membrane for efficient organic dye removal from wastewater, RSC Adv. 10 (2020) 8540–8547, https://doi.org/10.1039/d0ra01110j.
[27] N. Ma, J. Wei, S. Qi, Y. Zhao, Y. Gao, C.Y. Tang, Nanocomposite substrates for controlling internal concentration polarization in forward osmosis membranes, J. Membr. Sci. 441 (2013) 54–62, https://doi.org/10.1016/j.memsci.2013.04.004.
[28] S. Lim, M.J. Park, S. Phuntsho, L.D. Tijing, G.M. Nisola, W.-G. Shim, W.-J. Chung, H.K. Shon, Dual-layered nanocomposite substrate membrane based on polysulfone/graphene oxide for mitigating internal concentration polarization in forward osmosis, Polymer 110 (2017) 36–48, https://doi.org/10.1016/j.polymer.2016.12.066.
[29] T. Sirinupong, W. Youravong, D. Tirawat, W. Lau, G. Lai, A. Ismail, Synthesis and characterization of thin film composite membranes made of PSF-TiO_2/GO

nanocomposite substrate for forward osmosis applications, Arab. J. Chem. 11 (2018) 1144–1153, https://doi.org/10.1016/j.arabjc.2017.05.006.
[30] S.K. Elikaiy, K. Lari, M.T. Azad, A.S. Jahromi, A.M. Arasteh, Investigation and evaluation of salinity gradient power in Arvand River estuary using pressure retarded osmosis (PRO) method, Int. J. Environ. Sci. Technol. 18 (2020) 463–470, https://doi.org/10.1007/s13762-020-02993-6.
[31] N. Alzainati, H. Saleem, A. Altaee, S.J. Zaidi, M. Mohsen, A. Hawari, G.J. Millar, Pressure retarded osmosis: advancement, challenges and potential, J. Water Process Eng. 40 (2021) 101950, https://doi.org/10.1016/j.jwpe.2021.101950.
[32] J. Kim, K. Jeong, M. Park, H. Shon, J. Kim, Recent advances in osmotic energy generation via pressure-retarded osmosis (PRO): a review, Energies 8 (2015) 11821–11845, https://doi.org/10.3390/en81011821.
[33] P.G. Ingole, K.H. Kim, C.H. Park, W.K. Choi, H.K. Lee, Preparation, modification and characterization of polymeric hollow fiber membranes for pressure-retarded osmosis, RSC Adv. 4 (2014) 51430–51439, https://doi.org/10.1039/c4ra07619b.
[34] T.M. Salehi, M. Peyravi, M. Jahanshahi, W.-J. Lau, A.S. Rad, Impacts of zeolite nanoparticles on substrate properties of thin film nanocomposite membranes for engineered osmosis, J. Nanopart. Res. 20 (2018), https://doi.org/10.1007/s11051-018-4154-1.
[35] M. Tian, R. Wang, K. Goh, Y. Liao, A.G. Fane, Synthesis and characterization of high-performance novel thin film nanocomposite PRO membranes with tiered nanofiber support reinforced by functionalized carbon nanotubes, J. Membr. Sci. 486 (2015) 151–160, https://doi.org/10.1016/j.memsci.2015.03.054.
[36] M. Anjum, R. Miandad, M. Waqas, F. Gehany, M. Barakat, Remediation of wastewater using various nano-materials, Arab. J. Chem. 12 (2019) 4897–4919, https://doi.org/10.1016/j.arabjc.2016.10.004.
[37] O. Agboola, O.S.I. Fayomi, A. Ayodeji, A.O. Ayeni, E.E. Alagbe, S.E. Sanni, E.E. Okoro, L. Moropeng, R. Sadiku, K.W. Kupolati, B.A. Oni, A review on polymer nanocomposites and their effective applications in membranes and adsorbents for water treatment and gas separation, Membranes 11 (2021) 139, https://doi.org/10.3390/membranes11020139.
[38] P. Goh, C. Ong, B. Ng, A.F. Ismail, Applications of emerging nanomaterials for oily wastewater treatment, in: Nanotechnology in Water and Wastewater Treatment, 2019, pp. 101–113, https://doi.org/10.1016/b978-0-12-813902-8.00005-8.
[39] T. Lü, S. Zhang, D. Qi, D. Zhang, G.F. Vance, H. Zhao, Synthesis of pH-sensitive and recyclable magnetic nanoparticles for efficient separation of emulsified oil from aqueous environments, Appl. Surf. Sci. 396 (2017) 1604–1612, https://doi.org/10.1016/j.apsusc.2016.11.223.
[40] S. Song, H. Yang, C. Zhou, J. Cheng, Z. Jiang, Z. Lu, J. Miao, Underwater superoleophobic mesh based on $BiVO_4$ nanoparticles with sunlight-driven self-cleaning property for oil/water separation, Chem. Eng. J. 320 (2017) 342–351, https://doi.org/10.1016/j.cej.2017.03.071.
[41] T.D. Kusworo, Qudratun, D.P. Utomo, Performance evaluation of double stage process using nano hybrid PES/SiO_2-PES membrane and PES/ZnO-PES membranes for oily waste water treatment to clean water, J. Environ. Chem. Eng. 5 (2017) 6077–6086, https://doi.org/10.1016/j.jece.2017.11.044.
[42] P. Das, S. Dutta, K.K. Singh, Insights into membrane crystallization: a sustainable tool for value added product recovery from effluent streams, Sep. Purif. Technol. 257 (2021) 117666, https://doi.org/10.1016/j.seppur.2020.117666.
[43] J.F. Kim, G. Szekely, M. Schaepertoens, I.B. Valtcheva, M.F. Jimenez-Solomon, A.G. Livingston, In situ solvent recovery by organic solvent nanofiltration, ACS Sustain. Chem. Eng. 2 (2014) 2371–2379, https://doi.org/10.1021/sc5004083.

[44] Z. Yin, Y. Ma, B. Tanis-Kanbur, J.W. Chew, Fouling behavior of colloidal particles in organic solvent ultrafiltration, J. Membr. Sci. 599 (2020) 117836, https://doi.org/10.1016/j.memsci.2020.117836.
[45] S. Abdikheibari, L.F. Dumée, V. Jegatheesan, Z. Mustafa, P. Le-Clech, W. Lei, K. Baskaran, Natural organic matter removal and fouling resistance properties of a boron nitride nanosheet-functionalized thin film nanocomposite membrane and its impact on permeate chlorine demand, J. Water Process. Eng. 34 (2020) 101160, https://doi.org/10.1016/j.jwpe.2020.101160.
[46] Y. Han, Y. Jiang, C. Gao, High-flux graphene oxide nanofiltration membrane intercalated by carbon nanotubes, ACS Appl. Mater. Interfaces 7 (2015) 8147–8155, https://doi.org/10.1021/acsami.5b00986.
[47] R. Ding, H. Zhang, Y. Li, J. Wang, B. Shi, H. Mao, J. Dang, J. Liu, Graphene oxide-embedded nanocomposite membrane for solvent resistant nanofiltration with enhanced rejection ability, Chem. Eng. Sci. 138 (2015) 227–238, https://doi.org/10.1016/j.ces.2015.08.019.
[48] S. Sorribas, P. Gorgojo, C. Téllez, J. Coronas, A.G. Livingston, High flux thin film nanocomposite membranes based on metal–organic frameworks for organic solvent nanofiltration, J. Am. Chem. Soc. 135 (2013) 15201–15208, https://doi.org/10.1021/ja407665w.
[49] C. Echaide-Górriz, S. Sorribas, C. Téllez, J. Coronas, MOF nanoparticles of MIL-68 (Al), MIL-101(Cr) and ZIF-11 for thin film nanocomposite organic solvent nanofiltration membranes, RSC Adv. 6 (2016) 90417–90426, https://doi.org/10.1039/c6ra17522h.
[50] P. Chuntanalerg, S. Bureekaew, C. Klaysom, W.-J. Lau, K. Faungnawakij, Nanomaterial-incorporated nanofiltration membranes for organic solvent recovery, in: Advanced Nanomaterials for Membrane Synthesis and its Applications, 2019, pp. 159–181, https://doi.org/10.1016/b978-0-12-814503-6.00007-0.
[51] A. Houas, Photocatalytic degradation pathway of methylene blue in water, Appl. Catal. B Environ. 31 (2001) 145–157, https://doi.org/10.1016/s0926-3373(00)00276-9.
[52] C.-Y. Chen, Photocatalytic degradation of azo dye reactive orange 16 by TiO_2, Water Air Soil Pollut. 202 (2009) 335–342, https://doi.org/10.1007/s11270-009-9980-4.
[53] A. Gupta, A. Pal, C. Sahoo, Photocatalytic degradation of a mixture of crystal violet (basic violet 3) and methyl red dye in aqueous suspensions using Ag doped TiO2, Dyes Pigments 69 (2006) 224–232, https://doi.org/10.1016/j.dyepig.2005.04.001.
[54] E. Pramauro, A.B. Prevot, M. Vincenti, G. Brizzolesi, Photocatalytic degradation of carbaryl in aqueous solutions containing TiO_2 suspensions, Environ. Sci. Technol. 31 (1997) 3126–3131, https://doi.org/10.1021/es970072z.
[55] X. Vargas, E. Tauchert, J.-M. Marin, G. Restrepo, R. Dillert, D. Bahnemann, Fe-doped titanium dioxide synthesized: photocatalytic activity and mineralization study for azo dye, J. Photochem. Photobiol. A Chem. 243 (2012) 17–22, https://doi.org/10.1016/j.jphotochem.2012.06.001.
[56] Y. Li, C. Zhang, Z. Hu, Selective removal of pharmaceuticals and personal care products from water by titanium incorporated hierarchical diatoms in the presence of natural organic matter, Water Res. 189 (2021) 116628, https://doi.org/10.1016/j.watres.2020.116628.
[57] J.-C. Sin, S.-M. Lam, A.R. Mohamed, K.-T. Lee, Degrading endocrine disrupting chemicals from wastewater by photocatalysis: a review, Int. J. Photoenergy 2012 (2012) 1–23, https://doi.org/10.1155/2012/185159.
[58] H. Sofiah, A. Nora'aini, M. Marinah, The influence of polymer concentration on performance and morphology of asymmetric ultrafiltration membrane for lysozyme separation, J. Appl. Sci. 10 (2010) 3325–3330, https://doi.org/10.3923/jas.2010.3325.3330.

[59] G.A. Kallawar, B.A. Bhanvase, Nanomaterial-based photocatalytic membrane for organic pollutants removal, in: Handbook of Nanomaterials for Wastewater Treatment, 2021, pp. 699–737, https://doi.org/10.1016/b978-0-12-821496-1.00007-6.
[60] H. Dzinun, M.H.D. Othman, A.F. Ismail, M.H. Puteh, M.A. Rahman, J. Jaafar, Fabrication of dual layer hollow fibre membranes for photocatalytic degradation of organic pollutants, Int. J. Chem. Eng. Appl. 6 (2015) 289–292, https://doi.org/10.7763/ijcea.2015.v6.499.
[61] Y. Ahn, D. Lee, M. Kwon, I.-H. Choi, S.-N. Nam, J.-W. Kang, Characteristics and fate of natural organic matter during UV oxidation processes, Chemosphere 184 (2017) 960–968, https://doi.org/10.1016/j.chemosphere.2017.06.079.
[62] V.K. Gupta, R. Jain, A. Nayak, S. Agarwal, M. Shrivastava, Removal of the hazardous dye—Tartrazine by photodegradation on titanium dioxide surface, Mater. Sci. Eng. C 31 (2011) 1062–1067, https://doi.org/10.1016/j.msec.2011.03.006.
[63] V.K. Gupta, R. Kumar, A. Nayak, T.A. Saleh, M. Barakat, Adsorptive removal of dyes from aqueous solution onto carbon nanotubes: a review, Adv. Colloid Interf. Sci. 193–194 (2013) 24–34, https://doi.org/10.1016/j.cis.2013.03.003.
[64] F. Liu, S. Chung, G. Oh, T.S. Seo, Three-dimensional graphene oxide nanostructure for fast and efficient water-soluble dye removal, ACS Appl. Mater. Interfaces 4 (2012) 922–927, https://doi.org/10.1021/am201590z.
[65] C.C. Striemer, T.R. Gaborski, J.L. Mcgrath, P.M. Fauchet, Charge- and size-based separation of macromolecules using ultrathin silicon membranes, Nature 445 (2007) 749–753, https://doi.org/10.1038/nature05532.
[66] S. Filice, D. D'Angelo, S. Libertino, I. Nicotera, V. Kosma, V. Privitera, S. Scalese, Graphene oxide and titania hybrid Nafion membranes for efficient removal of methyl orange dye from water, Carbon 82 (2015) 489–499, https://doi.org/10.1016/j.carbon.2014.10.093.
[67] N. Wang, S. Ji, G. Zhang, J. Li, L. Wang, Self-assembly of graphene oxide and polyelectrolyte complex nanohybrid membranes for nanofiltration and pervaporation, Chem. Eng. J. 213 (2012) 318–329, https://doi.org/10.1016/j.cej.2012.09.080.
[68] R. Mukherjee, S. De, Adsorptive removal of phenolic compounds using cellulose acetate phthalate–alumina nanoparticle mixed matrix membrane, J. Hazard. Mater. 265 (2014) 8–19, https://doi.org/10.1016/j.jhazmat.2013.11.012.
[69] N.M. Wara, L.F. Francis, B.V. Velamakanni, Addition of alumina to cellulose acetate membranes, J. Membr. Sci. 104 (1995) 43–49, https://doi.org/10.1016/0376-7388(95)00010-a.
[70] L. Yan, Y.S. Li, C.B. Xiang, Preparation of poly(vinylidene fluoride)(pvdf) ultrafiltration membrane modified by nano-sized alumina (Al2O3) and its antifouling research, Polymer 46 (2005) 7701–7706, https://doi.org/10.1016/j.polymer.2005.05.155.
[71] M.L. Tran, C.-C. Fu, T.-H. Wei, C.-T. Hsieh, R.-S. Juang, Surface coating of titania and graphene oxide onto plasma-activated polymer membranes as efficient photocatalysts for organics removal from water, J. Water Process Eng. 37 (2020) 101488, https://doi.org/10.1016/j.jwpe.2020.101488.
[72] L. Aoudjit, P. Martins, F. Madjene, D. Petrovykh, S. Lanceros-Mendez, Photocatalytic reusable membranes for the effective degradation of tartrazine with a solar photoreactor, J. Hazard. Mater. 344 (2018) 408–416, https://doi.org/10.1016/j.jhazmat.2017.10.053.
[73] L. Guo, Y. Yang, F. Xu, Q. Lan, M. Wei, Y. Wang, Design of gradient nanopores in phenolics for ultrafast water permeation, Chem. Sci. 10 (2019) 2093–2100, https://doi.org/10.1039/c8sc03012j.
[74] H. Ali, E. Khan, I. Ilahi, Environmental chemistry and ecotoxicology of hazardous heavy metals: environmental persistence, toxicity, and bioaccumulation, J. Chem. 2019 (2019) 1–14, https://doi.org/10.1155/2019/6730305.

[75] K. Rehman, F. Fatima, I. Waheed, M.S.H. Akash, Prevalence of exposure of heavy metals and their impact on health consequences, J. Cell. Biochem. 119 (2017) 157–184, https://doi.org/10.1002/jcb.26234.
[76] M.R. Raviya, M.V. Gauswami, H.D. Raval, A novel polysulfone/iron-nickel oxide nanocomposite membrane for removal of heavy metal and protein from water, Water Environ. Res. 92 (2020) 1990–1998, https://doi.org/10.1002/wer.1356.
[77] S. Mishra, A.K. Singh, J.K. Singh, Ferrous sulfide and carboxyl-functionalized ferroferric oxide incorporated PVDF-based nanocomposite membranes for simultaneous removal of highly toxic heavy-metal ions from industrial ground water, J. Membr. Sci. 593 (2020) 117422, https://doi.org/10.1016/j.memsci.2019.117422.
[78] A. Saeedi-Jurkuyeh, A.J. Jafari, R.R. Kalantary, A. Esrafili, A novel synthetic thin-film nanocomposite forward osmosis membrane modified by graphene oxide and polyethylene glycol for heavy metals removal from aqueous solutions, React. Funct. Polym. 146 (2020) 104397, https://doi.org/10.1016/j.reactfunctpolym.2019.104397.
[79] S. Bandehali, A. Moghadassi, F. Parvizian, J. Shen, S.M. Hosseini, Glycidyl POSS-functionalized ZnO nanoparticles incorporated polyether-imide based nanofiltration membranes for heavy metal ions removal from water, Korean J. Chem. Eng. 37 (2020) 263–273, https://doi.org/10.1007/s11814-019-0441-5.
[80] X. Liu, B. Jiang, X. Yin, H. Ma, B.S. Hsiao, Highly permeable nanofibrous composite microfiltration membranes for removal of nanoparticles and heavy metal ions, Sep. Purif. Technol. 233 (2020) 115976, https://doi.org/10.1016/j.seppur.2019.115976.
[81] G. Abdi, A. Alizadeh, S. Zinadini, G. Moradi, Removal of dye and heavy metal ion using a novel synthetic polyethersulfone nanofiltration membrane modified by magnetic graphene oxide/metformin hybrid, J. Membr. Sci. 552 (2018) 326–335, https://doi.org/10.1016/j.memsci.2018.02.018.
[82] Z.U. Rehman, B. Khojah, T. Leiknes, S. Alsogair, M. Alsomali, Removal of bacteria and organic carbon by an integrated ultrafiltration—nanofiltration desalination pilot plant, Membranes 10 (2020) 223, https://doi.org/10.3390/membranes10090223.
[83] R. Singh, R. Bhadouria, P. Singh, A. Kumar, S. Pandey, V.K. Singh, Nanofiltration technology for removal of pathogens present in drinking water, in: Waterborne Pathogens, 2020, pp. 463–489, https://doi.org/10.1016/b978-0-12-818783-8.00021-9.
[84] T. Fujioka, A.T. Hoang, T. Ueyama, L.D. Nghiem, Integrity of reverse osmosis membrane for removing bacteria: new insight into bacterial passage, Environ. Sci. Water Res. Technol. 5 (2019) 239–245, https://doi.org/10.1039/c8ew00910d.
[85] K.R. Zodrow, E. Bar-Zeev, M.J. Giannetto, M. Elimelech, Biofouling and microbial communities in membrane distillation and reverse osmosis, Environ. Sci. Technol. 48 (2014) 13155–13164, https://doi.org/10.1021/es503051t.
[86] J.P.S. Cabral, Water microbiology. Bacterial pathogens and water, Int. J. Environ. Res. Public Health 7 (2010) 3657–3703, https://doi.org/10.3390/ijerph7103657.
[87] J.S. Gruber, A. Ercumen, J.M. Colford, Coliform bacteria as indicators of diarrheal risk in household drinking water: systematic review and meta-analysis, PLoS One 9 (2014), https://doi.org/10.1371/journal.pone.0107429.
[88] D.Y. Koseoglu-Imer, B. Kose, M. Altinbas, I. Koyuncu, The production of polysulfone (PS) membrane with silver nanoparticles (AgNP): physical properties, filtration performances, and biofouling resistances of membranes, J. Membr. Sci. 428 (2013) 620–628, https://doi.org/10.1016/j.memsci.2012.10.046.
[89] J. Zhang, Y. Zhang, Y. Chen, L. Du, B. Zhang, H. Zhang, J. Liu, K. Wang, Preparation and characterization of novel polyethersulfone hybrid ultrafiltration membranes bending with modified halloysite nanotubes loaded with silver nanoparticles, Ind. Eng. Chem. Res. 51 (2012) 3081–3090, https://doi.org/10.1021/ie202473u.
[90] M. Ben-Sasson, K.R. Zodrow, Q. Genggeng, Y. Kang, E.P. Giannelis, M. Elimelech, Surface functionalization of thin-film composite membranes with copper

nanoparticles for antimicrobial surface properties, Environ. Sci. Technol. 48 (2013) 384–393, https://doi.org/10.1021/es404232.
[91] F. Perreault, M.E. Tousley, M. Elimelech, Thin-film composite polyamide membranes functionalized with biocidal graphene oxide nanosheets, Environ. Sci. Technol. Lett. 1 (2013) 71–76, https://doi.org/10.1021/ez4001356.
[92] A. Ojha, Nanomaterials for removal of waterborne pathogens, in: Waterborne Pathogens, 2020, pp. 385–432, https://doi.org/10.1016/b978-0-12-818783-8.00019-0.
[93] H. Meng, T. Xia, S. George, A.E. Nel, A predictive toxicological paradigm for the safety assessment of nanomaterials, ACS Nano 3 (2009) 1620–1627, https://doi.org/10.1021/nn9005973.
[94] A. García, L. Delgado, J.A. Torà, E. Casals, E. González, V. Puntes, X. Font, J. Carrera, A. Sánchez, Effect of cerium dioxide, titanium dioxide, silver, and gold nanoparticles on the activity of microbial communities intended in wastewater treatment, J. Hazard. Mater. 199–200 (2012) 64–72, https://doi.org/10.1016/j.jhazmat.2011.10.057.
[95] E.Z. Gomaa, Silver nanoparticles as an antimicrobial agent: a case study on Staphylococcus aureus and Escherichia coli as models for gram-positive and gram-negative bacteria, J. Gen. Appl. Microbiol. 63 (2017) 36–43, https://doi.org/10.2323/jgam.2016.07.004.
[96] D.L. Zhao, S. Japip, Y. Zhang, M. Weber, C. Maletzko, T.-S. Chung, Emerging thin-film nanocomposite (TFN) membranes for reverse osmosis: a review, Water Res. 173 (2020) 115557, https://doi.org/10.1016/j.watres.2020.115557.
[97] . www.Lagchem.com/product.
[98] W. Lau, S. Gray, T. Matsuura, D. Emadzadeh, J.P. Chen, A. Ismail, A review on polyamide thin film nanocomposite (TFN) membranes: history, applications, challenges and approaches, Water Res. 80 (2015) 306–324, https://doi.org/10.1016/j.watres.2015.04.037.
[99] M.B. Dixon, D. Kim-Hak, Improving second-pass permeate quality using thin film nanocomposite (TFN) membranes, Desalin. Water Treat. (2014) 1–5, https://doi.org/10.1080/19443994.2014.939867.
[100] G.M. Urper-Bayram, B. Sayinli, R. Sengur-Tasdemir, T. Turken, E. Pekgenc, O. Gunes, E. Ates-Genceli, V.V. Tarabara, I. Koyuncu, Nanocomposite hollow fiber nanofiltration membranes: fabrication, characterization, and pilot-scale evaluation for surface water treatment, J. Appl. Polym. Sci. 136 (2019) 48205, https://doi.org/10.1002/app.48205.
[101] L. Barrera, R.B. Chandran, Harnessing Photoelectrochemistry for Wastewater Nitrate Treatment Coupled with Resource Recovery, 2020, https://doi.org/10.26434/chemrxiv.13182881.
[102] T. Varadavenkatesan, S. Pai, R. Vinayagam, A. Pugazhendhi, R. Selvaraj, Recovery of value-added products from wastewater using aqueous two-phase systems—a review, Sci. Total Environ. 778 (2021) 146293, https://doi.org/10.1016/j.scitotenv.2021.146293.
[103] Z. Ye, Y. Shen, X. Ye, Z. Zhang, S. Chen, J. Shi, Phosphorus recovery from wastewater by struvite crystallization: property of aggregates, J. Environ. Sci. 26 (2014) 991–1000, https://doi.org/10.1016/s1001-0742(13)60536-7.
[104] W. Rongwong, K. Goh, Resource recovery from industrial wastewaters by hydrophobic membrane contactors: a review, J. Environ. Chem. Eng. 8 (2020) 104242, https://doi.org/10.1016/j.jece.2020.104242.
[105] P. Suresh, C.E. Duval, Poly(acid)-functionalized membranes to sequester uranium from seawater, Ind. Eng. Chem. Res. 59 (2020) 12212–12222, https://doi.org/10.1021/acs.iecr.0c01090.

[106] K. Saito, S. Asai, Recovery of rare metals using nucleic acid bases and extractants immobilized by grafted polymer chains, Bunseki kagaku 66 (2017) 771–782, https://doi.org/10.2116/bunsekikagaku.66.771.
[107] L.F. Villalobos, T. Yapici, K.-V. Peinemann, Poly-thiosemicarbazide membrane for gold recovery, Sep. Purif. Technol. 136 (2014) 94–104, https://doi.org/10.1016/j.seppur.2014.08.027.
[108] R. Sujanani, M.R. Landsman, S. Jiao, J.D. Moon, M.S. Shell, D.F. Lawler, L.E. Katz, B.D. Freeman, Designing solute-tailored selectivity in membranes: perspectives for water reuse and resource recovery, ACS Macro Lett. 9 (2020) 1709–1717, https://doi.org/10.1021/acsmacrolett.0c00710.
[109] S. Guo, Y. Wan, X. Chen, J. Luo, Loose nanofiltration membrane custom-tailored for resource recovery, Chem. Eng. J. 409 (2021) 127376, https://doi.org/10.1016/j.cej.2020.127376.
[110] Y. Shirazi, A. Ghadimi, T. Mohammadi, Recovery of alcohols from water using polydimethylsiloxane-silica nanocomposite membranes: characterization and pervaporation performance, J. Appl. Polym. Sci. 124 (2011) 2871–2882, https://doi.org/10.1002/app.35313.
[111] A. Kamtsikakis, S. Mcbride, J.O. Zoppe, C. Weder, Cellulose nanofiber nanocomposite pervaporation membranes for ethanol recovery, ACS Appl. Nano Mater. 4 (2021) 568–579, https://doi.org/10.1021/acsanm.0c02881.
[112] X. Liu, H. Jin, Y. Li, H. Bux, Z. Hu, Y. Ban, W. Yang, Metal–organic framework ZIF-8 nanocomposite membrane for efficient recovery of furfural via pervaporation and vapor permeation, J. Membr. Sci. 428 (2013) 498–506, https://doi.org/10.1016/j.memsci.2012.10.028.
[113] X. Liu, Y. Li, Y. Liu, G. Zhu, J. Liu, W. Yang, Capillary supported ultrathin homogeneous silicalite-poly(dimethylsiloxane) nanocomposite membrane for bio-butanol recovery, J. Membr. Sci. 369 (2011) 228–232, https://doi.org/10.1016/j.memsci.2010.11.074.
[114] F. Soyekwo, Q. Zhang, Y. Qu, Z. Lin, X. Wu, A. Zhu, Q. Liu, Tetraamminezinc complex integrated interpenetrating polymer network nanocomposite membrane for phosphorous recovery, AICHE J. 65 (2018) 755–765, https://doi.org/10.1002/aic.16463.
[115] L.E. Peng, Z. Yao, J. Chen, H. Guo, C.Y. Tang, Highly selective separation and resource recovery using forward osmosis membrane assembled by polyphenol network, J. Membr. Sci. 611 (2020) 118305, https://doi.org/10.1016/j.memsci.2020.118305.
[116] M. Namvar-Mahboub, M. Pakizeh, S. Davari, Preparation and characterization of UZM-5/polyamide thin film nanocomposite membrane for dewaxing solvent recovery, J. Membr. Sci. 459 (2014) 22–32, https://doi.org/10.1016/j.memsci.2014.02.014.
[117] M. Mehrabian, A. Kargari, Prospects of nanocomposite membranes for the recovery of hydrogen and production of syngas, in: Nanocomposite Membranes for Water and Gas Separation, 2020, pp. 397–437, https://doi.org/10.1016/b978-0-12-816710-6.00016-x.

CHAPTER 6

Sustainable membranes with functionalized nanomaterials (FNMs) for environmental applications

Nur Hashimah Alias[a], Mohamad Nor Nor Azureen[b], Nur Hidayati Othman[a], Fauziah Marpani[a], Woei Jye Lau[c], and Munawar Zaman Shahruddin[a]

[a]Department of Oil and Gas Engineering, School of Chemical Engineering, College of Engineering, Universiti Teknologi MARA, Shah Alam, Selangor, Malaysia
[b]Department of Chemical Sciences, Faculty of Science & Technology, The National University of Malaysia (UKM), Bangi, Selangor, Malaysia
[c]Advanced Membrane Technology Research Centre (AMTEC), School of Chemical and Energy Engineering, Faculty of Engineering, Universiti Teknologi Malaysia, Skudai, Johor, Malaysia

6.1 Introduction

Over the past few decades, the concept of applying membrane technology for environmental applications is well known in various fields. Membrane processes are considered as potential methods useful in clean technologies that minimize the use of raw materials, rationalize energy consumption, and reduce waste production. This technology is capable of solving many environmental problems and gaining a research interest from various industries such as water treatment for domestic and industrial water supply, chemical, pharmaceutical, biotechnological, beverages, food metallurgy, and other separation processes [1]. In particular, membrane separation has been widely used for liquid/gas separation, filtration, distillation, ion exchange, and chemical treatment systems, which are closely related to environmental challenges. The pivotal discovery that transformed membrane separation from a laboratory to an industrial process was the development, in the early 1960s, of the Loeb-Sourirajan process for making defect-free, high-flux, anisotropic reverse osmosis membranes [2]. A membrane can be defined as a selective barrier that allows specific things such as molecules, ions, or other small particles to pass through while stopping others. The membrane interface could be molecularly homogeneous that owns uniform structure and composition or else it can be heterogeneous in physical and chemical

Membranes with Functionalized Nanomaterials
https://doi.org/10.1016/B978-0-323-85946-2.00003-5

properties. Besides that, it may contain pores or holes in finite dimension, or it may as well contain a layered structure. By convention, for a filter to be defined as a membrane, it is usually limited to the structures that perform a separation with particulate suspensions larger than 1–10 μm [3]. Therefore, the selection of a suitable membrane is very important in determining the success of the specific applications.

Nowadays, immense intensive efforts have been carried out to explore the new membrane materials and processes by altering traditionally used materials. In general, type materials used to fabricate the membrane layer can be categorized as organic membranes, inorganic membranes, and inorganic-organic hybrid membrane materials [4]. Until recently, most all organic membranes have been tailored using polymeric materials, although inorganic membranes receive more attention than organic ones. In general, polymer materials provide a vast variety of material structures and properties. Numerous membrane materials and processes have been investigated to address environmental concerns such as polyethersulfone (PES) and polyacrylonitrile (PAN) membranes for dye removal [5–7], heavy metal removal [8], polyphenylsulfone (PPSF) membrane for fuel cell applications [9], polysulfone (PS) and membrane for gas separation [10], hemodialysis [11], supercapacitor [12], polyvinylidene fluoride (PVDF) membrane for bisphenol A removal [13], alumina ceramic and polyacrylonitrile (PAN) membranes for produced water treatment [5,6,14,15], alumina-yttria stabilized zirconia, alumina ceramic, polyethersulfone (PES), and polyvinylidene fluoride (PVDF) membranes for oily wastewater [16–20]. Using membrane technology, the separation can be carried out continuously under mild conditions with relatively low energy consumption and without incorporation of any additives. Moreover, this membrane technology can be combined with other separation processes that will form the hybrid processes. This advanced process has greatly influenced research toward developing new membrane materials with intriguing physical and chemical properties by doping or hybridizing nanomaterials into membrane materials. The intrinsic properties of inorganic nanomaterials as a filler such as size, type, structure, and interactions with the polymer matrix can significantly affect the properties of the resultant matrix [21].

In the past two decades, the evolution of nanotechnology represents an ever-improving process in the design, discovery, creation, and novel utilization of artificial nanoscale materials. To meet the major challenges in membrane sustainability, this nanomaterial in various hierarchical fashions is stimulating various important practical applications in the environmental sector. The design, synthesis, and modification of novel nanomaterials allow for enhanced performance for environment-related applications [22]. Composite materials are a multiphase combination material of two or more component materials with different properties and different forms through compounding processes; it not only maintains the main characteristics of the original component, but also shows a new character which is not possessed by any of the original components [23]. While preparing the composite membrane with nanomaterials, some factors need to be considered depending on the type of nanomaterials. The first consideration while preparing the composite membranes is the degree of nanomaterials doping. The optimum degree of nanomaterials loading should be achieved to prevent deterioration in the mechanical integrity of the membranes. Next is the filler distribution in the polymer matrix. The nonuniform distribution will lead to particle agglomeration that results in failure due to a disruption in the homogeneity of the membrane film. The distribution of nanomaterials is affected by the chemical interactions between the polymer and the nanoparticles, the particle dimensions, and the dispersion of the fillers in the casting solvent [24].

Membranes with functionalized nanomaterials (FNMs) are effective hybrid components that attract a great deal of attention for increasing particle miscibility in polymer matrices [25]. They are manufactured by doping FNMs into the membrane materials matrix. The development of novel nanotechnologies suggests that functionalization on the nanomaterials surface offers an appropriate solution to tune their properties in order to be compatible with the membrane materials. Surface functionalization refers to the introduction of functional units or chemical groups on the surface of nanomaterials to create specific surface sites. Objectives for the development of FNMs enable control of the nanomaterials surface and can be used to integrate with the membrane matrix to produce hybrid systems with advanced features. Therefore, this chapter has been constructed to describe the limitations of membranes with functionalized materials in environmental applications as well as to afford a glimpse into the various syntheses and modifications of FNMs. Finally, it concludes with the future challenges of sustainable membranes with FNMs in environmental applications.

6.2 Limitations of membranes with functionalized nanomaterials (FNMs) in environmental applications

Nowadays, membrane processes are attracting a great deal of interest in many industrial applications such as wastewater treatment, power generation, air purification, and others. Membrane technology is relatively simple in concept and operation, flexible and compatible with integrated systems in many environmental applications. Over the past decades, numerous trials have been devoted to manufacturing of advanced membranes for particular applications that have appropriate features such as permeability, selectivity, and specific physicochemical properties [26]. Various organic and inorganic materials have been used for membrane preparation. The development of nanocomposite membranes by doping or hybridizing the membrane matrix with nanomaterials is an efficient approach for improving the properties of the pristine membrane. Generally, nanocomposite membranes are a multiphase combination of nanomaterials and the membrane matrix with different properties and different forms through compounding processes [27]. Nanomaterials may be either coated onto the membrane surface or dispersed in the polymer solution before membrane casting.

The most attractive nanomaterials for environmental applications are derived from silica, noble metals, semiconductors, metal oxides, polymers, and carbonaceous materials. Various environmental treatments and remediation from air, water, and green energy-related applications are being investigated [28–30]. The categories of treatment techniques include nanoadsorption, membrane and membrane processes, nanophotocatalyst, and nanosensing. There are some challenges that have been addressed from the applications of nanomaterials for environmental applications which are the potential impact of the nanomaterials toward human health and ecosystems, the reusability of the nanomaterials, and limited activity toward various applications that affect many other factors such as cost and environmental circumstance on nanomaterials itself [31]. Typically, the addition of the nanomaterials tends to change the surface properties of membranes that will influence the membrane behavior.

Surface functionalization of nanomaterials has been extensively explored and has proven to play a critical role in nanoparticle development for practical applications such as improving the miscibility and dispersion of nanoparticles in the membrane matrix [32]. The nonuniform distribution will lead to particle agglomeration that results in mechanical failure due to a disruption in the homogeneity of the membrane. Furthermore, nanoparticle distribution is also affected by the chemical interactions between the polymer and the particles, the particle dimensions, and the dispersion of the fillers in the casting solvent [33]. Although a variety of membranes with FNMs have been extensively explored for various environmental applications, there are still some restrictions that have limited the application of membranes with FNMs.

The selection of the types of membranes with FNMs for every environmental application depends on the physical state of the materials and the chemical nature of the functional moieties borne by the materials. The development of FNMs adopts various types of chemicals which are either hazardous or nonhazardous to the environment. The method to incorporate the FNMs onto/into the membrane matrix is sometimes a complex process and the sustainability of the FNMs in the membrane matrix is hard to define. Moreover, during prolonged filtration and the repeating cycle of membrane reuse, FNMs could leach from the membrane and potentially enter the environment. The leaching of the nanomaterials into environments can occur during a longtime operation, during the membrane process as well as due to inappropriate disposal of used membranes. The nanomaterials released can undergo an environmental transformation that could potentially pose a risk to both human health and the environment system. Furthermore, nonbiodegradable membranes with FNM material wastes are difficult to dispose of and require a complex process that is costly, which contributes to environmental concerns. Therefore, the selection of the membrane materials, nanomaterials, and functionalization method as well as the chemicals used for FNM synthesis are a great alarm toward environmental applications.

6.3 Synthesis and modification of membranes with functionalized nanomaterials (FNMs)

Due to their unique features, nanomaterials have taken their place among the most preferred materials in various branches of applications. It is well accepted that the incorporation of nanoparticles into a membrane matrix can significantly improve the physicochemical properties of the composite membrane. Unfortunately, the large interface between nanomaterials with the membrane materials increases the unfavorable enthalpic interaction between the hydrophobic and hydrophilic properties of both materials that often lead to aggregations [34]. In order to increase the dispersion of the nanomaterials in the membrane matrix, it is essential to create a compatible interface of the nanomaterials by functionalizing with either the ligand or polymers onto the nanomaterials surfaces. The nanomaterials can be decorated by a variety of functionalities. Functionalization of nanomaterials is defined as the insertion of chemical functional groups in the surface of materials regardless of their crystalline or amorphous nature [33]. Functionalization reactions can be conducted in metals, ceramics, and synthetic or natural polymers. The aim of functionalization is to modify, tune, or add surface properties to a material once they play a key role in the overall performance of materials, especially those with nanometric sizes [35]. However, the inert nature of most hybrid membrane materials and nanomaterials limits their development for specific applications. Therefore, the surface modification of nanomaterials must be carried out to improve the adhesion, printing, and wetting by bringing a variety of polar and other functional groups on the nanomaterials surface. In this section, the commonly used methods in the production and functionalization of nanomaterials are discussed.

6.3.1 In situ surface functionalization

The common functionalization strategy can be defined as an addition of a chemical functionality on the surfaces of nanomaterials. A wide range of functional groups can be used for direct functionalization of the nanomaterials depending on the application and compatibility with the membrane materials matrix. Understanding and controlling the bond breaking as well as formation processes of direct functionalization of a materials involved multistep complex processes. Direct functionalization refers to the modification of nanomaterials by the addition of various functional group compounds during the synthesis process [36]. In situ surface functionalization is a one-pot synthesis method in which the synthesis of nanomaterials and surface functionalization are carried out within a single step. This method was also used to reduce the particle size and narrowing particle size distribution along with the functionalization [37]. Precursors for synthesizing the nanomaterials as well as the functionalization materials are simultaneously used in the reaction mixture. One limitation of this functionalization method is the incompatibility of the functional groups with the preparation process such as a modifying group showing a fast reaction with the nanoparticles surface [38]. Carboxylate, phosphonates, and thiols groups are commonly groups that have been used for this method.

6.3.2 Postsynthesis surface functionalization

In this method, the procedure was totally separated into two parts which are the synthesis of nanomaterials and surface modification. This method is very general and a large variety of functional groups can be introduced to the surface of the nanomaterials. It is carried out by grafting functional materials onto the surface of the nanomaterials after undergoing the synthesis process. The main benefits of the method are that large numbers of coupling agents are commercially available which will be chosen based on the compatibility between two materials. Meanwhile, the main concern is that the functional group must possess a high affinity toward the nanomaterials surface, and it should not exist in the form of isolated clusters.

6.3.2.1 Polymer grafted nanoparticles

Surface polymerization can be used to improve the interfacial reactions of the nanomaterials for enhancing the stability and dispersion of the nanoparticles by combining the unique properties of the polymers and the nanoparticles. Grafting strategies have been adapted for anchoring the polymers onto the nanoparticles [39]. Various parameters were found to influence the dispersion of the nanoparticles including polymer molecular weight, polymer graft density, polymer architecture, and polymer composition. The grafting technique can be divided into three categories which are grafting "to," grafting "from," and grafting "through." Grafting "to" represents a facile technique that involves chemical bonding between polymers that contained end group functionalities of polymers anchoring specific functional group moieties and the active site of the nanoparticles [40]. This technique can lead to low grafting densities due to steric crowding reactive sites of the polymers and the nanoparticles. Meanwhile, grafting "from" represents the grafting of polymers from nanoparticles that initiated from the radical initiating group that was previously introduced onto the nanoparticle surface. It permits a high level of control over functionality and the density of the polymer chain [41]. The initiation of graft polymerization can be achieved by using small monomers introduced onto the active sites of nanomaterials. In addition, grafting "through" is accomplished by the copolymerization of the initiator and free monomers in solution followed by polymerization that results in the formation of a polymer chain directly onto the nanomaterials surface [42]. Sadri et al. [43] proposed a technique to functionalize the graphene nanoplatelets (GNPs) to improve the solubility and dispersion of the GNPs in the aqueous media to develop the nanofluids for heat transfer applications. The technique proposed is the grafting "through" method where hydrogen peroxide as the initiator and gallic acid (GA) were grafted on the GNPs surface. They present a successful finding as superior thermal conductivity enhancement of about 24.2% was achieved at 45°C. Fig. 6.1 shows the grafting approaches for synthesizing functionalized nanoparticles.

6.3.2.2 Covalent functionalization

Covalent functionalization is the attachment of the chemical moieties to the nanomaterials structure via the formation of covalent bonds which share at

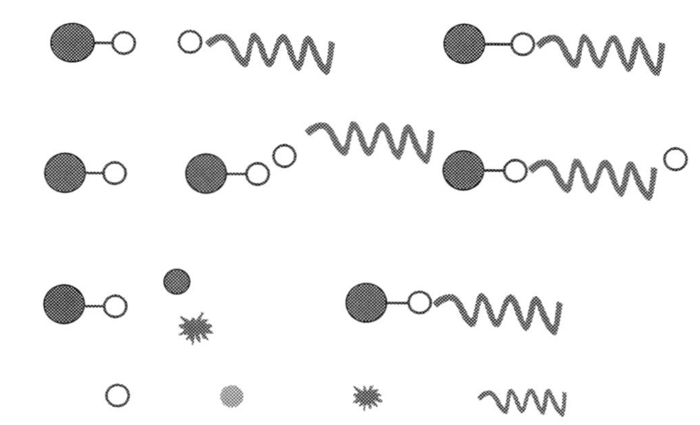

Fig. 6.1 Grafting approaches for synthesizing functionalized nanoparticles (A) grafting "to", (B) grafting "from," and (C) grafting "through". *(No permission required.)*

least one pair of electrons between the nanomaterials. The typical objective of covalent functionalization is to increase the dispersion, processability, and reactivity of the nanomaterials [44]. As a result, the physicochemical properties of the nanomaterials can be fine-tuned for specific interactions with the membrane matrix. Generally, the covalent attachments of molecular entities are more challenging which requires the use of highly reactive species [45]. There are a variety of functional groups that can be introduced onto the surface of the nanomaterials by covalent functionalization. Consequently, numerous studies have established several procedures for functionalized nanomaterials such as carboxylation, hydroxylation, hydrogenation, amination, sulfonation, and halogenation which depend on the suitability and applicability of the nanomaterials. Fig. 6.2 shows the several procedures

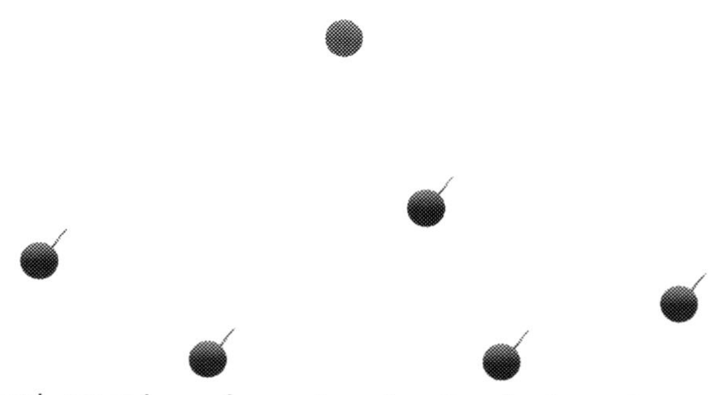

Fig. 6.2 Several procedures for surface functionalization of nanoparticles. *(No permission required.)*

that have been reported to modify the surface of the nanomaterials for covalent functionalization.

Carboxylation

Carboxylation refers to the chemical reaction in which carboxylic acid groups are produced by treating the substrate with carbon dioxide. It is a standard conversion in organic chemistry that was also applied to a functionalized variety of nanoparticles [46]. The carboxylate-oxide nanoparticles interactions have been extensively studied with some even produced commercially. The identity of the organic substituents on the carboxylate groups can control the physical properties of the functionalized nanoparticles. Alexander et al. [47] developed the functionalized alumina nanoparticles by carboxylation with carboxylic acid. The function of carboxylic acid is to enhance solubility as well as facilitate further reactivity. Meanwhile, Titkov et al. [48] synthesized silver nanoparticles functionalized with the carboxyl group to improve the stabilization of silver nanoparticles against aggregation.

Hydroxylation

Hydroxylation describes the chemical process to introduce the hydroxyl groups (—OH) into the substrate by oxidation reaction. The degree of hydroxylation refers to the number of OH groups in a molecule. Various strategies were examined for hydroxylation of nanoparticles to sponsor efficient functionalization through maximizing surface hydroxyl groups [49]. Li et al. [50] efficiently developed hydroxylated barium titanate that was treated using hydrogen peroxide. The study shows that adequate hydroxyl groups are vital parameters to improve the nanoparticle's surface reactivity with the dispersant. Ngo et al. [51] reported that the amount of the hydroxyl group of synthesized zinc oxide nanoparticles was modulated by the type of solvent and the annealing temperature will have an impact on the surface hydrophilicity.

Hydrogenation

Hydrogenation is a chemical reaction between molecular hydrogen and another element in the presence of a catalyst synthesized at normal temperature and pressure. This process typically constitutes the addition of pairs of the hydrogen atom to a molecule. Noncatalytic hydrogenation can only take place at a very high temperature [52]. The process required three components, unsaturated substrate, hydrogen sources, and a catalyst, that is carried out at different temperatures and pressure depending on the substrate and activity of the catalyst [53]. The obvious source of hydrogen is hydrogen gas itself, which is typically available commercially within the storage medium or pressurized cylinder. Another source of hydrogen comes from the donor molecules such as formic acid, isopropanol, and dihydroanthracene. Zhou and Zhou [54] developed hydrogenation of iridium nanoparticles catalyzed using formic acid as a hydrogen source. Platinum, palladium, rhodium, and ruthenium are commonly the highly active catalysts that are used for hydrogenation that can be operated at lower temperature and pressure. Meanwhile, nonprecious metal catalysts, especially those that are based on nickel, have also been commonly used as an alternative to economical materials but often result in a slower reaction that required high synthesizing temperature [55].

Amination/sulfonation

Amination is the process by which the amine group is introduced into the substrate. Amino groups permit the most versatile reaction for the surface modification of nanoparticles through either the formation of using amides or reductive amination. Wang et al. [56] reported that the amine group enhances the reactivity of the silica nanoparticles as well as the hydrophobicity. Meanwhile, the concept of sulfonation is similar to amination as sulfonic acid groups were added to the substrate. Sulfonic acid can promote the hydrophilicity of the substrate that is essential in several applications as well as improve the compatibility with other organic molecules. Bagheri et al. [57] reported that the addition of sulfonated silica oxide nanoparticles has improved proton conductivity as the SO_3H groups of sulfonic acid promote proton transportation. It has also been reported that highly sulfonated polyoctahedral silsesquioxane improves the water uptake as well as proton conductivity of the nanocomposite membrane [9].

6.3.2.3 Noncovalent functionalization

Noncovalent functionalization is different from covalent functionalization in that it is not involved in the sharing of electrons but rather involves more in the dispersed variation of electromagnetic interactions between or within nanoparticles and other molecules. The interaction between nanoparticles and other molecules are through physical adsorption, either the electrostatic interaction, Van der Waals interaction of cation-π interaction [58]. Noncovalent functionalization has the potential to be used for adsorption or wrapping of various functional groups like amphiphilic surfactants, polymers, or biomolecules that will be used to enhance the solubility and dispersibility of the nanoparticles [59]. Nanoparticles decorated with surfactants are involved in medical applications, structural materials, energy conversion processes, catalysts, cleaning and purification systems [60]. Noncovalent functionalized multiwalled carbon nanotubes (MWCNTs) with imidazolium amine-terminated ionic liquid (AIL) were facilely prepared through cation-π stacking interaction to enhance the thermal conductivity of epoxy composites. It shows that the MWCNTs functionalized with AIL achieve about 211% improvement in thermal conductivity as compared to modified acidified MWCNTs [61]. Mukherjee et al. [62] decorated silver nanoparticles with a mixture of Tween 20(Tw-20) surfactant and human serum albumin and hemoglobin proteins for the biological cell. While Emadi et al. [63] also reported the applications of nanoparticles for medical purposes using functionalized graphene oxide with chitosan for protein nanocarriers to protect against enzymatic cleavage and retain collagenase activity. The diversity of the modification of various nanoparticles via surface functionalization enables conjugation of various functional materials that have widely been applied in membrane technology. Table 6.1 shows the example of a membrane with FNMs in various applications that have been reported previously.

Table 6.1 Example of membrane with FNMs in various applications.

Applications	Membrane with FNMs	Remarks	Ref.
Reverse osmosis	PI thin layer/ Oxidized MWCNTs	The functionalized MWCNTs with hydrophilic groups enhanced the hydrophilicity, water flux and fouling resistance	[64]
Antifouling membrane	PVDF/Ag-GO	Silver functionalized GO was significantly reduced the fouling resistance of the PVDF membrane due to the increasing of membrane hydrophilicity	[65]
Nanofiltration	PI thin layer/ Modified Hydrous MO	The plasma coating of modified manganese oxide nanoparticles using PECVD method results in improve the dispersion stability of polar nanoparticles in organic phase during interfacial polymerization process	[66]

Continued

Table 6.1 Example of membrane with FNMs in various applications—cont'd

Applications	Membrane with FNMs	Remarks	Ref.
Protein adsorption	Electrospun PVA/(Poly (MVE/MA), PMA)	(Poly(MVE/MA), PMA) act as selective adsorption of proteins. The large specific surface area and abundant available carboxyl group of (Poly(MVE/MA), PMA) contributed to high adsorption performance	[67]
PEMFC	SPPSU/SPOSS	The functionalized POSS with sulfonic acid group has improve the connectivity between inorganic nanoparticles with SPPSU membrane as well as improve proton conductivity	[68]
Vacuum membrane distillation	PTFE/Oleic acid functionalized CNT	The oleic acid functionalized CNT showing better performance in terms of high permeation flux and salt rejection at lower CNT loading	[69]
Phenol and benzene removal	PSU-PES/ Functionalized CNT	Functionalized CNT has improved the dispersion of the CNT in the PSF/PES membrane that contribute to decrease the membrane pore size that attribute to improve the membrane rejection	[70]
CO_2 separation	PI/ Functionalized MWCNT	Functionalized MWCNT were homogenously dispersed in the PI matrix and the gas separation performance showed that CO_2 permeability coefficient increased 292%.	[71]
Separation of lignin	PSf/Fe-ZSM-5 and PSf/Cu-ZSM-5	Functionalized zeolite influences the lignin rejection from 85.2% to 88.5%. Incorporation of Fe and Cu as a functionalized material has provide a novel way of	[22]

Table 6.1 Example of membrane with FNMs in various applications—cont'd

Applications	Membrane with FNMs	Remarks	Ref.
		obtaining high dispersion within zeolite framework.	
Heavy metal removal	Alumina / Thiourea functionalized Si	Thiourea has creating a selective layer for functional coating on the surface of tubular ceramic membranes that has ensure better fixation of the silica on the ceramic membrane.	[72]

SPPSU: Sulfonated Polyphenylsulfone, SPOSS: Sulfonated Polyhedral Silsesquioxane, CNT: Carbon Nanotubes, PTFE: Polytetrafluoroethylene, Psf: Polysulfone, FE-ZSM-5: Ferrous functionalized zeolite, Cu-ZSM-5: Copper functionalized zeolite, Si: Silica, PI: Polyimide, MWCNTs: Multiwalled carbon nanotubes PVDF: Polyvinylidene Fluoride, Ag: Silver, GO: Graphene Oxide, PES: Polyethersulfone, (Poly(MVE/MA), PMA): Poly(methyl vinyl ether-alt-maleic anhydride), MO: Manganese Oxide, PECVD: Plasma Enhanced Chemical Vapour Deposition.

6.4 Future challenges for membranes with functionalized nanomaterials (FNMs)

In the future, membrane technology will continue to play a vital role in addressing global environmental challenges, such as water scarcity, climate change, and energy shortages. Engineering new high-performance membranes with targeted applications and understanding the limiting factors and their mechanisms in membrane separation are two key research directions that should be paid more attention in future research. In all scenarios, membrane technology can play a fundamental role in achieving the objective of sustainable production. The potential of membrane technology in the process industry is evident from the fact that membrane operations have outclassed their traditional counterparts on the basis of process intensification metrics developed to quantify their sustainability with respect to conventional operations. From a global perspective of the industrial scale process, membranes have succeeded in being alternate technologies to the conventional separation technique allowing more compact designs. Membranes can be fitted in various places in a production facility and can be synergistically combined with other processes leading to hybrid technologies. Membrane technology is currently the best available technology in many environmental applications, and an immense number of advanced membrane systems are actively being explored and developed.

The concept of sustainable membranes in various environmental applications has been of concern in the scientific community in recent years. The sustainability of membrane technology while bearing an advanced materials is still challenging over the year. In recent years, there have been many studies reporting on the development of membranes using green chemistry principles, which are driving researchers to propose new solutions able to promote the development of green membrane material. The green materials that are able to keep a cleaner environment while utilizing an abandoned material on earth are a great challenge to be established. A lot of effort and research needs to be put in to counter the problems. They are being used for the development and innovation in a range of industrial sectors. However, the use of functionalized nanomaterials is still in its infancy in many industrial settings. Functionalized nanomaterials have the potential to create cheaper and more effective consumer products and industrial processes. However, they could also have adverse effects on the environment, human health, and safety, and their sustainability is questionable, if used incorrectly. In spite of the progress in the application of nanomaterials-based membrane technology for various environmental applications, it is necessary to justify the environmental threats that may be associated with this technology. The assessment of membrane stability and understanding the dynamics of nanomaterials leaching out from the membrane matrix pose a great challenge for long-term membrane operations. The leaching effect of the nanomaterials could adversely affect the membrane performance and its life span. Yet other challenges for sustainable membranes with FNMs are to sustain the membrane performance in all operating conditions without sacrificing the membrane lifetime and preventing the membrane fouling, nanomaterials leaching, membrane aging as well as slowing down the growth of large-scale industrial use.

6.5 Conclusions

This chapter provides an overview of the membrane bearing functionalized nanomaterials (FNMs) for various environmental applications. The limitation of this advanced membrane technology lies in the leaching of the nanomaterials into the environment that hindered the longtime membrane operation. Various approaches of FNMs development have also been discussed in order to improve the adhesion, printing, and wetting by bringing a variety of polar and other functional groups on the nanomaterials surface toward compatibility with the membrane matrix. The future challenges of sustainable membranes with FNMs have also been presented. However, more intensive research needs to be carried out to reach an efficient advanced membrane technology that supports FNMs in the applied environment technology. It is believed that more focused and deep research needs to be employed on what has been achieved so far to come up with feasible applications along with the search for novel technologies.

References

[1] Y. Wen, J. Yuan, X. Ma, S. Wang, Y. Liu, Polymeric nanocomposite membranes for water treatment: a review, Environ. Chem. Lett. 17 (2019) 1539–1551.

[2] L. Nuang, J. Gu, H.G.L. Coster, A.G. Fane, Quantitative determination of the electrical properties of RO membranes during fouling and cleaning processes using electrical impedance spectroscopy, Desalination 379 (2016) 126–136.

[3] C. Ursino, R. Castro-Munoz, E. Drioli, L. Gzara, M.H. Albeirutty, A. Figoli, Progress of nanocomposite membranes for water treatment, Membranes 8 (18) (2018), 8020018.

[4] P.G. Ingole, M.I. Baig, W.K. Choi, H.K. Lee, Synthesis and characterization of polyamide/polyester thin-film nanocomposite membrane achieved by functionalized TiO_2 nanoparticles for water vapor separation, J. Mater. Chem. A 1–3 (2012) 1–15.

[5] N.H. Alias, J. Jaafar, S. Samitsu, A.F. Ismail, N.A.M. Nor, N. Yusof, F. Aziz, Mechanistic insight of the formation of visible-light responsive nanosheet graphitic carbon nitride embedded polyacrylonitrile nanofibres for wastewater treatment, J. Water Process Eng. 33 (2020), 101015.

[6] N.H. Alias, J. Jaafar, S. Samitsu, A.F. Ismail, M.H.D. Othman, M.A. Rahman, N.H. Othman, N. Yusof, F. Aziz, T.A.T. Mohd, Efficient removal of partially hydrolysed polyacrylamide in polymer-flooding produced water using photocatalytic graphitic carbon nitride nanofibres, Arab. J. Chem. 13 (2020) 4341–4349.

[7] S. Zinadini, S. Rostami, V. Vatanpour, E. Jalilian, Preparation of antibiofouling polyethersulfone mixed matrix NF membrane using photocatalytic activity of ZnO/MWCNTs nanocomposite, J. Membr. Sci. 529 (2017) 133–141.

[8] X. Fang, J. Li, X. Li, S. Pan, X. Zhang, X. Sun, J. Shen, W. Han, L. Wang, Internal pore decoration with polydopamine nanoparticle on polymeric ultrafiltration membrane for enhanced heavy metal removal, Chem. Eng. J. 314 (2017) 38–49.

[9] N.A.M. Nor, J. Jaafar, J.D. Kim, Improved properties of sulfonated octaphenyl polyhedral silsequioxane crosslink with highly sulfonated polyphenylsulfone as proton exchange membrane, J. Solid State Electrochem. 24 (2020) 1185–1195.

[10] N.A.H.M. Norddin, A.F. Ismail, A. Mustafa, R.S. Murali, T. Matsuura, Utilizing low ZIF-8 loading for an asymmetric PSf/ZIF-8 mixed matrix membrane for CO_2/CH_4 separation, RSC Adv. 5 (2015) 30206–30215.

[11] A. Roy, P. Dadhich, S. Dhara, S. De, In vitro cytocompatibility and blood compatibility of polysulfone blend, surface-modified polysulfone and polyacrylonitrile membranes for hemodialysis, RSC Adv. 5 (2015) 7023–7034.

[12] S.T. Gunday, E. Cevik, A. Yusuf, A. Bozkurt, Synthesis, characterization and supercapacitor application of ionic liquid incorporated nanocomposite based on SPSU/silicon dioxide, J. Phys. Chem. Solid 137 (2020), 109209 (1-8).

[13] N.A.M. Nor, J. Jaafar, A.F. Ismail, M.A. Mohamed, M.H.D. Othman, M.A. Rahman, N. Yusof, W.J. Lau, Preparation and performance of PVDF-based nanocomposite membrane consisting of TiO_2 nanofibers for organic pollutant decomposition in wastewater under UV irradiation, Desalination 391 (2016) (2016) 89–97.

[14] N.H. Alias, J. Jaafar, S. Samitsu, N. Yusof, M.H.D. Othman, M.A. Rahman, A.F. Ismail, F. Aziz, W.N.W. Salleh, N.H. Othman, Photocatalytic degradation of oilfield produced water using graphitic carbon nitride embedded in electrospun polyacrylonitrile nanofibers, Chemosphere 204 (2018) 79–86.
[15] N.H. Alias, J. Jaafar, S. Samitsu, T. Matsuura, A.F. Ismail, S. Huda, N. Yusof, F. Aziz, Photocatalytic nanofiber-coated alumina hollow fiber membranes for highly efficient oilfield produced water treatment, Chem. Eng. J. 360 (2019) 1437–1446.
[16] N.F.D. Junaidi, N.H. Othman, M.Z. Shahruddin, N.H. Alias, F. Marpani, W.J. Lau, A.F. Ismail, Fabrication and characterization of graphene-polyethersulfone (GO-PES) composite flat sheet and hollow fiber membranes for oil-water separation, J. Chem. Technol. Biotechnol. 95 (2020) 1308–1320.
[17] N.H. Othman, N.H. Alias, M.Z. Shahruddin, S.N.C.M. Hussein, A. Dollah, Supported graphene oxide hollow fibre membrane for oily wastewater treatment, AIP Conf. Proc. 1901 (1) (2017), 020008.
[18] S.H. Paiman, M.A. Rahman, T. Uchikoshi, N.A.H.M. Nordin, N.H. Alias, N. Abdullah, K.H. Abas, M.H.D. Othman, J. Jaafar, A.F. Ismail, In situ growth of α-Fe_2O_3 on Al_2O_3-YSZ hollow fiber membrane for oily wastewater, Sep. Purif. Technol. 236 (2020), 116250.
[19] U.N. Rusli, N.H. Alias, M.Z. Shahruddin, N.H. Othman, Photocatalytic degradation of oil using polyvinylidene fluoride/titanium dioxide composite membrane for oily wastewater treatment, MATEC Web Conf. 69 (05003) (2016) 1–5.
[20] M.Z. Shahruddin, N.H. Othman, N.H. Alias, S.N.A. Ghani, Desalination of produced water using bentonite as pre-treatment and membrane separation as main treatment, Procedia. Soc. Behav. Sci. 195 (2015) 2094–2100.
[21] R. Deng, W. Han, K.L. Yeung, Confined PFSA/MOF composite membranes in fuel cells for promoted water management and performance, Catal. Today 331 (2019) 12–17.
[22] G. Saranya, G. Arthanareeswaran, A.F. Ismail, N.L. Reddy, M.V. Shankar, J. Kweon, Efficient rejection of organic compounds using functionalized ZSM-5 incorporated PPSU mixed matrix membrane, RSC Adv. 7 (2017) 15536–15552.
[23] V.R. Pereira, A.M. Isloor, U.K. Bhat, A.F. Ismail, A. Obaid, H.-K. Fun, Preparation and performance studies of polysulfone-sulfonated nano-titania (S-TiO_2) nanofiltration membranes for dye removal, RSC Adv. 5 (2015) 53874–53885.
[24] F. Sun, L.L. Qin, J. Zhou, Y.K. Wang, J.Q. Rong, Y.J. Chen, S. Ayaz, Y.U. Hai-Yin, L. Liu, Friedel-crafts self-crosslinking of sulfonated poly(etheretherketone) composite proton exchange membrane doped with phosphotungstic acid and carbon-based nanomaterials for fuel cell applications, J. Membr. Sci. 611 (118381) (2020) 1–9.
[25] R. Vani, S. Ramaprabhu, P. Haridoss, Mechanically stable and economically viable polyvinyl alcohol-based membranes with sulfonated carbon nanotubes for PEMFC, Sustain. Energy Fuels 3 (4) (2020) 1372–1382.
[26] A.K. Shukla, J. Alam, M. Alhoshan, L.A. Dass, M.R. Muthumareeswaran, Development of a nanocomposite ultrafiltration membrane based on polyphenylsulfone blended with graphene oxide, Sci. Rep. 7 (2017), 41976 (1–12).
[27] P. Bhavani, D. Sangeetha, Proton conducting composite membranes for fuel cell application, Int. J. Hydrogen Energy 36 (22) (2011) 14858–14865.
[28] N. Jullok, R.V. Hooghten, P. Luis, A. Volodin, C.V. Haesendonck, J. Vermant, B.V. der Bruggen, Effect of silica nanoparticles in mixed matrix membranes for pervaporation dehydration of acetic acid aqueous solution: plat-inspired dewatering systems, J. Clean. Prod. 112 (2016) 4879–4889.
[29] M. Kumar, T.S. Rao, A.M. Isloor, G.P.S. Ibrahim, Inamuddin, N. Ismail, A.F. Ismail, A.M. Asiri, Use of cellulose acetate/polyphenylsulfone derivatives to fabricate ultrafiltration hollow fiber membranes for the removal of arsenic from drinking water, Int. J. Biol. Macromol. 129 (2019) 715–727.

[30] A. Naderi, T.S. Chung, M. Weber, C. Maletzko, High performance dual-layer hollow fiber membrane of SPPSU/polybenzimidazole for hydrogen purification, J. Membr. Sci. 591 (2019) 1–10.
[31] J. Dai, S. Li, J. Liu, J. He, J. Li, L. Wang, L. Wang, J. Lei, Fabrication and characterization of a defect-free mixed matrix membrane by facile mixing PPSU with ZIF-8 core-shell microspheres for solvent-resistant nanofiltration, J. Membr. Sci. 589 (2019) 1–11.
[32] J. Weingart, P. Vabbilisetty, X.L. Sun, Membrane mimetic surface functionalization of nanoparticles: method and applications, Adv. Colloid Interface Sci. 197–198 (2013) 68–84.
[33] A.K. Mishra, S. Bose, T. Kuila, N.H. Kim, J.H. Lee, Silicate-based polymer-nanocomposite membranes for polymer electrolyte membrane fuel cells, Prog. Polym. Sci. 37 (6) (2012) 842–869.
[34] A.K. Sahu, K. Ketpang, S. Shanmugam, O. Kwon, S. Lee, H. Kim, Sulfonated graphene-nafion composite membranes for polymer electrolyte fuel cells operating under reduced relative humidity, J. Phys. Chem. 120 (29) (2016) 15855–15866.
[35] R. Scipioni, D. Gazzoli, F. Teocoli, O. Palumbo, A. Paolone, N. Ibris, S. Brutti, M.A. Navarra, Preparation and characterization of nanocomposite polymer membranes containing functionalized SnO_2 additives, Membranes 4 (1) (2014) 123–142.
[36] A. Liberman, N. Mendez, W.C. Trogler, A.C. Kummel, Synthesis and surface functionalization of silica nanoparticles for nanomedicine, Surf. Sci. Rep. 69 (2–3) (2014) 132–158.
[37] I. Khan, K. Saeed, I. Khan, Nanoparticles: properties, applications and toxicities, Arab. J. Chem. 12 (7) (2019) 908–931.
[38] I.A. Rahman, V. Padavettan, Synthesis of silica nanoparticles by sol-gel: size-dependant properties, surface modification, and applications on silica-polymer nanocomposites—a review, J. Nanomater. 2012 (2012), 132424.
[39] A.A. Tashvigh, L. Luo, T.S. Chung, M. Weber, C. Maletzko, A novel ionically cross-linked sulfonated polyphneylsulfone (sPPSU) membrane for organic solvent nanofiltration (OSN), J. Membr. Sci. 545 (2018) 221–228.
[40] A.J. Chancellor, B.T. Seymour, B. Zhao, Characterizing polymer-grafted nanoparticles: from basic defining parameters to behaviour in solvents and self-assembled structures, Anal. Chem. 91 (10) (2019) 6391–6402.
[41] L. Li, C. Han, D. Xu, J.Y. Xing, Y.H. Xue, H. Liu, Polymer-grafted nanoparticles prepared via a grafting-from strategy: a computer simulation study, Phys. Chem. Chem. Phys. 20 (2018) 18400–18409.
[42] A. Jayaraman, Polymer grafted nanoparticles: effect of chemical and physical heterogeneity in polymer grafts on particle assembly and dispersion, J. Polym. Sci. B Polym. Phys. 51 (7) (2013) 524–534.
[43] R. Sadri, M. Hosseini, S.N. Kazi, S. Bagheri, N. Zubir, G. Ahmadi, M. Dahari, T. Zaharinie, A novel, eco-friendly technique for covalent functionalization of graphene nanoplatelets and the potential of their nanofluids for heat transfer applications, Chem. Phys. Lett. 675 (2017) 92–97.
[44] G. Graffius, F. Bernardoni, A.Y. Fadeev, Covalent functionalization of silica surface using "inert" poly(dimethylsiloxanes), Langmuir 30 (49) (2014) 14797–14807.
[45] V. Georgakilas, M. Otyepka, A.B. Bourlinos, V. Chandra, N. Kim, K.C. Kemp, P. Hobza, R. Zboril, K.S. Kim, Functionalization of graphene: covalent and non-covalent approaches, derivatives and applications, Chem. Rev. 112 (11) (2012) 6156–6214.
[46] A. Shukla, J. Alam, M. Alhoshan, L.A. Dass, F.A.A. Ali, M.R. Muthumareeswaran, U. Mishra, M.A. Ansari, Removal of heavy metal ions using carboxylated graphene oxide-incorporated polyphenylsulfone nanofiltration membrane, Environ. Sci. Water Res. Technol. 4 (2018) 438–448.
[47] S. Alexander, V. Gomez, A.R. Barron, Carboxylation and decarboxylation of aluminium oxide nanoparticles using bifunctional carboxylic acids and octylamine, J. Nanomater. 2016 (2016), 7950876.

[48] A.I. Titkov, O.A. Logutenko, E.Y. Gerasimov, I.K. Shundrina, E.V. Karpova, N.Z. Lyakho, Synthesis of silver nanoparticles stabilized by carboxylated methoxypoyethylene glycols: the role of carboxyl terminal groups in the particle size and morphology, J. Incl. Phenom. Macrocycl. Chem. 94 (2019) 287–295.
[49] T.F. O'Mahony, M.A. Morris, Hydroxylation methods for mesoporous silica and their impact of surface functionalisation, Microporous Mesoporous Mater. 317 (2021), 110989.
[50] C.C. Li, S.J. Chang, J.T. Lee, W.S. Liao, Efficient hydroxylation of $BaTiO_3$ nanoparticles by using hydrogen peroxide, Colloids Surf. A Physicochem. Eng. Asp. 361 (1–3) (2010) 143–149.
[51] G.V. Ngo, A. Margaillan, S. Villain, C. Leroux, C. Bressy, Synthesis of ZnO nanoparticles with tuneable size and surface hydroxylation, J. Nanopart. Res. 15 (2013) 1332.
[52] H. Liu, S. Fan, Z. Wang, K. Chen, A. Guo, Temperature and pressure effects on the catalytic performance of metalloporphyrins during hydrogenation of naphthalene, ChemistrySelect 2 (4) (2017) 1613–1619.
[53] K. Dobrezbenger, J. Bosters, N. Moser, N. Yigit, A. Nagl, K. Fottinger, D. Lennon, G. Ruppechter, Hydrogenation of palladium nanoparticles supported by graphene nanoplatelets, J. Phys. Chem. C 124 (43) (2020) 23674–23683.
[54] X. Zhou, M. Zhou, Polyvinylpyrrolidone-stabilized iridium nanoparticles catalyzed the transfer hydrogenation of nitrobenzene using formic acid as the source of hydrogen, Chemistry 2 (4) (2020) 960–968.
[55] E.G. Shubina, N.S. Filimonov, R.V. Shafigulin, A.V. Bulanova, I.V. Shishkovskii, Y.G. Morozov, Effect of size of nickel nanoparticles in hydrogenation of benzene, Pet. Chem. 57 (5) (2017) 410–414.
[56] J. Wang, L. Yang, J. Xie, Y. Wang, T.J. Wang, Surface amination of silica nanoparticles using tris(hydroxymethyl)aminomethane, Ind. Eng. Chem. Res. 59 (49) (2020) 21383–21392.
[57] A. Bagheri, P. Salarizadeh, M.S.A. Hazer, P. Hosseinabadi, S. Kashefi, H. Beydaghi, The effect of adding sulfonated SiO_2 nanoparticles and polymer blending on properties and performance of sulfonated poly ether sulfone membrane: fabrication and optimization, Electrochim. Acta 295 (2019) 875–890.
[58] M. Sardan, A. Yildirim, D. Mumcuoglu, A.B. Tekinay, M.O. Gule, Noncovalent functionalization of mesoporous silica nanoparticles with amphiphilic peptides, J. Mater. Chem. B 5 (2) (2014) 2168–2174.
[59] D. Ponnamma, S.H. Sung, J.S. Hong, K.H. Ahn, K.T. Varughese, S. Thomas, Influence of non-covalent functionalization of carbon nanotubes on the rheological behavior of natural rubber latex nanocomposites, Macromol. Biosci. 53 (2014) 147–159.
[60] H. Heinz, C. Pramanik, O. Heinz, Y. Ding, R.K. Mishra, D. Marchon, R.J. Flatt, I. Estrela-Lopis, J. Llop, S. Moya, R.F. Ziolo, Nanoparticle decoration with surfactants: molecular interactions, assembly, and applications, Surf. Sci. Rep. 77 (1) (2017) 1–58.
[61] C. Chen, X. Li, Y. Wen, J. Liu, X. Li, H. Zeng, Z. Xue, X. Zhou, X. Xie, Noncovalent engineering of carbon nanotube surface by imidazolium ionic liquids: a promising strategy for enhancing thermal conductivity of epoxy composites, Compos. A Appl. Sci. Manuf. 125 (2019), 105517.
[62] M. Mukherjee, K. Gangopadhyay, R. Das, P. Purkayastha, Development of non-ionic surfactant and protein-coated ultra small silver nanoparticles: increased viscoelasticity enables potency in biological applications, ACS Omega 5 (15) (2020) 8999–9006.
[63] F. Emadi, A. Amini, A. Gholami, Y. Ghasemi, Functionalized graphene oxide with chitosan for protein nanocarriers to protect against enzymatic cleavage and retain collagenase activity, Sci. Rep. 10 (7) (2017) 42258.
[64] J. Farahbakhsh, M. Delnavaz, V. Vatanpour, Investigation of raw and oxidized multiwalled carbon nanotubes in fabrication of reverse osmosis polyamide membranes for improvement in desalination and antifouling properties, Desalination 410 (2017) 1–9.

[65] K. Ko, Y. Yu, M.J. Kim, J. Kweon, H. Chung, Improvement in fouling resistance of silver-graphene oxide coated polyvinylidene fluoride membrane prepared by pressurized filtration, Sep. Purif. Technol. 194 (2018) 161–169.
[66] G.S. Lai, W.J. Lau, P.S. Goh, M. Karaman, M. Gursoy, A.F. Ismail, Development of thin film nanocomposite membrane incorporated with plasma enhanced chemical vapor deposition-modified hydrous manganese oxide for nanofiltration process, Compos. Part B 176 (2019), 107328.
[67] M. Najafi, J. Chery, M.M. Frey, Functionalized electrospun poly(vinyl alcohol) nanofibrous membranes with poly(methyl vinyl ether-alt-maleic anhydride) for protein adsorption, Materials 11 (6) (2018) 1–12.
[68] N.A.M. Nor, J. Jaafar, J.D. Kim, A.F. Ismail, M.H.D. Othman, A. Rahman, M., Effect of polyhedral silsesquioxane functionalized sulfonic acid groups incorporated into highly sulfonated polyphenylsulfone as proton conducting membrane, Arab. J. Sci. Eng. 29 (2020) 1–9.
[69] Z.A. Pouya, M.A. Tofighy, T. Mohammadi, Synthesis and characterization of polytetrafluoroethylene/oleic acid-functionalized carbon nanotubes composite membrane for desalination by vacuum membrane distillation, Desalination 503 (2021), 114931.
[70] M.S. Rameetse, O. Aberefa, M.O. Daramola, Effect of loading and functionalization of carbon nanotube on the performance of blended polysulfone/polyethersulfone membrane during treatment of wastewater containing phenol and benzene, Membranes 10 (54) (2020) 1–13.
[71] H. Sun, T. Wang, Y. Xu, W. Gao, P. Li, J. Niu, Fabrication of polyimide and functionalized multi-walled carbon nanotubes mixed matrix membranes by in-situ polymerization for CO_2 separation, Sep. Purif. Technol. 177 (2017) 327–336.
[72] V.V. Tomina, G.I. Nazarchuk, I.V. Melnyk, Modification of ceramic membranes by silica nanoparticles with thiourea functions, J. Nanomater. 4 (2019) 1–8. 2534934.

CHAPTER 7

Sustainable membranes with FNMs for biomedical applications

Zinnia Chowdhury[b], Sanjib Barma[b], Aparna Ray Sarkar[a], and Dwaipayan Sen[a]
[a]Department of Chemical Engineering, Heritage Institute of Technology, Kolkata, West Bengal, India
[b]Department of Chemical Technology, University of Calcutta, Kolkata, West Bengal, India

7.1 Introduction

Membrane science and technology has been one of the most coveted technologies for several decades, which has proven easy to implement in different application areas like air purification, water treatment, pharmaceutical manufacturing, food processing, biomedical applications, etc. Eco-friendliness and energy efficiency are the two major factors for which membrane technology has drawn attention from worldwide researchers. Nowadays, biomedical applications are one of the important areas, where membrane science has been introduced with enough opportunity in the development of advanced healthcare systems. An example is hemodialysis that involves the purification of blood at the time of erroneous kidney functioning [1]. However, one of the complexities in making the dialysis process less efficient is the fouling of the membrane, which results in a deviation in the permeation rate from the normal glomerular filtration rate (GFR). Secondly, biofilm growth over the membrane might pose an additional issue to the application of membrane technology in hemodialysis and several other biomedical applications. Nowadays, such issues have been resolved through fabricating the membrane after casting polymer mixed with other materials such as nanoparticles on the support. Such a mixed system helps to modulate the membrane property in several ways with a solution to the key issues related to the membrane; for example, in case with the silver-lined membrane, where silver is doped with the basic polymer in order to inhibit the growth of the biofilm over the membrane and also to increase the hydrophilicity of the membrane. Among several such technologies associated with the mixed matrix, doping of functionalized nanomaterials is one of the emerging techniques, where the property of the nanomaterials is also being modulated according to the applications. Nanosized materials

Membranes with Functionalized Nanomaterials
https://doi.org/10.1016/B978-0-323-85946-2.00008-4

(~100 nm) with large specific surface area, high reactivity, and temperature resistance are some of the significant properties that can be controlled through functionalization of the nanomaterials, which subsequently control the membrane property after being mixed with the basic polymeric solution. Functionalization of nanomaterials enhances their therapeutic effectiveness and thus results in better drug delivery preceded by disease diagnosis. From tissue engineering to drug delivery, membranes modified with functionalized nanomaterials has shown its immense potential in resolving many unaddressed issues continuing for several decades and thus, introduces a new era to healthcare. The chapter will thus try to illustrate different applications of such functionalized nanomaterial-based membranes preceded by a brief on the basics of membrane technology, nanomaterials' preparation, and their derivatization.

7.2 Overview on membrane technology

Membrane technology is a tertiary treatment process that has been considered as one of the imminent separation schemes in several application areas. A membrane is a selective barrier allowing selective passage of solutes without making any change either in chemical or mechanical properties of the barrier. Such things may be molecules, ions, or other small particles. The primary driving force for the permeation of the solvent/solute through the membrane depends on the nature of the process. For example, pressure difference across the membrane is mainly attributed to microfiltration (MF), ultrafiltration (UF), nanofiltration (NF), and reverse osmosis (RO). Likewise, the separation of ions during electro-osmosis is mainly manifested because of the electrokinetic difference across the membrane. On the contrary, separation of salts primarily follows the concentration difference across the membrane in the dialysis process. Further, the temperature difference across the membrane manifests as the separation of solutes in case of the membrane distillation [2]. Solvents/solutes passed through the membrane are known as permeates, while solutes rejected by the membrane are called the retentates. Membranes are mainly classified as the organic and inorganic membranes, depending on the materials from which they are fabricated. Organic polymers used for membrane fabrication are generally polyethersulfone (PES), polyamide-imide (PAI), polysulfone (PSf), polytetrafluoroethylene (PTFE), polyacrylonitrile (PAN), polyvinylidene fluoride (PVDF), etc. However, inorganic membranes are fabricated from ceramic, metallic, and zeolite sources. The properties and the corresponding separation efficacy greatly depend on such nature of the membrane.

7.2.1 Fabrication of polymeric membrane

Among several methods, the following fabrication methods are mostly used, especially when there will be a possibility of doping other materials with the polymer solution in order to fabricate functionalized membranes.

7.2.1.1 Phase inversion method

This is the most popular and elementary method for membrane fabrication and is being widely practiced. In this process, a homogeneous polymer solution is prepared in an appropriate solvent, which is further processed in order to have a transformation of dissolved polymer to a solid structure, called the membrane. The phase inversion method can be further classified into various types:

Immersion precipitation

In this process, a homogeneous polymer solution is prepared and casted over a chosen support resting on a glass plate using a blade (also called a Doctor Blade). The casted solution is then immersed in a coagulation bath of nonsolvent, preferably water, where the solvent from the polymeric solution gets dissolved into the nonsolvent. Such an escape of the solvent into the nonsolvent phase from the casted slurry makes the polymer to get precipitated over the support as a thin membrane film, which is further dried in order to get a desired morphology. Hence, the solvent for the polymer should have some polarity so that it will also be dissolved in the polar nonsolvent within the coagulation bath. In general, the fabricated membrane morphology with this method largely depends on the type of the polymer, its concentration within the solution, and the subsequent drying temperature. A schematic representation of the method has been shown in Fig. 7.1.

Evaporation-induced phase separation

This process involves the preparation of a homogeneous solution of polymer, nonvolatile solvent, and a volatile solvent. The volatile solvent is continuously evaporated and the polymer solubility decreases. The solution gets separated into two phases, namely the polymer-rich phase and the polymer-lean phase. The polymer-lean phase mainly consists of the nonvolatile solvent present at membrane pores and the polymer-rich phase forms

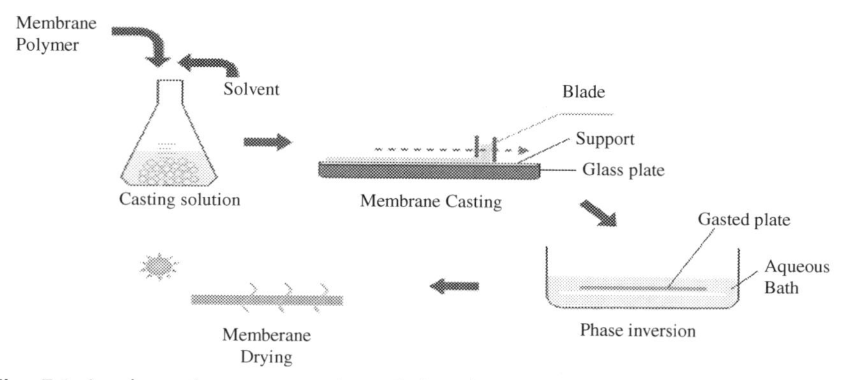

Fig. 7.1 A schematic representation of the phase inversion method. *(From N.Z.K. Shaari, N. Rahman, N.A. Sulaiman, R.M. Tajuddin, Thin film composite membranes: mechanical and antifouling properties, in: EDP Sciences, International Symposium on Civil and Environmental Engineering 2016 (ISCEE 2016), MATEC Web of Conferences, 06005, vol. 103, 2017, pp. 1–10. https://doi.org/10.1051/matecconf/201710306005.)*

the matrix of the membrane. Finally, the nonvolatile solvent is removed, which finally gives rise to a porous membrane [3]. The morphological structure of the membrane can be influenced by varying solvents based on its boiling point. Fig. 7.2 shows a membrane prepared using the method.

Vapor-induced phase separation

In this process, a casting solution of polymer in appropriate solvent is prepared, which is subsequently casted over the support. After the casting process, it is exposed to a humid atmosphere for a specific interval of time, where the water vapor present in the surrounding atmosphere dissolves into the casted thin film to make the porous structure. The porous casted plate is immersed in a nonsolvent bath to have the polymer precipitation as seen in case with the immersion precipitation followed by drying of the membrane.

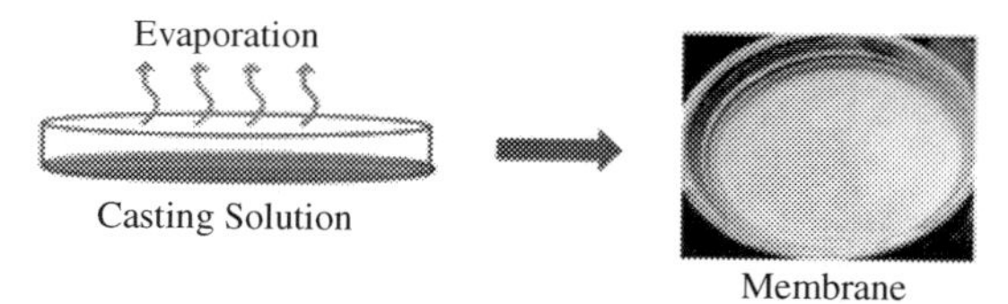

Fig. 7.2 Casting of membrane using evaporation-induced phase separation. *(From R. Pervin, P. Ghosh, M.G. Basavaraj, Tailoring pore distribution in polymer films via evaporation induced phase separation, RSC Adv. 9 (27) (2019) 15593–15605. https://doi.org/10.1039/C9RA01331H.)*

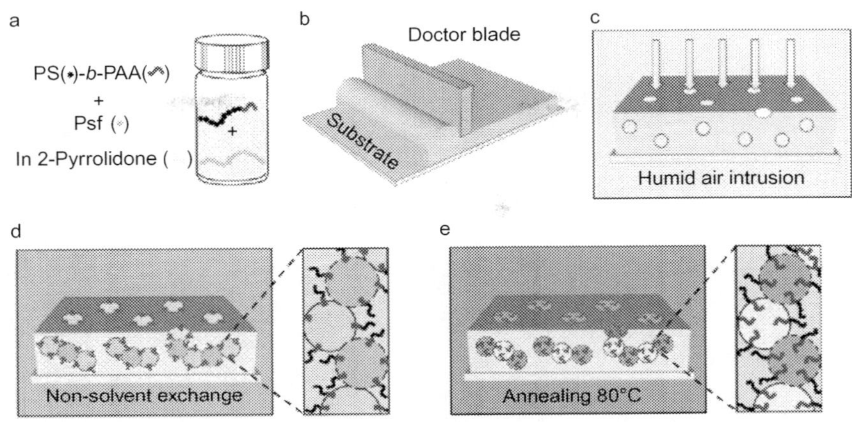

Fig. 7.3 Schematic of the surface-segregation and vapor-induced phase separation (SVIPS) membrane fabrication process. (A) The polymer solution was prepared by dissolving the polysulfone (PSf) and PS-b-PAA in 2-pyrrolidone. (B) The polymer solution was drawn into a uniform thin film on a glass substrate. (C) The casting solution thin film was exposed in a humid environment (with a relative humidity ~95%) for a predetermined amount of time. The intrusion of water vapor from the humid air into the casting solution contributes to the formation of a uniform cross-sectional architecture comprised of spongy cells. (D) The film was subsequently plunged into a nonsolvent water bath that caused the hydrophobic polymers to precipitate and vitrify the membrane nanostructure. Simultaneously, due to their hydrophilic nature, the PAA brushes preferentially segregate toward the surface of the pore wall. (E) The composite membrane was annealed in a bath of DI water at 80°C to allow the PAA brushes to extend toward the center of the pore. *PAA*, poly(acrylic acid). *(From Y. Zhang, J.R. Vallin, J.K. Sahoo, F. Gao, B.W. Boudouris, M.J. Webber, W.A. Phillip, High-affinity detection and capture of heavy metal contaminants using block polymer composite membranes, ACS Cent. Sci. 4 (12) (2018) 1697–1707. https://doi.org/10.1021/acscentsci.8b00690.)*

Fig. 7.3 shows the vapor-induced phase separation method for the fabrication of composite membrane [4].

Thermally induced phase separation

This method relies on the demixing of solution, where the polymer dissolved in the diluent gets separated from the diluent based on the affinity of the solvent for either polymer or diluent (Fig. 7.4). One of the primary advantages with the method is its easy operability in preparing a well-interconnected polymeric network attributing to a wide scope of fabrication of membrane with varied morphology and enough mechanical strength [5].

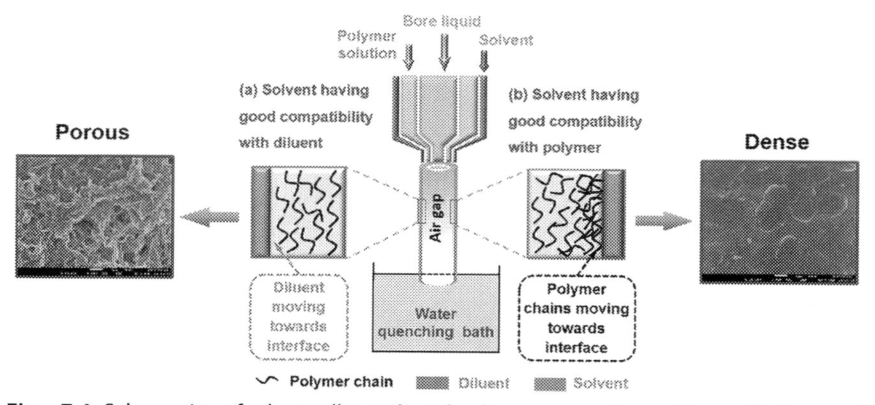

Fig. 7.4 Schematic of thermally induced phase separation. *(From C. Fang, S. Rajabzadeh, W. Liu, H.C. Wu, N. Kato, Y. Sun, S. Jeon, H. Matsuyama, Effect of mixed diluents during thermally induced phase separation process on structures and performances of hollow fiber membranes prepared using triple-orifice spinneret, J. Membr. Sci. 596 (2020). https://doi.org/10.1016/j.memsci.2019.117715.)*

7.2.1.2 Interfacial polymerization method

Interfacial polymerization is one of the mostly adopted commercial methods mainly for the fabrication of RO and NF thin film composite (TFC) membrane. Here, in this method primarily two different monomers are being dissolved in two different immiscible solvents. Initially, one of the solutions will be distributed over the support layer over which the other solution will be dispersed. The polymerization reaction takes place at the interface that subsequently forms a two-layer membrane on the support called TFC [6], which has been shown in Fig. 7.5.

7.2.1.3 Track-etching

This method is one of the most significant techniques used for the fabrication of membranes with several biomedical applications. In this process, heavy ion beams are generated to irradiate polymer films, which results in the formation of damaged tracks in the film (Fig. 7.6). The pore size and pore density of the membrane depends on the irradiation duration, temperature, and duration of etching. This strategy is famous for its exact control over the pore size and its distribution, where the size can be varied between few nanometers to micrometers. Polycarbonate and polyethytlene naphthalate membranes are commonly fabricated using the track-etching method due to their strong mechanical properties and resistance toward organic solvents and acids [5].

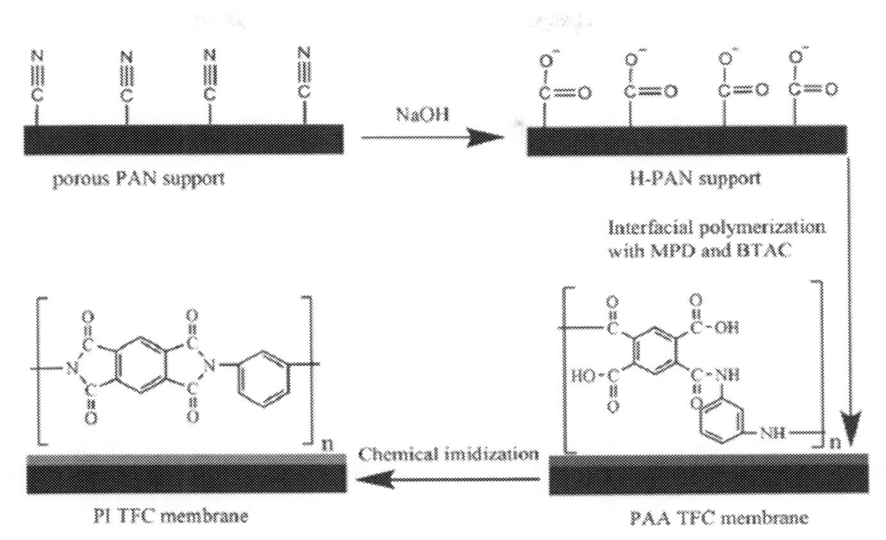

Fig. 7.5 Preparation of polyimide TFC membrane (*BTAC*, 1,2,4,5-benzenetetra acyl chloride; *MPD*, m-phenylenediamine; *PAN*, polyacrylonitrile). *(From S. Yang, H. Zhen, B. Su, Polyimide thin film composite (TFC) membranes: via interfacial polymerization on hydrolyzed polyacrylonitrile support for solvent resistant nanofiltration, RSC Adv. 7 (68) (2017) 42800–42810. https://doi.org/10.1039/c7ra08133b.)*

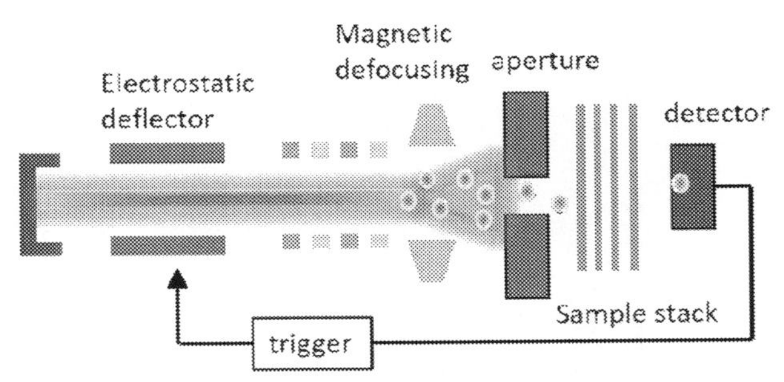

Fig. 7.6 Image of a single irradiation track etching setup. *(From A. Shiohara, B. Prieto-Simon, N.H. Voelcker, Porous polymeric membranes: fabrication techniques and biomedical applications, J. Mater. Chem. B (2020) 1–39. https://doi.org/10.1039/D0TB01727B.)*

7.2.1.4 Electrospinning method

Electrospinning is a relatively new procedure to fabricate membranes aiming for different process areas. A high potential difference is generated between a polymeric solution at the tip of the extruder and a grounded collector on which the polymeric deposition takes place. When the electrostatic potential turns out to be adequately high enough compared to the surface tension of the polymeric solution with the extruder surface, a jet of charged liquid expels out over the collector forming a fibrous membrane (Fig. 7.7). The aspect ratio, pore size distribution, and hydrophobicity of the membrane can be controlled by varying flow rate of the solution, viscosity of the solution, and applied electrostatic potential [7].

7.2.2 Application of membrane technology in the biomedical field

One of the noteworthy applications of membrane technology can be found in several biomedical fields, ranging from tissue engineering to hemodialysis. Tissue engineering aims to develop replacements for crippled organs and tissues due to infections. Tissue engineering has originated from transplantation of a donor tissue to replace a damaged or declined tissue. However, transmission of pathogens, shortage of donor organs, and rejection of donor organs are some of the intricacies that limit the application of direct tissue transplantation. To overcome these constraints, an alternative method has been developed, where tissues are built outside the host body. One of the processes in this method is "scaffold generation." A scaffold is a three-dimensional construction that acts temporarily as a support for growing new tissue cells. After tissue is generated over the scaffold, it is implanted

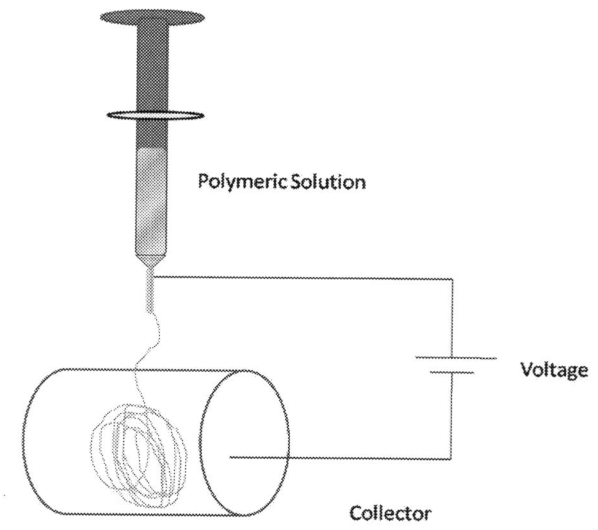

Fig. 7.7 A pictorial representation of the electrospinning process of membrane fabrication. *(No Permission Required.)*

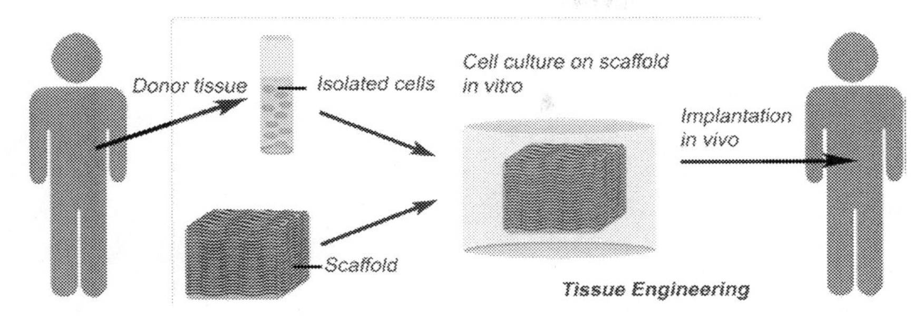

Fig. 7.8 A schematic representation of application of scaffold in tissue engineering. *(From D.F. Stamatialis, B.J. Papenburg, M. Gironés, S. Saiful, S.N.M. Bettahalli, S. Schmitmeier, M. Wessling, Medical applications of membranes: drug delivery, artificial organs and tissue engineering, J. Membr. Sci. 308 (1–2) (2008) 1–34. https://doi.org/10.1016/j.memsci.2007.09.059.)*

into the host body. One of the required properties of the scaffold is good porosity along with highly conductive pores. A porous scaffold ensures sufficient nutrient transport toward the tissue cells and thereafter removal of waste products from the tissue cells. Fig. 7.8 represents a schematic of the method [8].

In this context, polymers are mainly used for the fabrication of scaffolds using similar fabrication methods described for polymeric membrane. Both synthetic and natural biodegradable polymer-based membranes are utilized for scaffold fabrication. Among them, membrane fabricated from chitin sources has showed exceptional results in the tissue replacement process. Both α-chitin and β-chitin membranes are widely used for tissue engineering purposes due to their property of optimized cell adhesion and bioactiveness [9]. However, chitin-based membranes have improvised mechanical property, which can be enhanced by incorporating foreign substance like hydroxyapatite [10]. Another important application of membrane technology is related to drug delivery, where the controlled release of an encapsulated drug at a targeted site can be achieved from membrane-like encapsulation material. One of the advantages with this encapsulation technology is the increase in the sustainability of drug during its travel toward the targeted site. Membrane-based drug delivery can be classified into two types—one, the osmotic membrane system and the other is the diffusion-controlled membrane system. The osmotic membrane system comprises of a polymeric membrane enclosure, which is selectively permeable to water but restricts permeation of the drug. The concentrated drug solution is kept inside the membrane reservoir. When the reservoir passes through the

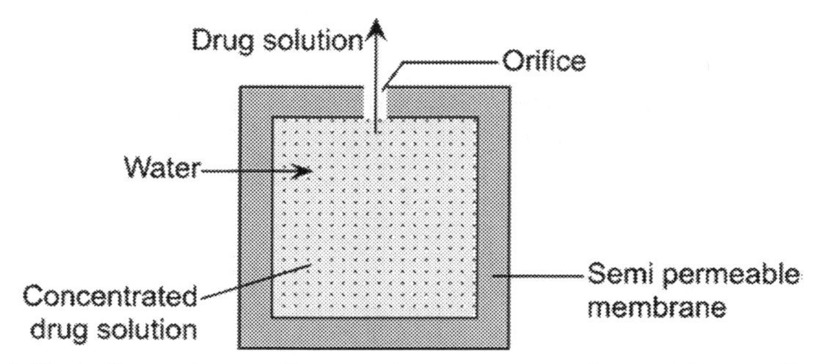

Fig. 7.9 Illustration of osmotic membrane system drug delivery. *(From D.F. Stamatialis, B.J. Papenburg, M. Gironés, S. Saiful, S.N.M. Bettahalli, S. Schmitmeier, M. Wessling, Medical applications of membranes: drug delivery, artificial organs and tissue engineering, J. Membr. Sci. 308 (1–2) (2008) 1–34. https://doi.org/10.1016/j.memsci.2007.09.059.)*

human body, water molecules passes through semipermeable membrane because of osmotic pressure, which in turn pumps out the drug through an orifice. Fig. 7.9 illustrates a schematic diagram of an osmotic membrane system drug delivery.

On the contrary, in diffusion-controlled membrane system, the delivery of drug occurs by transportation of the drug through the membrane. The passage of the drug greatly depends on the thickness of the membrane and also on its diffusivity through the membrane structure. The diffusion-controlled system can be very much observed in pills, patches, and implants. In case of pills, the drug is pressed into a form of a tablet and coated with a hydrophobic nondigestible membrane. When the pill passes within human body, it gets hydrated resulting in the formation of a gel layer through which drugs get released through diffusion. The illustration of the diffusion-controlled system in pills has been shown in Fig. 7.10.

In case of patches, the drug is kept in a reservoir that is held between two thin transparent layers of polymeric membrane, as shown in Fig. 7.11. The drug reservoir is surrounded by an annular border through which the drug is released to the specific site through diffusion. On the contrary, in implants, the drug is placed in a membrane reservoir. The drug can exist either in powder or in liquid form, which is released through the semipermeable membrane (Fig. 7.12). When the membrane is biodegradable in nature, drug release is accomplished by the degradation of the membrane. On the other hand, surgical removal of implants is done for nonbiodegradable membranes.

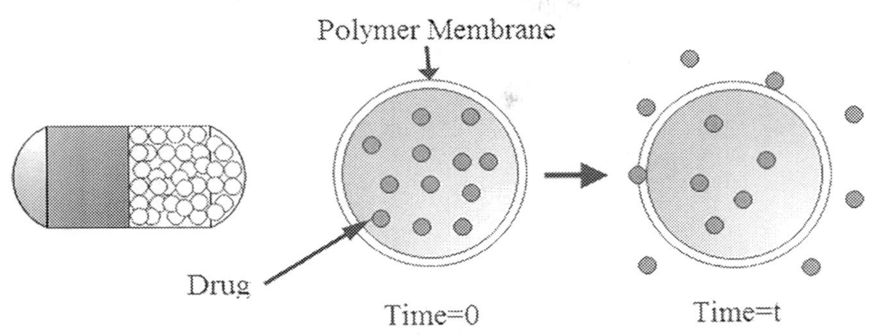

Fig. 7.10 Illustration of diffusion controlled membrane system drug delivery through pills. *(From Controlled (and sometimes local) drug release, Aleesha McDaniel, 2017. https://slideplayer.com/slide/10496095/. (Accessed 7 January 2021).)*

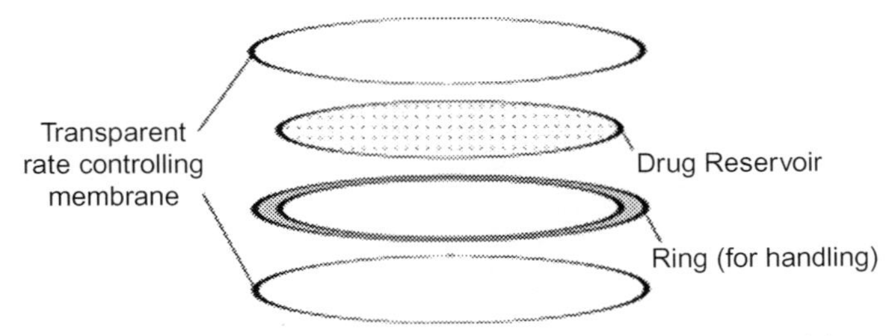

Fig. 7.11 Illustration of diffusion controlled membrane system drug delivery through patches. *(From D.F. Stamatialis, B.J. Papenburg, M. Gironés, S. Saiful, S.N.M. Bettahalli, S. Schmitmeier, M. Wessling, Medical applications of membranes: drug delivery, artificial organs and tissue engineering, J. Membr. Sci. 308 (1–2) (2008) 1–34. https://doi.org/10.1016/j.memsci.2007.09.059.)*

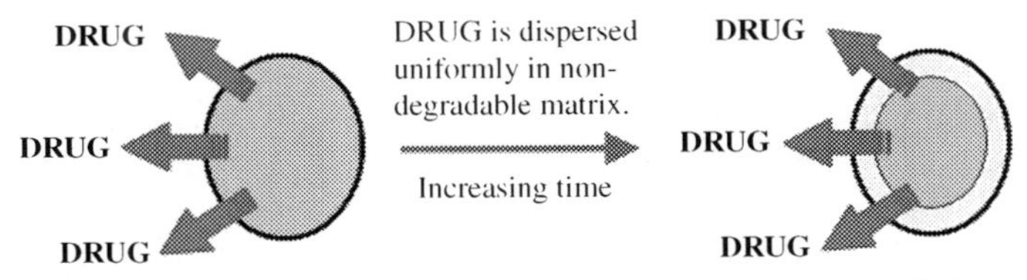

Fig. 7.12 Illustration of diffusion controlled membrane system drug delivery through nonbiodegradable membrane implants. *(From Implants, 2017. https://www.slideshare.net/sushirana/implants-73495771. (Accessed 7 January 2021).)*

One of the significant applications of membrane technology can be seen in the preparation of wearable biohybrid organs. Extra corporeal membrane oxygenator (ECMO) (also called oxygenator) in Fig. 7.13 is an example of such applications of membranes, when lungs fail to exchange gases with blood. Membrane oxygenators consist of a semipermeable membrane,

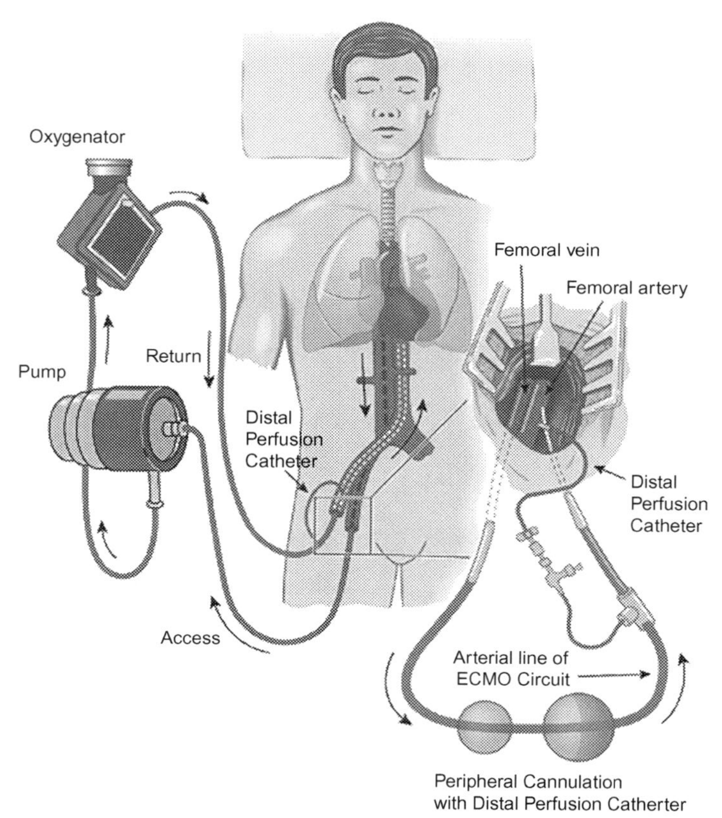

Fig. 7.13 ECMO arrangement. *(From M. Dennis, In depth extracorporeal cardiopulmonary resuscitation in adult out-of-hospital cardiac arrest, J. Am. Heart Assoc. 9 (10) (2020) 1–12. https://doi.org/10.1161/JAHA.120.016521.)*

which exchanges oxygen with blood. Since the membrane exposes blood to oxygen, the direct contact between blood and oxygen gets restricted and therefore probability of air embolism also gets reduced. However, solubility of the gas and its diffusivity greatly influences the activity of ECMO [8]. Bio-artificial liver support is a membrane-based bioreactor (Fig. 7.14), which is applied to substitute the function of a damaged and failed liver. The bioreactor consists of a high-density hepatocyte culture under subsequent oxygenation. The bioreactor consists of a gas-permeable membrane on which porcine hepatocytes are cultured along with nonparenchymal cells. In this scheme, the cells remain in polarized condition and help in maintaining normal liver functions. Polyethersulfone are the most commonly used membrane for human artificial liver preparation [8].

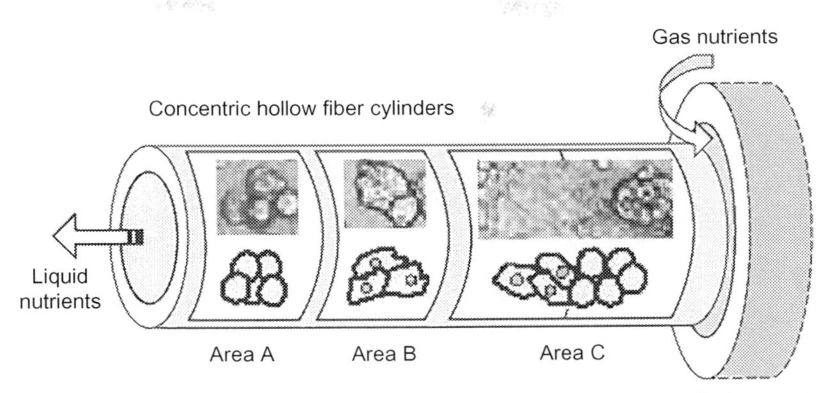

Fig. 7.14 Artificial liver: The outer surface of a hollow fiber cylinder with three unique surfaces used to exclusively activate distinctive receptor-ligand cell attachments. Area A indicates spheroids of hepatocytes attached onto Matrigel. Area B indicates hepatocytes that have spread into a monolayer when on type I collagen. Area C shows the interaction between both culture systems as different cell morphologies are brought together. *(From R.E. McClelland, L.M. Reid, Bioartificial livers, in: A. Atala, R. Lanza, J.A. Thomson, R.M. Nerem (Eds.), Principles of Regenerative Medicines. Academic Press, 2008, pp. 928–945. https://doi.org/10.1016/B978-012369410-2.50055-3.)*

Patients suffering from prolonged diabetes show abnormal pancreatic function and require insulin in type II diabetes to maintain proper glucose level in the body through maintaining the absorption of blood glucose. Long-term application of insulin injection may lead to other diabetic complications and kidney failure [8]. To eradicate this problem, the development of artificial pancreas is in utmost demand. Hollow-fiber- or flat-sheet membranes are used to integrate islets of Langerhans. This membrane allows glucose and insulin to pass through while restricting lymphocytes and immunoglobulins [11]. However, not restricted to the aforesaid applications, utilization of membrane technology can also be observed in the treatment of several ailments. One of the most common applications with membrane technology can be seen for hemodialysis process (Fig. 7.15). Erroneous kidney function increases the concentration of various metabolic wastes such as urea and creatinine within the blood that adversely affects human health. Therefore, chronic kidney disease (CKD) requires arrangement such as hemodialysis in order to remove such toxins. The principal mechanism of hemodialysis is blood flow over a polymeric membrane, where the permeate side is fed with a saline sweep in order to restrict the passage of ions from blood through the membrane. Different polymers have been used for fabricating hemodialysis membrane such as cellulose acetate,

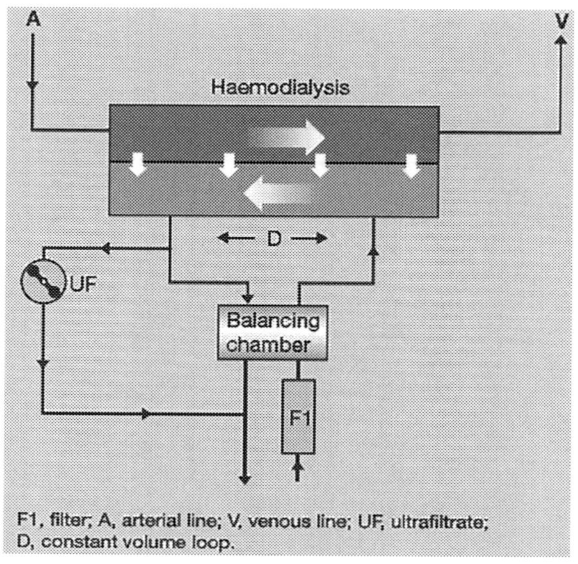

Fig. 7.15 Hemodialysis schematic. *(From O. Swift, E. Vilar, K. Farrington, Haemodialysis, Medicine 47 (9) (2019) 596–602. https://doi.org/10.1016/j.mpmed.2019.06.004.)*

polyethersulfone, etc. Hydrophilicity, biocompatibility, high porosity, protein retention capacity, and strong mechanical stability are some of the basic fundamental characteristics required for dialysis membrane. However, the fundamental intricacy associated with the hemodialysis process is the accumulation of protein-bound uremic toxins in the blood, which may cause artificial fibrillation (AF) in patients leading to fatal condition. Satisfactory removal of uremic toxins has not yet been achieved using the hemodialysis process and requires serious investigation [12].

7.2.3 Complexities: Membrane fabrication and application

The fundamental intricacy associated with the application of membrane technology is the selection of the casting material, which further influences the selectivity and permeability of the membrane [13]. Hence, modification of the membrane according to the application requires a proper control of the preparation of casting solution through mixing modifiers with the polymers during membrane fabrication. Since modifiers determine the property of the membrane, selection of such material is an intricate and delicate part during fabrication. For example, in case of tissue engineering, as discussed in the previous section, chitin membrane exhibits poor mechanical strength and requires certain modifiers such as hydroxyapatite to overcome such limitations. Further, during hemodialysis biocompatibility as well as recyclability of the membrane are the major issues that might be compensated through the usage of synthetic membranes such as PES, poly(methyl methacrylate), etc., compared to conventional cellulose acetate membrane. The other major intricacy associated with the effective reusability of the membrane is fouling, which might be either reversible or irreversible. During the separation for a long-run membrane, pores are getting blocked by the solutes to be separated and is manifested by a steady decline in the permeate flux. Hence, to continue the process membranes are taken for washing intermittently that eventually increases the net cost of the process and reduces the membrane life with chemical wash. Fig. 7.16 represents an illustration of the fouling process along with its cleaning methods.

Therefore, several studies had been done so far in order to fabricate the membrane through casting modifiers with the base polymer so that the fouling of the membrane could either be eliminated or minimized, ensuring the membrane usage for a long run. Engineered nanomaterials are nowadays a

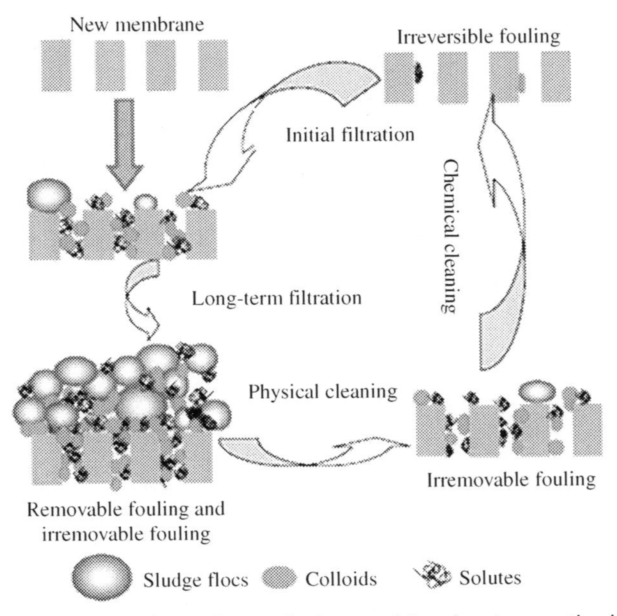

Fig. 7.16 An illustration of membrane fouling and its cleaning methods. *(From P.K. Gkotsis, D.C. Banti, E.N. Peleka, A.I. Zouboulis, P.E. Samaras, Fouling issues in membrane bioreactors (MBRs) for wastewater treatment: major mechanisms, prevention and control strategies, Processes 2 (4) (2014) 795–866. https://doi.org/10.3390/pr2040795.)*

mostly discussed topic along with membranes, where the incorporation of such materials with membranes will manifest a new dimension to the membrane science through alleviating the cons of the membrane application. In the subsequent portions of this chapter a thorough discussion will be made on the basic chemistry of nanomaterials, their preparation, and inclusion within the membrane to substantiate its breakthrough in biomedical fields.

7.3 Identification of sources for nanomaterial synthesis

The property of the nanomaterials primarily relies upon the nature of the parental components from where they are fabricated. Therefore, selection of feedstock has substantial impact on the properties of the prepared nanomaterials and its further derivatives. For example, nanomaterials like chitin have several applications in tissue engineering and drug delivery. Chitin is naturally found in the exoskeletons of shrimps, crayfish, and crabs along with fungal cell walls. During its application in a biomedical field, it is mainly converted into its deacytelated form chitosan as shown in Fig. 7.17.

Now, depending on the source of chitin, it exists as two allomorphs namely α-chitin and β-chitin. In case of β-chitin, the polymer chain occurs in a parallel form, while an antiparallel arrangement is found in α-chitin as shown in Fig. 7.18. However, the parallel morphology in β-chitin provides more flexibility in structure as compared to α-chitin [14]. On the other hand, α-chitin is more rigidly crystalline as compared to β-chitin and is most commonly applied for scaffold generation, which is an imminent step in tissue engineering.

One of the most used nanomaterials nowadays is either graphene or its different moieties, which are fabricated from carbonaceous sources. Graphene represents a hexagonal arrangement of sp^2 hybridized carbon atoms arranged either in single layer or in multilayer as shown in Fig. 7.19. Gene

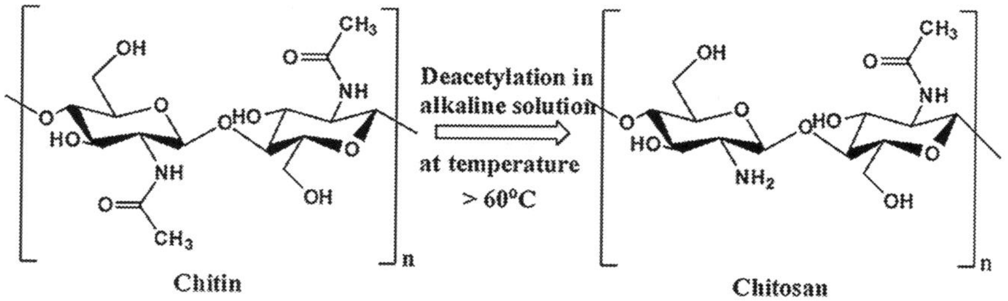

Fig. 7.17 Deacetylation of chitin to chitosan. *(From P.A. Alaba, N.A. Oladoja, Y.M. Sani, O.B. Ayodele, I.Y. Mohammed, S.F. Olupinla, W.M.W. Daud, Insight into wastewater decontamination using polymeric adsorbents, J. Environ. Chem. Eng. 6 (2) (2018) 1651–1672. https://doi.org/10.1016/j.jece.2018.02.019.)*

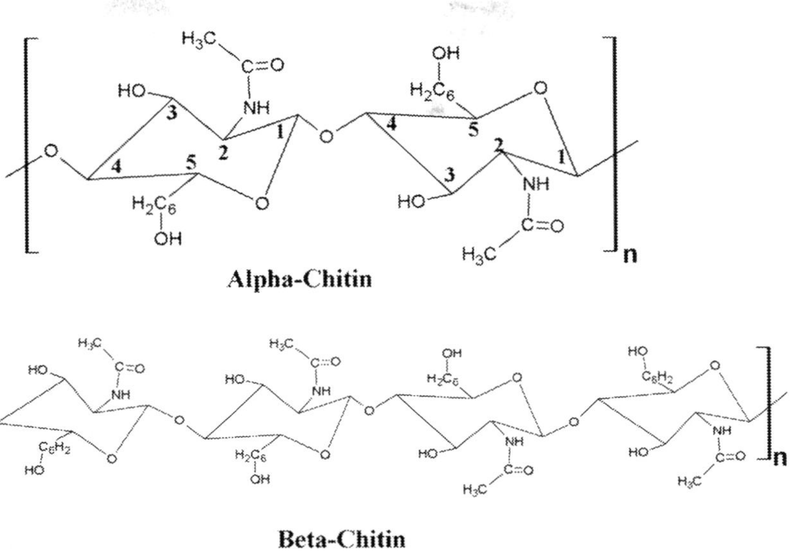

Alpha-Chitin

Beta-Chitin

Fig. 7.18 Molecular structure of α-chitin and β-chitin. *(From K.B. Rufato, J.P. Galdino, K.S. Ody, A.G.B. Pereira, E. Corradini, A.F. Martins, A.T. Paulino, A.R. Fajardo, F.A. Aouada, F.A.L. Porta, A.F. Rubira, E.C. Muniz, Hydrogels based on chitosan and chitosan derivatives for biomedical applications, in: L. Popa, M.V. Ghica, C. Dinu-Pirvu (Eds.), Hydrogels-Smart Materials for Biomedical Applications, InTechOpen, 2018, pp. 59–98. https://doi.org/10.5772/intechopen.81811.)*

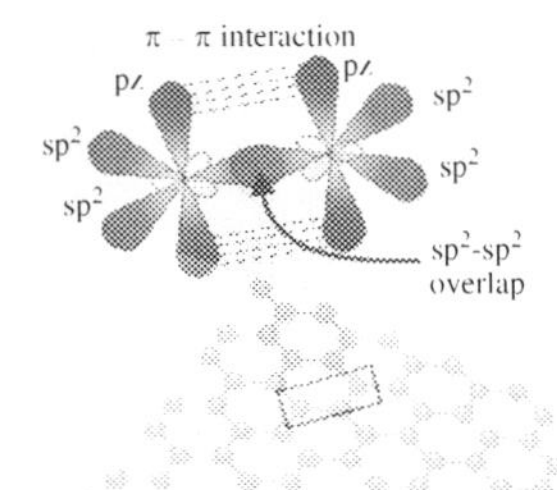

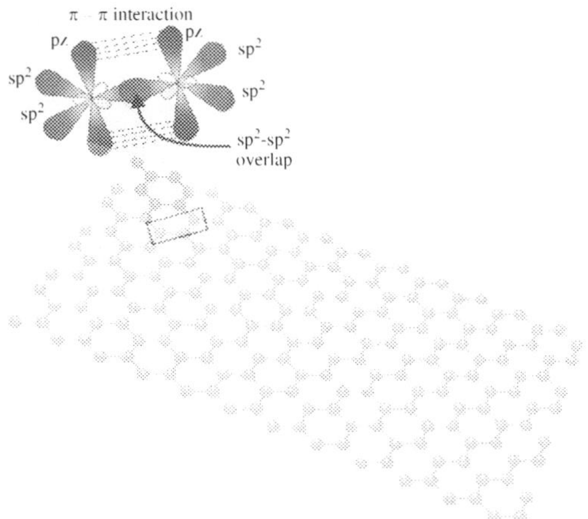

Fig. 7.19 Structure of single graphene layer with sp^2 hybridization. *(From H.-J. Lee, J.-G. Yook, Graphene nanomaterials-based radio-frequency/microwave biosensors for biomaterials detection, Materials 12 (6) (2019) 1–13. https://doi.org/10.3390/ma12060952.)*

Table 7.1 Different sources of graphene and its properties.

Source	Specific surface area ($m^2 g^{-1}$)	Layer	References
Glucose	820	Monolayered	[15]
Rice husk	2200	Monolayered	[15]
Waste tires	644.8	Monolayered	[16]
Peanut shells, walnut shells and almond shells	2070	Few-Layered	[17]
Sugarcane bagasse agricultural waste	320	Few-Layered	[18]
Printed Circuit Boards (PCB)	–	–	[19]
Elm tree waste	1085	–	[20]

therapy, drug delivery, cancer therapy, biosensors, and biomedical imaging are some of the remarkable areas where graphene-based nanomaterials are extensively used. Depending on its source, properties of graphene will also vary as shown in Table 7.1.

Nanomaterial synthesis is mainly done through two approaches—the top-down approach and the bottom-top approach. In case of top-down approach, fabrication starts with a bulk material followed by a gradual reduction into nano-scale products, while in the bottom-top approach agglomeration of atoms forms nano-scale materials. Based on these approaches, several methods have been developed for the preparation of nanomaterials and are described in the subsequent section.

7.4 Preparation of nanomaterials

7.4.1 Sol-gel method

The sol-gel method is a chemical process that involves a solution of precursors such as metal oxides or metal chlorides or colloidal particles to produce an integrated network through gel formation with the particles at the nanoscale, as shown in Fig. 7.20.

The name itself suggests that it transforms a colloidal liquid solution ("sol") to a gel formation. In general, the sol-gel occurs in two steps: hydrolysis and polycondensation. Metal alkoxides and metal chlorides are typical antecedents, which are being hydrolyzed followed by polycondensation to form a colloidal framework made out of nanoparticles scattered in a solvent. Subsequently, the solvent is removed either by evaporation or by drying and

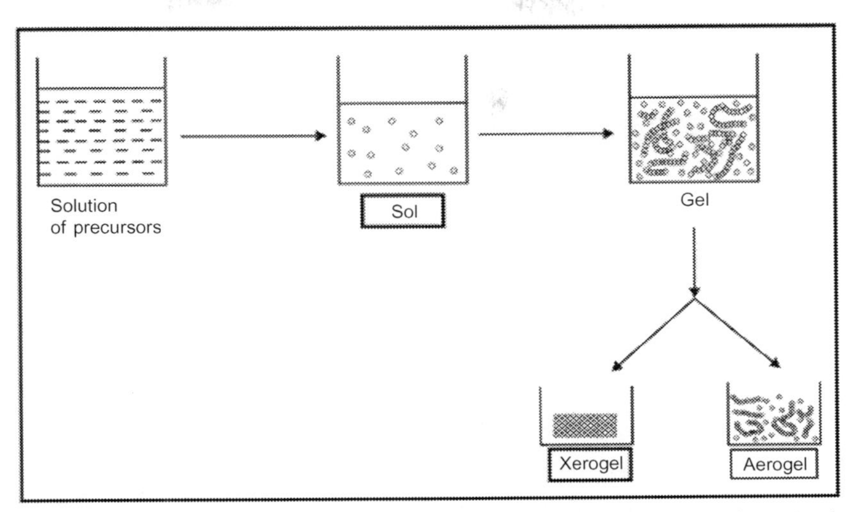

Fig. 7.20 Schematic diagram of sol-gel process. *(From Heterogeneous Catalysis: Catalyst preparation—Sol-gel, 2017. http://eacharya.inflibnet.ac.in/data-server/eacharya-documents/55daa452e41301c73a2cb5ac_INFIEP_208/337/ET/1.html. (Accessed 9 January 2021).)*

leaving behind a gel in the form of particles' network in a liquid phase. The gel formed after evaporation of the solvent is known as "xerogel" and the gel formed through drying process is called "aerogel." Fig. 7.20 shows the schematic representation of the method.

Initially, a solution is prepared by dissolving the precursor into a suitable solvent, usually ethanol. Hydrolysis of the material is done by adding small amounts of water with some acidic medium. The material produced on hydrolysis undergoes polycondensation to form a gel whose pores are filled with liquid phase. Finally, it undergoes heat treatment or drying to remove liquid from the gel and form aerogel or xerogel. Iron-based nanomaterial magnetite was successfully synthesized using the sol-gel process, and this process was found to be capable of synthesizing other metal oxide nanoparticles after oxidation of iron nanoparticles [21].

7.4.2 Chemical vapor deposition method

Fig. 7.21 shows a basic schematic for the chemical vapor deposition (CVD) method. It involves a reaction between a predetermined substrate (wafer) supported on a susceptor and volatile precursor containing the required material for nanoscale production, which finally gets deposited on the surface of the wafer. The whole process occurs under vacuum at high

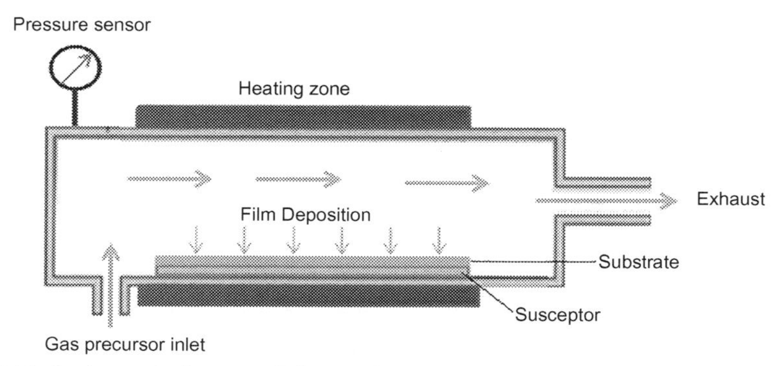

Fig. 7.21 A schematic diagram of chemical vapor deposition process. *(From Q. Zhang, D. Sando, V. Nagarajan, Chemical route derived bismuth ferrite thin films and nanomaterials, J. Mater. Chem. C 4 (19) (2016) 4092–4124. https://doi.org/10.1039/c6tc00243a.)*

temperature. Some volatile byproducts are also produced during the reaction, which are removed from the system by a continuous gas stream (usually N_2). The whole process occurs in five sequential steps: (i) transport of the reacting component to the wafer surface; (ii) absorption of the target species on the substrate surface; (iii) heterogeneous surface reaction between vapor phase and substrate; (iv) desorption of byproducts of the vapor phase reaction; and (v) transportation of byproducts away from the surface of the substrate [22].

7.4.3 Laser pyrolysis method

Laser pyrolysis consists of a chamber, where the vapor stage of the precursor is acquainted by a carrier gas (e.g., argon) to get into contact with the laser beam, as shown in the Fig. 7.22. Due to the high power of the laser beam (e.g., 2400 W), temperature (10,000°C) has been increased in the chamber, which triggers the nucleation rate to form nanoparticles. The nanoparticles are gathered by a "sample collector filter." If the reactant exists in liquid phase, initially it is vaporized and then introduced in the chamber. The laser beam is introduced on a continuous basis as the entire process depends on the thermal energy generated by the laser beam [23].

7.4.4 Flame spray pyrolysis method

In flame laser pyrolysis, an aqueous solution of metal salt is evaporated and decomposed on a predetermined substrate. The metal solution is sprayed

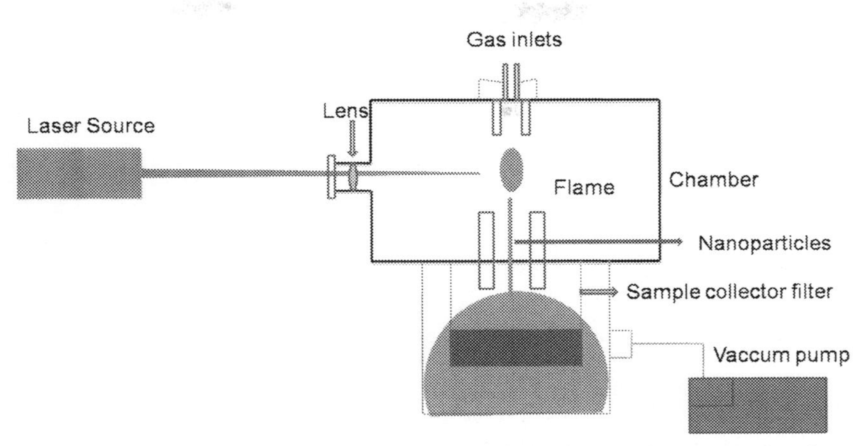

Fig. 7.22 A schematic diagram of laser pyrolysis method. *(From S. Wang, L. Gao, Laser-driven nanomaterials and laser-enabled nanofabrication for industrial applications, in: S. Thomas, Y. Grohens, Y.B. Pottathara (Eds.), Industrial Applications of Nanomaterials, 2019, pp. 181–203. Elsevier. https://doi.org/10.1016/B978-0-12-815749-7.00007-4.)*

through a capillary and comes in contact with a flame as shown in Fig. 7.23. When the solution is sprayed through the capillary, small drops are formed, which is exposed to a flame. The solvent burns out in the flame and metallic salts gets pyrolyzed to metal oxide atoms. The produced metal oxide atoms agglomerate to form nanoparticles through a bottom-top approach. The formulated nanomaterials are collected over the substrate [24]. One of the major advantages associated with this method is that the particle size as well as porosity of the film can be easily controlled [24].

7.5 Economic analysis for different methods

However, selection of the method for the synthesis of nanomaterials is greatly influenced by the total cost of the process. Starting with the sol-gel process, the capital cost of the process is not very high because of its less sophisticated technology. However, the precursor applied for the sol-gel process is quite expensive and researchers are sincerely focusing on minimizing the cost of such precursors in order to make the process more economically viable [25]. On the contrary, in the CVD method requirement of high temperature (the temperature ranges from 450°C to 1050°C) increases its operational cost. Furthermore, a comparative study of laser pyrolysis and flame spray pyrolysis showed that flame spray pyrolysis is much more economical compared to laser spray pyrolysis. High-end instruments such as a

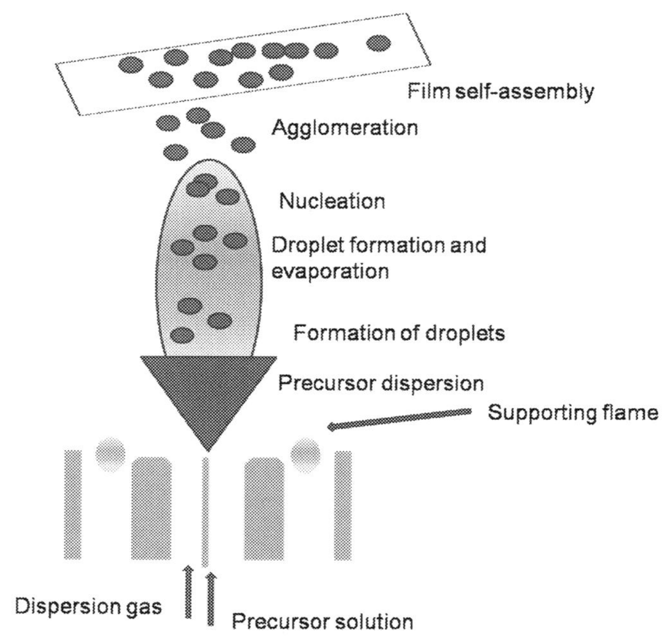

Fig. 7.23 A schematic diagram of flame spray pyrolysis method. *(From D. Nunes, A. Pimentel, L. Santos, P. Barquinha, L. Pereira, E. Fortunato, R. Martins, Synthesis, properties and applications, in: G. Korotcenkov (Ed.), Metal Oxide Nanostructures, first ed., Elsevier, 2019, pp. 21–57. https://doi.org/: https://doi.org/10.1016/B978-0-12-811512-1.00002-3.)*

vacuum chamber and laser arrangement along with costly precursors are some of the factors responsible for making laser pyrolysis an expensive method [26]. However, functionalization of the nanomaterials specific to the application is a big challenge, when the economy of the process will be considered. Therefore, in the next few sections a thorough discussion will be made on the functionalization of the nanomaterials to gain a notable understanding of the economy on application-specific production of functionalized nanomaterials.

7.6 Functionalization of nanomaterials

7.6.1 Chemistry behind the functionalization

Light dispersion and scattering properties of nanomaterials, called "plasmonic properties," as well as its chemistry with other materials plays a major role in its application for biomedical purposes. Nanomaterials confront certain limitations with respect to their plasmonic characteristics and hence, the interaction with the surrounding environment may also have some restrictions. Functionalization of nanomaterials leads to their stabilization and enhances the optical properties of the nanomaterials [27]. Additionally, nanomaterials which are not functionalized often undergo agglomeration and show certain toxic behaviors. To alleviate such possibilities toward agglomeration and loss of inherent property, functionalization of nanomaterials is essential prior to its application. In case of the surface functionalization, certain suitable foreign materials, such as polymers, biomolecules, etc., are incorporated within the nanomaterials that prevent its agglomeration, preparing it appropriately for several further applications [28]. Applications of functionalized nanomaterial are quite common in drug delivery, sensor fabrication, bioinstrumentations, and other industrial processes. Fabrication of functionalized nanomaterials is primarily done in two methods—presynthesis and postsynthesis. In presynthesis, functionalization is done during the preparation of nanomaterials. Here, materials adopted for functionalization are added during the synthesis of nanomaterials. Gold nanomaterials are prepared in the presynthesis method after reduction of aqueous phase choloauric acid [27]. However, the most common method for functionalization is the postsynthesis method. In this method, foreign materials for functionalization are coated on the surface of the nanomaterials without nurturing the parental properties of the nanomaterials. The postsynthesis method can be further classified into various types depending on the nature of the functionalization. These are noncovalent functionalization, covalent functionalization, physical adsorption, and grafting, as described below.

7.6.1.1 Noncovalent functionalization

In this method, functionalization is done based on the surface charge of the nanomaterial. A functional component gets attached to the surface of the nanomaterial through noncovalent bonding like pi-pi interactions, Van der Waal's bond, electrostatic charge, and hydrogen bonding. Gold nanoparticles coated with polymer polyethylene glycol were incorporated with amino-functionalized silicon phthalocyanine through noncovalent hydrophobic interactions. The modified nanoparticle was applied for drug loading [29]. Tabakman et al. in 2009 modified single-walled carbon nanotubes with polyethylene glycol and were applied as intravenous injection [30].

7.6.1.2 Covalent functionalization

In covalent functionalization, various chemical species like biomolecules, small organic and inorganic molecules, and polymers are attached to nanoparticles by virtue of the covalent bond. Here, initially one functional group of a long chain is attached to the surface of the nanoparticles, while the other end gets coupled to another functional material. The process can be further classified into direct and indirect method. In the direct method, the surface charge of the nanomaterials forms strong covalent bonds with the species used for functionalization. Coupling of cysteine with gold nanoparticles is an example of direct covalent functionalization, where after functionalization it is applied in biochemical analysis. Cysteine is an amino acid, which consists of thiol group protrusion in between two terminals containing amine on one side and carboxylic group on the other side. The electrostatic attraction between the thiol group and metal nanoparticles mobilizes the conjugation of cysteine over the gold nanoparticles' surface. On the other hand, in indirect methods oxygen-containing groups are primarily attached to the surface of the nanoparticles, which is further coupled with desired component for functionalization. In an example, primarily the thiol group is coordinated with a carboxyl group on a polymer resin support, which is further incubated in a butanethiol-resistant gold nanoparticle solution. After a one-to-one phase exchange reaction between a thiol-attached polymer support and nanomaterials, functionalized nanoparticles associated with carboxylic group were separated out from the medium [31]. Fig. 7.24

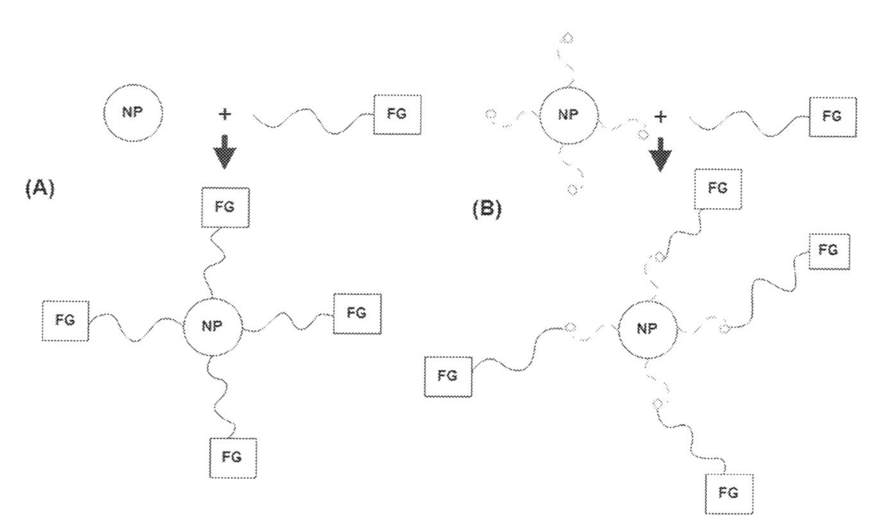

Fig. 7.24 A schematic diagram of (A) direct method and (B) indirect method (*FG*, functional group; *NP*, nanoparticle; *O*, oxygenated carbon group). *(No Permission Required.)*

illustrates the formation of functionalized nanoparticles through direct and indirect covalent functionalization.

7.6.1.3 *Physical adsorption*

Like the noncovalent functionalization method, here the functional groups bind onto the surface of the nanomaterials through adsorption due to affinity based on charge difference over the atom. Fig. 7.25 shows a representation of the physisorption, where functionalization has been done due to charge difference between the surface and the material. The coating acts as a passive layer and restricts agglomeration of the nanomaterials [31].

7.6.1.4 *Grafting*

Grafting is similar to covalent functionalization, where polymers are mainly attached on the surface of the nanomaterials during their functionalization. Nanomaterials functionalized through this method are mainly used as sensors as well as for several adsorption purposes. Fig. 7.26 illustrates the grafting of polyethylenimine into graphene.

7.6.2 Biocompatibility with functionalized nanomaterials

One of the important criteria for the application of functionalized nanomaterials in the biomedical field is their compatibility with the living cell tissues, known as biocompatibility. It must be ensured that when nanoparticles are in contact with human cells and tissues, no toxic response along with follow-up inflammation may occur. Shi et al. [32] studied the functionalization of

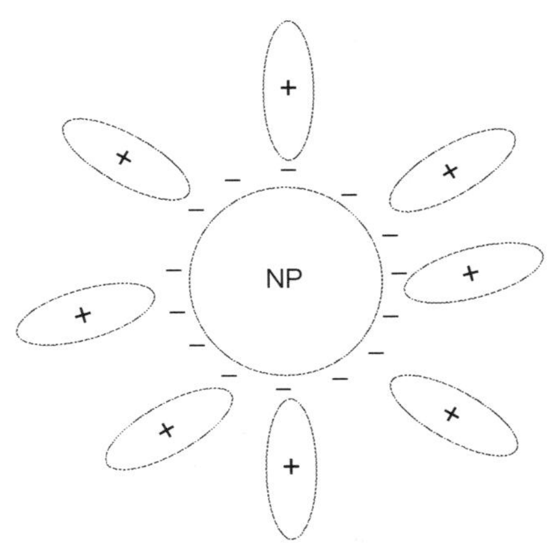

Fig. 7.25 An illustration of physisorption method of functionalization of nanomaterials. *(No Permission Required.)*

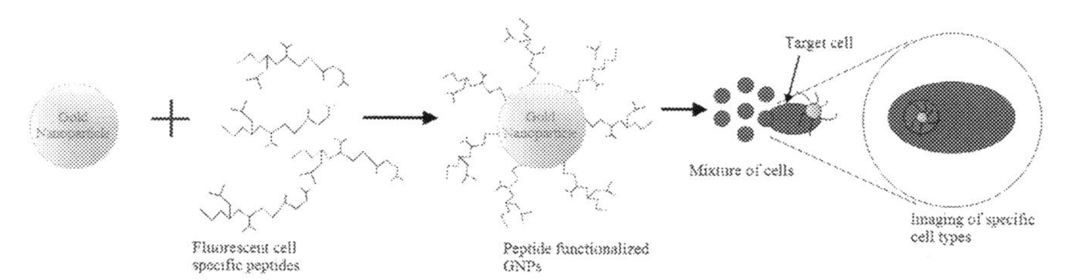

Fig. 7.26 A schematic diagram of imaging by peptides functionalized gold nanoparticles. *(From T. Pooja, V. Komal, D. Vida, S. Shree, Functionalized gold nanoparticles and their biomedical applications, Nanomaterials 1 (2011) 31–63. https://doi.org/10.3390/nano1010031.)*

gold nanoparticles with poly(amidoamine) dendrimers followed by an analysis of the cytotoxic effect of functionalized gold nanoparticles on the human body during drug delivery. According to their results, it was nontoxic to the human epithelial carcinoma cell line [32]. A comparative study on the toxic effect of dextran-functionalized reduced graphene oxide and conventional reduced graphene oxide was done on mammalian cell lines by Kim et al. [33]. The study revealed that the toxicity of dextran-functionalized reduced graphene oxide was comparatively much less compared to conventional reduced graphene oxide [33]. In 2013, Thorat et al. synthesized rare-earth-based magnetic nanoparticles $La_{0.7}Sr_{0.3}MnO_3$ functionalized with dextran for treating cancer through magnetic fluid hyperthermia. The study revealed that the biocompatibility of dextran-coated nanoparticles showed less cell toxicity as compared to bare nanomaterials [34]. Nowadays, carbon nanotubes have attracted much attention in targeted drug delivery. However, the low biocompatibility of carbon nanotubes restricts their application urging surface functionalization. Carbon nanotubes functionalized with succinylated derivative of β-lactoglobuline was much more biocompatible as compared to normal multiwalled carbon nanotubes. However, there was no significant difference in biocompatibility between β-lactoglobuline-incorporated carbon nanotubes and polyethylene glycol-incorporated carbon nanotubes [35]. Therefore, functionalization of nanomaterials not only inhibits the morphological changes of the nanoparticles, but also increases its biocompatibility compared to its pristine form, which makes its applications much wider in various biomedical purposes.

7.6.3 Imaging enrichment with functionalized nanomaterials

In order to better understand various inter- and intra-cellular biochemical processes, imaging is one of the most dependable techniques for proper diagnosis of any health-related issues. In this regard, functionalized nanomaterials have shown satisfactory performances in the advancement of imaging applications. Graphene-based functionalized nanomaterials are widely used in cancer cell imaging to help identify the actual target for faster recovery. Yin et al. [36] synthesized fluorescent silver nanoclusters and functionalized them with cancer-targeted DNA aptamers. The study revealed that the functionalized nanomaterials effectively specified lymphoblastic leukemia cells in the human body with immense luminescence [36]. A cobalt-ferrite-based nanoparticle was enclosed with rhodamine that was fluorescent in nature and the entire arrangement was encapsulated within a silica shell grid. The nanomaterial was further attached with AS1411 aptamer targeting specific proteins expressed by cancerous cells. This nanomaterial was studied for imaging cancerous cell by Hwang et al. The study revealed that successful imaging of cancerous cells was achieved with these nanomaterials, which enhances the diagnosis and proper treatment of cancer patients [37]. Functionalized magnetic nanoparticles are also widely applied for magnetic resonance imaging (MRI) techniques in cancer cell identification. Among various magnetic nanoparticles, iron oxide nanoparticles are majorly used due to their biocompatibility and low toxicity. Yang et al. [38] studied the effect of iron oxide-reduced graphene oxide nanoparticles grafted with polyethylene glycol polymer for magnetic resonance imaging of cancerous cells. According to them, polyethylene glycol-attached iron oxide-reduced graphene oxide nanoparticles showed efficient imaging for cancerous tumor as compared to normal graphene oxide nanoparticles [38]. Gold nanoparticles functionalized with specific biomolecules like RNA, DNA, drugs, peptides, proteins, and other such molecules are very much effective for imaging tumorous cells [39].

7.6.4 Targeted drug delivery with functionalized nanomaterials

One of the major applications of functionalized nanomaterials can be seen in designing drugs toward their targeted delivery. The primary requirements for any drug are its biocompatibliity, high surface area, selectivity, and large surface to volume ratio. During antibiotic formulation, carbon-based nanoparticles are used as an encapsulant, where the drug exists as the dispersed phase within the core for controlled drug delivery without any collateral side effects [40]. Graphene-based nanomaterials like graphene oxide and reduced graphene oxide have proven effective in formulating nanodrugs. Drug molecules are usually hydrophobic in nature, where either they are attached to graphene oxide through pi-pi interactions or the targeting molecules are

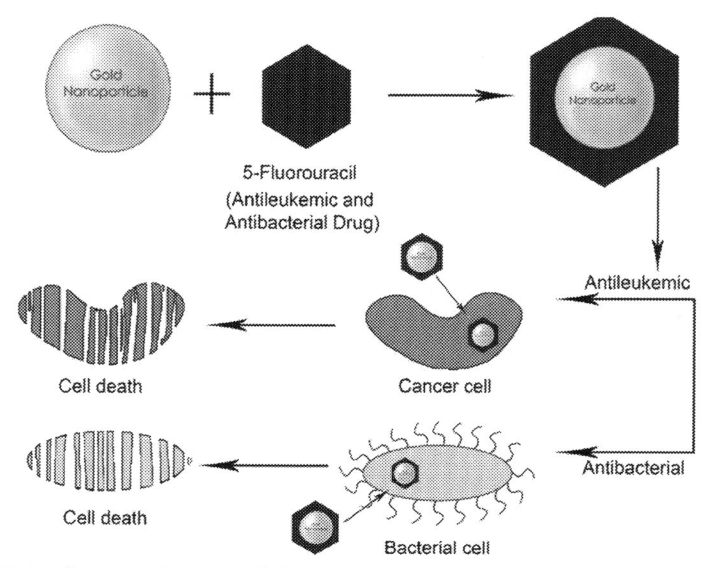

Fig. 7.27 A schematic diagram of drug delivery by functionalized gold nanoparticles. *(From T. Pooja, V. Komal, D. Vida, S. Shree, Functionalized gold nanoparticles and their biomedical applications, Nanomaterials 1 (2011) 31–63. https://doi.org/10.3390/nano1010031.)*

attached to graphene oxide through covalent bonds [41]. On the contrary, functionalized gold nanoparticles are also applied in nanodrug formulation for targeted delivery. Functionalization of gold nanoparticles with required biomolecules can effectively damage bacteria as well as cancerous cells, as shown in Fig. 7.27 [39].

7.6.5 Theranostics enhancement with functionalized nanomaterials

Theranostics is an amalgamated technique of therapeutics and diagnostics image-based treatment aiding in identifying the treatment consequences earlier [42]. Gold nanoparticles functionalized with heparin molecules for specific detection of metastatic cancerous cells with its subsequent destruction is an example of a nano-based theranostics approach. Metastatic cancerous cells secrete heparinase/heparanase enzymes, which enhances fluorescence signals that detect cancerous cells [43]. Photothermal therapy consists of a photo-absorbing material, which converts optical energy to heat energy and utilizes the heat energy to burn out cancerous cell. In this regard, functionalized iron oxide-reduced graphene oxide nanoparticles grafted with polyethylene glycol polymer were applied for photothermal therapy of cancerous cells, where efficient ablation of cancerous tumor can be achieved at a low laser power density compared to that of gold nanomaterials [38]. However, the potential of functionalized graphene-based nanomaterials for theranostic activity is yet to be fully achieved even after several decades of research on it [44]. Li et al. [45] functionalized polypyrrole nanoparticles with PEGylated indocyanine green for photothermal tumor detection and their effective ablation compared to bare nanoparticles [45].

7.6.6 Microbe detection with functionalized nanomaterials

In medical circumstances, diagnosis of a particular disease can be achieved by detection of microorganisms associated with the disease. Recent studies reveal that application of functionalized nanomaterials in the form of biosensors is quite effective in bacterial strain detection. Biosensors convert affinity of binding of a targeted analyte into an electrical signal through a transduction process. However, the accuracy and precision of the sensors depend much on the specificity of the binding capability of the analyte. In order to enhance the performance of the biosensors, nanomaterials are incorporated within the biosensors, where their high surface area enhances the binding propensity of the analyte ensuring proper signal resolution. Gold nanoparticles functionalized with poly (para-phenyleneethynylene) were efficiently applied for detecting three different strains of *Escherichia coli* responsible for intestinal diseases [46]. Graphene-based nanomaterials functionalized with DNA aptamers detected antibiotic-resistant *Staphylococcus aureus*, a point of concern nowadays as the superbug in nosocomial infections. Moreover, the technique showed high selectivity, simplicity, and accuracy [47]. Further, gold nanoparticles are functionalized with different ligands and antibodies for the detection of microbial strains. Moreover, microbes of a particular strain can be detected by attachment of sequences of nucleic acid, which is complementary to DNA or RNA of that strain. ssDNA modified by thiol can be associated with gold nanoparticles for detecting microbial strains [48]. Jayagopal et al. [49] functionalized gold nanoparticles with hair-pin DNA to detect living mRNA. The formulated nanoparticles successfully detected living mRNA without any toxic effect. The schematic diagram of gold functionalization with hair-pin DNA is shown in Fig. 7.28 [49].

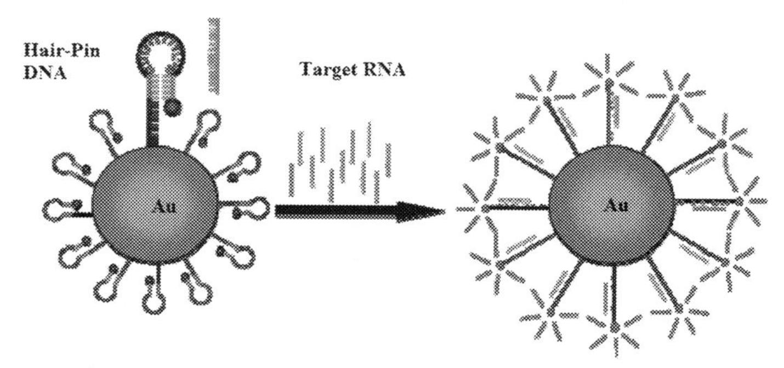

Fig. 7.28 A schematic diagram of gold functionalization by hairpin DNA and its attachment with mRNA. *(From A. Jayagopal, K.C. Halfpenny, J.W. Perez, D.W. Wright, Hairpin DNA-functionalized gold colloids for the imaging of mRNA in live cells, J. Am. Chem. Soc. 132 (28) (2010) 9789–9796. https://doi.org/10.1021/ja102585v.)*

7.7 Membrane modified with functionalized nanomaterials

As mentioned earlier, owing to the requirement of the biological process, certain modifications are made during membrane fabrication to increase its applicability in wider aspects. Incorporation of functionalized nanomaterials within the membrane is one of such surface modification techniques through which one can modulate the property of the membrane. Various properties like hydrophilicity, permeability, selectivity, porosity, mechanical stability, and surface charge density are modified over its pristine form with the functionalized nanomaterials inclusion [50]. During the membrane fabrication with the phase inversion method, functional nanomaterials and polymers are mixed at a certain proportion homogeneously to form the casting solution. Sometimes, dispersants like polymers and surfactants are also added to the solution for evenly dispersing nanomaterials into the casting solution [51]. Nanoparticles are also coated over the membrane surface for its texture modification. Ionita et al. [52] developed graphene oxide and carbon nanotube-doped cellulose acetate membrane through phase inversion method for hemodialysis. Cellulose acetate was initially hydrolyzed with sodium hydroxide, which in turn increases the availability of hydroxyl groups, as shown in Fig. 7.29.

Since hydrolysis of cellulose acetate increases the hydroxyl groups, hydrogen bond formation between cellulose acetate and graphene oxide becomes easier. Graphene oxide nanoparticles and carbon nanotubes functionalized with the amine group get attached to cellulose acetate's —OH

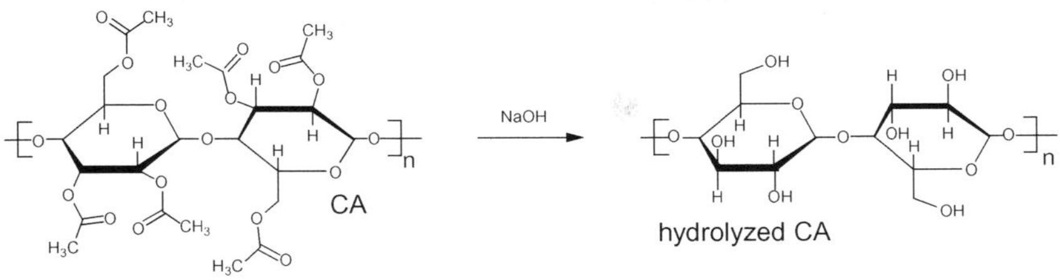

Fig. 7.29 A schematic diagram of hydrolysis of cellulose acetate using sodium hydroxide. *(From S.I. Voicu, R.M. Condruz, V. Mitran, A. Cimpean, F. Miculescu, C. Andronescu, M. Miculescu, V.K. Thakur, Sericin covalent immobilization onto cellulose acetate membrane for biomedical applications, ACS Sustain. Chem. Eng. 4 (3) (2016) 1765–1774. https://doi.org/10.1021/acssuschemeng.5b01756.)*

group through condensation. The schematic diagram has been illustrated in Fig. 7.30. The study revealed that retention of protein, especially bovine serum albumin was maximum, when cellulose acetate membrane was modified with both graphene and carbon nanotubes. Furthermore, graphene incorporation in cellulose acetate membrane shows much improved biocompatibility [52]. One of the major issues with the membrane is biofouling because of the water-repellent nature of the polymeric membrane and this can be reduced once the hydrophilicity of the membrane gets increased. In a study made by Zhu et al. [53], polysulfone membrane was coated with heparin (anticoagulant) immobilized copper hydroxide nanofibers. They found that the water contact angle became decreased attributing to an increase in the hydrophilicity of the membrane. Copper hydroxide nanofibers are produced through the reaction between 2-aminoethanol and copper nitrate solution at 4°C for a reaction time of 3 days. Over the fibers heparin solution was added dropwise followed by its filtration with a polysulfone membrane, which was further oven dried at 25°C [53]. Fig. 7.31 shows a schematic of

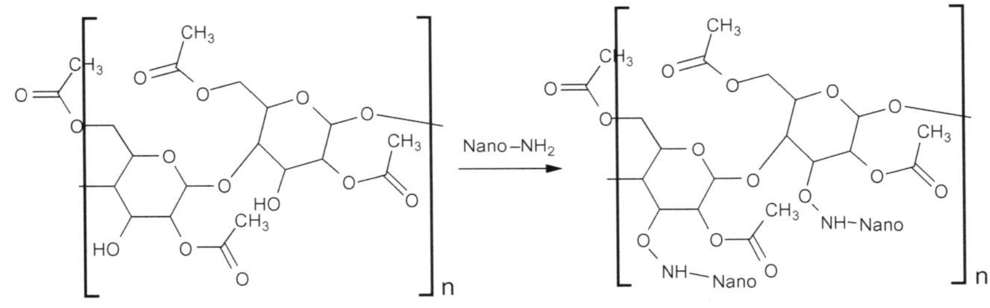

Fig. 7.30 A schematic diagram of cellulose acetate with amine-functionalized carbon nanomaterials. *(No Permission Required.)*

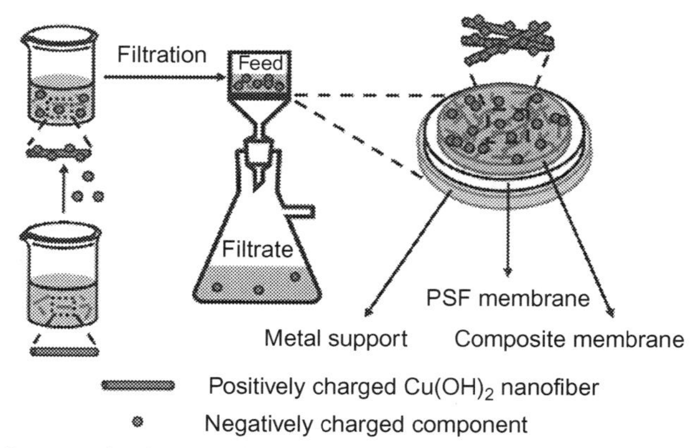

Fig. 7.31 Schematic for the preparation of polysulfone membrane coated with heparin immobilized on copper hydroxide nanofibers. *(From L.J. Zhu, L.P. Zhu, Z. Yi, J.H. Jiang, B.K. Zhu, Y.Y. Xu, Hemocompatible and antibacterial porous membranes with heparinized copper hydroxide nanofibers as separation layer, Colloids and Surfaces B: Biointerfaces 110 (2013) 36–44. https://doi.org/10.1016/j.colsurfb.2013.04.020.)*

the process and Fig. 7.32 shows the SEM images of the membrane after immobilization.

In a very recent study, Nan et al. [54] modified cellulose acetate membrane with ammonia functionalized graphene nanoparticles as shown in Fig. 7.33 to develop electroactive polymer (EAP) actuators, which has applications in the preparation of artificial muscles and other biomedical instruments. Furthermore, incorporation of functionalized nanoparticles along with polyvinylidenedifluoride (PVDF) in cellulose acetate membrane increased its durability, bending performance, and electrical signal response in the fabrication of artificial muscles [54]. Kumar et al. [55] studied a modified chitin scaffold, much applied for tissue engineering purposes, with silver nanoparticles, where silver nanoparticles were added to β-chitin hydrogel. After incorporation of silver nanoparticles into the hydrogel, it was cooled and dehydrated under vacuum for 48 hours to obtain a silver nanoparticle-induced chitin scaffold [55]. Here, silver ions get physically adsorbed within the chitin hydrogel through Van der Waals force [56]. Since silver shows excellent antibacterial property, the silver nanoparticles with β-chitin scaffold resist bacterial growth as well as showed no toxic effect. In 2009, Madhumathi et al. prepared a silica-incorporated chitin scaffold for bone tissue engineering applications, where biocompatibility was found to be higher with the modified membrane as compared to the normal chitin

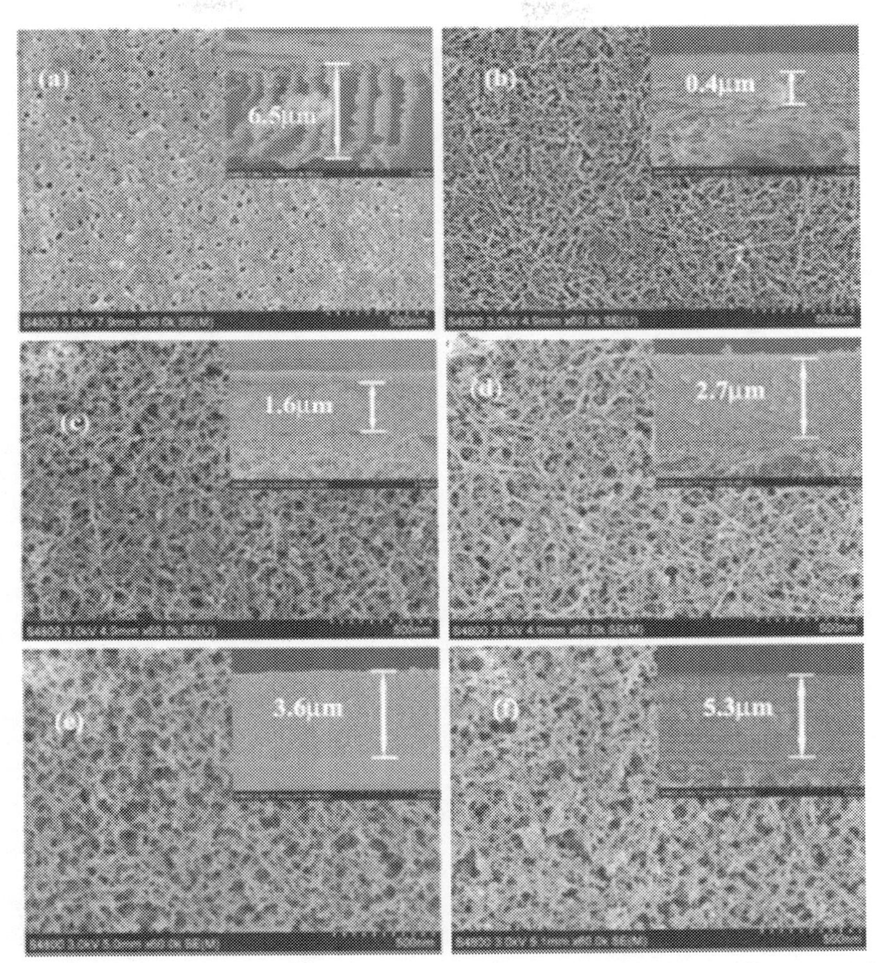

Fig. 7.32 (A) Pristine polusulfone membrane; (B) copper hydroxide nanofibers-coated polusulfone membrane; (C–F) polysulfone membrane coated with heparin immobilized on copper hydroxide nanofibers at different concentrations (65.9, 164.4, 212.0, 408.4 $\mu g/cm^2$, respectively). *(From L.J. Zhu, L.P. Zhu, Z. Yi, J.H. Jiang, B.K. Zhu, Y.Y. Xu, Hemocompatible and antibacterial porous membranes with heparinized copper hydroxide nanofibers as separation layer, Colloids and Surfaces B: Biointerfaces 110 (2013) 36–44. https://doi.org/10.1016/j.colsurfb.2013.04.020.)*

scaffold. Moreover, incorporation of silica nanoparticles enhances the porosity of the scaffold.

Li et al. [45] fabricated graphene oxide-functionalized collagen membrane loaded with *N*-acetyl cysteine for wound healing and the structure has been shown in Fig. 7.34. High capacity of water retention, porosity,

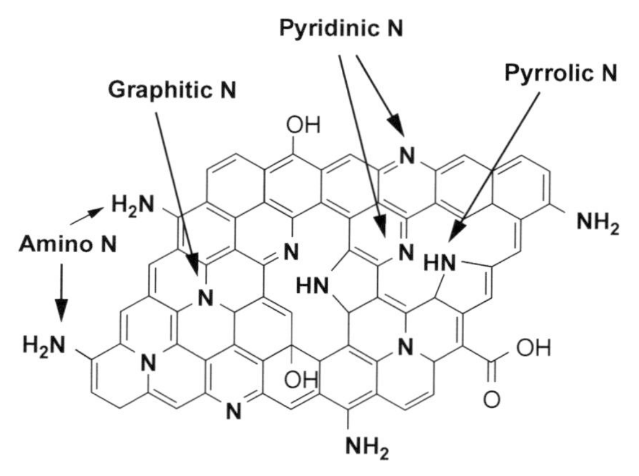

Fig. 7.33 A schematic diagram of amino-functionalized graphene. *(From C. Zhang, R. Hao, H. Liao, Y. Hou, Synthesis of amino-functionalized graphene as metal-free catalyst and exploration of the roles of various nitrogen states in oxygen reduction reaction, Nano Energy 2 (1) (2013) 88–97. https://doi.org/10.1016/j.nanoen.2012.07.021.)*

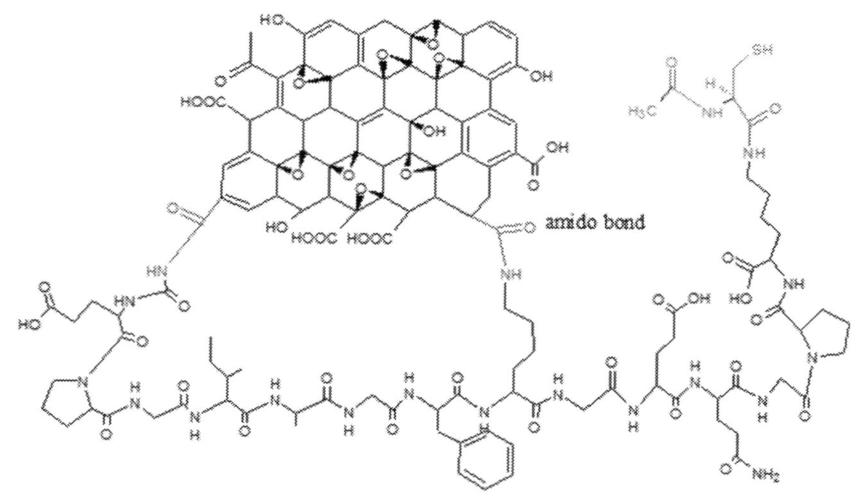

Fig. 7.34 Illustration of structure of graphene oxide-functionalized collagen membrane loaded with *N*-acetyl cysteine. *(From J. Li, C. Zhou, C. Luo, B. Qian, S. Liu, Y. Zeng, J. Hou, B. Deng, Y. Sun, J. Yang, Q. Yuan, A. Zhong, J. Wang, J. Sun, Z. Wang, N-acetyl cysteine-loaded graphene oxide-collagen hybrid membrane for scarless wound healing, Theranostics 9 (20) (2019) 5839–5853. https://doi.org/10.7150/thno.34480.)*

biocompatibility, and high antioxidative capacity are some of the characteristics required with a wound-healing scaffold. Fabricated graphene oxide-functionalized collagen membrane loaded with *N*-acetyl cysteine showed high water retention capacity along with enhanced biocompatibility.

Further, the antioxidative capacity of the membrane was also high due to the presence of graphene oxide nanoparticles. Functional groups present in graphene oxide nanoparticles facilitate its interaction with various biomolecules assisting in carrying antioxidant *N*-acetyl cysteine [57].

Yu et al. [58] modified poly(lactic acid) membrane to enhance its hydrophilicity, hemocompatibility, and antibacterial properties, where silver nanoparticles stabilized by dextran sulfate along with chitosan were deposited on the aminolyzed poly(lactic acid) membrane in a layer-by-layer arrangement (Fig. 7.35). One of the crucial concerns in hemodialysis is platelets' adhesion on the membrane surface that must be resisted for smooth circulation of blood. The time required for blood coagulation was prolonged with the modified membrane as compared to conventional poly(lactic acid) membrane making it suitable for application in the hemodialysis process. Moreover, human cell viability and proliferation were enhanced by silver incorporation on poly(lactic acid) membrane. Additionally, the modified membrane showed antibacterial effect against Methicilin-resistant *S. aureus* [58]. Ma et al. [59] incorporated graphene oxide

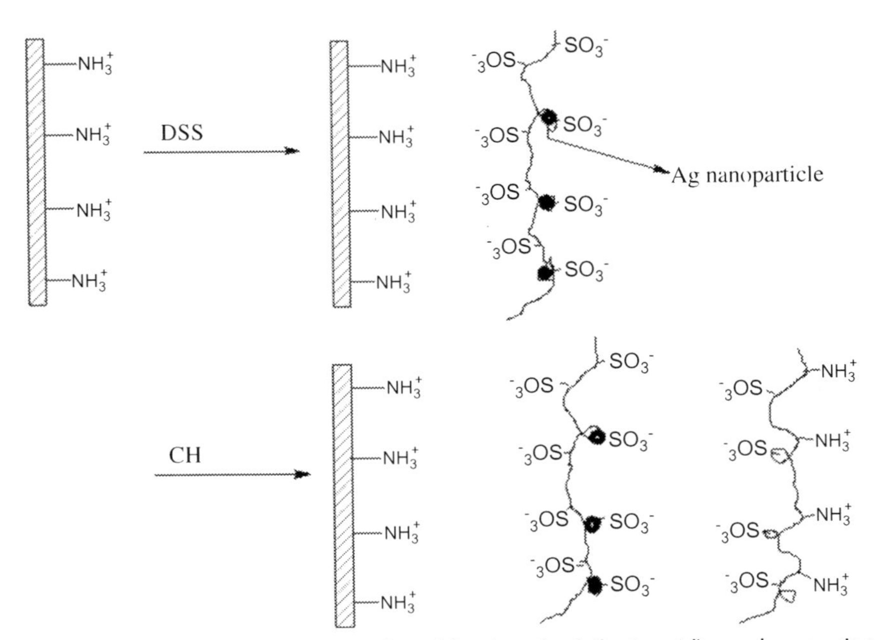

Fig. 7.35 Schematic representation of modification of poly(lactic acid) membrane using silver nanoparticles. *(From D.G. Yu, W.C. Lin, M.C. Yang, Surface modification of poly(L-lactic acid) membrane via layer-by-layer assembly of silver nanoparticle-embedded polyelectrolyte multilayer, Bioconjug. Chem. 18 (5) (2007) 1521–1529. https://doi.org/10.1021/bc060098s.)*

nanoparticles along with hydroxyapatite on polylactic acid membrane through electrospinning process to increase its biocompatibility. Moreover, the porous three-dimensional fibrous structure of the modified membrane assisted in better cell proliferation during tissue generation [59].

7.8 Summary

So far, the descriptions given in this chapter may not encompass the gamut of research that had been carried out on the FNM membrane until now. From drug delivery to microbial strain detection, functionalized nanomaterials are widely applied in different sectors of biomedical applications. Incorporation of functionalized nanomaterials within the membrane not only enhances their inherent strength and intrinsic properties, but also increases their biocompatibility essential for devising any biomedical devices.

References

[1] B. Seifert, G. Mihanetzis, T. Groth, W. Albrecht, K. Richau, Y. Missirlis, D. Paul, G. Von Sengbusch, Polyetherimide: a new membrane-forming polymer for biomedical applications, Artif. Organs 26 (2) (2002) 189–199, https://doi.org/10.1046/j.1525-1594.2002.06876.x.

[2] A. Asad, D. Sameoto, M. Sadrzadeh, Overview of membrane technology, in: M. Sadrzadeh, T. Mohammadi (Eds.), Nanocomposite Membranes for Water and Gas Separation, Elsevier, 2019, pp. 1–28, https://doi.org/10.1016/C2018-0-00150-4.

[3] T. Pasman, D. Baptista, S. van Riet, R.K. Truckenmüller, P.S. Hiemstra, R.J. Rottier, D. Stamatialis, A.A. Poot, Development of porous and flexible ptmc membranes for in vitro organ models fabricated by evaporation-induced phase separation, Membranes 10 (11) (2020) 1–19, https://doi.org/10.3390/membranes10110330.

[4] Y. Zhang, J.R. Vallin, J.K. Sahoo, F. Gao, B.W. Boudouris, M.J. Webber, W.A. Phillip, High-affinity detection and capture of heavy metal contaminants using block polymer composite membranes, ACS Cent. Sci. 4 (12) (2018) 1697–1707, https://doi.org/10.1021/acscentsci.8b00690.

[5] B.S. Lalia, V. Kochkodan, R. Hashaikeh, N. Hilal, A review on membrane fabrication: structure, properties and performance relationship, Desalination 326 (2013) 77–95, https://doi.org/10.1016/j.desal.2013.06.016.

[6] J. Cadotte, R. Forester, M. Kim, R. Petersen, T. Stocker, Nanofiltration membranes broaden the use of membrane separation technology, Desalination 70 (1–3) (1988) 77–88, https://doi.org/10.1016/0011-9164(88)85045-8.

[7] P.Y. Apel, S.N. Dmitriev, Micro-and nanoporous materials produced using accelerated heavy ion beams, Adv. Nat. Sci. Nanosci. Nanotechnol. 2 (1) (2011) 1–8, https://doi.org/10.1088/2043-6262/2/1/013002.

[8] D.F. Stamatialis, B.J. Papenburg, M. Gironés, S. Saiful, S.N.M. Bettahalli, S. Schmitmeier, M. Wessling, Medical applications of membranes: drug delivery, artificial organs and tissue engineering, J. Membr. Sci. 308 (1–2) (2008) 1–34, https://doi.org/10.1016/j.memsci.2007.09.059.

[9] H. Nagahama, V.V.D. Rani, K.T. Shalumon, R. Jayakumar, S.V. Nair, S. Koiwa, T. Furuike, H. Tamura, Preparation, characterization, bioactive and cell attachment

studies of α-chitin/gelatin composite membranes, Int. J. Biol. Macromol. 44 (4) (2009) 333–337, https://doi.org/10.1016/j.ijbiomac.2009.01.006.

[10] Z.H. Huang, Y.S. Dong, C.L. Chu, P.H. Lin, Electrochemistry assisted reacting deposition of hydroxyapatite in porous chitosan scaffolds, Mater. Lett. 62 (19) (2008) 3376–3378, https://doi.org/10.1016/j.matlet.2008.03.045.

[11] J. Beck, R. Angus, B. Madsen, D. Britt, B. Vernon, K.T. Nguyen, Islet encapsulation: strategies to enhance islet cell functions, Tissue Eng. 13 (3) (2007) 589–599, https://doi.org/10.1089/ten.2006.0183.

[12] M.Z. Fahmi, M. Wathoniyyah, M. Khasanah, Y. Rahardjo, S. Wafiroh, Abdulloh., Incorporation of graphene oxide in polyethersulfone mixed matrix membranes to enhance hemodialysis membrane performance, RSC Adv. 8 (2) (2018) 931–937, https://doi.org/10.1039/c7ra11247e.

[13] T.A. Saleh, V.K. Gupta, Nanomaterial and polymer membranes: synthesis, characterization, and applications, in: Nanomaterial and Polymer Membranes: Synthesis, Characterization, and Applications, Elsevier Inc., 2016, pp. 1–272, https://doi.org/10.1016/C2013-0-19381-6.

[14] D. Elieh-Ali-Komi, M.R. Hamblin, Chitin and chitosan: production and application of versatile biomedical nanomaterials, Int. J. Adv. Res. 4 (3) (2016) 411–427.

[15] N. Raghavan, S. Thangavel, G. Venugopal, A short review on preparation of graphene from waste and bioprecursors, Appl. Mater. Today 7 (2017) 246–254.

[16] C. Wang, et al., Direct conversion of waste tires into three-dimensional graphene, Energy Storage Mater. 23 (2019) 499–507.

[17] T. Purkait, et al., Large area few-layer graphene with scalable preparation from waste biomass for high-performance supercapacitor, Sci. Rep. 7 (1) (2017) 1–14.

[18] T. Somanathan, et al., Graphene oxide synthesis from agro waste, Nanomaterials 5 (2) (2015) 826–834.

[19] R. Rajarao, et al., Novel approach for processing hazardous electronic waste, Procedia Environ. Sci. 21 (2014) 33–41.

[20] S.M. Kharrazi, et al., A novel post-modification of powdered activated carbon prepared from lignocellulosic waste through thermal tension treatment to enhance the porosity and heavy metals adsorption, Powder Technol. 366 (2020) 358–368.

[21] J. Xu, H. Yang, W. Fu, K. Du, Y. Sui, J. Chen, Y. Zeng, M. Li, G. Zou, Preparation and magnetic properties of magnetite nanoparticles by sol-gel method, J. Magn. Magn. Mater. 309 (2) (2007) 307–311, https://doi.org/10.1016/j.jmmm.2006.07.037.

[22] Y.B. Pottathara, Y. Grohens, V. Kokol, N. Kalarikkal, S. Thomas, Synthesis and processing of emerging two-dimensional nanomaterials, in: Y.B. Pottahara, Y. Grohens, V. Kokol, N. Kalarikkal, S. Thomas (Eds.), Nanomaterials Synthesis: Design, Fabrication and Applications, Elsevier, 2019, pp. 1–25, https://doi.org/10.1016/B978-0-12-815751-0.00001-8.

[23] S. Wang, L. Gao, Laser-driven nanomaterials and laser-enabled nanofabrication for industrial applications, in: S. Thomas, Y. Grohens, Y.B. Pottathara (Eds.), Industrial Applications of Nanomaterials, Elsevier, 2019, pp. 181–203, https://doi.org/10.1016/B978-0-12-815749-7.00007-4.

[24] D. Nunes, A. Pimentel, L. Santos, P. Barquinha, L. Pereira, E. Fortunato, R. Martins, Synthesis, properties and applications, in: G. Korotcenkov (Ed.), Metal Oxide Nanostructures, first ed., Elsevier, 2019, pp. 21–57, https://doi.org/10.1016/B978-0-12-811512-1.00002-3.

[25] A.C. Marques, Sol-Gel Process: An Overview Gel Process: An Overview, 2007. https://www.lehigh.edu/imi/teched/LecBasic/Marques_Sol_gel.pdf. (Accessed 9 January 2021).

[26] A.I.Y. Tok, F.Y.C. Boey, X.L. Zhao, Novel synthesis of Al2O3 nano-particles by flame spray pyrolysis, J. Mater. Process. Technol. 178 (1-3) (2006) 270–273, https://doi.org/10.1016/j.jmatprotec.2006.04.007.

[27] S. Zeng, K.-T. Yong, I. Roy, X.-Q. Dinh, X. Yu, F. Luan, A review on functionalized gold nanoparticles for biosensing applications, Plasmonics 6 (2011) 491–506, https://doi.org/10.1007/s11468-011-9228-1.
[28] G. Jinhao, G. Hongwei, X. Bing, Multifunctional magnetic nanoparticles: design, synthesis, and biomedical applications, Acc. Chem. Res. 42 (8) (2009) 1097–1107, https://doi.org/10.1021/ar9000026.
[29] D.N. Grant, J. Benson, M.J. Cozad, O.E. Whelove, S.L. Bachman, B.J. Ramshaw, D.A. Grant, S.A. Grant, Conjugation of gold nanoparticles to polypropylene mesh for enhanced biocompatibility, J. Mater. Sci. Mater. Med. 22 (12) (2011) 2803–2812, https://doi.org/10.1007/s10856-011-4449-6.
[30] G. Prencipe, S.M. Tabakman, K. Welsher, Z. Liu, A.P. Goodwin, L. Zhang, J. Henry, H. Dai, PEG branched polymer for functionalization of nanomaterials with ultralong blood circulation, J. Am. Chem. Soc. 131 (13) (2009) 4783–4787, https://doi.org/10.1021/ja809086q.
[31] P.C. Kanth, V.S. Kumar, G. Nidhi, Functionalized nanomaterials for biomedical and agriculture industries, in: C.M. Hussain (Ed.), Handbook of Functionalized Nanomaterials for Industrial Applications, Elsevier, 2020, pp. 231–265, https://doi.org/10.1016/b978-0-12-816787-8.00010-7.
[32] X. Shi, S. Wang, H. Sun, J.R. Baker, Improved biocompatibility of surface functionalized dendrimer-entrapped gold nanoparticles, Soft Matter 3 (1) (2007) 71–74, https://doi.org/10.1039/b612972b.
[33] Y.K. Kim, M.H. Kim, D.H. Min, Biocompatible reduced graphene oxide prepared by using dextran as a multifunctional reducing agent, Chem. Commun. 47 (11) (2011) 3195–3197, https://doi.org/10.1039/c0cc05005a.
[34] N.D. Thorat, V.M. Khot, A.B. Salunkhe, R.S. Ningthoujam, S.H. Pawar, Functionalization of La0.7Sr0.3MnO3 nanoparticles with polymer: studies on enhanced hyperthermia and biocompatibility properties for biomedical applications, Colloids Surf. B Biointerfaces 104 (2013) 40–47, https://doi.org/10.1016/j.colsurfb.2012.11.028.
[35] S. Jain, S.M. Dongave, T. Date, V. Kushwah, R.R. Mahajan, N. Pujara, T. Kumeria, A. Popat, Succinylated β-lactoglobuline-functionalized multiwalled carbon nanotubes with improved colloidal stability and biocompatibility, ACS Biomater Sci. Eng. 5 (7) (2019) 3361–3372, https://doi.org/10.1021/acsbiomaterials.9b00268.
[36] J. Yin, X. He, K. Wang, Z. Qing, X. Wu, H. Shi, X. Yang, One-step engineering of silver nanoclusters-aptamer assemblies as luminescent labels to target tumor cells, Nanoscale 4 (1) (2012) 110–112, https://doi.org/10.1039/c1nr11265a.
[37] D.W. Hwang, H.Y. Ko, J.H. Lee, H. Kang, S.H. Ryu, I.C. Song, D.S. Lee, S. Kim, A nucleolin-targeted multimodal nanoparticle imaging probe for tracking cancer cells using an aptamer, J. Nucl. Med. 51 (1) (2010) 98–105, https://doi.org/10.2967/jnumed.109.069880.
[38] K. Yang, L. Hu, X. Ma, S. Ye, L. Cheng, X. Shi, C. Li, Y. Li, Z. Liu, Multimodal imaging guided photothermal therapy using functionalized graphene nanosheets anchored with magnetic nanoparticles, Adv. Mater. 24 (14) (2012) 1868–1872, https://doi.org/10.1002/adma.201104964.
[39] T. Pooja, V. Komal, D. Vida, S. Shree, Functionalized gold nanoparticles and their biomedical applications, Nanomaterials 1 (2011) 31–63, https://doi.org/10.3390/nano1010031.
[40] N. Tangboriboon, in: S.S. Mohapatra, S. Ranjan, N. Dasgupta, R.K. Mishra, S. Thomas (Eds.), Synthesis, Design, and Morphology of Metal Oxide Nanostructures, Elsevier, 2019, pp. 451–467, https://doi.org/10.1016/B978-0-12-814033-8.00015-1.
[41] G. Reina, J.M. González-Domínguez, A. Criado, E. Vázquez, A. Bianco, M. Prato, Promises, facts and challenges for graphene in biomedical applications, Chem. Soc. Rev. 46 (15) (2017) 4400–4416, https://doi.org/10.1039/c7cs00363c.

[42] R.S. Kalash, V.K. Lakshmanan, C.S. Cho, I.K. Park, Theranostics, in: M. Ebara (Ed.), Biomaterials Nanoarchitectonics, William Andrew Applied Science, 2016, pp. 197–215, https://doi.org/10.1016/B978-0-323-37127-8.00012-1.

[43] K. Lee, H. Lee, K.H. Bae, T.G. Park, Heparin immobilized gold nanoparticles for targeted detection and apoptotic death of metastatic cancer cells, Biomaterials 31 (25) (2010) 6530–6536, https://doi.org/10.1016/j.biomaterials.2010.04.046.

[44] T. He, F. Li, Y. Huang, T. Sun, J. Lin, P. Huang, Graphene as 2D nano-theranostic materials for cancer, in: J. Conde (Ed.), Handbook of Nanomaterials for Cancer Theranostics, Elsevier, 2018, pp. 97–124, https://doi.org/10.1016/B978-0-12-813339-2.00004-9.

[45] J. Li, C. Zhou, C. Luo, B. Qian, S. Liu, Y. Zeng, J. Hou, B. Deng, Y. Sun, J. Yang, Q. Yuan, A. Zhong, J. Wang, J. Sun, Z. Wang, N-acetyl cysteine-loaded graphene oxide-collagen hybrid membrane for scarless wound healing, Theranostics 9 (20) (2019) 5839–5853, https://doi.org/10.7150/thno.34480.

[46] R.L. Phillips, O.R. Miranda, C.C. You, V.M. Rotello, U.H.F. Bunz, Rapid and efficient identification of bacteria using gold-nanoparticle- poly(para-phenyleneethynylene) constructs, Angew. Chem. Int. Ed. 47 (14) (2008) 2590–2594, https://doi.org/10.1002/anie.200703369.

[47] R. Hernández, C. Vallés, A.M. Benito, W.K. Maser, F. Xavier Rius, J. Riu, Graphene-based potentiometric biosensor for the immediate detection of living bacteria, Biosens. Bioelectron. 54 (2014) 553–557, https://doi.org/10.1016/j.bios.2013.11.053.

[48] M.A. Syed, S.H.A. Bokhari, Gold nanoparticle based microbial detection and identification, J. Biomed. Nanotechnol. 7 (2) (2011) 229–237, https://doi.org/10.1166/jbn.2011.1281.

[49] A. Jayagopal, K.C. Halfpenny, J.W. Perez, D.W. Wright, Hairpin DNA-functionalized gold colloids for the imaging of mRNA in live cells, J. Am. Chem. Soc. 132 (28) (2010) 9789–9796, https://doi.org/10.1021/ja102585v.

[50] M.A.A. Shahmirzadi, A. Kargari, Nanocomposite membranes, in: V.G. Gude (Ed.), Emerging Technologies for Sustainable Desalination Handbook, Elsevier, 2018, pp. 285–330, https://doi.org/10.1016/B978-0-12-815818-0.00009-6.

[51] L.Y. Ng, A.W. Mohammad, C.P. Leo, N. Hilal, Polymeric membranes incorporated with metal/metal oxide nanoparticles: a comprehensive review, Desalination 308 (2013) 15–33, https://doi.org/10.1016/j.desal.2010.11.033.

[52] M. Ioniță, L.E. Crică, S.I. Voicu, S. Dinescu, F. Miculescu, M. Costache, H. Iovu, Synergistic effect of carbon nanotubes and graphene for high performance cellulose acetate membranes in biomedical applications, Carbohydr. Polym. 183 (2018) 50–61, https://doi.org/10.1016/j.carbpol.2017.10.095.

[53] L.J. Zhu, L.P. Zhu, Z. Yi, J.H. Jiang, B.K. Zhu, Y.Y. Xu, Hemocompatible and antibacterial porous membranes with heparinized copper hydroxide nanofibers as separation layer, Colloids Surf. B Biointerfaces 110 (2013) 36–44, https://doi.org/10.1016/j.colsurfb.2013.04.020.

[54] M. Nan, D. Bang, S. Zheng, G. Gd, B.A. Darmawan, S. Kim, H. Li, C.-S. Kim, A. Hong, F. Wang, J.-O. Park, E. Choi, High-performance biocompatible nanobiocomposite artificial muscles based on ammonia-functionalized graphene nanoplatelets-cellulose acetate combined with PVDF, Sens. Actuators B 323 (2019), https://doi.org/10.1016/j.snb.2020.128709.

[55] P.T.S. Kumar, S. Abhilash, K. Manzoor, S.V. Nair, H. Tamura, R. Jayakumar, Preparation and characterization of novel β-chitin/nanosilver composite scaffolds for wound dressing applications, Carbohydr. Polym. 80 (3) (2010) 761–767, https://doi.org/10.1016/j.carbpol.2009.12.024.

[56] K. Madhumathi, P.T. Sudheesh Kumar, S. Abhilash, V. Sreeja, H. Tamura, K. Manzoor, S.V. Nair, R. Jayakumar, Development of novel chitin/nanosilver composite

scaffolds for wound dressing applications, J. Mater. Sci. Mater. Med. 21 (2) (2010) 807–813, https://doi.org/10.1007/s10856-009-3877-z.
[57] W. Li, X. Wang, J. Wang, Y. Guo, S.Y. Lu, C.M. Li, Y. Kang, Z.G. Wang, H.T. Ran, Y. Cao, H. Liu, Enhanced photoacoustic and photothermal effect of functionalized polypyrrole nanoparticles for near-infrared theranostic treatment of tumor, Biomacromolecules 20 (1) (2019) 401–411, https://doi.org/10.1021/acs.biomac.8b01453.
[58] D.G. Yu, W.C. Lin, M.C. Yang, Surface modification of poly(L-lactic acid) membrane via layer-by-layer assembly of silver nanoparticle-embedded polyelectrolyte multilayer, Bioconjug. Chem. 18 (5) (2007) 1521–1529, https://doi.org/10.1021/bc060098s.
[59] H.B. Ma, W.X. Su, Z.X. Tai, D.F. Sun, X.B. Yan, B. Liu, Q.J. Xue, Preparation and cytocompatibility of polylactic acid/hydroxyapatite/graphene oxide nanocomposite fibrous membrane, Chin. Sci. Bull. 57 (23) (2012) 3051–3058, https://doi.org/10.1007/s11434-012-5336-3.

CHAPTER 8

Sustainable membranes with FNMs for energy generation and fuel cells

K. Khoiruddin[a], G.T.M. Kadja[b], and I.G. Wenten[a]
[a]Department of Chemical Engineering, Faculty of Industrial Technology, Institut Teknologi Bandung, Bandung, Indonesia
[b]Division of Inorganic and Physical Chemistry, Faculty of Mathematics and Natural Sciences, Institut Teknologi Bandung, Bandung, Indonesia

8.1 Introduction

Membranes are a relatively new technology but have been considered to play important roles in various strategic sectors, including water and wastewater treatment [1–3], gas separation [4,5], food and beverage processing [6,7], medical and pharmaceutical industries [8,9], biotechnology and biorefinery [10,11], and energy conversion and storage [12,13]. For example, membrane technology has been used in the desalination process with a total capacity of more than 65% of global capacity [14]. Interest in using membranes to replace conventional processes is increasing because membranes can produce high-quality water, require lower energy, consume fewer chemicals, are compact and require less footprint, and are easy to scale up [15,16].

Membranes are also widely applied in the field of energy conversion and storage, such as fuel cells, redox flow batteries (RFBs), and reverse electrodialysis (RED). These technologies utilize ion-exchange membrane (IEM) as a key component to the separation process. Fuel cells are compact energy converting devices that can generate higher energy efficiency with low environmental impact [17]. One of the most recent fuel cells is the microbial fuel cell (MFB), which allows the conversion of microbial activity into electrical energy while simultaneously performing the wastewater treatment [18]. RFB is an energy storage device that is potentially applied for large-scale applications having the advantages of independent power output and storage capacity, durable and long life cycle, and high tolerance to environmental conditions [19]. Such a storage device is crucially needed to ensure a stable energy supply and demand under intermittent characteristics of renewable energy sources. RED can harness electrical energy directly from a salinity

https://doi.org/10.1016/B978-0-323-85946-2.00010-2

gradient of two solutions, such as seawater and river water. This process generates clean energy with almost zero pollutants and carbon emissions [20]. Salinity gradient power is a potential energy source since it has enormous sources on the Earth.

The performances and cost of these membrane-based energy devices depend on the properties of the membrane. Commercially available membranes are usually made of polymeric materials as they are easy to process with sufficient stability and flexibility [21]. Despite these advantages, current available IEMs still need more improvement to increase process efficiency and more development to obtain low-cost and sustainable membranes. Therefore, research on IEM for fuel cells, RFB, and RED has gained increasing interest as shown in Fig. 8.1A and B.

One of the most promising methods to improve the performance of IEM is the incorporation of nanomaterials and functionalized nanomaterials (FNMs). The introduction of FNMs is expected to achieve the desired characteristics of IEMs due to the collaborative impacts of organic–inorganic interaction within the fabricated nanocomposite membrane. They do not only enhance the mechanical, thermal, and chemical stability of IEM but also augment the process efficiency. The use of nanomaterials and functionalized nanomaterials (FNMs) for IEM modification is currently an interesting subject of numerous studies [22–24]. This chapter summarizes recent developments of IEM with FNMs, including nanomaterial functionalization, incorporation of FNMs in IEM matrix, and applications of IEM/FNMs in fuel cells, RFB, and RED.

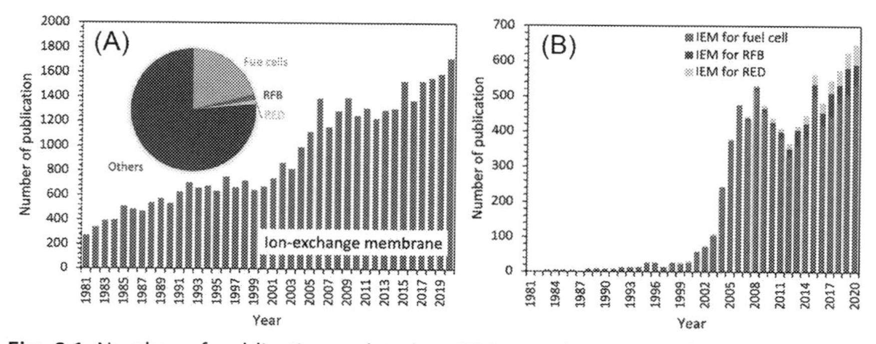

Fig. 8.1 Number of publications related to (A) ion-exchange membrane and (B) ion exchange membrane for fuel cells, redox flow battery (RFB), and reverse electrodialysis (RED). Number of publications indexed by SCOPUS [attained on April 10th, 2021; queries TITLE-ABS-KEY(term)].

8.2 Preparation of IEM containing FNMs

Preparation of IEMs generally depends on membrane type, either homogeneous or heterogeneous. Approaches for the preparation of homogeneous membranes include polymerization of monomer-containing functional groups, fabrication of membrane from polymers having functional groups, and functionalization of the polymeric film with charged groups [25]. Meanwhile, the preparation of heterogeneous membranes comprises two main steps, i.e., mixing the polymeric binder with charged polymer and casting it [26–30]. Techniques of IEM fabrication on a commercial scale have been explained in detail by Tanaka [31]. This section summarizes the recent preparation of FNMs used in the modification of IEMs and the preparation of IEM/FNMs, especially on the introduction of FNMs into the membrane matrix.

8.2.1 Functionalization of nanomaterials

Various nanomaterials have been used to improve the properties of IEM, including two-dimensional carbon-based materials (graphene and graphene oxide). Graphene consists of sp^2-bonded carbon atoms organized in a honeycomb-like pattern with one atomic thickness. Meanwhile, graphene oxide has a construction similar to that of graphene, except that a fraction of carbon atoms has an sp^3 hybridized orbitals with various oxygen-based functional groups (epoxide, hydroxyl, carbonyl, and carboxyl). These groups can be readily tuned or further functionalized to meet the desired physicochemical properties of the membranes. In addition, graphene oxide possesses high mechanical strength, high surface area, and high proton conductivity, which are highly beneficial for the application of IEM [32].

Acid group, such as phosphate, is a favorable functional group for the functionalization of nanomaterials since it will act as a conductive pathway for proton transport. Phosphonated graphene oxide can be obtained via oxidation, exfoliation, and phosphonation [33]. The first and the second steps are for obtaining graphene oxide from graphite, while the last step is for introducing the phosphate group. Furthermore, graphene oxide can be modified by introducing conductive polymers, such as polyaniline, to enhance its electrochemical properties [34]. In situ oxidative chemical polymerization can facilitate the functionalization of graphene oxide with polyaniline resulting in a uniform distribution of polyaniline on graphene oxide [34]. Sulfonated poly(arylene thioether sulfone) (SATS), another polymeric functional groups, can be attached to graphene oxide by reacting graphene oxide with the thiolate group of the SATS [35].

Graphene oxide may also be modified with multifunctional groups. An example is imidized graphene oxide containing sulfonic and carboxylic groups, which are obtained through a multistep chemical reaction of graphene oxide [36]. Graphene oxide may also have both basic and acidic groups on its surface as demonstrated in the preparation of sulfonated polytriazole-grafted graphene oxide through the click reaction (Fig. 8.2A) [37]. Another intriguing functionalized graphene oxide is quaternized graphene oxide, which can improve the hydroxide transport in the IEM matrix [38,39]. Fig. 8.2B and C shows the typical reactions for obtaining quaternized graphene oxide. Graphene oxide functionalization may be conducted through the sol–gel process, such as for the synthesis of graphene oxide-zirconium phosphate nanocomposite [40]. Zirconium phosphate is able to provide ion-exchange groups to graphene oxide that leads to the increased ion-exchange capacity of the modified membrane and enhances the electrochemical properties [40].

Carbon nanotube (CNT) is a carbon-based material having a cylindrical morphology, which consists of rolled-up sheets of graphene with a diameter in the nanoscale. Carbon nanotubes can be formed by a single sheet or multiple sheets, so-called single-walled carbon nanotubes (SWCNT) and multi-walled carbon nanotubes (MWCNT), respectively. CNT is often used as an additive for preparing the composite membrane due to its low density, and high electrical conductivity, and robust mechanical stability [41,42]. CNT may have various functional groups obtained by various functionalization methods, such as acid treatment, grafting, in situ polymerization, and hydrothermal treatment [41,42]. Acid treatment can be used to introduce sulfonic groups on the CNT surface (Fig. 8.3A). For instance, immersion of CNTs in acetyl sulfuric acid could result in sulfonated CNTs [43]. The sulfonic group can also be adhered to the CNT surface by mixing carboxylated CNTs with ammonium sulfate solution, under relatively high temperature [44]. Sulfonated CNTs have a better dispersion in solvent compared to pristine CNT (Fig. 8.3B).

Moreover, CNT may also be anchored by phosphotungstic acid through the electrostatic self-assembly using poly(diallyldimethylammonium chloride) (PDDA) [45], as shown in Fig. 8.3C. First, the CNT undergoes attachment of PDDA, by dispersing CNT in PDDA solution. The PDDA-CNT is then dispersed in phosphotungstic acid solution and the electro-statical interaction with phosphotungstic acid shall be formed [45]. Besides, an even distribution of CNT in the cationic membrane matrix can be obtained by

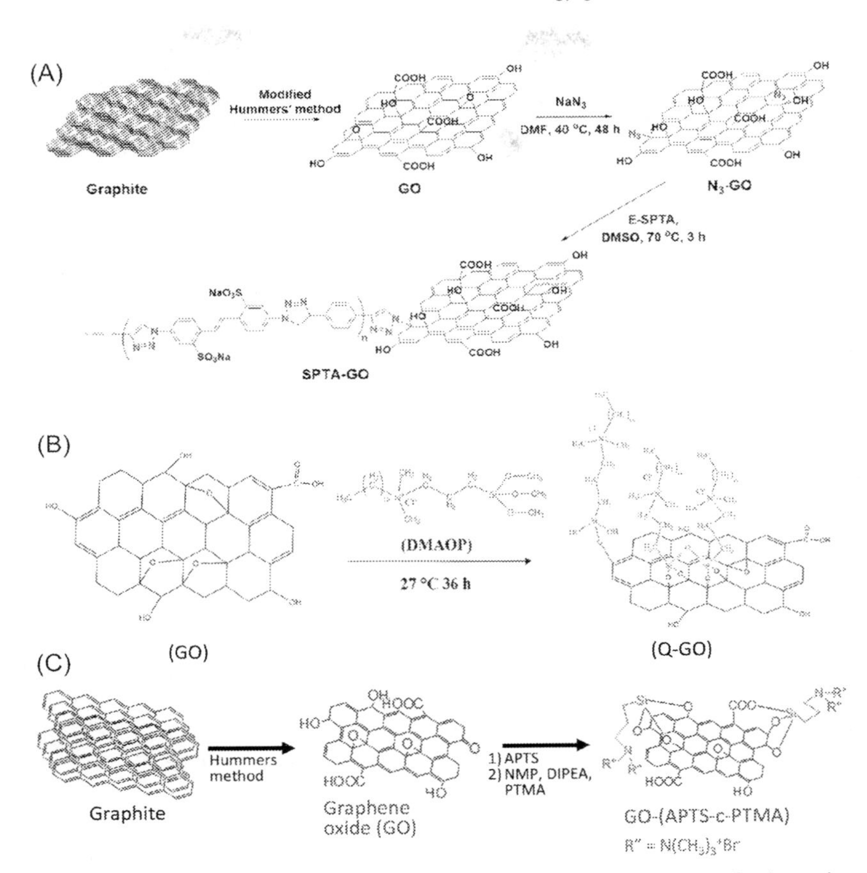

Fig. 8.2 Functionalized graphene oxide. (A) Sulfonated polytriazole-grafted graphene oxide (SPTA-GO). (B) Quaternized graphene oxide. (C) Quaternized graphene oxide. *(A) Reprinted with permission from J. Han, H. Lee, J. Kim, S. Kim, H. Kim, E. Kim, Y.-E. Sung, K. Kim, J.-C. Lee, Sulfonated poly(arylene ether sulfone) composite membrane having sulfonated polytriazole grafted graphene oxide for high-performance proton exchange membrane fuel cells, J. Membr. Sci. 612 (2020) 118428. https://doi.org/10.1016/j.memsci.2020.118428, Elsevier. (B) Reprinted with permission from S. Changkhamchom, P. Kunanupatham, K. Phasuksom, A. Sirivat, Anion exchange membranes composed of quaternized polybenzimidazole and quaternized graphene oxide for glucose fuel cell, Int. J. Hydrogen Energy 46 (2021) 5642–5652. https://doi.org/10.1016/j.ijhydene.2020.11.043, Elsevier. (C) Reprinted with permission from J.Y. Chu, K.H. Lee, A.R. Kim, D.J. Yoo, Improved electrochemical performance of composite anion exchange membranes for fuel cells through cross linking of the polymer chain with functionalized graphene oxide, J. Membr. Sci. 611 (2020) 118385. https://doi.org/10.1016/j.memsci.2020.118385, Elsevier.*

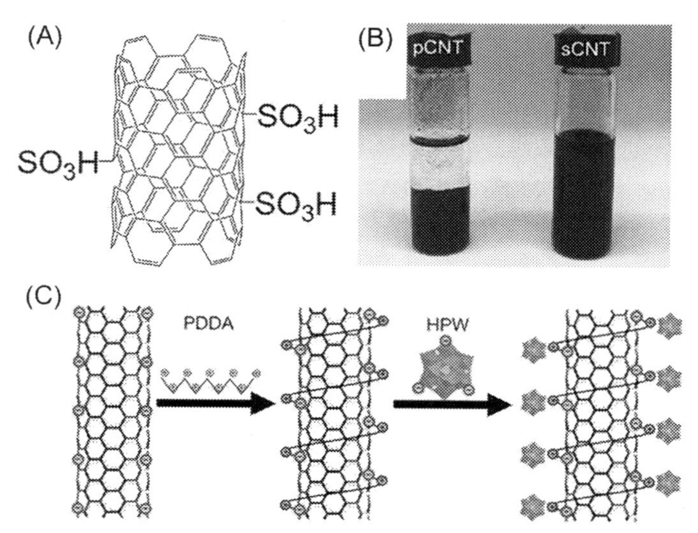

Fig. 8.3 Functionalized CNTs. (A) Sulfonated CNT. (B) Pristine CNT (pCNT) and sulfonated CNT (sCNT) in solvent after 10h. (C) HPW-PDDA-CNTs. *(A, B) Reprinted with permission from H. Fan, Y. Huang, N.Y. Yip, Advancing the conductivity-permselectivity tradeoff of electrodialysis ion-exchange membranes with sulfonated CNT nanocomposites, J. Membr. Sci. 610 (2020) 118259. https://doi.org/10.1016/j.memsci.2020.118259, Elsevier. (C) Reprinted with permission from Y. Li, H. Wang, Q. Wu, X. Xu, S. Lu, Y. Xiang, A poly(vinyl alcohol)-based composite membrane with immobilized phosphotungstic acid molecules for direct methanol fuel cells, Electrochim. Acta 224 (2017) 369–377. https://doi.org/10.1016/j.electacta.2016.12.076, Elsevier.*

utilizing electrostatic interaction between basic and acidic groups of the filler and the membrane matrix, respectively. For this purpose, CNT can be functionalized by two-step reactions: attaching the carboxylic group to CNT and then substituting it with the amine group [46]. Another emerging carbon-based material for IEM modification is carbon nanofiber. It has a high specific area, high electrical conductivity, and overall negative charge [47]. With such features, it is possible to improve the electrochemical properties of IEM with carbon fiber. However, functionalization of carbon fiber for modifying IEMs is rarely reported.

Inorganic oxide nanoparticle is another class of nanomaterials which is frequently used to modify polymeric membrane, including IEMs [48]. Improving the water uptake of IEM is one of the main purposes of incorporating hygroscopic metal oxides, such as SiO_2 and TiO_2. The use of inorganic oxide nanoparticles is crucial, especially when the IEM should be operated at high temperature, such as in high-temperature proton exchange fuel cells. Incorporating SiO_2 nanoparticles in IEMs has the function of either improving water sorption or retaining water content of the membrane [49]. Furthermore, the composite polymeric/SiO_2 nanoparticle membrane has shown increasing mechanical strength and thermal resistance [50]. Sulfonation leads to SiO_2 nanoparticle function extended by providing a strong conductive way for proton diffusion. Sulfonated silica can be obtained by the sol–gel method using tetraethyl orthosilicate and chlorosulfuric acid as the raw materials [51].

Sulfonation of TiO_2 is also reported in the literature [52], and can be conducted by dispersing titania in sulfuric acid solution. In addition to better hydration state, the sulfonic group also creates a pathway for the proton transport [52]. TiO_2 can be in a form of nanotube called as titanate nanotube, which could provide larger surface area than its nanoparticle form. The higher surface of titanate nanotube, about 225 m^2/g [53], can contain more functional sites. For instance, it can form covalent bonding with silane ionic liquid through a condensation reaction between the hydroxyl group of the nanotube and the methoxy group of ionic liquid [53].

Iron(III) oxide or Fe_2O_3 nanoparticles is another example of metal oxide that has potential for improving the properties of IEMs. Fe_2O_3 nanoparticles display high adsorption capacity, high electrical conductivity, and substantial hydrophilicity [54]. These advantages can be further improved by introducing functional groups. Sulfate functional groups can be added to the surface of Fe_2O_3 nanoparticles by sulfuric acid solution [54,55]. It may serve as additional functional groups facilitating the transport of cation within the membrane matrix.

In addition to the interesting features of functionalized single inorganic filler, combining two or more inorganic materials has attracted considerable interest from the researchers to explore the advantages. Combining the synergetic properties of mordenite zeolite, graphene oxide, and silane has been reported [56]. To prepare such composite materials, acid-treated mordenite is first dispersed in graphene oxide suspension to form mordenite-graphene oxide particles because of the covalent ester linkage resulted from the reaction between hydroxylic group on mordenite surface and carboxylic group on graphene oxide surface. Subsequently, silanization with (3-mercaptopropyl)triethoxysilane (MPTES) is performed resulting in the modernite-graphene oxide functionalized with organosilane-bearing thiol groups [56]. Preventing nanofiller accumulation in the membrane matrix is the aim of combining titanium oxide with reduced graphene oxide nanoribbons [57]. This is to assist the formation of uniform nanofiller distribution in a polymeric membrane matrix. Reduced graphene oxide nanoribbons can be obtained from the longitudinal unzipping of CNTs. Then, the

TiO_2 nanoparticles are wrapped by reduced graphene oxides nanoribbons via the hydrothermal process, which allows the dispersion and thermal treatment of TiO_2 nanoparticles and reduced graphene oxides are dispersed in water-ethanol solution [57].

8.2.2 IEM/FNMs preparation methods

Various methods for IEM/FNM preparation have been reported, including additive blending, sol–gel process, solution casting, and others. These methods are discussed in the following subsection in detail and are summarized in Table 8.1.

8.2.2.1 Additive blending

One of the facile methods for incorporating FNMs into the IEM matrix is additive blending. In this method, nanomaterials are dispersed in the membrane solution prior to the casting. The nanomaterials should be homogeneously mixed for obtaining a good distribution in the synthesized membrane. Generally, neither chemical bonding nor molecular interaction occurs between the inorganic nanoparticles and the polymer binder; hence, the inorganic fillers can easily agglomerate [42,58].

Additive blending is a favorable method for incorporating FNMs due to its simplicity. This method, for example, has been used to incorporate ZnO nanoparticles and carbon nanofibers in the matrix of heterogeneous IEMs [47,60,64]. Although additive blending can produce better membrane properties, it is quite challenging to distribute the inorganic nanoparticles in the polymer solution and the membrane structure. The incompatibility between the inorganic filler and the organic membrane matrix usually causes defect formation, leading to poor permselectivity [65]. To break up the agglomeration of nanoparticles, the polymeric solution containing nanomaterials may undergo sonication [60]. Ultra-sonication of the solution can facilitate better nanoparticle distribution in the membrane matrix. Another approach to obtain better nanoparticle distribution is using FNMs, as demonstrated in the introduction of polyaniline-modified graphene oxide nanomaterials [34].

8.2.2.2 Sol–gel method

The sol–gel process, which involves transforming a solution into a gel, is another alternative method for introducing nanomaterials into the membrane matrix. In this method, nanomaterials can be introduced into the membrane structure while simultaneously creating a cross-linking between

Table 8.1 Preparation of IEM with FNMs.

Method	FNMs	Membrane matrix	Remarks	Ref.
Additive blending	Polyaniline/graphene oxide	PVC	Improved electrochemical properties	[34]
Additive blending	Sulfonated imidized graphene oxide	SPEEK	Improved electrochemical properties	[36]
Additive blending	Sulfate-functionalized Fe_2O_3 nanoparticles	PVC/HIPS/ABS	Improved electrochemical properties	[54]
Additive blending	Sulfonated CNT	SPPO	Improved electrochemical properties	[43]
Additive blending	Sulfonated SiO_2	Nafion	Improved electrochemical properties	[51]
Additive blending	Graphene oxide-zirconium phosphate nanocomposite	PVC	Improved electrochemical properties	[40]
Sol–gel	Sulfonated silica	SPAES	Decreased methanol permeability	[58]
Solution casting	Oxidized MWCN (CNTs-COO −) in PPO solution	Ralex AM-PP	Improved hydrophilicity, improved fouling resistance	[59]
Solution casting	Sulfonated iron oxide nanoparticles (Fe_2O_3-$SO_4{}^{2-}$)	Ralex AM-PP	Improved hydrophilicity, improved fouling resistance	[59]
In-situ formation	ZnO	SPPO/SPVC	Improved electrochemical properties	[60]
Electro-deposition	Graphene oxide	Commercial heterogeneous membrane	Improved hydrophilicity, improved fouling resistance	[61]
Electro-deposition	Sulfonated reduced graphene oxide/polydopamine	Commercial JAM-II-05 AEM	Improved monovalent permselectivity and antifouling property	[62]
Dip-coating	Zirconium phosphate nanoparticles	RALEX CM	Improved monovalent permselectivity	[63]

ABS, acrylonitrile-butadiene-styrene; CNT, carbon nanotube; HIPS, high impact grade polystyrene; MWCNT, multiwalled carbon nanotube; PVC, polyvinyl chloride; SPAES, sulfonated poly(arylene ether sulfone); SPEEK, sulfonated poly(ether ether ketone); SPVC, sulfonated polyvinyl chloride.

the inorganic filler with the organic matrix [58]. Therefore, agglomeration of the filler, which is usually observed in the additive blending method, can be avoided.

For instance, this method has been used to synthesize proton exchange membranes comprising sulfonated poly(arylene ether sulfone) and sulfonated silica [58]. Besides, the silica and membrane matrix were covalently bonded by reacting (3-isocyanatopropyl) triethoxysilane and 3-(trihydroxysilyl) propane-1-sulfonic acid [58]. The formation of cross-linkage improved the compatibility between the inorganic filler and the organic membrane components and resulted in a uniform membrane structure [58].

8.2.2.3 Solution casting method

Solution casting comprises polymer dissolution, nanomaterials dispersion, and the solution casting on the membrane surface. It creates a new layer on the existing membrane resulting in a composite membrane. This method is used to modify the properties of the membrane surface, such as membrane hydrophilicity.

Some studies have reported the use of the solution casting method for incorporating FNMs into IEMs. For instance, sulfonated iron oxide nanoparticles blended with sulfonated poly (2,6-dimethyl-1,4-phenylene oxide) (SPPO) were casted on a commercial AEM [59]. The Fe_2O_3-$SO_4{}^{2-}$/SPPO layer was successfully deposited on the commercial membrane with a thickness of 19 μm. Besides, the new layer displayed a homogeneous surface and lowered water contact angle (34°).

8.2.2.4 Other methods

There are also other methods for introducing nanomaterials in the membrane matrix, such as electro-deposition, in situ formation, and dip-coating. Electro-deposition is a method for the deposition of nanomaterials by utilizing electrical direct current. This method has been used to coat graphene oxide on commercial IEMs [61]. In brief, graphene oxide was dispersed in pure water assisted by sonication. The graphene oxide suspension was mixed with NaCl solution with a certain concentration. IEMs were then arranged in a simple ED unit where CEMs and AEMs were placed in an alternating pattern. The homogeneous solution of graphene oxide was circulated into the cell while applying a specific current density. As a result, the graphene oxide was attracted and deposited on the membrane surface. The graphene oxide was successfully coated on IEMs' surface with 0.72 μm thickness [61].

In situ formation has been demonstrated in the preparation of SPPO/sulfonated polyvinyl chloride (SPVC)/ZnO CEM [60]. The ZnO precursor, $Zn(NO_3)_2.4H_2O$, was mixed with a polymer solution, and NH_3 was added dropwise to obtain a pH ~12. Suitable pH value induces the in situ formation of ZnO nanoparticles. From XRD analysis, the ZnO nanoparticle had a 25 nm crystallite size [60].

Another facile method for the incorporation of FNMs is dip-coating. This method has been used to deposit sulfonated reduced graphene oxide on the AEM surface [66]. First, the sulfonated reduced graphene oxide was dispersed in NaCl solution. Two unmodified AEMs were placed in a cell where a compartment separated the AEMs. The solution was poured into this compartment and stirred (100 rpm) for 24 h. Afterward, a sulfonated reduced graphene oxide layer was created on the membrane surface indicated by a yellow layer. This deposition may occur due to electrostatic interaction between the FNMs and the membrane charges. The dip-coating method was also used to coat zirconium phosphate nanoparticles on commercial CEMs [63].

8.3 IEM/FNMs for fuel cell applications

The performance of IEM/FNMs has been demonstrated in various types of fuel cells (Table 8.2). The role of FNMs in improving the performance of fuel cells is indicated by some aspects. One of the crucial aspects is enhancing membrane conductivity. FNMs can enhance a fuel cell's power density by improving membrane conductivity. For example, SPTA-GO, which has both acidic and basic groups, enhances the proton conductivity in the SPAES membrane matrix by providing more pathways for the proton transport and an additional protonation-deprotonation loop [37]. As a result, a higher energy density is obtained. Sulfonic groups on FNMs also give the same effect, especially for the pathway of the proton [52]. In a relatively low humidity environment, phosphonic acid shows higher proton conductivity than sulfonic and carboxylic groups, making it a potential for functionalization of nanomaterials for high-temperature fuel cell application [67]. Some possible pathways for proton conduction by FNMs are shown in Fig. 8.4.

The suppressed fuel crossover is another effect of incorporating FNMs in IEM used in fuel cells. Fuel crossover decreases power efficiency and power density over time. HPW-PDDA-CNT particles when introduced into the PVA membrane decrease methanol permeability [45]. This is proportional to the decreasing swelling ratio of the HPW-PDDA-CNT/PVA

Table 8.2 Performances of IEM/FNMs in fuel cell.

Membrane	Types	Fuel cell	Conductivity	Remarks	Ref.
Am-SPAES/I-SiO_2(10%)	PEM	DMFC	110 mS/cm at 100°C[a]	Methanol permeability = 4.76×10^{-7} $Cm^2\,s^{-1}$	[58]
PVA/HPW-PDDA-CNTs (2%)	PEM	DMFC	5.27 mS/cm at 30°C[a]	Power density = 16 mW/cm^2 at 60°C	[45]
Nafion/silanated-mordenite-GO(0.05%)	PEM	DMFC	86.5 mS/cm at 70°C[a]	Power density = 29.55 mW/cm^2 at 70°C	[56]
SPAES/SPTA-GO	PEM	PEMFC	412.5 mS/cm at 90% RH[a]	Power density = 1.21 W/cm^2 (50%);1.58 W/cm^2 (100% RH)	[37]
PVdF-co-HFP/sulfonated TiO_2	PEM	PEMFC	3.6 mS/cm at 79.85°C[a]	Power density = 85 mW/cm^2 at 80°C	[52]
Py-PBI/phosphonated GO	PEM	PEMFC	76.4 mS/cm at 140°C[a]	Power density = 359 mW/cm^2 at 120°C under anhydrous condition	[67]
PBI/sulfonated TiO_2(2%)	PEM	PEMFC	96 mS/cm at 150°C[a]	Power density = 621 mW/cm^2	[68]
PBI/IL-GO	AEM	AFC	40–50 mS/cm at 60°C[b]	–	[69]
QPPO/PSF/ZnO(2%)	AEM	AFC	52.34 mS/cm at 80°C[b]	Power density = 69 mW/cm^2	[70]
QPAE/APTS-co-PTMA-GO(0.7%)	AEM	AFC	114.2 mS/cm at 90°C[b]	Power density = 135.8 mW/cm^2 at 70°C	[39]
PIPPO/ASU-GO	AEM	AFC	73.7 mS/cm at 80°C[b]	Power density = 102 mW/cm^2 at 80°C	[71]
QAPSU/IL-TnT(5%)	AEM	AFC	20 mS/cm at 30°C[b]	Power density = 315 mW/cm^2 at 60°C	[53]
PVDF-g-PSSA/sulfonated TiO_2(5%)	CEM	MFC	67 mS/cm[a]	Power density = 130.54 mW/cm^2	[72]
Aniline-treated polysulfone/sulfonated CNT	CEM	MFC	190 mS/cm[a]	Power density = 304.2 mW/cm^2 at 30°C	[73]
SPBI/sulfonated GO(3%)	CEM	MFC	16 mS/cm at room temperature[a]	Power density = 472.46 mW/cm^2.	[74]

[a] Proton conductivity.
[b] Hydroxide ion conductivity.

AFC, alkaline fuel cell; DMFC, direct methanol fuel cell; GO, graphene oxide; IL, ionic liquid; MFC, microbial fuel cell; PBI, polybenzimidazole; PEM, proton exchange membrane; PSF, polysulfone; PVdF-co-HFP, poly(vinylidene fluoride-hexa fluoro propylene); PVA, polyvinyl alcohol; Py-PBI, 2,6-pyridine functionalized polybenzimidazole; QAPSU, quaternary ammonium functionalized polysulfone; QPAE, quaternized poly(arylene ether); QPPO, quaternized poly(2.6 dimethyl-1.4 phenylene oxide); RH, relative humidity; SPAES, sulfonated poly(arylene ether sulfone); SPBI, sulfonated polybenzimidazole; TnT, titanate nanotube.

membrane. In addition, the use of HPW-PDDA-CNT particles results in the blockage of methanol transport through the membrane. As the methanol crossover decreases, the power density of the DMFC increases [45].

Functionalization may alter the interaction of nanomaterials with each other as well as in the membrane matrix. Graphene oxide containing 6-azonia-spiro[5.5]undecane groups blended in a membrane matrix, shows higher conductivity than of the pristine graphene oxide [71]. The alteration of functionalized graphene oxide stacking was considered as the possible cause of this improvement (shown in Fig. 8.4A). The functionalized graphene oxide sheets show loose stacking in the membrane matrix compared to the pristine graphene oxide. The functional group on graphene oxide

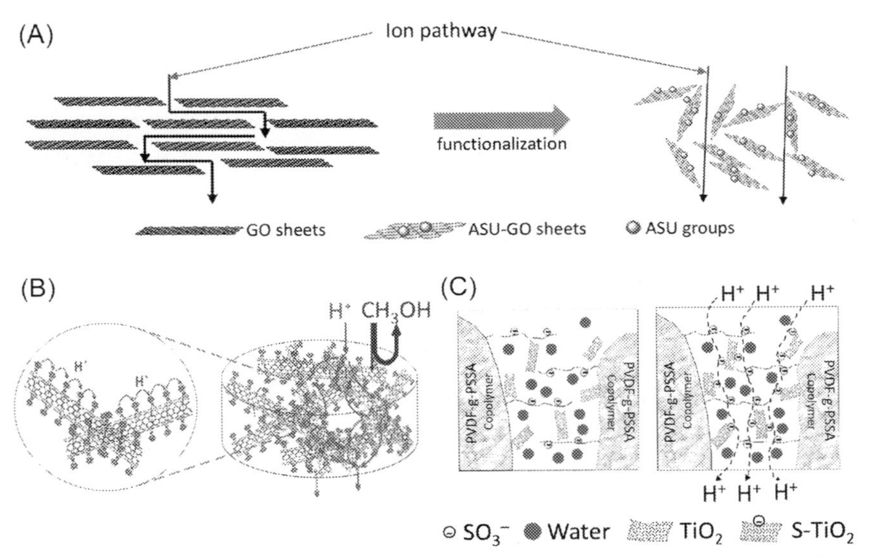

Fig. 8.4 Proton transport improvement by FNMs. (A) GO sheet and functionalized GO sheets. (B) HPW-PDDA-CNTs. (C) Sulfonated TiO_2. *(A) Reprinted with permission from C. Long, C. Lu, Y. Li, Z. Wang, H. Zhu, N-spirocyclic ammonium-functionalized graphene oxide-based anion exchange membrane for fuel cells, Int. J. Hydrogen Energy 45 (2020) 19778–19790. https://doi.org/10.1016/j.ijhydene.2020.05.085, Elsevier. (B) Reprinted with permission from Y. Li, H. Wang, Q. Wu, X. Xu, S. Lu, Y. Xiang, A poly(vinyl alcohol)-based composite membrane with immobilized phosphotungstic acid molecules for direct methanol fuel cells, Electrochim. Acta 224 (2017) 369–377. https://doi.org/10.1016/j.electacta.2016.12.076, Elsevier. (C) Reprinted with permission from C. Li, Y. Song, X. Wang, Q. Zhang, Synthesis, characterization and application of S-TiO2/PVDF-g-PSSA composite membrane for improved performance in MFCs, Fuel 264 (2020) 116847. https://doi.org/10.1016/j.fuel.2019.116847, Elsevier.*

may weaken the interlayer interaction between the graphene oxide sheet, which then results in a shorter transport pathway for ions [71].

The improved properties of IEM/FNMs may be due to the interaction between FNMs and the polymeric matrix of the membrane. Interconnection between quaternary ammonium groups of functionalized graphene oxide and the chloromethyl groups of quaternized poly(arylene ether) (QPAE) forms ionic clusters in the membrane matrix, which results in high chemical resistance toward a high pH environment [39]. Another example is the interaction between sulfonated CNT and sulfonated block copolymer (SDBC) based on poly(ether ether ketone) blocks copolymerized with partially fluorinated poly(arylene ether sulfone) [75]. Hydrogen bonding and π-π interactions between sulfonated CNT and SDBC yield in the unseparated phases of filler and membrane matrix, which eventually results in higher mechanical stability [75].

For microbial fuel cells, in addition to high electrochemical properties, the membrane should be resistant to organic fouling and biofouling, since this technology utilizes microbial activity for harvesting energy from wastewater containing organic matters [76]. Organic substances in the feed solution may form fouling on the membrane surface and create higher resistance to ionic or proton transport. The fouling (organic and biofouling) formation may cause detrimental effects on the long-term operation of microbial fuel cell, especially the generated energy. To prevent severe fouling formation, the membrane needs to be highly hydrophilic. Incorporating hydrophilic FNMs is one of the approaches of producing highly hydrophilic IEM. For example, sulfonated TiO_2 can be used to modify the hydrophilicity of the polyvinylidene fluoride-poly sodium styrene sulfonate (PVDF-g-PSSA) membrane [72]. The water contact angle of the PVDF-g-PSSA membrane decreases from $\sim 80°$ to $\sim 61°$ due to the presence of 5% sulfonated TiO_2 [72]. The modified membrane also displays less bacterial communities, more loose and porous fouling layer than those observed on commercial membrane after 2 months of operation [72]. The results indicate that the modified PVDF-g-PSSA membrane shows improved biofouling resistance, leading to stable MFC performance. Oxygen diffusion from the anode to the cathode chamber adversely affects columbic efficiency and redox potential in MFC. Therefore, IEM used in MFC should have low oxygen diffusivity in order to generate high power density. It is evidenced that a cationic aniline-treated polysulfone membrane with sulfonic CNTs displays lower oxygen diffusivity than the unmodified membrane, almost by one order [73]. This decreasing oxygen transfer yields higher Coulombic efficiency as well as generates power density [73].

8.4 IEM/FNMs in redox flow battery (RFB)

RFB converts electrical energy into chemical energy using reduction–oxidation (redox) reactions. These reactions change the oxidation state of anolyte and catholyte during charging and discharging process of RFB. These electrolytes are placed in two chambers separated by IEM (see Fig. 8.5A and B). Here, IEM act as a separator that facilitates the transfer of charged carrier and avoids crossover of electrolyte from one chamber to another [77,78]. RFB technologies offer an attractive feature of independence between energy capacity and power output. In addition, RFB

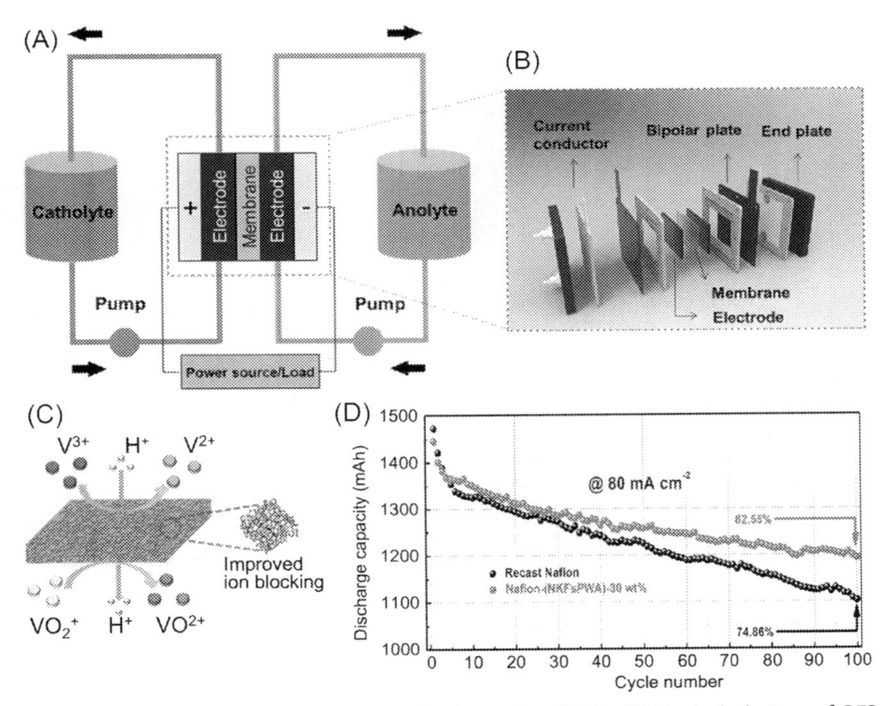

Fig. 8.5 RFB with IEM/FNMs. (A) Typical schematic of RFB. (B) Exploded view of RFB module. (C) Blocking mechanism of vanadium ions by PWA-NFK additive. (D) Discharge capacity of the vanadium RFB with recast Nafion and recast Nafion containing PWA-NFK additive. *(A, B) Reprinted with permission from Yue Zhang, Y. Zhong, W. Bian, W. Liao, X. Zhou, F. Jiang, Robust proton exchange membrane for vanadium redox flow batteries reinforced by silica-encapsulated nanocellulose, Int. J. Hydrogen Energy 45 (2020b) 9803–9810. https://doi.org/10.1016/j.ijhydene.2020.01.188, Elsevier. (C, D) Reprinted with permission from X.-B. Yang, L. Zhao, X.-L. Sui, L.-H. Meng, Z.-B. Wang, Phosphotungstic acid immobilized nanofibers-Nafion composite membrane with low vanadium permeability and high selectivity for vanadium redox flow battery, J. Colloid Interface Sci. 542 (2019) 177–186. https://doi.org/10.1016/j.jcis.2019.02.001, Elsevier.*

technologies generally exhibit a long life cycle, high efficiency, and good scalability, making them suitable for large-scale energy storage [79]. The amount of energy to be stored depends on the volume of the electrolyte and, thus, the electrolyte tank. As a consequence, the energy can be stored as much as the available electrolyte.

The energy conversion rate (charging) and power output (discharging) depend on the membrane area or the module dimension. To achieve efficient charging-discharging cycles, IEM with high electrochemical properties should be used. At higher membrane conductivity, charge carrier transport is faster, leading to more efficient charging-discharging cycles. Therefore, IEM for RFB should be conductive to ions, especially protons. Meanwhile, to maintain its capacity overtime, RFB needs IEM that is highly resistant to electrolyte crossover, such as the crossover of vanadium ions in all-vanadium RFB. It is because electrolyte crossover induces self-discharge of the battery that decreases the battery capacity.

All-vanadium RFB is a preferred RFB type since it uses the same element in both electrolytes, i.e., vanadium ions. This gives the advantage of no cross contamination between anolyte and catholyte. Hence, the electrolyte can be used repeatedly without purification [80]. Nafion-based membranes are usually employed in RFB, because they are proton conductive, selective, as well as chemically and mechanically stable. However, they display high vanadium permeability, which drives efforts to Nafion modification or development of new membrane material.

Some reported studies of IEM/FNMs are tabulated in Table 8.3. The role of FNMs in decreasing the permeability of vanadium has been demonstrated by examining the performance of SPAES membrane modified by silica-encapsulated nanocellulose in all-vanadium RFB [85]. Silica-encapsulated nanocellulose—which has particle size around 500 nm, is trapped in the membrane matrix and blocks the permeation of vanadium ions. The lower vanadium permeability then leads to the higher retained battery capacity after 160 discharging cycles, outperforming Nafion 212. Besides, the silica-encapsulated nanocellulose also reinforces the tensile strength of the membrane from 16.3 to 54.5 MPa. It is ascribed to the effect of a strong hydrogen-bonding in the nanocellulose network. Amino-silica nanoparticles embedded in Nafion 117 matrix also restrict the permeation of vanadium ions [81]. Electrostatic attraction between the amine groups of amino-silica and sulfonic groups of Nafion makes the amino-silica nanoparticle localized in the charge cluster of the membrane. Hence, vanadium diffusion is decreased by the expulsion effect provided by the positive charge of amine groups.

Table 8.3 Performances of IEM and IEM/FNMs in RFB from selected studies.

Membrane	Vanadium permeability	Anolyte	Catholyte	Remarks	Ref.
Nafion 117	$VO^{2+} = 8.65 \times 10^{-7}\ cm^2\ min^{-1}$; $VO_2^{+} = 2.73 \times 10^{-7}\ cm^2\ min^{-1}$	1.6 M V^{2+}/V^{3+}	1.6 M VO^{2+}/VO_2^{+}	–	[81]
Nafion 117/ amino-silica	$VO^{2+} = 2.32 \times 10^{-7}\ cm^2\ min^{-1}$; $VO_2^{+} = 0.85 \times 10^{-7}\ cm^2\ min^{-1}$	1.6 M V^{2+}/V^{3+}	1.6 M VO^{2+}/VO_2^{+}	The capacity was 4.5% higher than the pristine Nafion 117 membrane; Retained capacity after 150 discharged cycle = ~99%	[81]
Nafion 115	$VO^{2+} = 3.9 \times 10^{-6}\ cm^2\ min^{-1}$	1.0 M V(VI)	1.0 M V(III)	Energy efficiency = 81% (at 40 mA/cm^2)	[82]
Nafion 115/ AATMS-silica	$VO^{2+} = 8.5 \times 10^{-7}\ cm^2\ min^{-1}$	1.0 M V(VI)	1.0 M V(III)	Energy efficiency = 85% (at 40 mA/cm^2)	[82]
Nafion 212	$VO^{2+} = 8.23 \times 10^{-7}\ cm^2\ min^{-1}$	1.5 M V^{2+}/V^{3+}	1.5 M VO^{2+}/VO_2^{+}	–	[83]
Nafion 212/ MWCNT-OH	$VO^{2+} = 1.93 \times 10^{-7}\ cm^2\ min^{-1}$	1.5 M V^{2+}/V^{3+}	1.5 M VO^{2+}/VO_2^{+}	Retained capacity after 400 discharged cycle = ~80% Retained capacity after 950 discharged cycle = 50%	[83]
Nafion 115 (recast)	$VO^{2+} = 20.2 \times 10^{-7}\ cm^2\ min^{-1}$	–	–	Retained capacity after 100 discharged cycle = 74.86%	[84]
Nafion 115/ PWA-NKFs (30 wt%)	$VO^{2+} = 2.46 \times 10^{-7}\ cm^2\ min^{-1}$	–	–	Retained capacity after 100 discharged cycle = 82.55%	[84]
Nafion 212	$VO^{2+} = \sim 4.4 \times 10^{-7}\ cm^2\ min^{-1}$	1.5 M V^{2+}/V^{3+}	1.5 M VO^{2+}/VO_2^{+}	Retained capacity after 160 discharged cycle = 68.1%	[85]
SPAES/silica encapsulated nanocellulose	$VO^{2+} = 3.67 \times 10^{-7}\ cm^2\ min^{-1}$	1.5 M V^{2+}/V^{3+}	1.5 M VO^{2+}/VO_2^{+}	Retained capacity after 160 discharged cycle = 97.2%	[85]

Continued

Table 8.3 Performances of IEM and IEM/FNMs in RFB from selected studies—cont'd

Membrane	Vanadium permeability	Anolyte	Catholyte	Remarks	Ref.
SPEEK	$VO^{2+} = 8.18 \times 10^{-7}$ cm^2 min^{-1}	–	–	Retained capacity after 100 discharged cycle = 92%	[86]
SPEEK/ethylenediamine-GO (2 wt%)	$VO^{2+} = 2.04 \times 10^{-7}$ cm^2 min^{-1}	–	–	Retained capacity after 100 discharged cycle = ~30%	[86]

Note: AATMS, [N-(2-aminoethyl)-3-aminopropyl]trimethoxysilane; SPAES, sulfonated poly(arylene ether sulfone); SPEEK, sulfonated poly(ether ether ketone); PWA, phosphotungstic acid; NKFs, nano-Kevlar fibers.

Recast Nafion membrane doped by phosphotungstic acid (PWA)-nano-Kevlar fibers (NKFs) or PWA-NFKs also experienced decreasing vanadium permeability and gaining more stable battery capacity [84]. The complex additive of PWA-NFKs blocks vanadium permeation but with unaffected conductivity for protons (Fig. 8.5C). The unaffected proton transport is associated with its smaller Stokes radius than vanadium [84]. By suppressing the vanadium transport, the discharge capacity remained high after a 100 discharging cycle test, outperforming the unmodified recast Nafion 115 (Fig. 8.5D).

Finding a new membrane material is a strategy to decrease the cost of RFB. Sulfonated poly(ether ether ketone) (SPEEK) is a potential candidate as the substitute of Nafion. Unfortunately, SPEEK exhibits high vanadium permeability, which gives detrimental effects to RFB performance. The vanadium permeability of the SPEEK membrane has been successfully suppressed by embedding ethylenediamine-functionalized graphene oxide, or NH_2-GO [86]. When 2%wt NH_2-GO was added, the vanadium permeability was dramatically decreased by almost 4 times lower than the pristine SPEEK. This led to a remarkable capacity retention after a 100 charging-discharging test. Interactions between amine and sulfonic groups that tighten the pathway for vanadium transport is ascribed as the possible mechanism of vanadium permeability reduction.

Three types of amine-functionalized graphene oxide, namely primary amine, ethylenediamine, and 1,6-hexanediamine, have been used to modify SPEEK membranes [87]. All of them can increase the efficiency of RFB and retain the battery capacity higher than pristine SPEEK and Nafion membranes. Among them, ethylenediamine-functionalized graphene oxide shows the highest proton conductivity, selectivity, and mechanical properties. This could be due to the better interaction of ethylene diamine with the membrane matrix. Another alternative material is SPPO. The SPPO membrane exhibits lower vanadium permeability (VO^{2+}) of 2.5×10^{-8} cm^2 min^{-1} [88]. This value can be further suppressed to 4.76×10^{-9} cm^2 min^{-1} by utilizing sulfonated silica as a filler [88].

8.5 IEM/FNMs in reverse electrodialysis (RED)

RED is a reverse process of conventional electrodialysis that can harness the electrical energy from the gradient salinity of two solutions. RED employs both CEM and AEM arranged alternately between a pair of electrodes (Fig. 8.6A and B). Here, IEMs facilitate ion migration from concentrated

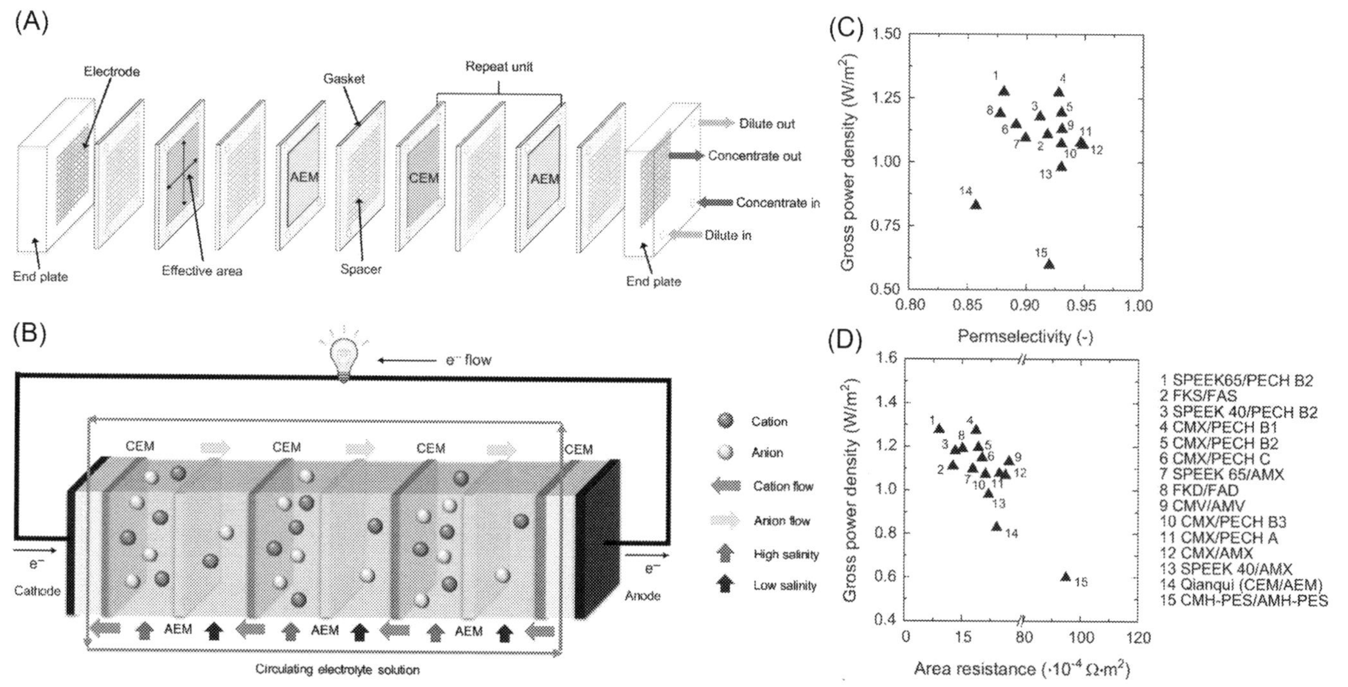

Fig. 8.6 Reverse electrodialysis. (A) Stack arrangement. (B) Schematic illustration of RED process. (C) Gross power density vs permselectivity. (D) Gross power density vs area resistance. *(A, B) Reprinted with permission from J. Jang, Y. Kang, J.-H. Han, K. Jang, C.-M. Kim, I.S. Kim, Developments and future prospects of reverse electrodialysis for salinity gradient power generation: influence of ion exchange membranes and electrodes, Desalination 491 (2020), https://doi.org/10.1016/j.desal.2020.114540, Elsevier. (C, D) Reprinted with permission from E. Güler, R. Elizen, D.A. Vermaas, M. Saakes, K. Nijmeijer, Performance-determining membrane properties in reverse electrodialysis, J. Membr. Sci. 446 (2013) 266–276. https://doi.org/10.1016/j.memsci.2013.06.045, Elsevier.*

solution (CS) into dilute solution (DS) and separate cations and anions for obtaining electrical potential difference at the electrodes. When the electrode is connected to an external load, the generated potential difference drives the electrical current. IEMs need to be conductive toward the counterion to achieve high current density as well as selective toward coions to obtain high electrical potential. Combining these desired properties will result in high power density. However, available commercial membranes, which are generally not specially designed for RED, exhibit lower power density and as well as high production cost [89]. Therefore, there is an increasing effort devoted to developing IEMs that can improve the performance of RED. Table 8.4 summarizes selected studies that demonstrated the use of IEM/FNMs for generating energy in RED technology.

The gross power output of RED is closely related to the membrane resistance [90] (Fig. 8.6C and D) and the ion-exchange capacity of the membrane [91]. These parameters are the major target of the extensive research in preparing IEM for RED applications. The addition of FNMs in the IEM matrix helps to achieve these desired electrochemical properties. Sulfonated SiO_2 increased the power density of sulfonated poly(2,6-dimethyl-1,4-phenyleneoxide) CEM in RED [92]. The improvement was attributed to the enhanced permselectivity and reduced area resistance.

Table 8.4 RED performance with IEM/FNMs.

Membrane		NaCl concentration (in molar)		Power density (W/m^2)	Ref.
CEM	AEM	CS	DS		
FKS (from Fumatech)	FAS (Fumatech)	0.5	0.017	~1.1	[92]
SPPO	FAS (Fumatech)	0.5	0.017	~0.9	[92]
SPPO/sulfonated SiO_2 (0.5%, 70nm)	FAS (Fumatech)	0.5	0.017	1.3	[92]
CSO (Selemion)	ASV (Selemion)	0.5	0.017	~1.2	[55]
SPPO/sulfonated Fe_2O_3 (0.7%)	ASV (Selemion)	0.5	0.017	1.4	[55]
Nafion/sulfonated-SWCNT	FAA-3-based membrane	0.6	0.006	0.079	[98]

CS, high concentration solution; DS, low concentration solution; Ps, permselectivity; SPPO, sulfonated poly(2,6-dimethyl-1,4-phenyleneoxide); SWCNT, single-wall carbon nanotube.

The void faction in the membrane matrix created by FNMs allows higher permeation of counterions. In addition, the sulfonic group improves membrane ion-exchange capacity and ion transport. Ion-exchange capacity is one of the important properties that regulates ionic separation by IEM. Higher ion-exchange capacity means more available charge ions in the membrane matrix for facilitating counterion transport and excluding coions. A similar improvement was obtained by sulfonated iron oxide particles [55].

Usually, RED performance is examined by pure sodium chloride solution. If 10% of pure sodium chloride solution is replaced by magnesium chloride, membrane resistance increases by 50% [93]. The effect of divalent ions on RED power output is then modeled, showing the trend of decreasing power density with the concentration of divalent ion, which is in agreement with experimental data [94]. In practice, the feed solution, such as seawater and river water, contains bivalent or multivalent ions, therefore, monovalent permselective IEMs are needed. A lab test of RED with monovalent permselective IEM demonstrates the improvement of power density by 42% compared to unmodified membrane when multiion solution is used [95]. As also shown in a pilot test RED, the absence of multivalent ions could suppress concentration polarization near the membrane surface resulting in a higher gross power output [96].

The development of IEM with high permselectivity toward monovalent ions is one strategy to address the negative effect of divalent ions. FNMs such as sulfonated reduced GO (S-rGO) can facilitate the improvement of monovalent permselectivity [62]. A commercial AEM with S-rGO showed an improved monovalent permselectivity of Cl^-/SO_4^{2-}, about 2.5 higher than the unmodified ones. FNMs can improve the monovalent permselectivity by increasing electrostatic interaction between the membrane and the multivalent ion or by increasing the sieving effect. The sulfonic groups on reduced graphene oxide provide charges that give electrostatic repulsion to divalent ions [62,66]. The sulfonic group is negatively charged that will affect the exclusion of anions. However, divalent or multivalent ions are more repulsed than monovalent ions due to their higher valence. Another improvement is attributed to the formation of anion channels created by reduced graphene oxide [62,66]. As the hydrated size of divalent ions is larger, the anion channel lowers divalent ion migration. Hence, monovalent permselectivity can be enhanced.

The actual feed solution of RED may contain organic or inorganic contaminants. These contaminants could be attracted on the membrane surface due to electrostatic interaction, leading to fouling layer formation. This layer becomes an additional barrier that decreases the ionic migration rate. Therefore, chemical regeneration will be needed to recover the membrane properties. In operation, chemical regeneration will result in higher chemical consumption and membrane replacement cost. To be sustained in long-term operation, the membrane needs an improvement of its properties, such as increasing membrane hydrophilicity. This way, the membrane can be resistant to the fouling propensity. Membrane hydrophilicity improvement has been successfully performed by incorporating oxidized multiwalled carbon nanotubes (O-MWCNT) [97]. The SPPO cationic membrane hydrophilicity was decreased from 81.5° to 50.8° when the membrane contained 0.5% O-MWCNT. Moreover, the membrane with O-MWCNT displayed better electrochemical properties than the unmodified ones. Sulfonated MWCNT also displays a similar function that attracts more water molecules to the hydrophilic sites of sulfonic groups and results in higher membrane hydrophilicity [98].

8.6 Concluding remarks

Engineering electrochemical properties of membranes is a crucial aspect in the development of energy conversion and storage technologies based on IEM. These technologies not only produce high-energy output and efficiency but also are economically viable if they use highly conductive, permselective, and durable membranes. The conductive membrane can promote better ionic transport so that the resulting electric current is high. Meanwhile, a selective membrane will produce a larger energy capacity. In some instances, the membrane also requires specific characteristics, such as antifouling, because the technology may use solutions that can cause fouling formation. Fouling usually decreases the membrane performance resulting in poor membrane performance stability. Also, to withstand a long-term operation, the membrane should be mechanically, chemically, and thermally stable.

Numerous studies show that those objectives can be achieved by employing FNMs in the membrane preparation. Several FNMs based on graphene oxide, carbon nanotubes, and metal oxide have been widely used to modify the characteristics of IEM. These modified membranes were tested on fuel cells, RFB, and RED. In addition to their intrinsic properties, functionalization adds more attractive and advantageous features to nanomaterials, such as (1) the nanomaterial can be distributed evenly in the membrane matrix, (2) functional groups help facilitate the transport of ions or charge carriers so that the membrane is more conductive, (3) the functional group provides an electrostatic repulsion effect that decreases the permeability of fuels, e.g., vanadium ion (4), new chemical bonds between the nanomaterial and the polymer binder strengthening the membrane matrix, and (5) functional groups create better water-membrane interaction leading to increased membrane hydrophilicity. Therefore, fuel cells, RFB, and RED equipped with IEM/FNMs exhibit higher power output and more stable performance than unmodified IEMs or even commercial membranes.

References

[1] F.E. Ahmed, R. Hashaikeh, N. Hilal, Hybrid technologies: the future of energy efficient desalination—a review, Desalination 495 (2020), https://doi.org/10.1016/j.desal.2020.114659, 114659.
[2] A.N. Hakim, K. Khoiruddin, D. Ariono, I.G. Wenten, Ionic separation in electrodeionization system: mass transfer mechanism and factor affecting separation performance, Sep. Purif. Rev. 49 (2020) 294–316, https://doi.org/10.1080/15422119.2019.1608562.
[3] N.K. Khanzada, M.U. Farid, J.A. Kharraz, J. Choi, C.Y. Tang, L.D. Nghiem, A. Jang, A.K. An, Removal of organic micropollutants using advanced membrane-based water and wastewater treatment: a review, J. Membr. Sci. 598 (2020), https://doi.org/10.1016/j.memsci.2019.117672, 117672.
[4] I.G.B.N. Makertihartha, K.S. Kencana, T.R. Dwiputra, K. Khoiruddin, G. Lugito, R.R. Mukti, I.G. Wenten, SAPO-34 zeotype membrane for gas sweetening, Rev. Chem. Eng. (2020), https://doi.org/10.1515/revce-2019-0086.
[5] X. Zou, G. Zhu, Microporous organic materials for membrane-based gas separation, Adv. Mater. 30 (2018) 1700750, https://doi.org/10.1002/adma.201700750.
[6] C. Bhattacharjee, V.K. Saxena, S. Dutta, Fruit juice processing using membrane technology: a review, Innov. Food Sci. Emerg. Technol. 43 (2017) 136–153, https://doi.org/10.1016/j.ifset.2017.08.002.
[7] I.G. Wenten, K. Khoiruddin, R. Reynard, G. Lugito, H. Julian, Advancement of forward osmosis (FO) membrane for fruit juice concentration, J. Food Eng. 290 (2021), https://doi.org/10.1016/j.jfoodeng.2020.110216, 110216.
[8] J. Vanneste, D. Ormerod, G. Theys, D. Van Gool, B. Van Camp, S. Darvishmanesh, B. Van der Bruggen, Towards high resolution membrane-based pharmaceutical separations, J. Chem. Technol. Biotechnol. 88 (2013) 98–108, https://doi.org/10.1002/jctb.3848.
[9] I.G. Wenten, P.T.P. Aryanti, K. Khoiruddin, A.N. Hakim, N.F. Himma, Advances in polysulfone-based membranes for hemodialysis, J. Membr. Sci. Res. 2 (2016) 78–89.
[10] L. Handojo, A.K. Wardani, D. Regina, C. Bella, M.T.A.P. Kresnowati, I.G. Wenten, Electro-membrane processes for organic acid recovery, RSC Adv. 9 (2019) 7854–7869, https://doi.org/10.1039/C8RA09227C.
[11] E. Ratnaningsih, R. Reynard, K. Khoiruddin, I.G. Wenten, R. Boopathy, Recent advancements of UF-based separation for selective enrichment of proteins and bioactive peptides—a review, Appl. Sci. (2021), https://doi.org/10.3390/app11031078.
[12] H. Chang, Y. Zou, R. Hu, H. Feng, H. Wu, N. Zhong, J. Hu, Membrane applications for microbial energy conversion: a review, Environ. Chem. Lett. 18 (2020) 1581–1592, https://doi.org/10.1007/s10311-020-01032-7.

[13] Y. Wang, D.F. Ruiz Diaz, K.S. Chen, Z. Wang, X.C. Adroher, Materials, technological status, and fundamentals of PEM fuel cells – a review, Mater. Today 32 (2020) 178–203, https://doi.org/10.1016/j.mattod.2019.06.005.
[14] M.A. Abdelkareem, M. El Haj Assad, E.T. Sayed, B. Soudan, Recent progress in the use of renewable energy sources to power water desalination plants, Desalination 435 (2018) 97–113, https://doi.org/10.1016/j.desal.2017.11.018.
[15] E. Drioli, E. Curcio, Membrane engineering for process intensification: a perspective, J. Chem. Technol. Biotechnol. 82 (2007) 223–227, https://doi.org/10.1002/jctb.1650.
[16] I.G. Wenten, K. Khoiruddin, P.T.P. Aryanti, A.N. Hakim, Scale-up strategies for membrane-based desalination processes: a review, J. Membr. Sci. Res. 2 (2016) 42–58.
[17] M.A. Abdelkareem, K. Elsaid, T. Wilberforce, M. Kamil, E.T. Sayed, A. Olabi, Environmental aspects of fuel cells: a review, Sci. Total Environ. 752 (2021), https://doi.org/10.1016/j.scitotenv.2020.141803, 141803.
[18] S.Z. Abbas, M. Rafatullah, Recent advances in soil microbial fuel cells for soil contaminants remediation, Chemosphere 272 (2021), https://doi.org/10.1016/j.chemosphere.2021.129691, 129691.
[19] Y.H. Wan, J. Sun, H.R. Jiang, X.Z. Fan, T.S. Zhao, A highly-efficient composite polybenzimidazole membrane for vanadium redox flow battery, J. Power Sources 489 (2021), https://doi.org/10.1016/j.jpowsour.2021.229502, 229502.
[20] A. Zoungrana, M. Çakmakci, From non-renewable energy to renewable by harvesting salinity gradient power by reverse electrodialysis: a review, Int. J. Energy Res. 45 (2021) 3495–3522, https://doi.org/10.1002/er.6062.
[21] I.G. Wenten, K. Khoiruddin, A.K. Wardani, I.N. Widiasa, Synthetic polymer-based membranes for heavy metal removal, in: A.F. Ismail, W.N.W. Salleh, N. Yusof (Eds.), Synthetic Polymeric Membranes for Advanced Water Treatment, Gas Separation, and Energy Sustainability, Elsevier, 2020, pp. 71–101, https://doi.org/10.1016/B978-0-12-818485-1.00005-8.
[22] D. Ariono, Khoiruddin, Improving ion-exchange membrane properties by the role of nanoparticles, AIP Conf. Proc. 1788 (2017) 030003, https://doi.org/10.1063/1.4968256.
[23] M. Son, K.H. Cho, K. Jeong, J. Park, Membrane and electrochemical processes for water desalination: a short perspective and the role of nanotechnology, Membranes (Basel). (2020), https://doi.org/10.3390/membranes10100280.
[24] I.G. Wenten, K. Khoiruddin, G.T.M. Kadja, R.R. Mukti, P.D. Sutrisna, Modified zeolite-based polymer nanocomposite membranes for pervaporation, in: S. Thomas, S.C. George, T. Jose (Eds.), Polymer Nanocomposite Membranes for Pervaporation, Elsevier, 2020, pp. 263–300, https://doi.org/10.1016/B978-0-12-816785-4.00011-2.
[25] R.K. Nagarale, G.S. Gohil, V.K. Shahi, Recent developments on ion-exchange membranes and electro-membrane processes, Adv. Colloid Interface Sci. 119 (2006) 97–130, https://doi.org/10.1016/j.cis.2005.09.005.
[26] S.M. Hosseini, N. Rafiei, A. Salabat, A. Ahmadi, Fabrication of new type of barium ferrite/copper oxide composite nanoparticles blended polyvinylchloride based heterogeneous ion exchange membrane, Arab. J. Chem. 13 (2020) 2470–2482, https://doi.org/10.1016/j.arabjc.2018.06.001.
[27] S.M. Hosseini, M.M.B. Usefi, M. Habibi, F. Parvizian, B. Van der Bruggen, A. Ahmadi, M. Nemati, Fabrication of mixed matrix anion exchange membrane decorated with polyaniline nanoparticles to chloride and sulfate ions removal from water, Ionics (Kiel). 25 (2019) 6135–6145, https://doi.org/10.1007/s11581-019-03151-w.
[28] K. Khoiruddin, D. Ariono, S. Subagjo, I.G. Wenten, Structure and transport properties of polyvinyl chloride-based heterogeneous cation-exchange membrane modified by additive blending and sulfonation, J. Electroanal. Chem. 873 (2020), https://doi.org/10.1016/j.jelechem.2020.114304, 114304.

[29] K. Khoiruddin, D. Ariono, S. Subagjo, I.G. Wenten, Effect of hydrophilic additive and PVC polymerization degree on morphology and electrochemical properties of PVC-based heterogeneous cation-exchange membrane, J. Appl. Polym. Sci. (2018), https://doi.org/10.1002/app.46690.
[30] J. Křivčík, J. Vladařová, J. Hadrava, A. Černín, L. Brožová, The effect of an organic ion-exchange resin on properties of heterogeneous ion-exchange membrane, Desalin. Water Treat. 14 (2010) 179–184, https://doi.org/10.5004/dwt.2010.1025.
[31] Y. Tanaka, Ion Exchange Membranes: Fundamentals and Applications, second ed., Elsevier Science, Amsterdam, 2015.
[32] J. Dong, Z. Yao, T. Yang, L. Jiang, C. Shen, Control of superhydrophilic and superhydrophobic graphene Interface, Sci. Rep. 3 (2013) 1733, https://doi.org/10.1038/srep01733.
[33] M. Etesami, E. Abouzari-Lotf, A. Ripin, M. Mahmoud Nasef, T.M. Ting, A. Saharkhiz, A. Ahmad, Phosphonated graphene oxide with high electrocatalytic performance for vanadium redox flow battery, Int. J. Hydrogen Energy 43 (2018) 189–197, https://doi.org/10.1016/j.ijhydene.2017.11.050.
[34] S.M. Hosseini, E. Jashni, M. Habibi, B. van der Bruggen, Fabrication of novel electrodialysis heterogeneous ion exchange membranes by incorporating PANI/GO functionalized composite nanoplates, Ionics (Kiel). 24 (2018) 1789–1801, https://doi.org/10.1007/s11581-017-2319-z.
[35] H. Lee, J. Han, K. Kim, J. Kim, E. Kim, H. Shin, J.-C. Lee, Highly sulfonated polymer-grafted graphene oxide composite membranes for proton exchange membrane fuel cells, J. Ind. Eng. Chem. 74 (2019) 223–232, https://doi.org/10.1016/j.jiec.2019.03.012.
[36] G. Shukla, V.K. Shahi, Sulfonated poly(ether ether ketone)/imidized graphene oxide composite cation exchange membrane with improved conductivity and stability for electrodialytic water desalination, Desalination 451 (2019) 200–208, https://doi.org/10.1016/j.desal.2018.03.018.
[37] J. Han, H. Lee, J. Kim, S. Kim, H. Kim, E. Kim, Y.-E. Sung, K. Kim, J.-C. Lee, Sulfonated poly(arylene ether sulfone) composite membrane having sulfonated polytriazole grafted graphene oxide for high-performance proton exchange membrane fuel cells, J. Membr. Sci. 612 (2020), https://doi.org/10.1016/j.memsci.2020.118428, 118428.
[38] S. Changkhamchom, P. Kunanupatham, K. Phasuksom, A. Sirivat, Anion exchange membranes composed of quaternized polybenzimidazole and quaternized graphene oxide for glucose fuel cell, Int. J. Hydrogen Energy 46 (2021) 5642–5652, https://doi.org/10.1016/j.ijhydene.2020.11.043.
[39] J.Y. Chu, K.H. Lee, A.R. Kim, D.J. Yoo, Improved electrochemical performance of composite anion exchange membranes for fuel cells through cross linking of the polymer chain with functionalized graphene oxide, J. Membr. Sci. 611 (2020), https://doi.org/10.1016/j.memsci.2020.118385, 118385.
[40] V. Kumari, R. Badru, S. Singh, S. Kaushal, P.P. Singh, Synthesis and electrochemical behaviour of GO doped ZrP nanocomposite membranes, J. Environ. Chem. Eng. 8 (2020), https://doi.org/10.1016/j.jece.2020.103690, 103690.
[41] M.T. Musa, N. Shaari, S.K. Kamarudin, Carbon nanotube, graphene oxide and montmorillonite as conductive fillers in polymer electrolyte membrane for fuel cell: an overview, Int. J. Energy Res. 45 (2021) 1309–1346, https://doi.org/10.1002/er.5874.
[42] M. Sianipar, S.H. Kim, Khoiruddin, F. Iskandar, I.G. Wenten, Functionalized carbon nanotube (CNT) membrane: progress and challenges, RSC Adv. 7 (2017) 51175–51198, https://doi.org/10.1039/C7RA08570B.
[43] H. Fan, Y. Huang, N.Y. Yip, Advancing the conductivity-permselectivity tradeoff of electrodialysis ion-exchange membranes with sulfonated CNT nanocomposites, J. Membr. Sci. 610 (2020), https://doi.org/10.1016/j.memsci.2020.118259, 118259.

[44] R. Vani, S. Ramaprabhu, P. Haridoss, Mechanically stable and economically viable polyvinyl alcohol-based membranes with sulfonated carbon nanotubes for proton exchange membrane fuel cells, Sustain. Energy Fuels 4 (2020) 1372–1382, https://doi.org/10.1039/C9SE01031A.
[45] Y. Li, H. Wang, Q. Wu, X. Xu, S. Lu, Y. Xiang, A poly(vinyl alcohol)-based composite membrane with immobilized phosphotungstic acid molecules for direct methanol fuel cells, Electrochim. Acta 224 (2017) 369–377, https://doi.org/10.1016/j.electacta.2016.12.076.
[46] A.R. Kim, M. Vinothkannan, M.H. Song, J.-Y. Lee, H.-K. Lee, D.J. Yoo, Amine functionalized carbon nanotube (ACNT) filled in sulfonated poly(ether ether ketone) membrane: effects of ACNT in improving polymer electrolyte fuel cell performance under reduced relative humidity, Compos. Part B Eng. 188 (2020), https://doi.org/10.1016/j.compositesb.2020.107890, 107890.
[47] E. Jashni, S.M. Hosseini, J.N. Shen, B. Van der Bruggen, Electrochemical characterization of mixed matrix electrodialysis cation exchange membrane incorporated with carbon nanofibers for desalination, Ionics (Kiel). 25 (2019) 5595–5610, https://doi.org/10.1007/s11581-019-03068-4.
[48] D.V. Golubenko, R.R. Shaydullin, A.B. Yaroslavtsev, Improving the conductivity and permselectivity of ion-exchange membranes by introduction of inorganic oxide nanoparticles: impact of acid–base properties, Colloid Polym. Sci. 297 (2019) 741–748, https://doi.org/10.1007/s00396-019-04499-1.
[49] K. Akli, K. Khoiruddin, I.G. Wenten, Preparation and characterization of heterogeneous PVC-silica proton exchange membrane, J. Membr. Sci. Res. 2 (2016) 141–146, https://doi.org/10.22079/JMSR.2016.20312.
[50] P. Martina, R. Gayathri, M.R. Pugalenthi, G. Cao, C. Liu, M.R. Prabhu, Nanosulfonated silica incorporated SPEEK/SPVdF-HFP polymer blend membrane for PEM fuel cell application, Ionics (Kiel). 26 (2020) 3447–3458, https://doi.org/10.1007/s11581-020-03478-9.
[51] K. Oh, O. Kwon, B. Son, D.H. Lee, S. Shanmugam, Nafion-sulfonated silica composite membrane for proton exchange membrane fuel cells under operating low humidity condition, J. Membr. Sci. 583 (2019) 103–109, https://doi.org/10.1016/j.memsci.2019.04.031.
[52] K.S. Kumar, S. Rajendran, M.R. Prabhu, A study of influence on sulfonated TiO2-poly (Vinylidene fluoride-co-hexafluoropropylene) nano composite membranes for PEM fuel cell application, Appl. Surf. Sci. 418 (2017) 64–71, https://doi.org/10.1016/j.apsusc.2016.11.139.
[53] V. Elumalai, D. Sangeetha, Synergic effect of ionic liquid grafted titanate nanotubes on the performance of anion exchange membrane fuel cell, J. Power Sources 412 (2019) 586–596, https://doi.org/10.1016/j.jpowsour.2018.11.096.
[54] M. Namdari, T. Kikhavani, S.N. Ashrafizadeh, B. Van der Bruggen, Improvements in heterogeneous cation exchange membranes by incorporation of Fe2O3 nanoparticles, Ionics (Kiel). 25 (2019) 4953–4968, https://doi.org/10.1007/s11581-019-03045-x.
[55] J. Gi Hong, Y. Chen, Evaluation of electrochemical properties and reverse electrodialysis performance for porous cation exchange membranes with sulfate-functionalized iron oxide, J. Membr. Sci. 473 (2015) 210–217, https://doi.org/10.1016/j.memsci.2014.09.012.
[56] P. Prapainainar, N. Pattanapisutkun, C. Prapainainar, P. Kongkachuichay, Incorporating graphene oxide to improve the performance of Nafion-mordenite composite membranes for a direct methanol fuel cell, Int. J. Hydrogen Energy 44 (2019) 362–378, https://doi.org/10.1016/j.ijhydene.2018.08.008.
[57] T. Roy, S.K. Wanchoo, K. Pal, Novel sulfonated poly(ether ether ketone)/rGONR@TiO2 nanohybrid membrane for proton exchange membrane fuel cells, Solid State Ion. 349 (2020), https://doi.org/10.1016/j.ssi.2020.115296, 115296.

[58] Y. Liu, P. Huo, J. Ren, G. Wang, Organic–inorganic hybrid proton-conducting electrolyte membranes based on sulfonated poly(arylene ether sulfone) and SiO2–SO3H network for fuel cells, High Perform. Polym. 29 (2016) 1037–1048, https://doi.org/10.1177/0954008316667790.
[59] C. Fernandez-Gonzalez, B. Zhang, A. Dominguez-Ramos, R. Ibañez, A. Irabien, Y. Chen, Enhancing fouling resistance of polyethylene anion exchange membranes using carbon nanotubes and iron oxide nanoparticles, Desalination 411 (2017) 19–27, https://doi.org/10.1016/j.desal.2017.02.007.
[60] F. Heidary, A.R. Khodabakhshi, D. Ghanbari, A novel sulfonated poly Phenylene oxide-poly Vinylchloride/ZnO cation-exchange membrane applicable in refining of saline liquids, J. Clust. Sci. 28 (2017) 1489–1507, https://doi.org/10.1007/s10876-017-1156-6.
[61] Y. Li, S. Shi, H. Cao, Z. Zhao, H. Wen, Modification and properties characterization of heterogeneous anion-exchange membranes by electrodeposition of graphene oxide (GO), Appl. Surf. Sci. 442 (2018) 700–710, https://doi.org/10.1016/j.apsusc.2018.02.166.
[62] Y. Jin, Y. Zhao, H. Liu, A. Sotto, C. Gao, J. Shen, A durable and antifouling monovalent selective anion exchange membrane modified by polydopamine and sulfonated reduced graphene oxide, Sep. Purif. Technol. 207 (2018) 116–123, https://doi.org/10.1016/j.seppur.2018.06.053.
[63] D.V. Golubenko, Y.A. Karavanova, S.S. Melnikov, A.R. Achoh, G. Pourcelly, A.B. Yaroslavtsev, An approach to increase the permselectivity and mono-valent ion selectivity of cation-exchange membranes by introduction of amorphous zirconium phosphate nanoparticles, J. Membr. Sci. 563 (2018) 777–784, https://doi.org/10.1016/j.memsci.2018.06.024.
[64] K. Khoiruddin, D. Ariono, S. Subagjo, I.G. Wenten, Improved anti-organic fouling of polyvinyl chloride-based heterogeneous anion-exchange membrane modified by hydrophilic additives, J. Water Process Eng. 41 (2021), https://doi.org/10.1016/j.jwpe.2021.102007, 102007.
[65] J. Pandey, A. Shukla, Synthesis and characterization of PVDF supported silica immobilized phosphotungstic acid (Si-PWA/PVDF) ion exchange membrane, Mater. Lett. 100 (2013) 292–295, https://doi.org/10.1016/j.matlet.2013.03.039.
[66] Y. Zhao, K. Tang, H. Ruan, L. Xue, B. Van der Bruggen, C. Gao, J. Shen, Sulfonated reduced graphene oxide modification layers to improve monovalent anions selectivity and controllable resistance of anion exchange membrane, J. Membr. Sci. 536 (2017) 167–175, https://doi.org/10.1016/j.memsci.2017.05.002.
[67] E. Abouzari-Lotf, M. Zakeri, M.M. Nasef, M. Miyake, P. Mozarmnia, N.A. Bazilah, N.F. Emelin, A. Ahmad, Highly durable polybenzimidazole composite membranes with phosphonated graphene oxide for high temperature polymer electrolyte membrane fuel cells, J. Power Sources 412 (2019) 238–245, https://doi.org/10.1016/j.jpowsour.2018.11.057.
[68] S. Lee, K. Seo, R.V. Ghorpade, K.-H. Nam, H. Han, High temperature anhydrous proton exchange membranes based on chemically-functionalized titanium/polybenzimidazole composites for fuel cells, Mater. Lett. 263 (2020), https://doi.org/10.1016/j.matlet.2019.127167, 127167.
[69] C. Wang, B. Lin, G. Qiao, L. Wang, L. Zhu, F. Chu, T. Feng, N. Yuan, J. Ding, Polybenzimidazole/ionic liquid functionalized graphene oxide nanocomposite membrane for alkaline anion exchange membrane fuel cells, Mater. Lett. 173 (2016) 219–222, https://doi.org/10.1016/j.matlet.2016.03.057.
[70] P.F. Msomi, P. Nonjola, P.G. Ndungu, J. Ramonjta, Quaternized poly(2.6 dimethyl-1.4 phenylene oxide)/polysulfone blend composite membrane doped with ZnO-nanoparticles for alkaline fuel cells, J. Appl. Polym. Sci. 135 (2018), https://doi.org/10.1002/app.45959.

[71] C. Long, C. Lu, Y. Li, Z. Wang, H. Zhu, N-spirocyclic ammonium-functionalized graphene oxide-based anion exchange membrane for fuel cells, Int. J. Hydrogen Energy 45 (2020) 19778–19790, https://doi.org/10.1016/j.ijhydene.2020.05.085.

[72] C. Li, Y. Song, X. Wang, Q. Zhang, Synthesis, characterization and application of S-TiO2/PVDF-g-PSSA composite membrane for improved performance in MFCs, Fuel 264 (2020), https://doi.org/10.1016/j.fuel.2019.116847, 116847.

[73] N. Shaik, J. Veeriah, S.K. Bhargava, S. Sridhar, Microbial fuel cell–aided processing of kitchen wastewater using high-performance nanocomposite membrane, J. Environ. Eng. 146 (2020) 4020073, https://doi.org/10.1061/(ASCE)EE.1943-7870.0001717.

[74] S. Mondal, F. Papiya, S.N. Ash, P.P. Kundu, Composite membrane of sulfonated polybenzimidazole and sulfonated graphene oxide for potential application in microbial fuel cell, J. Environ. Chem. Eng. 9 (2021), https://doi.org/10.1016/j.jece.2020.104945, 104945.

[75] A.R. Kim, J.C. Gabunada, D.J. Yoo, Amelioration in physicochemical properties and single cell performance of sulfonated poly(ether ether ketone) block copolymer composite membrane using sulfonated carbon nanotubes for intermediate humidity fuel cells, Int. J. Energy Res. 43 (2019) 2974–2989, https://doi.org/10.1002/er.4494.

[76] P. Choudhury, R.N. Ray, T.K. Bandyopadhyay, B. Basak, M. Muthuraj, B. Bhunia, Process engineering for stable power recovery from dairy wastewater using microbial fuel cell, Int. J. Hydrogen Energy 46 (2021) 3171–3182, https://doi.org/10.1016/j.ijhydene.2020.06.152.

[77] S.-H. Cha, Recent development of nanocomposite membranes for vanadium redox flow batteries, J. Nanomater. 2015 (2015), https://doi.org/10.1155/2015/207525, 207525.

[78] Y. Shi, C. Eze, B. Xiong, W. He, H. Zhang, T.M. Lim, A. Ukil, J. Zhao, Recent development of membrane for vanadium redox flow battery applications: a review, Appl. Energy 238 (2019) 202–224, https://doi.org/10.1016/j.apenergy.2018.12.087.

[79] H.R. Jiang, J. Sun, L. Wei, M.C. Wu, W. Shyy, T.S. Zhao, A high power density and long cycle life vanadium redox flow battery, Energy Storage Mater. 24 (2020) 529–540, https://doi.org/10.1016/j.ensm.2019.07.005.

[80] K. Lourenssen, J. Williams, F. Ahmadpour, R. Clemmer, S. Tasnim, Vanadium redox flow batteries: a comprehensive review, J. Energy Storage 25 (2019), https://doi.org/10.1016/j.est.2019.100844.

[81] C.-H. Lin, M.-C. Yang, H.-J. Wei, Amino-silica modified Nafion membrane for vanadium redox flow battery, J. Power Sources 282 (2015) 562–571, https://doi.org/10.1016/j.jpowsour.2015.02.102.

[82] M.S. Kondratenko, E.A. Karpushkin, N.A. Gvozdik, M.O. Gallyamov, K.J. Stevenson, V.G. Sergeyev, Influence of aminosilane precursor concentration on physicochemical properties of composite Nafion membranes for vanadium redox flow battery applications, J. Power Sources 340 (2017) 32–39, https://doi.org/10.1016/j.jpowsour.2016.11.045.

[83] M. Ding, X. Ling, D. Yuan, Y. Cheng, C. Wu, Z.-S. Chao, L. Sun, C. Yan, C. Jia, SPEEK membrane of ultrahigh stability enhanced by functionalized carbon nanotubes for vanadium redox flow battery, Front. Chem. 6 (2018), https://doi.org/10.3389/fchem.2018.00286, 286.

[84] X.-B. Yang, L. Zhao, X.-L. Sui, L.-H. Meng, Z.-B. Wang, Phosphotungstic acid immobilized nanofibers-Nafion composite membrane with low vanadium permeability and high selectivity for vanadium redox flow battery, J. Colloid Interface Sci. 542 (2019) 177–186, https://doi.org/10.1016/j.jcis.2019.02.001.

[85] Y. Zhang, Y. Zhong, W. Bian, W. Liao, X. Zhou, F. Jiang, Robust proton exchange membrane for vanadium redox flow batteries reinforced by silica-encapsulated nanocellulose, Int. J. Hydrogen Energy 45 (2020) 9803–9810, https://doi.org/10.1016/j.ijhydene.2020.01.188.

[86] S. Liu, D. Li, L. Wang, H. Yang, X. Han, B. Liu, Ethylenediamine-functionalized graphene oxide incorporated acid-base ion exchange membranes for vanadium redox flow battery, Electrochim. Acta 230 (2017) 204–211, https://doi.org/10.1016/j.electacta.2017.01.170.
[87] Y. Zhang, H. Wang, P. Qian, Y. Zhou, J. Shi, H. Shi, Sulfonated poly(ether ether ketone)/amine-functionalized graphene oxide hybrid membrane with various chain lengths for vanadium redox flow battery: a comparative study, J. Membr. Sci. 610 (2020), https://doi.org/10.1016/j.memsci.2020.118232, 118232.
[88] T. Sadhasivam, H.-T. Kim, W.-S. Park, H. Lim, S.-K. Ryi, S.-H. Roh, H.-Y. Jung, Low permeable composite membrane based on sulfonated poly(phenylene oxide) (sPPO) and silica for vanadium redox flow battery, Int. J. Hydrogen Energy 42 (2017) 19035–19043, https://doi.org/10.1016/j.ijhydene.2017.06.030.
[89] E. Guler, K. Nijmeijer, Reverse electrodialysis for salinity gradient power generation: challenges and future perspectives, J. Membr. Sci. Res. 4 (2018) 108–110, https://doi.org/10.22079/jmsr.2018.86747.1193.
[90] E. Güler, R. Elizen, D.A. Vermaas, M. Saakes, K. Nijmeijer, Performance-determining membrane properties in reverse electrodialysis, J. Membr. Sci. 446 (2013) 266–276, https://doi.org/10.1016/j.memsci.2013.06.045.
[91] J. Jang, Y. Kang, J.-H. Han, K. Jang, C.-M. Kim, I.S. Kim, Developments and future prospects of reverse electrodialysis for salinity gradient power generation: influence of ion exchange membranes and electrodes, Desalination 491 (2020), https://doi.org/10.1016/j.desal.2020.114540.
[92] J. Gi Hong, S. Glabman, Y. Chen, Effect of inorganic filler size on electrochemical performance of nanocomposite cation exchange membranes for salinity gradient power generation, J. Membr. Sci. 482 (2015) 33–41, https://doi.org/10.1016/j.memsci.2015.02.018.
[93] T. Rijnaarts, E. Huerta, W. van Baak, K. Nijmeijer, Effect of divalent cations on RED performance and cation exchange membrane selection to enhance power densities, Environ. Sci. Technol. 51 (2017) 13028–13035, https://doi.org/10.1021/acs.est.7b03858.
[94] L. Gómez-Coma, V.M. Ortiz-Martínez, J. Carmona, L. Palacio, P. Prádanos, M. Fallanza, A. Ortiz, R. Ibañez, I. Ortiz, Modeling the influence of divalent ions on membrane resistance and electric power in reverse electrodialysis, J. Membr. Sci. 592 (2019), https://doi.org/10.1016/j.memsci.2019.117385, 117385.
[95] R.A. Tufa, T. Piallat, J. Hnát, E. Fontananova, M. Paidar, D. Chanda, E. Curcio, G. di Profio, K. Bouzek, Salinity gradient power reverse electrodialysis: cation exchange membrane design based on polypyrrole-chitosan composites for enhanced monovalent selectivity, Chem. Eng. J. 380 (2020), https://doi.org/10.1016/j.cej.2019.122461.
[96] S. Mehdizadeh, Y. Kakihana, T. Abo, Q. Yuan, M. Higa, Power generation performance of a pilot-scale reverse electrodialysis using monovalent selective ion-exchange membranes, Membranes (Basel). 11 (2021) 1–25, https://doi.org/10.3390/membranes11010027.
[97] X. Tong, B. Zhang, Y. Chen, Fouling resistant nanocomposite cation exchange membrane with enhanced power generation for reverse electrodialysis, J. Membr. Sci. 516 (2016) 162–171, https://doi.org/10.1016/j.memsci.2016.05.060.
[98] S.A. Shah, S.-Y. Choi, S. Cho, M. Shahbabaei, R. Singh, D. Kim, Modified single-wall carbon nanotube for reducing fouling in perfluorinated membrane-based reverse electrodialysis, Int. J. Hydrogen Energy 45 (2020) 30703–30719, https://doi.org/10.1016/j.ijhydene.2020.08.111.

CHAPTER 9

Sustainable membranes with FNMs for pharmaceuticals and personal care products

Maher Darwish[a], A.A. Abuhabi[b], and Hanan Mohammad[a]

[a]Department of Pharmaceutical Chemistry and Drug Control, Faculty of Pharmacy, Wadi International University, Homs, Syria

[b]Water Technology PhD Joint Programme Between Islamic University of Gaza IUG & Al-Azhar University of Gaza AUG, Gaza, Palestine

9.1 Introduction

Recently, pharmaceuticals and personal care products (PPCPs), and endocrine disruptive compounds (EDCs), collectively denoted here as PhACs, have emerged as a new type of organic micropollutants (MPs). The widespread consumption of such MPs in the fields of healthcare and evolving lifestyles standards resulted in their omnipresence in the environment [1]. The persistency of PhACs in water and wastewater, albeit in trace amounts, has impacted numerously population health conditions and aquatic ecosystems, including premature birth and death, cancer, chronic bronchitis, respiratory tract infection, heart disease, and permanent reduction in lung capacity [2]. Most conventional water/wastewater treatment systems are challenged by inherent shortcomings in addressing this issue as they are not specifically intended or equipped to remove PPCPs; in fact, they are believed to worsen the problem by contributing to the formation of hazardous metabolites and transformation products alongside the parent compounds [3].

A diverse set of innovative systems, including adsorption, advanced oxidation procedures (AOPs), and novel membrane filtration, has been identified as including key and mainstream removal technologies. Attributed to their MPs outstanding removal, membrane technology embraces beyond-doubts potential to fill in the gaps in related water applications. In relation to filtration and rejection capacity, membranes can be diversely classified with respect to several parameters, including:

(1) pore size where material particle size passing and membrane screening process (sieving and repulsion related to both particles and membrane surface layer charge) determine membrane scale type; and

https://doi.org/10.1016/B978-0-323-85946-2.00005-9

(2) membrane fabrication material where membranes are either made of organic or inorganic materials, mostly polymeric and ceramic materialized membranes.

Considering pore size and scale, membranes are categorized as microfiltration (MF), ultrafiltration (UF), nanofiltration (NF), and reverse osmosis (RO) [4]. MF membranes have a particle removal capacity in the range of 0.025–10 mm from liquid solutions in a pure screening or sieving process. UF membranes depend on hydrostatic pressure to force a liquid to pass as a semipermeable-based membrane where suspended particles and solutes having high molecular weight are retained (0.01–0.05 μm or less) while lower molecular weight solutes pass [5]. NF and RO membranes have some similarity in their separation mechanism where uncharged components are removed by size exclusion depending on pore size in a so-called steric hindrance while charged components are removed by chemical repulsion or electrostatic effect [6]. Not far from such membrane categorization, MF, UF, NF, and RO membranes are fabricated from either polymeric organic materials or inorganic materials, commonly known as ceramic membranes [4]. Organic membranes are practically fabricated of various available polymers, including polysulfone, polyethersulfone, cellulose acetate, polymethylpentene, polyimide, polyetherimide, polycarbonate, polydimethylsiloxane, and polyphenylene oxide, while inorganic membranes include ceramics, carbon molecular sieves, nanoporous carbon, mixed conducting perovskite, zeolites, amorphous silica, and palladium alloys [4,7,8]. Presently, polymeric membranes are very dominant in water treatment and desalination despite some setbacks observed including poor long-term stability, tendency to fouling, and shortened lifespan, which limit their applications in many ways [9,10]. The industrial and academic research market of ceramic membranes tends to be limited compared to that for polymeric membranes. Yet interest in ceramic membranes' water and wastewater treatment applications is generally extending globally due to their remarkable properties and ceramic materials types. Ceramic membranes are applicable for MF, UF, and occasionally for NF and RO [11]. Characterized by good separation capacity, flux and porosity at high rates, clear pore size distribution, chemical, mechanical, and thermal stability, being highly hydrophilic with low tendency to fouling at low pressures, and finally long lifespan, ceramic membranes seem to be favorable over polymeric membranes [12–14]. Both polymeric and ceramic membranes have found numerous applications in water and wastewater industries, namely, treatment and desalination [15,16] in addition to other industries including food processing

[17–19], medical and pharmaceutical [20–22], biotechnology [23–25], and energy conversion [26,27]. Such wide applicability is attributed to several merits, such as low operational cost, flexible scalability, ease of operation, low energy consumption and requirements, and small footprint [28].

Focusing on pharmaceutical components removal, NF membranes have been widely applied due to their remarkable capacities in removing MPs in general from water and aqueous solutions at high flux; as such, substances are scaled to NF and RO membranes exclusively. This is attributed to their unique separation mechanism, which is a mix of size exclusion and chemical repulsion depending on particle size and substances charge. Characterized by low-pressure requirements and therefore lower energy requirements, NF membranes tend to be more favorable for MPs removal than RO membranes [4,7,8]. Both polymeric and ceramic materialized NF membranes have nanoscale pores either originally fabricated or blended after fabrication with various organic or inorganic nanoparticles (NPs) [21]. The blending process is usually meant to achieve better performance, as it enhances flux and rejection by improving selectivity and permeability along with fouling resistance [29]. NF membranes' fouling tendency and low rejection rate are major challenges that need to be overcome in order to achieve better performance in MPs' removal from water [30]. Accordingly, several attempts were made aiming to apply a surface modification of NF membranes to achieve improved performance toward MPs removal, benefiting from the molecular sieving capacity of these membranes [31–33]. Such modification counts mostly on functional nanomaterials potentially resulted either due to surface modification or due to NPs blending at the beginning of preparation process.

Based on the above, the aim of the current chapter is to discuss the limitations of nanomaterial-based membrane filtration of PPCPs and EDCs in water and to shed some light on the innovations and strategies of functionalization techniques that can assist in mitigating such limitations toward the realization of nano-enabled membrane capacities reshaping filtration research innovatively.

9.2 PhACs in the environment

9.2.1 Source of PhACs

PPCPs and EDCs have been classified as emerging contaminants due to their persistence in-ground, surface, and even in drinking waters at traceable of ng/L to low μg/L levels. These contaminants emanate mainly from

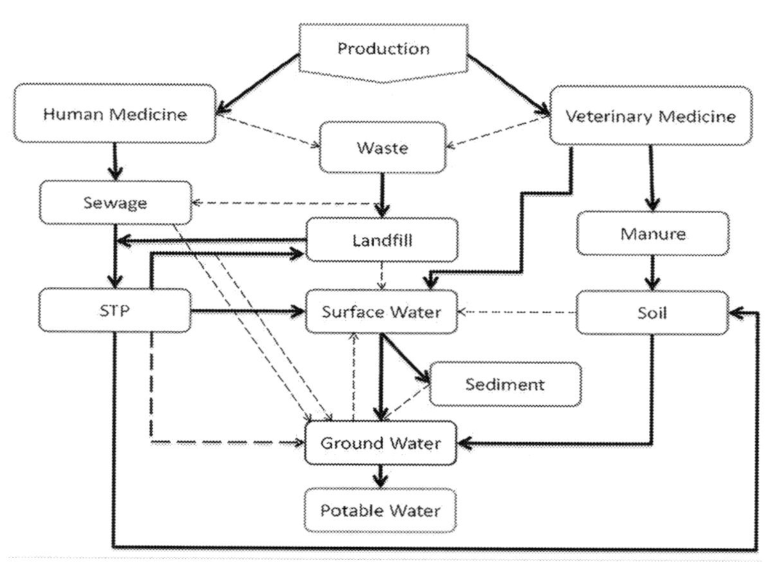

Fig. 9.1 Source, fate, and circulation of PhACs in the environment. *(From Z. Hoyett, Pharmaceuticals and Personal Care Products: Risks, Challenges, and Solutions, IntechOpen, 2018. https://doi.org/10.5772/intechopen.70799.)*

food-based additives and human and veterinary drugs [34]. Fig. 9.1 demonstrates the numerous pathways by which PhACs are introduced into the environment [35].

Excretion of prescribed or nonprescribed pharmaceuticals represents a major source of PhACs and other emerging contaminants in nature. These pharmaceuticals are mostly excreted in urine and feces as unchanged parent compounds or after minor metabolic changes. They transfer wastewater from domestic activities such as the use of soaps and sinks' water taps washing-off through water piping. The improper disposal of expired or unwanted drugs in the sink, toilet, or in household solid waste also contributes to conveying intact PhACs into the wastewater. Interestingly, some active metabolites and biotransformation products may form during treatment in wastewater treatment plants (WWTPs) due to various physical, chemical, and biological reactions taking place there. Hence, incomplete treatments count as the principal source of the PhACs and their metabolites occurrence in aquatic environments [36]. Furthermore, insufficient treatment of effluents from different industries and clinics and hospitals, biosolids and manure land applications, dead animal disposal, and unused medicines landfilling potentially discharge PhACs into surface water [37]. Of note, septic tanks are another key source of the occurrence of PhACs in environments, even on a small scale in rural and suburb areas where they are used to store sewage for the treatment of domestic waste. Due to their primitive design, they are prone to cracking and leaking their contents into groundwater aquifers and adjacent surface water bodies [38]. Landfills are likewise among the primary occurring sources of PhACs in the environment. Medication, soft drinks, and personal care products seem to be sourced from these sites. The latter semisolids and solids waste depository has a major amount of PhACs pollutants. Once a considerable amount of solid and semisolid wastes is dispositioned or leaked, PhACs will also either leak or transform metabolically into different forms by the landfill original microorganisms' colonies [18]. On the other hand, an increase in contaminates' concentration can be observed in groundwater due to the flowing of leachate into the underground aquifers. It is worth mentioning that veterinary drugs used for disease prevention and productivity promotion in the breeding livestock are considered as potential sources in the environment. PhACs ordinarily come from sewage in waste lagoons whereby leachate is a potential or major cause of PhACs release to the ecosystems and environment. In such a context, the PhACs contain various types of antibiotics, metabolites, and some xeno-hormones [39].

In developing countries, municipal wastewater tends to be the common PhACs source while landfill leachate and industries are the most common PPCPs source. Having said this, it is worth mentioning that agricultural industries and landfill leachate may produce PhACs in high concentrations in developed countries whereas municipal wastewater can also be a PhACs source for developing countries.

9.2.2 Occurrence of PhACs

The unique physicochemical characters of PPCPs and EDCs make them unsuitable for conventional water treatment processes, as confirmed by their being present in drinking water. Since PhACs are not completely removed when treating wastewater in the WWTP, aquatic organisms and public health tend to be jeopardized. The alarming evidence from surveillance studies indicates that PhACs can be observed in each type of aquatic environment and are omnipresent [40]. In water and wastewater, advancements in analytical techniques have paved the way for detecting and monitoring

PhACs and their metabolites' very low concentrations in terms of investigating the presence of these pollutants as well as how they influence our health and environment substantially [41]. In general, PhACs had a trace concentration with less than 100 ng/L in both groundwater and surface water, whereas such concentration could be 50 ng/L or less in treated water. Yet a concerning increase of these pollutants in waters can be observed worldwide [34].

Table 9.1 shows the presence of PPCPs and EDCs in aquatic environments of America, Asia, and Europe [42]. The data indicate that concentrations of the contaminants in WWTPs are always higher than those detected in groundwater and surface water and, unfortunately, many concentrations higher than their predicted no-effect concentration (PNEC) attributed to poor metabolism and slow biodegradation. Acetaminophen (AAP), which belongs to the analgesics group of pharmaceuticals, could be present in concentrations above the lowest acceptable PNEC value in different WWTPs of Asia and America, while ciprofloxacin, a common antibiotic, was found in a concentration higher than the PNEC in surface water and wastewater in Asia.

In addition to the above compounds, the use of insecticides such as diethyltoluamide (DEET) often occur in aquatic environments, mainly rural areas, as mosquito or other similar insects are considered as major health and/or sanitation problems. Despite the rare usage nowadays, DEET was detected in ranges from 540 to 1010 ng/L in groundwater [43]. The occurrence of triclocarban and triclosan (TCS) in selected WWTP has been reported in Gauteng Province (South Africa), showing these contaminates at high concentrations. TCS concentrations varied in rivers (0.880–8.72 μg/L), influent WW (2.01–17.6 μg/L), raw and treated sludge (3.65–15.0 μg/kg), and (2.08–7.81 μg/kg), respectively [44]. Compounds used in cosmetics have also been detected in considerable concentrations. Octocrylene and ethylhexyl methoxycinnamate could be detected in samples taken from nearby groundwater of the metropolitan city of Barcelona, Spain. In this location also, musk components, galaxolide (HHCB), and tonaline as well as other musk compounds could be recognized as resembling PPCPs' compounds, all of which were identified as hazardous compounds to ecological sustainability despite the fact that the amounts traced were at existence levels only [45].

Table 9.1 Occurrence and concentration of PhACs in WWTPs of different countries (literature published between 2010 and 2020).

Emerging organic contaminants (EOCs)	America			Asia			Europe			Lowest PNEC (ng/L)
	WWTP	Surface water	Groundwater	WWTP	Surface water	Groundwater	WWTP	Surface water	Groundwater	
Analgesics/antinflammatory										
Diclofenac	2363	1209		523	2.76		69.7	5.4	9.7	10,000
Ketoprofen				58.2		4.1	458	3.4	2.8	15.6×10^6
Ibuprofen	1983	730		268	4.3	19.7	1596	5.5		5000
Acetaminophen	11,600	3422	1890	51,900	2.6	0.647	2463	14.7	10.3	9200
Naproxen	2600	3990	41,900	12,500	0.1	67	741	3.5	1.2	37,000
Antibiotics										
Ciprofloxacin				246	112.4	0.519	221			20
Erythromycin				254.24	2.4	5.6	92.7			
Sulfamethoxazole	1143	173	170	2935.4	61.49	28.7	912	1.9	3	20,000
Beta-blocker										
Propranolol				8.08	0.13	0.098	8.98	1.2	1.8	500
Lipid regulators										
Bezafibrate	3105	1371		125			490	3.4		100,000
Gemfibrozil	178	49	1.2	14.5					165.3	100,000
Psychoactive drug										
Carbamazepine	99.1	6.1	420	14.5			565	13.9	10.4	25,000
Diazepam	4.8	6.1		0.422			6.5			
Hormones										
Estrone (E1)				11.8	4.2	1.08		0.5	0.7	
β E2				3.8	1.5	0.11		0.2	0.4	
αE2				1.1	14.7			0.2	0.7	

From A. Olasupo, F.B.M. Suah, Recent advances in the removal of pharmaceuticals and endocrine-disrupting compounds in the aquatic system: a case of polymer inclusion membranes, J. Hazard. Mater. 406 (2021). https://doi.org/10.1016/j.jhazmat.2020.124317.

9.2.3 Health concerns regarding PhACs

In light of the appreciable concentrations of PPCPs in diverse environmental media, the scientific community expressed significant concern about their potential toxicity [42]. This is attributed to the limited risk assessments achieved so far regarding the effects of PhACs on the environment despite their well-known biological activity. Yet it is evident that antibiotics, for instance, with continuous exposure by environmental microorganisms, caused genes' resistance to develop to the antibiotic, evidently indicated by the bacterial resistance discovered in some facilities used for livestock purposes and nearby environment. For example, six antibiotics (ciprofloxacin, tetracycline, ampicillin, trimethoprim, erythromycin, and trimethoprim/sulfamethoxazole) could be found in WWTP effluent in Australia which increased the resistance of two strains of natural bacteria found in the receiving waters [40,46]. Growing day by day, research supports the fact that genes' resistance in human and other species like pathogens potentially being transferred by microorganisms living in the environment nearby without any previous history of exposure to antibiotic contamination. Having said this, aquatic and terrestrial ecosystems seem to have clearly confirmed antibiotic influence [47]. The residual PPCPs in the environment could find their way to the human body through food as they are able to transfer during farming activities like manure modifications, irrigation by untreated wastewater, and finally disposing sludge into agricultural lands directly. Indeed, PPCPs have been found in breast milk, blood, and children's urine. Moreover, the presence of paraben and benzophenone-4 in placenta confirming their accumulation is taken as early evidence suggesting the possibility of mother-fetal transference [36].

Another great concern regarding PhACs in aquatic bodies is their endocrine-disrupting effects. Exposure to the synthetic hormone 17-ethinylestradiol was recognized to induce feminizing effects in fish at trace amounts. In contrast, 17-bestradiol stimulated a substantial rise in plasma vitellogenin (Vtg) in both male and female of the freshwater chub (*Leuciscus cephalus*) [41]. Waterborne nonsteroidal antiinflammatory drugs (NSAIDs) and their degradation products showed diverse effects on aquatic and aquatic-depending organisms. Diclofenac (DFC) has a harmful effect on the population and variety of birds, for example, steppe eagles and vultures, while at higher levels, it may affect fish in terms of kidney and gill integrity and selected immune parameters. Additionally, chronic human exposure to DFC may result in thyroid tumors and hemodynamic changes [36,48].

Exposure to ibuprofen (IBU) triggers serious side effects to the human body comprising vomiting, bleeding, kidney problems, and cardiovascular risks. On the other hand, 4-acetylbenzoic acid, the main degradation product of IBU, exhibited a noteworthy interference with the growth of micro-green algae, *Chlorella vulgaris*. In addition, photodegradation products of naproxen (NPX) were found to have more toxic effects than the parental compound on algae, rotifers, and microcrustaceans [2]. Waterborne gemfibrozil (GBZ) has a hazardous influence on marine organisms such as goldfish (*Carassius auratus*), represented by a reduction in plasma testosterone. This fibrates-type drug is pharmaceutically applied to assist in lowering the triglycerides in the blood but still causes unknown environmental risks to humans [40]. Moreover, it had been reported that the clofibric primary degradation product known as 4-chlorophenol was found to be very toxic for different organism types and humans as well [2]. Likewise, waterborne carbamazepine (CBZ), an antiepileptic drug, may inhibit human embryonic cell growth and bioaccumulates in freshwater and marine species, such as algae, bacteria, fish, and gastropods. Finally, exposure to 4-methyl benzylidene camphor and TCS can cause development of malformations (up to 3%) in frog embryos and adverse effects on frogs' early life stages, respectively [36]. In addition to parent PhACs toxicity, many of these contaminants are introduced to the environment after human or veterinary use in the form of a metabolite that may be more dangerous than their parent compounds. For example, some acetylated metabolites of antibiotics (such as N4-acetylsulfapyridine) found to have higher toxicity than the parental compound (sulfapyridine) in algae [41]. Given the aforementioned ecological implications, it is urgent to consider the removal of these MPs as a high priority.

9.3 Methods to remove PhACs from water

The wastewater and drinking-water conventional mechanical, chemical, biological, and physical wastewater treatment methods (or a combination of these) were not intended specially to remove pharmaceuticals [5]. Therefore, they are unable to eliminate or degrade most of these emerging components, and a considerable portion of them, intact or metabolized, are eventually discharged to aquatic environments via sewage effluents [49]. Innovative strategies for tertiary treatment of secondary effluents are continuously developed to tackle the hazardous effects of PPCPs and restore higher water quality. Advanced wastewater treatment methods present high elimination rates (above 99%) for intended pharmaceutical compounds. Particular importance was attached to the cost-effectiveness of such tertiary treatment methods [50]. Among the aforementioned techniques, some are considered as cost-effective promising techniques for the PhACs elimination as follows.

9.3.1 Adsorption

The adsorption process utilizing solid adsorbents is amongst the most proficient techniques for contaminants removal due to inexpensiveness, easiness of operation, diversity of adsorbents, and high capacity to abate pollutants concentrations drastically. The diverse types of adsorbents are generally categorized into natural (i.e., charcoal, clays, clay minerals, zeolites) and synthetic (carbonous materials, metal-organic frameworks, polymeric adsorbents, etc.). Each adsorbent has special characteristics such as porosity, pore structure, surface area, and nature of its adsorbing surfaces that affect properties such as selectivity, capacity, and adsorption kinetics. Markedly, nanosorbents are superior to their conventional counterparts in terms of surface area and active sites, and thus have shown a better capacity of adsorbing target contaminants. The major obstacle here is to retain the nanosorbent and ensure it is not released into the environment, which is considered secondary contamination. Fixing the nanosorbents on a support material or filter has been tested to solve such a problem. Among the various types of nanosorbents, both carbon nanotubes (CNTs) and zeolites are eliciting excellent adsorption features; henceforth, they have been broadly explored to remove PhACs from water and/or wastewater [51].

9.3.2 Advanced oxidation procedures

AOPs such as photocatalysis, UV photolysis, Fenton, ozonation, and sonolysis are considered powerful tools for water and wastewater treatment. AOPs have been well-recognized due to their great potential in converting photon/chemical energies into oxidizing power to instantly degrade and mineralize recalcitrant environmental organic pollutants. For instance, heterogeneous photocatalytic degradation employing titania (TiO_2) NPs is one of the most investigated candidates in this regard. Irradiating the colloidal suspension composed of the photocatalyst and the treated solution utilizing light of sufficient energy triggers a series of redox reactions leading subsequently to mineralization of organic contaminants [52].

An intriguing example of PPCPs decomposition via AOPs is the case of DFC and amoxicillin (AMX). DFC is highly degradable and was completely removed from aqueous solutions by UVC-photolysis, whereas AMX is not susceptible to direct photolysis by light [53]. Here, adding H_2O_2 to the AMX solution while using UV irradiation greatly improved the antibiotic degradation [54]. Still, low mineralization was achieved in both processes, showing a maximum removal of 50% for total organic carbon (TOC) with UV/H_2O_2 following reaction time of 80 min. On the other hand, the photocatalytic procedures fully decomposed DFC from aqueous solutions at high rates, and the mitigation of TOC was nearly 80% [55]. Some drawbacks to adopting photocatalysts are limited selectivity to organics, limited activity under visible light (TiO_2, ZnO), fast recombination of photogenerated charge carriers, and postrecovery issues [56].

9.3.3 Membrane separation processes

Membrane separation represents a mature significant technology toward the fresh water via desalination and reuse of industrial wastewater. Membranes work as separation barriers that separate two phases and restore the transparency of water by inhibiting the transport of various chemicals. Among the various properties of membranes is their aptitude to regulate the rate of permeation of different species. Many mechanisms are involved in the process, including size exclusion for MF and NF, rejection by the difference in solubility and diffusivity in RO, and separation by charge difference between species in electrodialysis membranes. Micropollutants comprising PhACs with molecular weight in the range of 200–400 Da have been amply removed via filtration. Some factors should be taken into account when applying such a technique, including membrane properties, a compound of interest physicochemical characteristics, transport mechanism, and matrix effect [57].

As aforementioned, membranes are classified generally into MF, UF, NF, and RO. However, when it comes to the removal of pharmaceuticals, bare MF and UF are rarely used due to the fact that the molecular weight cutoff (MWCO) of UF membranes (10–100 kDa) is at least one order of magnitude above the MW of most micropollutants (<1 kDa) and the same therefore applies for MF [58,59]. Currently, it is very common to modify the pores of MF and UF with some types of nanomaterials to mimic the NF pore size and scale. Here, we are focusing on the functionalization of NF in addition to nanomaterial-based UF/MF composites, especially with carbon nanomaterials and inorganic NPs.

9.3.3.1 Mechanism of rejection of PhACs in membrane filtration

Many mechanisms have been proposed to illustrate the rejection pathways of PhACs over membrane filters. Generally, PhACs may occur in water as charged molecules or neutral based on the medium pH. Consequently, pH alterations have a significant influence on PhACs ionic state and thereafter on their rejection by membranes [60,61]. The dissociation constant (pKa) of the dissolved PhACs is the basic determinant of their ionic state. When the pKa value is lower than studied pH range, the pollutant has a negative charge. Alternatively, the pollutant would either have a positive charge or a mixed of both neutral and positive charge. The three mechanisms suggested so far for the rejection process and illustrated in Fig. 9.2 include size exclusion, adsorption comprising physical adsorption and, less commonly, chemisorption, and electrostatic interaction [30].

Size exclusion or steric hindrance effect

This is the simplest mechanism in which PhACs molecules are sieved when their size is larger than pore size of membrane while the small molecules may pass unhindered. Consequently, large molecules are rejected efficiently, and small ones are slightly rejected (Fig. 9.2A) [62]. Hydrophobicity, on the

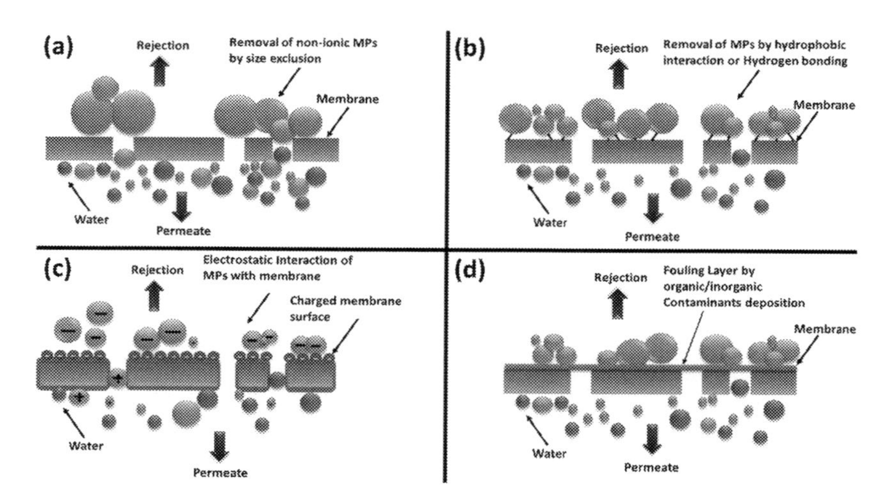

Fig. 9.2 Mechanism of PhACs removal by membrane filtration: (A) Size exclusion; (B) hydrophobic interaction; (C) electrostatic interaction; (D) adsorption. *(From N.K. Khanzada, M.U. Farid, J.A. Kharraz, J. Choi, C.Y. Tang, L.D. Nghiem, A. Jang, A.K. An, Removal of organic micropollutants using advanced membrane-based water and wastewater treatment: a review, J. Membr. Sci. 598 (2020). https://doi.org/10.1016/j.memsci.2019.117672.)*

other hand, is an important factor for rejection where hydrophobic PhACs may interact, adsorb onto, and partition into a hydrophobic membrane surface enabling diffusion through the membrane, therefore bringing about a lower rejection of hydrophobic solutes than expected based on the steric hindrance effect. The case study here is IBU, a hydrophobic pharmaceutical removed by NF. When the pH of the feed was lower than the pKa of the compound, it was highly adsorbed on the membrane, whereas at a higher pH, IBU became negatively charged, and adsorption was minimized; here, the rejection was considered electrostatic in nature [63,64].

MF and UF membranes are efficient in removing solid particles in wastewater, but as it is hard to sustain PPCPs by size exclusions, this contributed less than 10% removal efficiency. It was also noticed that hydrophobic compounds (log Kow > 3) were less likely to be removed by UF and MF membranes [65].

Hydrophobic/adsorptive mechanism

Earlier, we stated that UF membranes are inefficient when applied for MPs rejection based on size exclusion due to the large pore size. However, adsorption can contribute to a small extent of UF membranes removing MPs. This might be attributed to the fact that the adsorption of MPs in membrane filtration not only is limited to the membrane surface but also can develop in the porous membrane structure and is frequently directly correlated to pore radius. Typically, MPs may enter the membrane internal porous structure of UF, and the greater the porosity is, the higher the extent of adsorption [66]. A comparative study of NF and UF was carried out for the removal of 27 kinds of different PPCPs in water using the dead-end filtration process. An overall separation tendency via hydrophobic adsorption as a function of Kow between the hydrophobic compounds and porous hydrophobic membrane during the membrane filtration was noticed. The investigation results showed that both hydrophobic adsorption and size exclusion mechanisms were prevailing to keep PPCPs for the NF membrane, while the UF membrane kept characteristically hydrophobic PPCPs due primarily to hydrophobic adsorption [67].

Commonly, membranes of polymeric origin are hydrophobic in nature. Such a characteristic allows PPCPs with neutral/positive charge to be adsorbed onto these membranes (Fig. 9.2B and D). This process takes place and continues until saturation, after which the low molecular weight solutes diffuse through while the high molecular weight solute remains trapped due to the steric hindrance effect [68]. Some hydrophobic molecules are able to

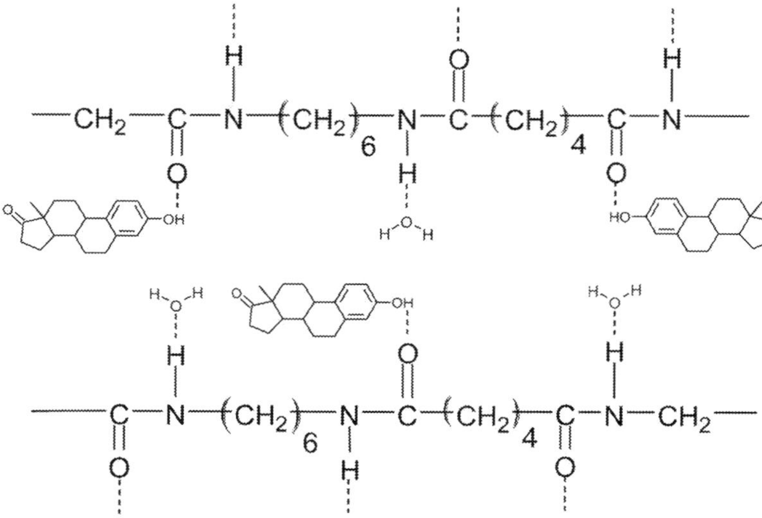

Fig. 9.3 Hydrogen bonding of estrone in polyamide microfiltration membrane. *(From J. Han, W. Qiu, J. Hu, W. Gao, Chemisorption of estrone in nylon microfiltration membranes: adsorption mechanism and potential use for estrone removal from water, Water Res. 46 (3) (2012) 873–881. https://doi.org/10.1016/j.watres.2011.11.066.)*

interact with the membrane surface through hydrogen bonding which may be assigned to the adsorptive mechanism. For instance, estrone, one of the major mammalian estrogens with aromatized C18 steroid and a 3-hydroxyl group and a 17-ketone, can form hydrogen bonds with the polyamide moiety and stay adsorbed on the membrane (Fig. 9.3) [69].

The adsorption process keeps going until the membrane is saturated; thereafter, diffusion through the membrane takes place and causes the rejection process to reduce no matter what the process of interaction is.

Electrostatic interaction

Currently, the most polymeric membrane has a negative charge at neutral pH. This is due to the functionalization of carboxylic acid during the synthesis process in order to boost the selectivity and permeability of membranes. The carboxyl group tends to deprotonate easily at the membrane surface, endowing it with a negative charge. This special character elevates the rejection rate between the surface of membrane and negatively charged PhACs solutes, which is attributed to electrostatic repulsion (Fig. 9.2C). Thus, the hydrophobic interaction/adsorption process is not likely to occur as it is difficult for the negatively charged solutes to come in proximity with the oppositely charged membrane surface, hence the PhACs solute is not capable of being adsorbed or diffused through the membrane [63,70]. Positively charged nonhydrophobic PhACs solutes, on the other hand, are mainly rejected by electrostatic attractions with a negatively charged membrane surface. This attraction may induce solute accumulation at the membrane surface and the formation of a "concentration polarization layer." The strong interaction between the oppositely charged solute/membrane assists the dissolution of the solutes and thereafter diffusion across the membrane matrix. Nonetheless, if the PhACs' molecular size is larger than the MWCO of the membrane, then rejection involves a size-exclusion mechanism. For instance, high rejection was noted for two β-beta-blockers, sotalol and metropolol, using NF-90 (MWCO 90 D) and NF-200 (MWCO 200 D) membranes due to the considerable difference in the NF membranes MWCO and these solutes MW [71].

9.3.3.2 Limitations of membrane filtration used for removal of PhACs

The application of membrane separation techniques in treatment to eventually eradicate the occurrence of PhACs in the water environment has been the focus of many studies. The results showed that the majority of PhACs was barely eliminated when passing through the MF and UF systems. NF and RO, on the other hand, showed a higher removal efficiency—up to 90%–99% in some cases. The degree of removal efficiency was directly related to the membrane characteristics and molecular properties associated with its targeted compounds [65]. However, even though NF has demonstrated such a promising efficacy, it suffers, in addition to the other three classes of membranes, from the following crucial challenges that hinder their full-scale applications in the environmental field to remove PhACs.

Membrane fouling

Membrane fouling is among the most important membrane operations obstacle; it is caused by the excess solute accumulation on the membrane surface and/or inside membrane pores. The membrane and adsorbent need to be regenerated or replaced frequently, and the pollutant is simply migrated to another phase and is not degraded [72,73]. This problem greatly affects membrane performance, as it causes productivity and effluent quality decline, energy-cost requirement increases, essentiality of proper disposal or rejected concentrate treatment, and, ultimately, a short lifespan of membranes. Up to 50% of the total operation cost is due to this specific issue, emphasizing the need for a proper solution to it [21].

Insufficient separation

For uncharged solutes, membranes are characterized by a sigmoidal rejection curve (rejection as a function of molar mass), which results in an insufficient separation between different compounds on the basis of molecular size. A typical sigmoidal rejection curve is given in Fig. 9.4 [9]. Moreover, hydrophobicity and charge interactions play an important role in separation process. As such, the permeate would contain variously sized molecules, either lower or higher than the claimed membrane pore size.

Permeability/selectivity tradeoff relations

Membranes with both high permeability and selectivity are always encouraged. A critical challenge facing the separation techniques is how to allow chemicals with similar sizes to be separated on the membrane while concurrently maintaining a high flux for only one chemical through a membrane. The concept of permeability/selectivity tradeoff relationship implies that as the infusion of one compound increases, the selectivity of the membrane will mostly decrease [74]. To resolve such a challenge, considerable research effort by research laboratories has produced in membranes that are both more permeable and more selective than first-generation materials. Polymers cross-linked with well-defined pores—for instance, two-dimensional (2D) materials, coated nanofibers, carbon nanotubes, metal NPs, or other nanomaterials—determine which compound can be separated [75].

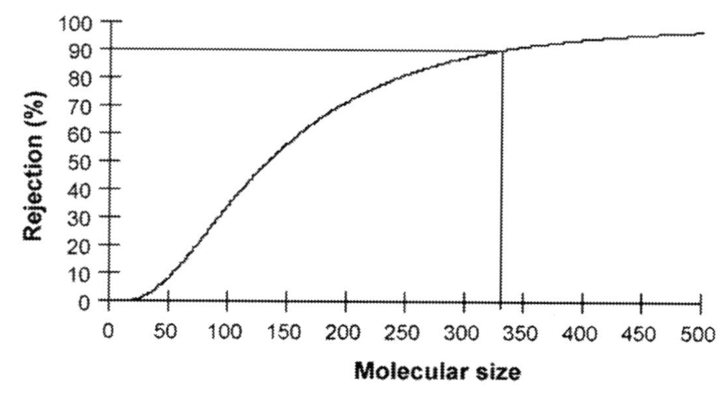

Fig. 9.4 Characteristic sigmoidal rejection curve attained for rejection of uncharged solutes with a nanofiltration membrane. *(From B. Van der Bruggen, M. Mänttäri, M. Nyström, Drawbacks of applying nanofiltration and how to avoid them: a review, Sep. Purif. Technol. 63 (2) (2008) 251–263. https://doi.org/10.1016/j.seppur.2008.05.010.)*

Progressive studies of membrane functionalization by introducing extra organic or inorganic moieties have been developed to overcome all the limitations. Such modifications help in controlling the physicochemical properties of the membranes and serve for the introduction of various reactive functional groups on their surfaces to allow for a tailored functionality toward the desired application. Functional groups introducing into membranes have been recently the focus of membrane modification research. Commonly, functional groups are —OH, —COOH, $—NH_2$, —SH, $—SO_3H$, $—CONH_2$, etc. If the proper activated group is associated with the functionalized membrane, the application range can be extended observably toward selective and tuned permeation as well as separation, capturing of toxic metal, and NPs immobilization for toxic organics degradation to biocatalyst.

9.4 Approaches of membrane functionalization

9.4.1 Functional groups bonding types

9.4.1.1 Functionalization by a noncovalent bond (physisorption process)

This functionalization type reversibly involves binding ligand toward nanomaterial-based membranes through forming relatively weak bonds. As such, several ways may lead to achieving physisorption, including electrostatic interactions, hydrogen bonding, hydrophobic interactions, and/or van der Waals forces. The noncovalent functionalization of carbon nanotube and graphene by π-π interactions represent an excellent example of this approach offering the chance of having functional groups, e.g., aromatic organic molecules attached to carbonaceous material at the same time while having electronic properties maintained and intact simultaneously. So-called steric stabilization is another common approach considered for noncovalent surface functionalization involving polymers or surfactants applied as a capping layer. The individual NPs are stabilized accordingly by surface coating while using steric repulsion to inhibit agglomeration may successfully preserve nanoparticle dispersion intact.

9.4.1.2 Functionalization by a covalent bond (chemisorption process)

One of the most dominating functionalization types involves chemisorption, in which covalent bonds formed between the ligand and membrane assist in achieving robust linkage as well as surface stabilization. The functional group reactivation made by ligand possession toward substrate materials may have the chemisorption capacity to the material surface as well as yielding self-assembled structures. Chemisorbed molecules are most frequently those with the bonding of organic molecules to metals, inorganic semiconductors, or their oxides involvement. Ligands possessing the corresponding functional groups are chosen and synthesized by the substrate material nature. Thiols, amines, silanes, and phosphates on metals and metal-based nanomaterials represent ideal chemisorption examples [76].

9.4.2 Functionalization techniques

Functionalization of the membrane is mainly attributed to either noncovalent or covalent attachment. As simple as it could be, modification techniques are conceptually based on in-situ polymerization, simple adsorption, or grafting to/on the existing support material. Some of the functionalization techniques discussed in this chapter apply to polymeric membranes and others to ceramics. They include coating, layer-by-layer assembly, grafting, immersion, sol-gel, and in-situ hydrogel cross-linking.

9.4.2.1 Surface coating

The coating is a simple physical functionalization method applied widely to diminish fouling while elevating the rejection rate and permeate flux [77]. Various materials are broadly available for functionalization by coating, including polymers, NPs, surfactants, and polyelectrolytes. The coating can be achieved through the following three techniques:

(1) adsorption/adhesion of hydrophilic or compatible materials on the base polymer support. The binding strength is potentially enhanced by functional groups in the molecule and membrane surface interaction;

(2) interpenetration at the interface between the macromolecules and base polymer support; and

(3) entanglement of the macromolecules and the membrane pore structure.

Modification application may alter surface properties, hydrophobicity, and hydrophilicity. However, the low stability attributed to adsorbed functional material leaching can be disadvantageous. There are several coating methods, including dip-coating, spin-coating, spray-coating, and filtration coating.

9.4.2.2 Layer-by-layer assembly (LBL)

Layer-by-layer assembly is considered a brand-new membrane functionalization method that involves the formation of NPs and polyelectrolytes into a well-defined network on the membrane surface [78]. It is a simple mechanism but a highly versatile method involving electrostatic interactions and covalent bonding, hydrogen bonding, hydrophobic interactions, and partitioning. Ideally, charged layer adsorption onto an opposite charged one might result in achieving LBL. This method can be performed through the following four steps:

(1) adsorption of positive components;
(2) washing;
(3) adsorption of negative components; and
(4) washing [79].

Employing coupling agents such as silane in the synthesis process may improve the electrostatic interaction and increase membrane stability. However, covalent bonding-based self-assembly is favored over electrostatic interaction in terms of stability and durability. The surface charge, film thickness, and composition are possibly tuned with the different adsorbed molecules, a number of layers deposited based on application and requirements and practice. The diversity of LBL makes broad applications allowable, such as other functionalization on NPs, polymers, proteins, and dye molecules.

9.4.2.3 Grafting

Grafting is known to be the most frequently applied technique for the functionalization of membranes. It is a chemical modification method potentially classified by four major types: chemical, radiation, photochemical and plasma-induced polymerization [80]. Compared with physical modification, such as coating, grafting considerably offers attachment covalently with membrane structure and maintains long-term stability compared to physical transformation (e.g., coating). Additionally, with grafting, membrane surface can be tailed with various monomers choice toward obtaining desired properties. Chemical grafting involves two significant types: free radical and ionic. Free radical polymerization contains three essential steps: initiation, propagation, and termination. In addition to general free radical polymerization, atom transfer radical polymerization and reversible free radical-fragmentation chain transfer polymerization have gained more interest due to their controllable reaction and narrow molecular weight distribution. Radiation grafting includes preirradiation by vacuum or inert gas to form free radicals, peroxidation with formed hydroperoxides or diperoxides in the presence of air or oxygen, and mutual irradiation technique. The main setback for this technique is related to the possibility of polymer support destruction due to irradiation. Plasma grafting is a well-established technique for membrane functionalization to form linear polymer chains in porous supports. The graft copolymerization is initiated by the macromolecule radicals formed via the cleavage of chemical bonds in the base structure [81]. Such technique is utilized to poly(acrylic acid) onto polypropylene grafting along with printability or adhesion enhancement. The weakness of this method is attributed to the various side reactions which are caused by solid interaction between plasma species (electrons, ions, etc.) and polymer supports base.

9.4.2.4 Immersion method

The most well-recognized method of functionalization. It is a facile and rapid technique that uses activated silane groups as grafts on the ceramic membrane. The process is initiated by mixing silane with the proper solvent to yield Si-OH groups on the surface. Afterwards, the membrane is immersed into the solution, where a covalent bond is formed between Si-OH, and the silane molecules occur on the membrane surface in the shape of the Si—O—Si bond [82]. As a consequence, chemical and mechanical stability and hydrophobicity of the membrane are enhanced. The high cost of silane agents is the only drawback of this method. However, the controlled pore size, permeability, and water contact angle will eventually assist in improving the antifouling resistance of the membrane.

9.4.2.5 Sol-gel method

This method also helps introduce a hydrophobic layer on the ceramic membrane surface. The functionalization is performed by treating the synthesized membrane with the alumina sol and then heating until the gel is obtained. The obtained gel is thereafter oxidized at a temperature range between 400°C and 900°C. The lone drawback here is crack formation in the gel drying stage. Organic functional groups can be added as binders to avoid the development of cracks in the drying step. In addition, preparing the sol-gel solution by the modified Stöber method may help shorten the time required for the process [83].

9.4.2.6 In-situ hydrogel cross-linking

Generally, the incorporation of cross-linked hydrogel inside the porous material is known to be an in-situ hydrogel cross-linking. Hydrogels impose a flexibility degree of toward having considerable volume change amount. However, mechanical strength lacking limits its separation applications to some extent. Such an issue can be dealt with by embedding hydrogels in a rigid and porous matrix (membrane, etc.) through cross-linking supported by in-situ polymerization and simultaneous cross-linking. The cross-linked polymer chains are entangled within the porous matrix and lead to high stability [84].

It is worth mentioning that membrane support performance determined by noncovalent modification in this technique is influenced by hydrogel concentration and loading amount. Generally, several advantages are provided by cross-linking including high stability, uniform distribution, and high concentration of functional groups. However, cross-linking exaggerating and repetition are very likely to lead to polymer chains movement limitation inside its matrix, which in turn lead to reasonable resistance of mass transfer. Too much cross-linking also severely limits the movement of the polymer chains in the matrix, resulting in mass transfer resistance.

9.5 Functionalized NF membranes for PhACs removal

NF membranes are efficient for water applications with the remarkable ability to remove organic MPs from water and aqueous solutions at high flux. This is attributed to lower space requirement, ease of operation and configuration, and relatively low energy consumption compared to other technologies. Additionally, the composite separation layer of the NF membrane is generally highly charged. Thus, this surface charge is observed to be reasonable assistance in conserving energy and improving separation efficiency toward small particles or those with opposite charge [22]. For nearly a decade, NF membranes have shown good removal capacities for natural organic matter (NOM) and MPs, including PhACs from various waters [30]. Practically, several processes are involved when applying NF membranes to eliminate these PhACs, including electrostatic and hydrophobic interactions, size exclusion, etc. [85]. In addition, the negative-charged surface of the membrane assists through electrostatic repulsion in the removal of several negatively charged PPCP components (i.e., IBU, dipyrone, and DFC) in a more efficient manner [86]. Observably, positively charged MPs tend to be removed with negatively charged membranes but less removal efficiency attributed to the electrostatic interaction [87]. For some uncharged MPs and PhACs components, a different membrane behavioral removal can be noticed, potentially or more dominantly related to size exclusion [88]. Additionally, NF membrane rejection (NF-270) reported for some of the MPs was relatively low, affected by two functional moieties on the linkage side of the sulfonamide [88]. The polar molecules of the MPs are also proven to have a significant role in altering membrane pores' polarity, leading to lower rejection obtaining compared to MPs components with lower polarity [89]. Solution temperature is another determining parameter for NF membrane rejection performance for PPCPs. Temperature increasing may lead to pore sizes increasing and, therefore, water flux increasing, increasing MPs diffusion capacity [90]. The membrane rejection rate for negatively charged MPs increased with temperature, while the opposite performance could be observed for positive and neutral ones.

Despite researchers' efforts to apply NF membranes for MPs removal, a few drawbacks presented themselves and hindered this important application. Membrane fouling tendency and low rejection rate are significant challenges that need to be overcome to achieve better performance of NF membranes in MPs removal from water. To do so, several attempts could be observed aiming to apply a surface modification of NF membranes to achieve improved performance toward MPs removal, benefiting from the molecular sieving capacity of these membranes as an advantage. Some examples of such attempts are as follows.

9.5.1 Enhancing antifouling property

The execution of an in-situ modification of NF membrane with hydrolyzed aluminum NPs resulted in significant improvement of membrane hydrophilicity and membrane positive-charge capability. Along with coagulation application to the feed, an improvement of MPs removal as high as 88% could be obtained compared to unmodified NF membrane in addition to fouling reduction improvement with 2.13 times compared to the unmodified membrane [29]. Another successful attempt was by applying chitosan (CS) as a novel antifouling agent for a thin-film composite-NF (TFC-NF) acrylic membrane [91]. CS, a polyaminosaccharide with many hydroxyl and amine functional groups, helped to reduce surface roughness and membrane contact angle, and to confirm significant hydrophilicity improvement. Additionally, an effective removal rate of two common drugs (diphenhydramine and mebeverine) of 97% and 98%, respectively, could be obtained with a lower fouling tendency than the unmodified original membrane.

9.5.2 Enhancing both removal efficiency and permeate flux

The use of graphene oxide (GO) coat of a catalytic NF membrane aiming to remove high-toxic MPs was reported [92]. The size exclusion issue treating different high-toxic MPs with such a combination was confirmed, achieving as high as 96%–99% removal efficiency. Generally, GO sheets, functionalized by epoxides, alcohols, and carboxylic acid groups, assist the membrane in establishing types of well-organized laminar nano-channels that encourage water to pass effectively through blocking other much larger molecules within a simplified and energy effective process [20–22]. In addition, a combined method of Fenton and Fe-based incorporation on polyacrylonite PAN film NF membrane was applied to improve both antifouling and some PhACs rejection [93]. The study revealed some promising results related to flux increasing, foulants degradation on the membrane surface, and reasonable PhACs rejection rates obtained.

Fig. 9.5 shows a radical graft polymerization to modify NF90 commercial membrane to improve six PPCPs elimination for water applications [94]. Another achievement could also be made while applying such a technique, whereby both organic and biofouling caused by humic acid (HA) and sodium alginate (SA) could also be mitigated accordingly. It is worth

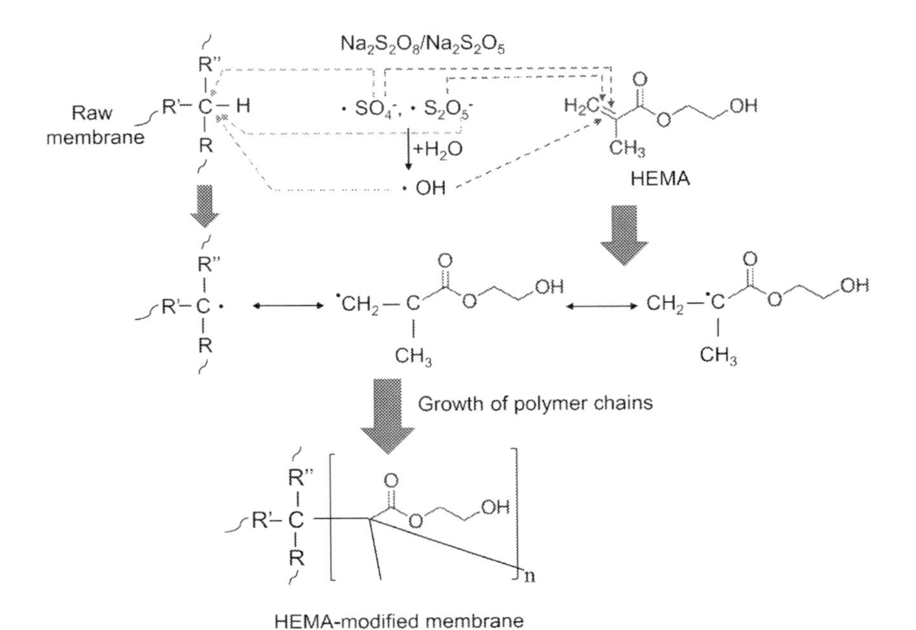

Fig. 9.5 Radical graft polymerization technique using HEMA for NF90 commercial membrane surface modification. *(From Y.L., Lin, C.C. Tsai, N.Y. Zheng, Improving the organic and biological fouling resistance and removal of pharmaceutical and personal care products through nanofiltration by using in situ radical graft polymerization, Sci. Total Environ. 635 (2018) 543–550. https://doi.org/10.1016/j.scitotenv.2018.04.131.)*

mentioning that size exclusion combined with electrostatic repulsion forming a synergistic effect led to having proper PPCPs removal under fouling condition caused by the SA + HA combination.

9.5.3 Enhance membrane rejection of hydrophobic PhACs

The rejection of hydrophobic EDCs by traditional TFC polyamide membranes is known to be poor and may pose a high risk in membrane-based water reclamation. This phenomenon is attributed to the sorption of EDS onto the membrane surface. Even though boosted sorption raises the initial rejection, it can quickly break through the membrane rejection layer and induce moderately low rejections at a steady state, according to the solution diffusion theory. Coating the membrane surface by hydrophilic functionality greatly enhanced this MPs rejection. For example, polydopamine (PDA) coating was introduced to improve the hydrophilicity for the NF90 membrane, as shown in Fig. 9.6 [95,96]. The PDA-modified membrane experienced enhanced performance toward many hydrophobic EDCs (ethyl-paraben, propyl-paraben, benzyl-paraben, and bisphenol A) with up to 75% minimization in the passage of bisphenol A (BPA) compared to the unmodified NF90 and a systematic increase in rejection of the parabens with an increase in PDA coating time. Meanwhile, no variation was observed in the rejection of neutral hydrophilic polyethylene glycol, signifying that the

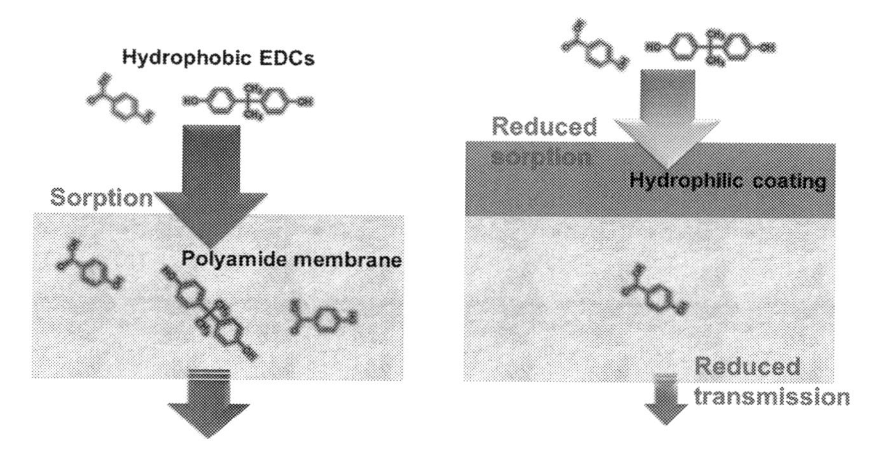

Fig. 9.6 Illustration of the sorption and transmission of hydrophobic EDCs through original and coated membranes. *(From H. Guo, Y. Deng, Z. Tao, Z. Yao, J. Wang, C. Lin, T. Zhang, B. Zhu, C.Y. Tang, Does hydrophilic polydopamine coating enhance membrane rejection of hydrophobic endocrine-disrupting compounds? Environ. Sci. Technol. Lett. 3 (9) (2016) 332–338. https://doi.org/10.1021/acs.estlett.6b00263.)*

impaired hydrophobic EDCs-membrane interactions and the decreased rate of sorption are the leading steps of improved performance.

For the same purpose, metal-organic frameworks (MOFs) were a suitable candidate to enhance polyamide membrane hydrophilic property. MIL-101(Cr) MOF, for instance, was incorporated to create water/EDC selective nano-channels. The water flux of the MOF-membrane was 2.3 times that of the control. The rejection rates against EDCs (methylparaben, propylparaben, benzyl paraben, and BPA) were likewise dramatically higher than the corresponding values of the control membrane, and the water/EDC doubled selectivity could be achieved for BPA [97].

Away from polyamide-based NF membranes, a nonpolyamide NF membrane was fabricated using a complex green tannic acid iron (TA-Fe) [7]. The membrane was applied aiming to achieve high rejection of EDCs, which was successfully achieved with a rejection rate of up to 99.7%. Moreover, PDA and quaternate CS dually charged polyelectrolyte NF membrane was also prepared as an effective approach for higher rejection (Fig. 9.7) [98]. The dually charged membrane surface showed

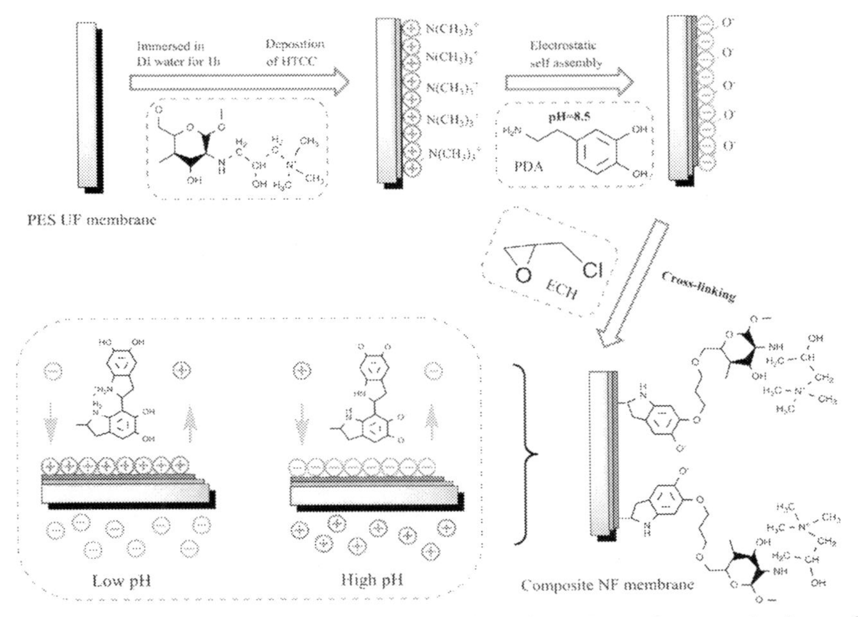

Fig. 9.7 Fabrication of the dually charged polyelectrolyte NF membrane and enhanced rejection of PhACs at various pH conditions. *(From Z. Ouyang, Z. Huang, X. Tang, C. Xiong, M. Tang, Y. Lu, A dually charged nanofiltration membrane by pH-responsive polydopamine for pharmaceuticals and personal care products removal, Sep. Purif. Technol. 211 (2019) 90–97. https://doi.org/10.1016/j.seppur.2018.09.059.)*

rejection improvement against atenolol (ATE) (81.67%), CBZ (92.50%), and IBU (94.5%) at optimum pHs.

Enhanced selectivity toward EDSs was also realized by in-situ immobilization of silver NPs over the preliminarily PDA coated membrane [99]. At this juncture, the higher detention of EDCs was attributed to the combined effect of both steric impediment and weak hydrophobic interactions.

9.5.4 Minimizing the permeability-selectivity tradeoff (enhanced selectivity of membrane against MPs)

Surface modification may also contribute to tackling the permeability-selectivity tradeoff issue. The polyamide NF membrane performance regarding permeability-selectivity tradeoff was enhanced by using TFN membranes attained by the intercalation and controlling the positioning of MOFs filler within the PA layer (MOF bi-layered TFC). More than 98% rejection of both DFC and NPX was obtained with 4.9- and 3.4-times improvements compared to the TFC and TFN membranes values, respectively. The improvement was ascribed to the PA layer thickness, MOF porosity, membrane hydrophilicity, and membrane roughness [100]. Another assessed endeavor was when preparing TFN NF on a polysulfone (PSF) support with in-situ embedded zeolite NPs followed by interfacial polymerization to form the polyamide layer. The produced membrane illustrated double water permeability and a similar rejection of negatively charged PhACs compared with the TFC control membrane. The demonstrated performance could be due to the internal pores of zeolite NPs, the increased membrane surface roughness, and the microporous defects between the NPs and the polyamide matrix [101]. In parallel and compared to the NF-90, a simultaneous enhancement of both permeability and rejection against tris(2-chloroethyl) phosphate, tris(1-chloro-2-propyl) phosphate, and tris(1,3-dichloro-2-propyl) phosphate molecules was realized from a TFN hollow-fiber membrane encompassing nanoporous SAPO-34 NPs. The NP exposed TFN membrane was prepared via the cosolvent assisted interfacial polymerization method (Fig. 9.8) [102]. In this NF membrane, larger nanoporosity and a higher cross-linking degree contributed to the enhanced water permeability. In contrast, the superior hydrophilicity and lower streaming potential contributed to having improved rejection of MPs.

Likewise, TFN membrane containing oleic acid modified SiO_2 NPs on PSF support fabricated via interfacial polymerization in Fig. 9.9 [103] has

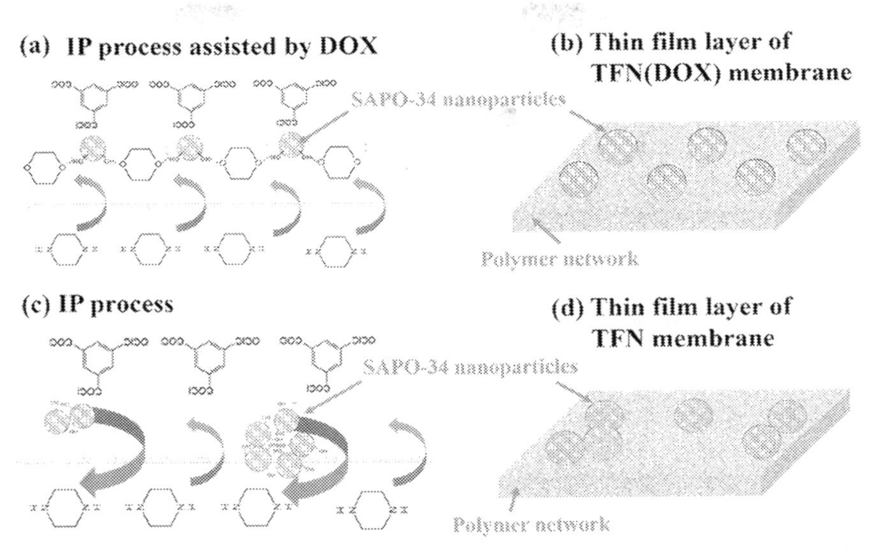

Fig. 9.8 Preparation of TFN hollow fiber membrane with SAPO-34 nanoparticles on the dual-layer (PES/PVDF) hollow fiber substrate. *(From T.Y., Liu, Z.H. Liu, R.X. Zhang, Y. Wang, B.V.d. Bruggen, X.L. Wang, Fabrication of a thin film nanocomposite hollow fiber nanofiltration membrane for wastewater treatment, J. Membr. Sci. 488 (2015) 92–102. https://doi.org/10.1016/j.memsci.2015.04.020.)*

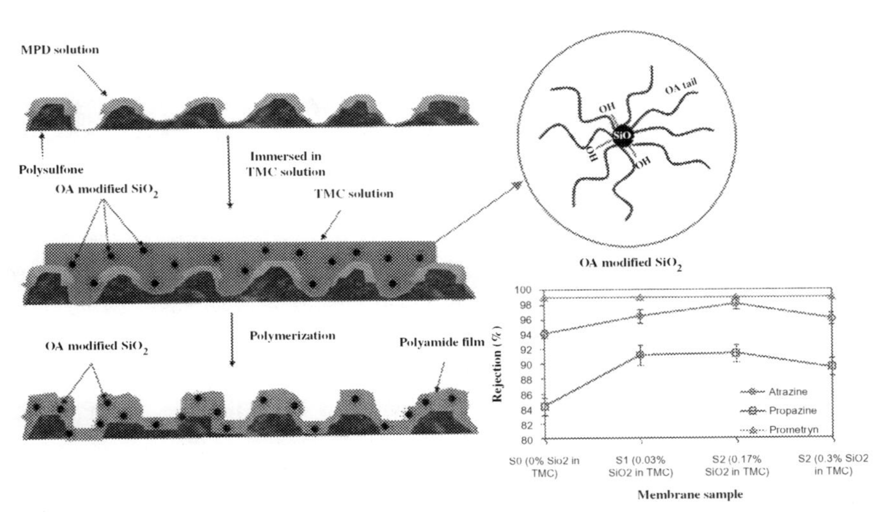

Fig. 9.9 Oleic acid modified SiO_2/polyamide membrane preparation and rejection performance. *(From N. Rakhshan, M. Pakizeh, Removal of triazines from water using a novel OA modified SiO2/PA/PSf nanocomposite membrane, Sep. Purif. Technol. 147 (2015) 245–256. https://doi.org/10.1016/j.seppur.2015.04.013.)*

shown increased permeate flux and enhanced rejection for three triazines including atrazine, propazine, and prometryn (PTN) when compared with a pristine membrane.

9.6 Nanocomposite membranes for PhACs removal

The nanocomposite membranes are among the pioneering options to address all the challenges of membranes application. Indeed, fabrication and investigating this type of functionality focus on nanotechnology in membranes for MPs removal. For example, nanofiber and NPs-based membranes have elicited advanced properties, such as low fouling, compared with pristine ones. On the other hand, carbon nanomaterials (CNM)-based membranes, including CNT and graphene, have many remarkable features placing them on top as candidates for water purification processes such as adsorption, filtration, and supports for AOPs [104]. A detailed description of the improved performance over these three types of composites in bare and functionalized forms is reviewed in this section.

9.6.1 NPs/functionalized NPs-based composite membranes

Inorganic membranes and organic polymers blending with some inorganic NPs toward enhancing membrane rejection against PhACs and other MPs can be seen as one of the recent breakthroughs. The NPs of interest can be embedded onto the membrane surface or otherwise dispensed in the solution prior to membrane casting [105]. Presently, several inorganic NPs available for potential blending with membrane, including zinc oxide, germanium dioxide, silver nitrate, silicon dioxide, aluminum oxide, magnesium oxide, and titanium dioxide. Having such particles blended with polymeric membranes, for instance, polyethersulfone (PES), positively impacts the membrane hydrophilicity and possibly converts this membrane from being hydrophobic to hydrophilic. Consequently, features such as high permeability, stable flux, excellent rejection against foulants, and better antifouling behavior can be realized and sustained through this technique.

Blended PES membrane with two types of NPs, namely, silica (SiO_2) and germanium dioxide (GeO_2), for PhACs removal from the water had a removal rate as high as 99%. Such a high removal rate is attributed to the increase of transmembrane pressure caused by NPs loading, resulting in higher water permeation and lower PhACs permeation. Additionally, PhACs components with high hydrophobicity and sorption potential tended to be removed at a higher rate than those of hydrophilic properties

[106]. Several polymeric membranes were modified by TiO_2, forming a promising nanocomposite membrane with high hydrophilicity properties and obtaining better permeability and higher flux. Enhancing such MPs' removal related properties also contributed to more enhanced removal rates [95]. Nanosize TiO_2 is also well-known photocatalytic material with high degrading potency for PPCPs compared to photolysis ones [107]. A poly (vinylidene fluoride) PVDF dual-layer hollow fiber membrane configured with TiO_2 NPs as a photocatalyst for potential PhACs elimination was fabricated and applied by Paredes et al. [108]. The use of catalyst photo-transformation capability enabled the photo-transformation of several PhACs 2.2 times the normal photolysis process, resulting in highly improved photocatalytic PhACs removal from different waters. Having said this, a visual glitch associated with applying TiO_2 as a photocatalyst specifically for the PPCPs removal is related to the issue of transformation products (TPs) formation, potentially toxic and/or bio-accumulative based. The occurrence of TPs is attributed to the separation mechanism of the TiO_2 from its reaction mixture prior to using it for PPCPs removal [109]. This issue needs to be dealt with appropriately to reduce the potential occurrence of TPs [107,110].

Characterized by several unique properties, including microporosity, nano-sizable particles, good adsorption capability, layering structure, and nano-clay particles, researchers are increasingly interested in water and wastewater contaminants removal. Despite the previously mentioned attractive properties, several obstacles are associated with nano-clay particles, such as surface hydration, lower molecules accessibility, and low attraction to organics [111]. Therefore, benefiting from the nano-clay merits is dependent on overcoming their drawbacks, which may be possible with the blending of these particles with membranes and so producing a clay-polymer nanocomposite membrane [112]. A nanocomposite poly(vinyl alcohol) PVA/porous PSF membrane combined/incorporated with montmorillonite (Mt) nano-clay had shown high but differentiated rejection rates (86%–98%) for several PhACs components. This has the potential to differentiate to the hydrophilicity level of each component [113].

In general, membrane capacities in removing organic MPs from water and aqueous solutions are known to be significant and reasonable [15,16]. Such capacities can also be enhanced/amplified by having the membrane decorated with nano-catalysts, producing a catalytic membrane [21]. Ceramic membranes fit better than polymeric ones as catalytic because ceramic membranes tend to be more structurally and chemically stable, with higher antioxidation capability [114], not to mention their remarkable thermal and mechanical stability, their antiswelling ability, and their long lifespan [101]. A catalytic ceramic membrane decorated with α-$Al_2O_3/CoFe_2O_4$, a gravity-driven membrane, combined with in-situ catalytic oxidation degrading material to eliminate organic MPs from water. This ceramic membrane has a controllable structure with pore size of 267.5 nm on average. The nano-catalyst coating caused a slight pure water flux decrease of no more than 2.2% but without significant alteration to the membrane structure [115]. Other studies focusing on removing PhACs using NPs and surface functionalized NPs-based ceramic and polymeric membranes are summarized in Table 9.2 [36].

9.6.2 Functionalized nanofibers-based composite membranes

Characterized by their unique properties in terms of high porosity, increased specific surface area, and distinctive interconnected construction, high water flux, low fouling tendency, and stable structure, nanofiber membranes are well-distinguished and established. Accordingly, these membranes have become rapidly advantageous and at the focus of water purification field but with the utilization of nanofibers growing, the bare nanofiber membrane is incapable of meeting the requirements of the relatively complicated process of PhACs removal. To overcome this issue, nanofibers functionalization through multifunctional constituents joining allows for filtration efficacy. For that, a wide range of preelectrospinning and postelectrospinning technologies has been adopted recently to modify the nanofibers. Fig. 9.10 shows the functionalization methods of the electrospun nanofiber membranes (ENMs) which are divided into three categories: nanofiber modification technology, nanofiber surface modification, and modification of TFC nanofiber membranes [96,123].

In general, nanofiber modification technologies can be categorized in relation to nanoparticle doping, polymer mixing, and coaxial electrospinning technology (Fig. 9.10A). Additionally, the intended groups may be anchored on the nanofiber surface by physisorption or chemisorption (Fig. 9.10B). Physical surface functionalization is basically conducted by submitting the target group or substance on the surface of ENMs through coating, deposition, electrostatic attraction, etc. However, the long-term stability during filtration process is always questionable in terms of physisorption, hence chemical modification via chemical reactions can make up for this inadequacy efficiently. TFC modification endows the ENMs with a

Table 9.2 Examples on the removal of PhACs using NPs or functionalized NPs-based ceramic and polymeric membranes.

Membrane support	Functionality	Functionalization method	PhAC	Removal mechanism	Removal/ degradation on control	Removal/ degradation on functionalized membrane	Ref.
Ceramic membranes							
α-Al_2O_3	N-doped TiO_2	Surface coating	CBZ	Electrostatic interaction and catalytic degradation	Less than 10%	~70%	[116]
Al_2O_3	CoO	In-situ self-sacrificed template method	Sulfamethoxazole	Size-exclusion and catalytic degradation via peroxymonosulfate (PMS) activation	25%	59%	[117]
α-Al_2O_3	V_2O_3	Impregnations	BPA	Retention and Fenton processes	The BSA retention level of one-layered simple ceramic membranes was above 96 Degradation: <5% BPA	Degradation up to 77% with high V_2O_3 and H_2O_2 levels	[118]
Titania-based microporous top layer and the α-Al_2O_3 macroporous support	Mesoporous maghemite (g-Fe_2O_3) thin layer	Sol-gel deposition	para-Chlorobenzoic acid	Electrostatic attraction and ozonation	Retention: (~50%) Degradation: R_{ct}, 1.49×10^{-9}	Retention: (~65%) Degradation: R_{ct}, 4.26×10^{-10}	[119]
Clay-alumina based macroporous support	CuO and TiO_2	Surface coating	Ciprofloxacin	Surface adsorption	–	99.50%	[120]

Continued

Table 9.2 Examples on the removal of PhACs using NPs or functionalized NPs-based ceramic and polymeric membranes—cont'd

Membrane support	Functionality	Functionalization method	PhAC	Removal mechanism	Removal/ degradation on control	Removal/ degradation on functionalized membrane	Ref.
Tubular shaped α-Al_2O_3 substrate	CeO_x	Citrate sol-gel assisted wet impregnation method	Mixture of BPA, benzotriazole (BTZ), and clofibric acid (CA)	Adsorption and catalytic ozonation	Adsorptions or ozonation alone less than 5%	~38% TOC removal	[117]
α-Al_2O_3	$CoFe_2O_4$	Impregnation technique combined with low temperature calcination	IBU	Adsorption and catalytic degradation via PMS activation	Adsorption of control and control +$CoFe_2O_4$ ~20%	99.5% removal	[21]
Tabular diatomite	MgO	Dip-coating	Tetracycline	Electrostatic adsorption effect (on the positively charged nano coating)	24.20%	98.40%	[121]
Polymer membranes							
PES	$Fe_3O_4@SiO_2$	Liquid-induced phase separation	Methylene blue	Sieving and Fenton-like reaction	30%	94.60%	[122]
PES	Hyperbranched polyethyleneimine HPIE and $BiVO_4$	Blending	TCS	Retention and photocatalytic degradation	–	Up to 86%	[44]
PES	SiO_2 and GeO_2 with sodium dodecyl sulfate (SDS) cross-linker	Blending	Caffeine (CAF) Galaxolide (HHCB)	–	~60%	97.85%–99.00% (for HHCB) 87.42%–93.00% (for CAF)	[106]

No Permission Required.

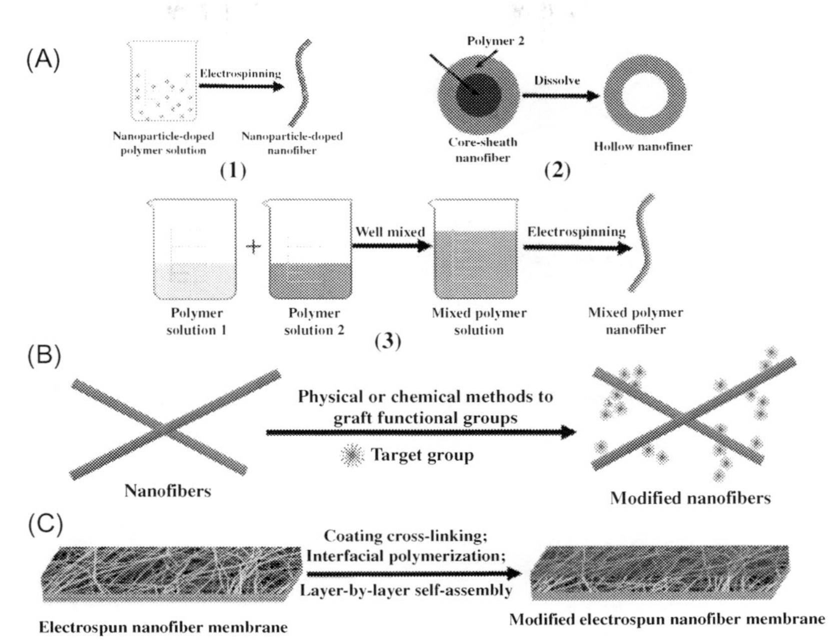

Fig. 9.10 Representation of ENMs modification methods: (A) Nanofiber modification technology; (B) nanofiber surface modification; and (C) thin film composite nanofiber membranes. *(From H. Chen, M. Huang, Y. Liu, L. Meng, M. Ma, Functionalized electrospun nanofiber membranes for water treatment: a review, Sci. Total Environ. 739 (2020) 139944. https://doi.org/10.1016/j.scitotenv.2020.139944.)*

selective layer that increases the porosity and reduces the pore size to allow for enhanced PhACs rejection and permeate flux. As shown in Fig. 9.10C, TFC is composed of the nanofiber support layer and a selective layer also called the active layer. High hydrophilicity, improved selectivity, high strength-to-weight ratio, low mass transfer resistance, and easy control of membrane thickness are among the additional advantages of such functionalization.

Recently, β-cyclodextrin (β-CD)/chitosan (CS)-surface functionalized PVA nanofiber membrane for the rapid removal of heavy metals and PhACs has been successfully fabricated with epichlorohydrin as a cross-linking agent [124]. The vital role of the cross-linker is emphasized in Fig. 9.11, where bead-free and uniform nanofibers formulate the uncross-linked β-CD/CS/PVA membrane. Observably and following the epichlorohydrin cross-linking process, the cross-linked β-CD/CS/PVA nanofiber

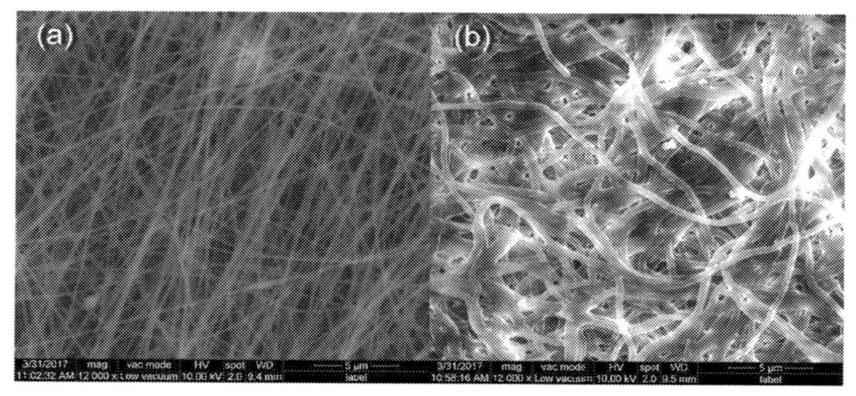

Fig. 9.11 SEM images of (A) uncross-linked β-CD/CS/PVA membrane and (B) cross-linked β-CD/CS/PVA membrane. *(From J.P. Fan, J.J. Luo, X.H. Zhang, B. Zhen, C.Y. Dong, Y.C. Li, J. Shen, Y.T. Cheng, H.P. Chen, A novel electrospun B-CD/CS/PVA nanofiber membrane for simultaneous and rapid removal of organic micropollutants and heavy metal ions from water, Chem. Eng. J. 378 (2019). https://doi.org/10.1016/j.cej.2019.122232.)*

membrane morphology (Fig. 9.11B) changed. This is potentially due to fiber swelling and slight adhesion increasing while the fibrous form is intact.

Similarly, electrospun PES nanofibers with β-CD-epichlorohydrin (βCDP) deposited on PES UF membranes were fabricated to remove some steroid hormones from the water [34]. Maximum E2 removal (80% vs ~40% for UF only) was achieved within 5 h in static adsorption, while it reached 99% in dynamic adsorption. The composite nanofiber membrane showed enhanced water permeability and a high capacity was estimated for micropollutant uptake.

Photocatalyst membranes for enhanced decolorization of MB and degradation of isoproturon were prepared by immobilization of Degussa P25 TiO_2 nanoparticles on both polymeric (polyamide 6) and ceramic (silica) nanofibrous supports [125]. NPs were added by inline functionalization and dip-coating process as pre- and postfunctionalization methods, respectively. Inline functionalization was shown to be the best functionalization method for polyamide 6, since dip-coating results in pore blocking. For silica, on the contrary, dip-coating was the preferred functionalization method, avoiding shielding of the TiO_2 by the silica shell. A similar trend was observed when PAN nanofibers were employed as a support for grafting (TiO_2-NH_2) nano-photocatalyst [126]. Prior to surface functionalization with nanoparticles, PAN was composited with MWCNTs to improve the tensile strength, Young's modulus, and chemical resistance.

PAN/MWCNT/TiO_2-NH_2 was used for the photodegradation of three pharmaceuticals, namely IBP, NPX, and cetirizine (CIZ), in aqueous media under UV irradiation and the results demonstrated that more than 99% degradation efficiency was obtained for IBP, CIZ, and NPX within 120, 40, and 25 min, respectively, whereas for PAN/TiO_2-NH_2, the photodegradation did not exceed 60%–70% and was negligible for PAN alone.

Laccase, a multicopper oxidase enzyme, is considered an attractive functionality and could be covalently immobilized on the surface of PVA/CS/f-MWNTs nanofibrous membranes with an average fiber diameter of 100–200 nm. Both enzyme loading and activity retention of the immobilized laccase were significantly higher on the nanofibrous membranes with MWNTs than those without. Immobilized laccase on the PVA/CS/MWNTs nanofibrous membranes exhibited high stabilities, reuse capabilities, and removal efficiency for DFC. Cyclic voltammetry measurements demonstrated that MWNTs enhanced the electrochemical capacity of the nanofibrous membrane. The assumed mechanism here is that CNTs enhance electron transfer between laccase and substrate molecules on the nanofibers. Fig. 9.12 displays the removal of DFC by laccase versus time under optimal conditions. The removal efficiency of DFC by the laccase-PVA/CS/MWNTs membrane was 18 times higher than that by the laccase-PVA/CS membrane under the same conditions [90].

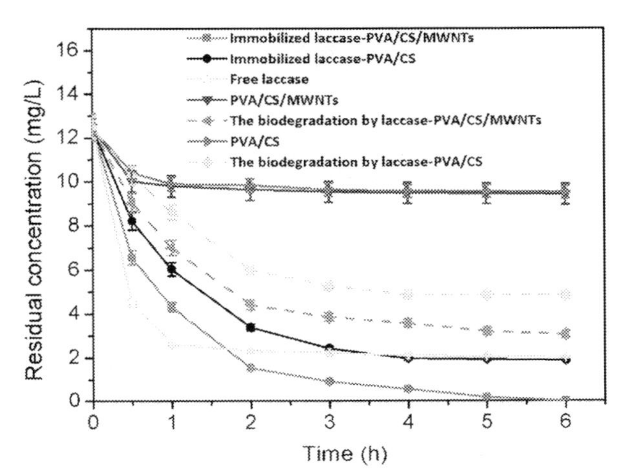

Fig. 9.12 Removal of DFC by laccase versus time. *(From R. Xu, R. Tang, Q. Zhou, F. Li, B. Zhang, Enhancement of catalytic activity of immobilized laccase for diclofenac biodegradation by carbon nanotubes, Chem. Eng. J. 262 (2015) 88–95. https://doi.org/10.1016/j.cej.2014.09.072.)*

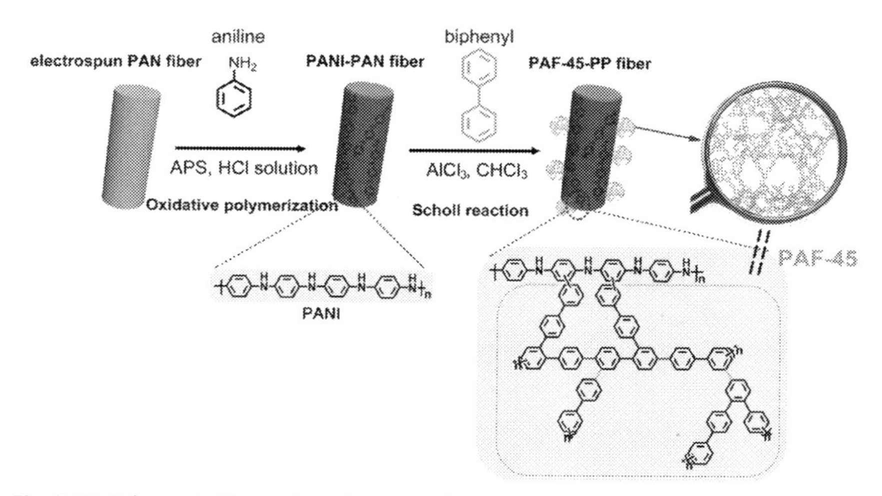

Fig. 9.13 Schematic illustration of coating electrospun PAN fiber membrane surface by PANI-PAF aromatic framework. *(From R. Zhao, T. Ma, S. Li, Y. Tian, G. Zhu, Porous aromatic framework modified electrospun fiber membrane as a highly efficient and reusable adsorbent for pharmaceuticals and personal care products removal, ACS Appl. Mater. Interfaces 11 (18) (2019) 16662–16673. https://doi.org/10.1021/acsami.9b04326.)*

To expand the adsorption capacity of ENMs, coating the PAN surface by a porous aromatic framework (PAF) was implemented as a rising strategy. Meanwhile the processability of PAFs could be positively amended with the assistance of aromatic seed layer, polyaniline (PANI) (Fig. 9.13) [26,27]. Benefiting from this composition, an increase in the surface area of PAN fiber membrane to 262.4 m^2/g from 9.2 m^2/g was obtained. The functionalized ENM presented decent adsorption performance toward three typical PPCPs: IBU, chloroxylenol (CLXN), and DEET. Hydrophobic interaction, π-π interaction, and pore capture mechanism contributed to the high adsorption capacity and good recycling ability.

9.6.3 CNMs/functionalized CNMs-based composite membranes

9.6.3.1 CNTs-based membranes for PhACs removal

In water filtration, the application of CNMs is widely spreading. This is because of their transport selectivity ability through water and their size reduction contributing to water permeability improvement outperforming polymeric membranes [127]. They also have the exclusion capacity to allow water to pass while blocking contaminants flow simultaneously [69], considering their strong affinity for organic matter characterized by nonpolar compounds [128,129]. The CNTs are graphite rooted material consisting of various sheets with a single atom in a hexagonal-shaped structure embedded in honeycomb lattice crystal, so-called graphene. These structures varied from cylindrical, giving rise to single, double, or multiwalled CNTs, respectively (SWCNTs, DWCNTs, and MWCNTs). Characterized by high adsorption capacity and blended as carbon nanocomposite membranes, CNTs can efficiently remove contaminants from both water and wastewater, especially when applying the vacuum filtration method [130,131]. Several water quality parameters may affect the removal capacity of these membranes, including pH, whereby chemical groups of both PhACs and CNTs are affected by pH solution variation positively/negatively, leading to alteration of the PhACs and CNTs electrostatic interaction, ionic strength and calcium concentration influencing PhAC/CNT interactions, and level of NOM in the solution which has the adsorption modification capacity of naturally existing solids in the water [132].

Despite all the above, the hydrophobic characteristic of CNT makes it harder for the tubes to disperse and increases their tendency to aggregate and decrease the distribution aptitude in the solvent. A reduction in the compatibility of a membrane structure may occur. Functionalization of CNT with hydrophilic moieties is anticipated to resolve those drawbacks. Furthermore, the antitradeoff between permeability and selectivity might be realized via the electrostatic repulsion concept offered by the available hydrophilic functional groups on CNTs. During this process, the pollutant will be driven out and be prevented from attachment to the membrane surface while water molecules are easily attracted and pass through. Hence, different tools have been explored to functionalize CNTs to accomplish a high dispensing of CNTs and high compatibility between CNTs and polymer matrix to modify membrane properties and performances [104].

9.6.3.2 Functionalized CNTs (f-CNTs)-based membranes for PhACs removal

Mechanical and chemical treatments are the common approaches employed to functionalize CNTs. As a usual mechanical method, sonication was used to solve the entanglement behavior of CNTs and enhance their distribution in solution. In the meantime, embedding hydrophilic functional moieties was carried out to help CNTs dispersing in a solvent solution as well as in polymer matrices. The introduced polar functional groups may contribute to elevated membrane characteristics due to higher hydrophilicity, larger pore size, higher porosity, and rougher/smoother surface structure. These properties, in turn, guarantee higher permeability, rejection, antitradeoff between permeability and selectivity, fouling control properties, and short equilibrium time relative to other carbonaceous materials on f-CNT membranes [104]. Fortunately, the unique structure of CNTs enables them to associate firmly with functional moieties via covalent and noncovalent bonds depending on different reaction mechanisms. This covalent functionalization process is achieved generally by strong acids such as sulfuric acid (H_2SO_4) and nitric acid (HNO_3), as shown in Fig. 9.14 [133].

The oxidant is utilized to deliberately grafting functional groups such as —COOH, —OH, (—C=O), and —NH_2 groups at the open ends, sidewalls, and defect sites of CNTs. Some of these covalent modifications may require harsh treatments and induce damage to the CNTs structure. Here,

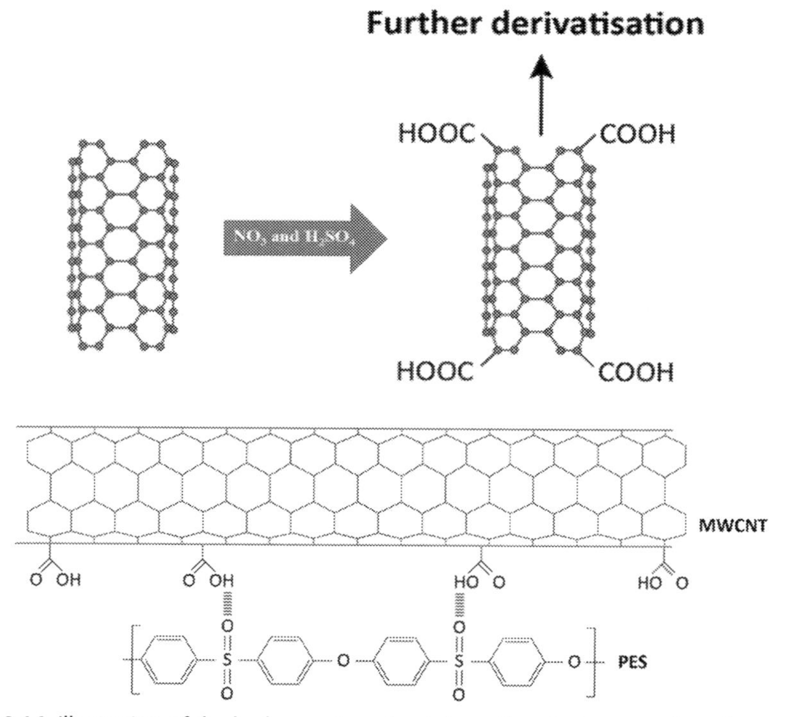

Fig. 9.14 Illustration of the hydrogen-bonding interactions between carboxylic groups of f-MWCNTs and PES support. *(From S. Kurwadkar, T.V. Hoang, K. Malwade, S.R. Kanel, W.F. Harper, G. Struckhoff, Application of carbon nanotubes for removal of emerging contaminants of concern in engineered water and wastewater treatment systems, Nanotechnol. Environ. Eng. 4 (1) (2019). https://doi.org/10.1007/s41204-019-0059-1.)*

noncovalent functionalization implies providing the additional moieties through adsorption, π-π interaction, hydrogen bonding, and electrostatic attraction to attach surfactants, polymers, or large organic molecules onto the surface of CNTs without any effects on the internal structure of CNTs. The functionalization methods of CNTs-based membranes are summarized in Table 9.3 (from Ref.[134]).

Furthermore, inorganic NPs and heteroatoms f-CNTs blended with organic polymers are a particular case of NPs based composite membranes. They have been reported as an efficient inorganic functionalization to improve the hydrophilicity of the hydrophobic commercial membranes. Some examples of the f-CNTs-based membranes with enhanced properties are as follows.

Table 9.3 Examples on functionalization of CNTs.

Composite membranes	Functionalization	Major aims
CNTs-mullite	HNO_3	Incorporate acidic functional groups Decrease impurities Increase membrane performances
CNTs-AAO	H_2O_2	Incorporate functional groups Increase hydrophilicity
CNTs-PES	H_2SO_4: HNO_3 (3:1)	Incorporate functional groups Shorten the length of CNTs Increase CNTs dispersion and interfacial bonding
CNTs-PA	O_3	Increase sidewall functionalities Shorten the length of CNTs Increase CNTs dispersion Reduce biofouling
CNTs-PVA	Chitosan	Maintain intact structure of CNTs Increase CNTs dispersion Improve the packing structure
CNTs-polyetherimide	Surfactant	Incorporate surfactants Maintain intact structure of CNTs Enhance membrane thermal stability and mechanical strength

From L. Ma, X. Dong, M. Chen, L. Zhu, C. Wang, F. Yang, Y. Dong, Fabrication and water treatment application of carbon nanotubes (CNTs)-based composite membranes: a review, Membranes 7 (1) (2017). https://doi.org/10.3390/membranes7010016.

Efficient removal of PhACs

An interesting study showed the variety of physical and chemical properties among selecting commercially available CNT, including pristine MWCNT, hydroxylated MWCNT (H-MWCNT), thin-walled MWCNT with a large inner diameter (LMWCNT), high-purity MWCNT (HP-MWCNT), and aminated MWCNT (A-MWCNT). Eight representative types of PPCPs, encompassing TCS, IBU, AAP, CBZ, 4-acetylamino-antipyrine (ATAP), CAF, PTN, and carbendazim (CBD), were filtered with membranes coated with one of five MWCNT different types under controlled solution chemical conditions. The removal ratios of different PPCPs by the pristine MWCNT followed a decreasing order of triclosan (0.93) > prometryn (0.71) > 4-acetylamino antipyrine (0.67) > carbendazim (0.65) > caffeine (0.42) > ibuprofen (0.34) > acetaminophen (0.29) at 100 min of filtration. The removal ratio of AAP was increased to 0.74 by using H-MWCNT [132]. In another study, fabrication of carboxylic CNT (C-CNT) intercalated reduced graphene oxide (RGO) composite electro-Fenton membrane was developed in an attempt to coenhance membrane permeability and rejection as well as catalytic oxidation for the PhACs removal [135]. Attributing to intercalation of C-CNT, the coenhancement had significantly increased membrane pore size and interlayer distance and conductivity and charge density. As such, both membrane permeability and rejection of PhACs could be increased.

Furthermore, a polyvinylidene fluoride nanocomposite membrane with multiwalled carbon nanotubes PVDF/MWCNT having covalently immobilized laccase enzyme was developed and applied for PPCPs removal from water in a biodegradation process [136]. The two PPCPs components, namely CBZ and DFC, were successfully removed at 27% in 48 h and 95% in 4 h. The laccase tended to act as an effective catalyst contributing to the high level of removal obtained.

Fabrication of heteroatom-doped CNTs/blend membranes was also investigated as a novel covalent functionalization method. This type of modified membrane provides a unique opportunity to eliminate the recalcitrant compounds from water systems. It is advantageous due to improved water flux, porosity, hydrophilicity, roughness, mechanical, and rejection properties. Nitrogen-doped carbon nanotubes/polyethersulfone (N-CNT/PES) blend membranes were fabricated using the modified phase-inversion method to remove PPCPs such as CBZ, HHCB, CAF, tonalide, technical 4-nonylphenol, and BPA from water. The water flux of the membranes was enhanced by the addition of N-CNTs to the pristine PES membranes. The

N-CNT/PES blend membranes were highly proficient in removing the selected PPCPs, with the highest and lowest removal efficiencies being 99.20%–99.92% (for HHCB) and 84.61%–87.21% (for CAF), respectively. The results show the superior capability of the N-CNT/PES blend membranes in removing PPCPs from water [137]. Furthermore, a membrane consisting of highly catalytic Fe-Cu bimetallic NPs, HPEI, and MWCNTs was used to degrade 2,4,6-trichlorophenol. An HPEI/MWCNTs/Fe-Cu nanocomposite was fabricated with a PES membrane via the phase inversion method, as shown in Fig. 9.15. The efficiency of the fabricated membranes was evaluated to dechlorinate 2,4,6-TCP. Higher pollutant removal (99.4%) was achieved with the 0.5 wt% HPEI/MWCNTs/Fe-Cu/PES membrane as well as improved hydrophilicity and permeability ($26.3 \pm 1.3\,L/m^2\,h$), compared to those of the 0.1 wt% HPEI/MWCNTs/Fe-Cu/PES membrane. Additionally, the antifouling performance of all prepared membranes was found to be more than 90% [138].

In some cases, f-CNTs showed less capability to remove PPCPs. Wang et al. prepared an oxidized CNTs-PVDF composite membrane for the filtration of TCS, AAP, and IBU to determine the capabilities and mechanisms of PPCPs removal. The removal rates ranged from approximately 10% to 95%. They increased with more aromatic rings in the tested molecules (AAP $\approx$ IBU $<$ TCS), CNTs with lesser surface oxygen content (oxidized MWCNTs $<$ MWCNTs) and with more specific surface area (MWCNTs $<$ SWCNTs). This result was interpreted in terms of zeta potentials. In the case of O-MWCNT, the presence of more oxygenated functional groups on CNT surfaces led to more negative zeta potential at pH 7 and, therefore increasing electrostatic repulsion between CNT and TCS/IBU. Consequently, the O-MWCNT membrane had the lowest TCS and IBU removals [139].

Enhanced antifouling capability of f-CNT membranes

Besides removing emerging pollutants in water or wastewater, f-CNT membranes to resist the fouling issue caused by accumulated dissolved organics in membrane matrices were developed. Many studies have been carried out for the utilization of f-CNT membranes to enhance the antifouling property. NPs such as titania and zinc oxide and electrochemically active f-CNT filters have been embedded into polymer structures to minimize membrane fouling by in-situ foulant destruction and biological inactivation. A reduction in maintenance and improvement in membrane permeability have been witnessed in this regard [140]. Pod-like nitrogen-doped CNTs

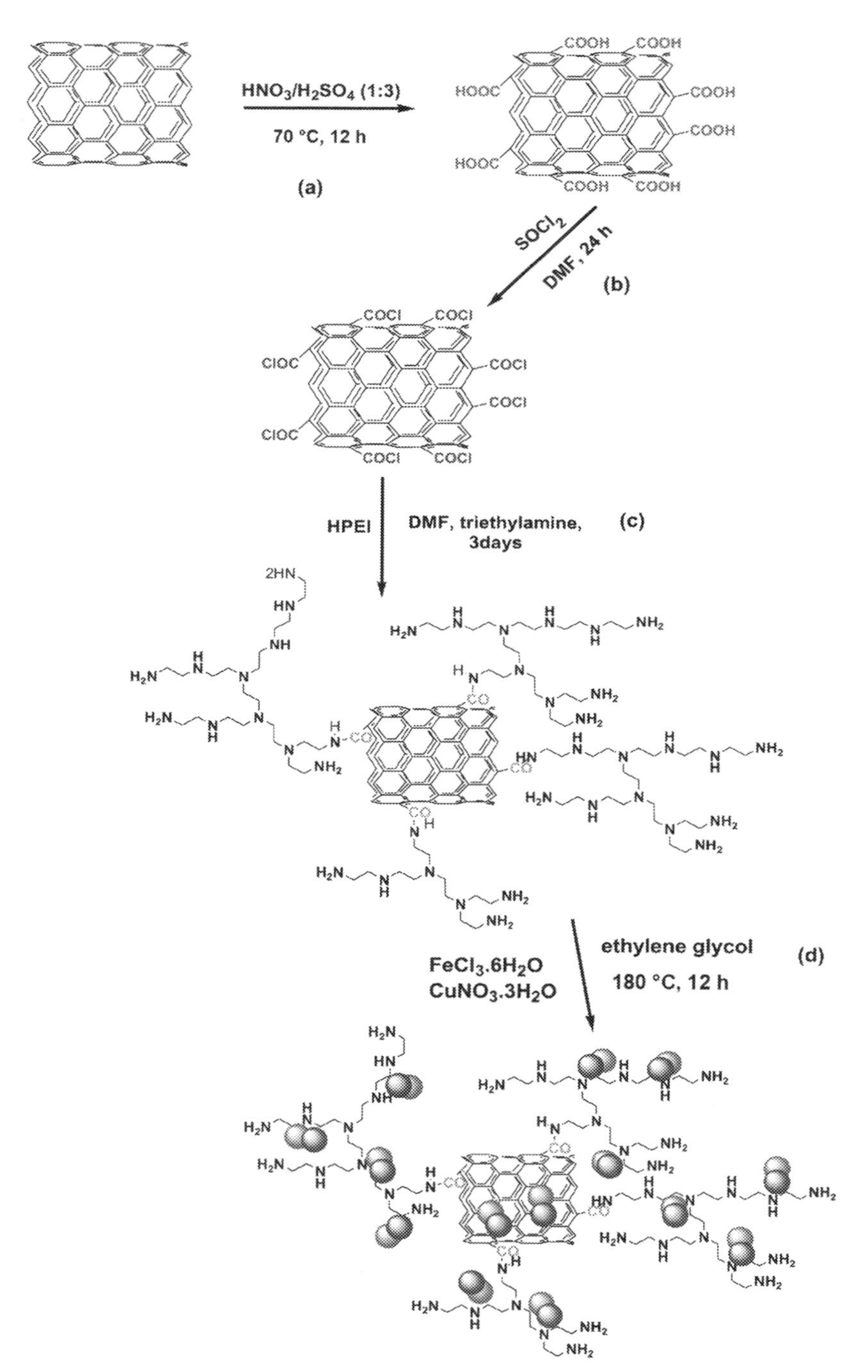

Fig. 9.15 Synthesis route for HPEI/MWCNTs/Fe-Cu nanocomposite. *(From S.T. Dube, R.M. Moutloali, S.P. Malinga, Hyperbranched polyethyleneimine/multi-walled carbon nanotubes polyethersulfone membrane incorporated with Fe-Cu bimetallic nanoparticles for water treatment, J. Environ. Chem. Eng. 8 (4) (2020) 103962. https://doi.org/10.1016/j.jece.2020.103962.)*

with encapsulation of Ni-Co alloy NPs were fabricated. The material was then applied in a flat Teflon membrane for PMS-mediated heterogeneous degradation of IBU. The synergistic effect of nitrogen doping and metal alloy encapsulation significantly enhanced the stability and catalytic activity, yielding a fast reaction rate of 0.31 min^{-1}, which was 23.4 and 5.8 times higher than that of pristine CNTs and monometallic (Ni or Co) encased CNTs, respectively [141].

The electrochemical applications of f-CNT filters are related to their excellent electrocatalytic properties for electrochemical reactions. A comparison was made between carboxylate f-CNTs and pristine CNTs to assert the boosted functioning of C-MWCNTs in the removal of IBU using an electrochemical filtration process under acidic conditions. Throughout the study, C-MWCNTs constantly displayed a better performance (almost complete removal) of IBU compared to pristine MWCNTs (which did not exceed 20% removal) under acidic conditions. They applied potentials of 2 and 3 V at lower flow rates. The C-CNTs have higher adsorption capacities for IBU molecules with oxygen-containing functional groups due to enhanced surface interactions with carbon nanotubes via hydrogen bonding and π-π hydrophobic interactions. Therefore, IBU molecules are efficiently trapped on the nanotubes for enhanced electrochemical degradation mediated by the generation of electroactive species as the primary contributor to the degradation of IBU, and its by-products, under these conditions [142].

Functionalization with cross-linkers to enhance blending with membranes

Applying the filtration process, GO membranes were prepared by a pressure-assisted self-assembly method whereby cross-linking of three diamine monomers on a PES support was applied. The different small molecular diamines, ethylenediamine, butane diamine, and *p*-phenylenediamine, were introduced as cross-linking agents to investigate the effect of diamine on the properties of GO membranes [143]. The hydrophobic substances IBU, GBZ, and TCS were selected as target PPCPs. The comparison made between GO membrane adsorption on different substances exhibited that cross-linked GO membranes capacity for both GBZ and TCS was nearly 1.5 times that of GO-0 GBZ and TCS. Having said this, GO-0 adsorption capacity for IBU tended to be slightly higher than that observed for GBZ and TCS.

9.7 Conclusions

This chapter has shown that various types of functionalization can efficiently control and enhance the water purification performance of nanomaterial-based membrane compared to the other state of the art. The results suggest that various hydrophilic, inorganic, enzymatic, etc. functionalities are vital for enhancing the water flux and rejection of PhACs.

Such a strategy paves the way for the preparation of membranes with tunable and sustainable behavior where specific membrane properties involved directly in PhACs removal and/or affected by PhACs types and concentrations are sustained and improved. Among these, flux and rejection as well as fouling tendency are the main properties relating to sufficient PhACs removal. These properties can be sustained and improved potentially by applying original fabricated and/or modified nanofiltration membranes, blended nanoparticle membranes, nanofibers, and nanotubes. In addition, functionalized nanomaterials can play an important role in such improvements not only by sustaining and enhancing membrane properties but also in terms of mitigating PhACs components types and concentration effects on membrane performance.

References

[1] G. Korekar, A. Kumar, C. Ugale, Occurrence, fate, persistence and remediation of caffeine: a review, Environ. Sci. Pollut. Res. 27 (28) (2020) 34715–34733, https://doi.org/10.1007/s11356-019-06998-8.

[2] S. Maldonado-Torres, R. Gurung, H. Rijal, A. Chan, S. Acharya, S. Rogelj, M. Piyasena, G. Rubasinghege, Fate, transformation, and toxicological impacts of pharmaceutical and personal care products in surface waters, Environ. Health Insights 12 (2018), https://doi.org/10.1177/1178630218795836. 117863021879583.

[3] R.R. Zepon Tarpani, A. Azapagic, Life cycle environmental impacts of advanced wastewater treatment techniques for removal of pharmaceuticals and personal care products (PPCPs), J. Environ. Manag. 215 (2018) 258–272, https://doi.org/10.1016/j.jenvman.2018.03.047.

[4] I. Koyuncu, R. Sengur, T. Turken, S. Guclu, M.E. Pasaoglu, Advances in water treatment by microfiltration, ultrafiltration, and nanofiltration, in: Advances in Membrane Technologies for Water Treatment: Materials, Processes and Applications, Elsevier Inc., 2015, pp. 83–128, https://doi.org/10.1016/B978-1-78242-121-4.00003-4.

[5] J.C. Crittenden, R.R. Trussell, D.W. Hand, K.J. Howe, G. Tchobanoglous, MWH's Water Treatment: Principles and Design, third ed., John Wiley and Sons, 2012, https://doi.org/10.1002/9781118131473.

[6] N. Hilal, H. Al-Zoubi, N.A. Darwish, A.W. Mohammad, M. Abu Arabi, A comprehensive review of nanofiltration membranes: treatment, pretreatment, modelling, and atomic force microscopy, Desalination 170 (3) (2004) 281–308, https://doi.org/10.1016/j.desal.2004.01.007.

[7] H. Guo, L.E. Peng, Z. Yao, Z. Yang, X. Ma, C.Y. Tang, Non-polyamide based nanofiltration membranes using green metal-organic coordination complexes: implications for the removal of trace organic contaminants, Environ. Sci. Technol. 53 (5) (2019) 2688–2694, https://doi.org/10.1021/acs.est.8b06422.
[8] K. Scott, Handbook of Industrial Membranes, Elsevier, 1995.
[9] B. Van der Bruggen, M. Mänttäri, M. Nyström, Drawbacks of applying nanofiltration and how to avoid them: a review, Sep. Purif. Technol. 63 (2) (2008) 251–263, https://doi.org/10.1016/j.seppur.2008.05.010.
[10] M. Zahid, A comprehensive review on polymeric nano-composite membranes for water treatment, J. Membr. Sci. Technol. 8 (2018) 1–20.
[11] Z. He, Z. Lyu, Q. Gu, L. Zhang, J. Wang, Ceramic-based membranes for water and wastewater treatment, Colloids Surf. A Physicochem. Eng. Asp. 578 (2019) 123513, https://doi.org/10.1016/j.colsurfa.2019.05.074.
[12] S. Basu, A.L. Khan, A. Cano-Odena, C. Liu, I.F.J. Vankelecom, Membrane-based technologies for biogas separations, Chem. Soc. Rev. 39 (2) (2010) 750–768, https://doi.org/10.1039/b817050a.
[13] A.K. Fard, G. McKay, A. Buekenhoudt, H. Al Sulaiti, F. Motmans, M. Khraisheh, M. Atieh, Inorganic membranes: preparation and application for water treatment and desalination, Materials 11 (1) (2018), https://doi.org/10.3390/ma11010074.
[14] S.M. Samaei, S. Gato-Trinidad, A. Altaee, The application of pressure-driven ceramic membrane technology for the treatment of industrial wastewaters—a review, Sep. Purif. Technol. 200 (2018) 198–220, https://doi.org/10.1016/j.seppur.2018.02.041.
[15] M. Bodzek, Membrane technologies for the removal of micropollutants in water treatment, in: Advances in Membrane Technologies for Water Treatment: Materials, Processes and Applications, Elsevier Inc., 2015, pp. 465–517, https://doi.org/10.1016/B978-1-78242-121-4.00015-0.
[16] M. Bodzek, Membrane techniques in wastewater treatment, Environ. Protect. Eng. 25 (1999) 153–192.
[17] M. Cheryan, Membranes in Food Processing, Springer Science and Business Media LLC, 1991, pp. 157–180, https://doi.org/10.1007/978-94-011-3682-2_11.
[18] F.P. Cuperus, H.H. Nijhuis, Applications of membrane technology to food processing, Trends Food Sci. Technol. 4 (9) (1993) 277–282, https://doi.org/10.1016/0924-2244(93)90070-Q.
[19] K.V. Peinemann, S.P. Nunes, L. Giorno, Membranes for Food Applications, John Wiley & Sons, 2011.
[20] S. Wang, J. Tian, L. Jia, J. Jia, S. Shan, Q. Wang, F. Cui, Removal of aqueous organic contaminants using submerged ceramic hollow fiber membrane coupled with peroxymonosulfate oxidation: comparison of CuO catalyst dispersed in the feed water and immobilized on the membrane, J. Membr. Sci. 618 (2021) 118707, https://doi.org/10.1016/j.memsci.2020.118707.
[21] X. Wang, Y. Li, H. Yu, F. Yang, C.Y. Tang, X. Quan, Y. Dong, High-flux robust ceramic membranes functionally decorated with nano-catalyst for emerging micropollutant removal from water, J. Membr. Sci. 611 (2020) 118281, https://doi.org/10.1016/j.memsci.2020.118281.
[22] X. Wei, X. Bao, J. Wu, C. Li, Y. Shi, J. Chen, B. Lv, B. Zhu, Typical pharmaceutical molecule removal behavior from water by positively and negatively charged composite hollow fiber nanofiltration membranes, RSC Adv. 8 (19) (2018) 10396–10408, https://doi.org/10.1039/c8ra00519b.
[23] M.B. Asif, F.I. Jegatheesan, W.E. Price, L.D. Nghiem, K. Yamamoto, Applications of membrane bioreactors in biotechnology processes, in: Current Trends and Future Developments on (Bio-) Membranes: Membrane Processes in the Pharmaceutical and Biotechnological Field, Elsevier, 2019, pp. 223–257 (Chapter 8).

[24] C. Charcosset, Membrane processes in biotechnology: an overview, Biotechnol. Adv. 24 (5) (2006) 482–492, https://doi.org/10.1016/j.biotechadv.2006.03.002.
[25] C. Charcosset, Membrane Processes in Biotechnology and Pharmaceutics, Elsevier, 2012.
[26] S. Nunes, K.V. Peinemann, Membranes for Energy Conversion, vol. 2, Wiley-VCH, 2008, pp. 1–286, https://doi.org/10.1002/9783527622146.
[27] X. Zhao, C. Lu, L. Yang, W. Chen, W. Xin, X.-Y. Kong, Q. Fu, L. Wen, G. Qiao, L. Jiang, Metal organic framework enhanced SPEEK/SPSF heterogeneous membrane for ion transport and energy conversion, Nano Energy 81 (2021) 105657, https://doi.org/10.1016/j.nanoen.2020.105657.
[28] V. Calabrò, A. Basile, Fundamental membrane processes, science and engineering, in: Advanced Membrane Science and Technology for Sustainable Energy and Environmental Applications, Elsevier Inc., 2011, pp. 3–21, https://doi.org/10.1533/9780857093790.1.3.
[29] P. Wang, F. Wang, H. Jiang, Y. Zhang, M. Zhao, R. Xiong, J. Ma, Strong improvement of nanofiltration performance on micropollutant removal and reduction of membrane fouling by hydrolyzed-aluminum nanoparticles, Water Res. 175 (2020) 115649, https://doi.org/10.1016/j.watres.2020.115649.
[30] N.K. Khanzada, M.U. Farid, J.A. Kharraz, J. Choi, C.Y. Tang, L.D. Nghiem, A. Jang, A.K. An, Removal of organic micropollutants using advanced membrane-based water and wastewater treatment: a review, J. Membr. Sci. 598 (2020), https://doi.org/10.1016/j.memsci.2019.117672.
[31] A. Azaïs, J. Mendret, E. Petit, S. Brosillon, Evidence of solute-solute interactions and cake enhanced concentration polarization during removal of pharmaceuticals from urban wastewater by nanofiltration, Water Res. 104 (2016) 156–167, https://doi.org/10.1016/j.watres.2016.08.014.
[32] C. Li, Y. Yang, Y. Liu, L. Hou, a., Removal of PhACs and their impacts on membrane fouling in NF/RO membrane filtration of various matrices, J. Membr. Sci. 548 (2018) 439–448, https://doi.org/10.1016/j.memsci.2017.11.032.
[33] J. Li, S.Y. Pang, Y. Zhou, S. Sun, L. Wang, Z. Wang, Y. Gao, Y. Yang, J. Jiang, Transformation of bisphenol AF and bisphenol S by manganese dioxide and effect of iodide, Water Res. 143 (2018) 47–55, https://doi.org/10.1016/j.watres.2018.06.029.
[34] A.M.E. Khalil, F.A. Memon, T.A. Tabish, D. Salmon, S. Zhang, D. Butler, Nanostructured porous graphene for efficient removal of emerging contaminants (pharmaceuticals) from water, Chem. Eng. J. 398 (2020), https://doi.org/10.1016/j.cej.2020.125440.
[35] Z. Hoyett, Pharmaceuticals and Personal Care Products: Risks, Challenges, and Solutions, IntechOpen, 2018, https://doi.org/10.5772/intechopen.70799.
[36] M. Zhang, J. Shen, Y. Zhong, T. Ding, P.D. Dissanayake, Y. Yang, Y.F. Tsang, Y.S. Ok, Sorption of pharmaceuticals and personal care products (PPCPs) from water and wastewater by carbonaceous materials: a review, Crit. Rev. Environ. Sci. Technol. (2020) 1–40, https://doi.org/10.1080/10643389.2020.1835436.
[37] I. Martínez-Alcalá, J.M. Guillén-Navarro, A. Lahora, Occurrence and fate of pharmaceuticals in a wastewater treatment plant from southeast of Spain and risk assessment, J. Environ. Manag. 279 (2021), https://doi.org/10.1016/j.jenvman.2020.111565.
[38] P.J. Phillips, C. Schubert, D. Argue, I. Fisher, E.T. Furlong, W. Foreman, J. Gray, A. Chalmers, Concentrations of hormones, pharmaceuticals and other micropollutants in groundwater affected by septic systems in New England and New York, Sci. Total Environ. 512–513 (2015) 43–54, https://doi.org/10.1016/j.scitotenv.2014.12.067.
[39] M. Padri, M. Sahrul Tamzil, Fate and transport of PPCPs in the environment: a review on occurrences, sources, and cases, Mater. Sci. Forum 967 (2019) 179–188, https://doi.org/10.4028/www.scientific.net/msf.967.179.

[40] A.J. Ebele, M. Abou-Elwafa Abdallah, S. Harrad, Pharmaceuticals and personal care products (PPCPs) in the freshwater aquatic environment, Emerg. Contam. (2017) 1–16, https://doi.org/10.1016/j.emcon.2016.12.004.

[41] M. Goel, A. Das, A Review on Treatment of Pharmaceuticals and Personal Care Products (PPCPs) in Water and Wastewater, Springer International Publishing, 2018, pp. 1–12, https://doi.org/10.1007/978-3-319-58538-3_41-1.

[42] A. Olasupo, F.B.M. Suah, Recent advances in the removal of pharmaceuticals and endocrine-disrupting compounds in the aquatic system: a case of polymer inclusion membranes, J. Hazard. Mater. 406 (2021), https://doi.org/10.1016/j.jhazmat.2020.124317.

[43] K.L. Del Rosario, S. Mitra, C.P. Humphrey, M.A. O'Driscoll, Detection of pharmaceuticals and other personal care products in groundwater beneath and adjacent to onsite wastewater treatment systems in a coastal plain shallow aquifer, Sci. Total Environ. 487 (1) (2014) 216–223, https://doi.org/10.1016/j.scitotenv.2014.03.135.

[44] K.M. Shaku, L.N. Dlamini, S.P. Malinga, Highly efficient photocatalytic hyperbranched polyethyleneimine/bismuth vanadate membranes for the degradation of triclosan, Int. J. Environ. Sci. Technol. 17 (6) (2020) 3297–3312, https://doi.org/10.1007/s13762-020-02699-9.

[45] Y. Cabeza, L. Candela, D. Ronen, G. Teijon, Monitoring the occurrence of emerging contaminants in treated wastewater and groundwater between 2008 and 2010. The Baix Llobregat (Barcelona, Spain), J. Hazard. Mater. 239–240 (2012) 32–39, https://doi.org/10.1016/j.jhazmat.2012.07.032.

[46] S.D. Costanzo, J. Murby, J. Bates, Ecosystem response to antibiotics entering the aquatic environment, Mar. Pollut. Bull. 51 (1–4) (2005) 218–223, https://doi.org/10.1016/j.marpolbul.2004.10.038.

[47] M. Darwish, A. Mohammadi, N. Assi, Integration of nickel doping with loading on graphene for enhanced adsorptive and catalytic properties of CdS nanoparticles towards visible light degradation of some antibiotics, J. Hazard. Mater. 320 (2016) 304–314, https://doi.org/10.1016/j.jhazmat.2016.08.043.

[48] M. Schriks, M.B. Heringa, M.M.E. van der Kooi, P. de Voogt, A.P. van Wezel, Toxicological relevance of emerging contaminants for drinking water quality, Water Res. 44 (2) (2010) 461–476, https://doi.org/10.1016/j.watres.2009.08.023.

[49] World Health Organization, Pharmaceuticals in Drinking-Water, 2012.

[50] K.S. Betts, Keeping drugs out of drinking water, Environ. Sci. Technol. 36 (19) (2002).

[51] A. Cincinelli, T. Martellini, E. Coppini, D. Fibbi, A. Katsoyiannis, Nanotechnologies for removal of pharmaceuticals and personal care products from water and wastewater. A review, J. Nanosci. Nanotechnol. 15 (5) (2015) 3333–3347, https://doi.org/10.1166/jnn.2015.10036.

[52] M. Darwish, A. Mohammadi, N. Assi, Partially decomposed PVP as a surface modification of ZnO, CdO, ZnS and CdS nanostructures for enhanced stability and catalytic activity towards sulphamethoxazole degradation, Bull. Mater. Sci. 40 (3) (2017) 513–522, https://doi.org/10.1007/s12034-017-1405-1.

[53] S.K. Alharbi, J. Kang, L.D. Nghiem, J.P. van de Merwe, F.D.L. Leusch, W.E. Price, Photolysis and UV/H2O2 of diclofenac, sulfamethoxazole, carbamazepine, and trimethoprim: identification of their major degradation products by ESI–LC–MS and assessment of the toxicity of reaction mixtures, Process Saf. Environ. Prot. 112 (2017) 222–234, https://doi.org/10.1016/j.psep.2017.07.015.

[54] Y.J. Jung, W.G. Kim, Y. Yoon, J.W. Kang, Y.M. Hong, H.W. Kim, Removal of amoxicillin by UV and UV/H2O2 processes, Sci. Total Environ. 420 (2012) 160–167, https://doi.org/10.1016/j.scitotenv.2011.12.011.

[55] J.F. García-Araya, F.J. Beltrán, A. Aguinaco, Diclofenac removal from water by ozone and photolytic TiO2 catalysed processes, J. Chem. Technol. Biotechnol. 85 (6) (2010) 798–804, https://doi.org/10.1002/jctb.2363.

[56] M. Darwish, A. Mohammadi, Functionalized nanomaterial for environmental techniques, in: Nanotechnology in Environmental Science, vols. 1–2, Wiley-VCH Verlag, 2018, pp. 315–350, https://doi.org/10.1002/9783527808854.ch10.
[57] L.L.S. Silva, C.G. Moreira, B.A. Curzio, F.V. da Fonseca, Micropollutant removal from water by membrane and advanced oxidation processes—a review, J. Water Resour. Prot. 411–431 (2017), https://doi.org/10.4236/jwarp.2017.95027.
[58] D. Dolar, K. Košutić, Removal of pharmaceuticals by ultrafiltration (UF), nanofiltration (NF), and reverse osmosis (RO), in: Comprehensive Analytical Chemistry, vol. 62, Elsevier B.V., 2013, pp. 319–344, https://doi.org/10.1016/B978-0-444-62657-8.00010-0.
[59] D. Jermann, W. Pronk, M. Boller, A.I. Schäfer, The role of NOM fouling for the retention of estradiol and ibuprofen during ultrafiltration, J. Membr. Sci. 329 (1–2) (2009) 75–84, https://doi.org/10.1016/j.memsci.2008.12.016.
[60] P. Berg, G. Hagmeyer, R. Gimbel, Removal of pesticides and other micropollutants by nanofiltration, Desalination 113 (2–3) (1997) 205–208, https://doi.org/10.1016/S0011-9164(97)00130-6.
[61] A.R.D. Verliefde, S.G.J. Heijman, E.R. Cornelissen, G. Amy, B. Van der Bruggen, J.-C. van Dijk, Influence of electrostatic interactions on the rejection with NF and assessment of the removal efficiency during NF/GAC treatment of pharmaceutically active compounds in surface water, Water Res. 41 (15) (2007) 3227–3240, https://doi.org/10.1016/j.watres.2007.05.022.
[62] Y. Yoon, P. Westerhoff, S.A. Snyder, E.C. Wert, Nanofiltration and ultrafiltration of endocrine disrupting compounds, pharmaceuticals and personal care products, J. Membr. Sci. 270 (1–2) (2006) 88–100, https://doi.org/10.1016/j.memsci.2005.06.045.
[63] C. Bellona, J.E. Drewes, The role of membrane surface charge and solute physico-chemical properties in the rejection of organic acids by NF membranes, J. Membr. Sci. 249 (1–2) (2005) 227–234, https://doi.org/10.1016/j.memsci.2004.09.041.
[64] V. Yangali-Quintanilla, A. Sadmani, M. McConville, M. Kennedy, G. Amy, Rejection of pharmaceutically active compounds and endocrine disrupting compounds by clean and fouled nanofiltration membranes, Water Res. 43 (9) (2009) 2349–2362, https://doi.org/10.1016/j.watres.2009.02.027.
[65] K.K. Ng, A.Y.C. Lin, T.H. Yu, C.F. Lin, Tertiary treatment of pharmaceuticals and personal care products by pretreatment and membrane processes, Sustain. Environ. Res. 21 (3) (2011) 173–180. https://link.springer.com/journal/42834.
[66] A.M. Comerton, R.C. Andrews, D.M. Bagley, P. Yang, Membrane adsorption of endocrine disrupting compounds and pharmaceutically active compounds, J. Membr. Sci. 303 (1–2) (2007) 267–277, https://doi.org/10.1016/j.memsci.2007.07.025.
[67] Y. Yoon, P. Westerhoff, S.A. Snyder, E.C. Wert, J. Yoon, Removal of endocrine disrupting compounds and pharmaceuticals by nanofiltration and ultrafiltration membranes, Desalination 202 (1–3) (2007) 16–23, https://doi.org/10.1016/j.desal.2005.12.033.
[68] L. Braeken, R. Ramaekers, Y. Zhang, G. Maes, B. Van Der Bruggen, C. Vandecasteele, Influence of hydrophobicity on retention in nanofiltration of aqueous solutions containing organic compounds, J. Membr. Sci. 252 (1–2) (2005) 195–203, https://doi.org/10.1016/j.memsci.2004.12.017.
[69] J. Han, W. Qiu, J. Hu, W. Gao, Chemisorption of estrone in nylon microfiltration membranes: adsorption mechanism and potential use for estrone removal from water, Water Res. 46 (3) (2012) 873–881, https://doi.org/10.1016/j.watres.2011.11.066.
[70] L.D. Nghiem, A.I. Schäfer, M. Elimelech, Pharmaceutical retention mechanisms by nanofiltration membranes, Environ. Sci. Technol. 39 (19) (2005) 7698–7705, https://doi.org/10.1021/es0507665.

[71] J. Radjenović, M. Petrović, F. Ventura, D. Barceló, Rejection of pharmaceuticals in nanofiltration and reverse osmosis membrane drinking water treatment, Water Res. 42 (14) (2008) 3601–3610, https://doi.org/10.1016/j.watres.2008.05.020.
[72] S.O. Ganiyu, E.D. Van Hullebusch, M. Cretin, G. Esposito, M.A. Oturan, Coupling of membrane filtration and advanced oxidation processes for removal of pharmaceutical residues: a critical review, Sep. Purif. Technol. 156 (2015) 891–914, https://doi.org/10.1016/j.seppur.2015.09.059.
[73] X. Qian, L. Xu, Y. Zhu, H. Yu, J. Niu, Removal of aqueous triclosan using TiO2 nanotube arrays reactive membrane by sequential adsorption and electrochemical degradation, Chem. Eng. J. (2020) 127615, https://doi.org/10.1016/j.cej.2020.127615.
[74] B.D. Freeman, Basis of permeability/selectivity tradeoff relations in polymeric gas separation membranes, Macromolecules 32 (2) (1999) 375–380, https://doi.org/10.1021/ma9814548.
[75] N.B. Bowden, Nanomaterials-based membranes increase flux and selectivity to enable chemical separations, ACS Appl. Nano Mater. 3 (10) (2020) 9538–9541, https://doi.org/10.1021/acsanm.0c02471.
[76] M. Rani, U. Shanker, Remediation of Organic Pollutants by Potential Functionalized Nanomaterials, Elsevier BV, 2020, pp. 327–398, https://doi.org/10.1016/b978-0-12-816787-8.00013-2.
[77] Y. Li, J.B. Richardson, R. Mark Bricka, X. Niu, H. Yang, L. Li, A. Jimenez, Leaching of heavy metals from E-waste in simulated landfill columns, Waste Manag. 29 (7) (2009) 2147–2150, https://doi.org/10.1016/j.wasman.2009.02.005.
[78] Z. Gao, S. Liu, Z. Wang, S. Yu, Composite NF membranes with anti-bacterial activity prepared by electrostatic self-assembly for dye recycle, J. Taiwan Inst. Chem. Eng. 106 (2020) 34–50, https://doi.org/10.1016/j.jtice.2019.10.020.
[79] H. Chen, M. Huang, Y. Liu, L. Meng, M. Ma, Functionalized electrospun nanofiber membranes for water treatment: a review, Sci. Total Environ. 739 (2020) 139944, https://doi.org/10.1016/j.scitotenv.2020.139944.
[80] S.J. Zaidi, K.A. Mauritz, M.K. Hassan, Membrane Surface Modification and Functionalization, Springer International Publishing, 2018, pp. 1–26, https://doi.org/10.1007/978-3-319-92067-2_11-1.
[81] Z. Xu, L. Wan, X. Huang, Surface modification by graft polymerization, Springer Science and Business Media LLC, 2009, pp. 80–149, https://doi.org/10.1007/978-3-540-88413-2_4.
[82] M. Dimitriadi, M. Zafiropoulou, S. Zinelis, N. Silikas, G. Eliades, Silane reactivity and resin bond strength to lithium disilicate ceramic surfaces, Dent. Mater. 35 (8) (2019) 1082–1094, https://doi.org/10.1016/j.dental.2019.05.002.
[83] A.L. Ahmad, N.F. Idrus, M.R. Othman, Preparation of perovskite alumina ceramic membrane using sol-gel method, J. Membr. Sci. 262 (1–2) (2005) 129–137, https://doi.org/10.1016/j.memsci.2005.06.042.
[84] J. Elisseeff, Hydrogels: structure starts to gel, Nat. Mater. 7 (4) (2008) 271–273, https://doi.org/10.1038/nmat2147.
[85] C. Bellona, J.E. Drewes, P. Xu, G. Amy, Factors affecting the rejection of organic solutes during NF/RO treatment—a literature review, Water Res. 38 (12) (2004) 2795–2809, https://doi.org/10.1016/j.watres.2004.03.034.
[86] K.P.M. Licona, L.R.d.O. Geaquinto, J.V. Nicolini, N.G. Figueiredo, S.C. Chiapetta, A.C. Habert, L. Yokoyama, Assessing potential of nanofiltration and reverse osmosis for removal of toxic pharmaceuticals from water, J. Water Process Eng. 25 (2018) 195–204, https://doi.org/10.1016/j.jwpe.2018.08.002.
[87] S. Gur-Reznik, I. Koren-Menashe, L. Heller-Grossman, O. Rufel, C.G. Dosoretz, Influence of seasonal and operating conditions on the rejection of pharmaceutical

active compounds by RO and NF membranes, Desalination 277 (1–3) (2011) 250–256, https://doi.org/10.1016/j.desal.2011.04.029.
[88] L.D. Nghiem, A. Manis, K. Soldenhoff, A.I. Schäfer, Estrogenic hormone removal from wastewater using NF/RO membranes, J. Membr. Sci. 242 (1–2) (2004) 37–45, https://doi.org/10.1016/j.memsci.2003.12.034.
[89] B. Van Der Bruggen, J. Schaep, D. Wilms, C. Vandecasteele, Influence of molecular size, polarity and charge on the retention of organic molecules by nanofiltration, J. Membr. Sci. 156 (1) (1999) 29–41, https://doi.org/10.1016/S0376-7388(98)00326-3.
[90] R. Xu, M. Zhou, H. Wang, X. Wang, X. Wen, Influences of temperature on the retention of PPCPs by nanofiltration membranes: experiments and modeling assessment, J. Membr. Sci. 599 (2020) 117817, https://doi.org/10.1016/j.memsci.2020.117817.
[91] M. Kamrani, A. Akbari, A. Yunessnia lehi, Chitosan-modified acrylic nanofiltration membrane for efficient removal of pharmaceutical compounds, J. Environ. Chem. Eng. 6 (1) (2018) 583–587, https://doi.org/10.1016/j.jece.2017.12.044.
[92] Y. Zhong, S. Mahmud, Z. He, Y. Yang, Z. Zhang, F. Guo, Z. Chen, Z. Xiong, Y. Zhao, Graphene oxide modified membrane for highly efficient wastewater treatment by dynamic combination of nanofiltration and catalysis, J. Hazard. Mater. 397 (2020) 122774, https://doi.org/10.1016/j.jhazmat.2020.122774.
[93] H. Karimnezhad, A.H. Navarchian, T. Tavakoli Gheinani, S. Zinadini, Amoxicillin removal by Fe-based nanoparticles immobilized on polyacrylonitrile membrane: individual nanofiltration or Fenton reaction, vs. engineered combined process, Chem. Eng. Res. Des. 153 (2020) 187–200, https://doi.org/10.1016/j.cherd.2019.10.031.
[94] Y.L. Lin, C.C. Tsai, N.Y. Zheng, Improving the organic and biological fouling resistance and removal of pharmaceutical and personal care products through nanofiltration by using in situ radical graft polymerization, Sci. Total Environ. 635 (2018) 543–550, https://doi.org/10.1016/j.scitotenv.2018.04.131.
[95] P. Anadão, A.M. Grumezescu, 15—Nanocomposite Filtration Membranes for Drinking Water Purification, Academic Press, 2017, pp. 517–549, https://doi.org/10.1016/B978-0-12-804300-4.00015-0.
[96] Y. Guo, B. Xu, F. Qi, A novel ceramic membrane coated with MnO2-Co3O4 nanoparticles catalytic ozonation for benzophenone-3 degradation in aqueous solution: fabrication, characterization and performance, Chem. Eng. J. 287 (2016) 381–389, https://doi.org/10.1016/j.cej.2015.11.067.
[97] R. Dai, H. Guo, C.Y. Tang, M. Chen, J. Li, Z. Wang, Hydrophilic selective nanochannels created by metal organic frameworks in nanofiltration membranes enhance rejection of hydrophobic endocrine-disrupting compounds, Environ. Sci. Technol. 53 (23) (2019) 13776–13783, https://doi.org/10.1021/acs.est.9b05343.
[98] Z. Ouyang, Z. Huang, X. Tang, C. Xiong, M. Tang, Y. Lu, A dually charged nanofiltration membrane by pH-responsive polydopamine for pharmaceuticals and personal care products removal, Sep. Purif. Technol. 211 (2019) 90–97, https://doi.org/10.1016/j.seppur.2018.09.059.
[99] H. Guo, Y. Deng, Z. Yao, Z. Yang, J. Wang, C. Lin, T. Zhang, B. Zhu, C.Y. Tang, A highly selective surface coating for enhanced membrane rejection of endocrine disrupting compounds: mechanistic insights and implications, Water Res. 121 (2017) 197–203, https://doi.org/10.1016/j.watres.2017.05.037.
[100] L. Paseta, D. Antorán, J. Coronas, C. Téllez, 110th anniversary: polyamide/metal-organic framework bilayered thin film composite membranes for the removal of pharmaceutical compounds from water, Ind. Eng. Chem. Res. 58 (10) (2019) 4222–4230, https://doi.org/10.1021/acs.iecr.8b06017.
[101] L.X. Dong, X.C. Huang, Z. Wang, Z. Yang, X.M. Wang, C.Y. Tang, A thin-film nanocomposite nanofiltration membrane prepared on a support with in situ embedded

zeolite nanoparticles, Sep. Purif. Technol. 166 (2016) 230–239, https://doi.org/10.1016/j.seppur.2016.04.043.
[102] T.Y. Liu, Z.H. Liu, R.X. Zhang, Y. Wang, B.V.d. Bruggen, X.L. Wang, Fabrication of a thin film nanocomposite hollow fiber nanofiltration membrane for wastewater treatment, J. Membr. Sci. 488 (2015) 92–102, https://doi.org/10.1016/j.memsci.2015.04.020.
[103] N. Rakhshan, M. Pakizeh, Removal of triazines from water using a novel OA modified SiO2/PA/PSf nanocomposite membrane, Sep. Purif. Technol. 147 (2015) 245–256, https://doi.org/10.1016/j.seppur.2015.04.013.
[104] M. Sianipar, S.H. Kim, Khoiruddin, F. Iskandar, I.G. Wenten, Functionalized carbon nanotube (CNT) membrane: progress and challenges, RSC Adv. 7 (81) (2017) 51175–51198, https://doi.org/10.1039/c7ra08570b.
[105] C. Ursino, R. Castro-Muñoz, E. Drioli, L. Gzara, M.H. Albeirutty, A. Figoli, Progress of nanocomposite membranes for water treatment, Membranes 8 (2) (2018), https://doi.org/10.3390/membranes8020018.
[106] E.M.M. Wanda, H. Nyoni, B.B. Mamba, T.A.M. Msagati, Application of silica and germanium dioxide nanoparticles/polyethersulfone blend membranes for removal of emerging micropollutants from water, Phys. Chem. Earth 108 (2018) 28–47, https://doi.org/10.1016/j.pce.2018.08.004.
[107] W.M.M. Mahmoud, T. Rastogi, K. Kümmerer, Application of titanium dioxide nanoparticles as a photocatalyst for the removal of micropollutants such as pharmaceuticals from water, Curr. Opin. Green Sustain. Chem. 6 (2017) 1–10, https://doi.org/10.1016/j.cogsc.2017.04.001.
[108] L. Paredes, S. Murgolo, H. Dzinun, M.H. Dzarfan Othman, A.F. Ismail, M. Carballa, G. Mascolo, Application of immobilized TiO2 on PVDF dual layer hollow fibre membrane to improve the photocatalytic removal of pharmaceuticals in different water matrices, Appl. Catal. B Environ. 240 (2019) 9–18, https://doi.org/10.1016/j.apcatb.2018.08.067.
[109] D. Fatta-Kassinos, M.I. Vasquez, K. Kümmerer, Transformation products of pharmaceuticals in surface waters and wastewater formed during photolysis and advanced oxidation processes—degradation, elucidation of byproducts and assessment of their biological potency, Chemosphere 85 (5) (2011) 693–709, https://doi.org/10.1016/j.chemosphere.2011.06.082.
[110] W.M.M. Mahmoud, A.P. Toolaram, J. Menz, C. Leder, M. Schneider, K. Kümmerer, Identification of phototransformation products of thalidomide and mixture toxicity assessment: an experimental and quantitative structural activity relationships (QSAR) approach, Water Res. 49 (1) (2014) 11–22, https://doi.org/10.1016/j.watres.2013.11.014.
[111] M. Borisover, Z. Gerstl, F. Burshtein, S. Yariv, U. Mingelgrin, Organic sorbate-organoclay interactions in aqueous and hydrophobic environments: sorbate-water competition, Environ. Sci. Technol. 42 (19) (2008) 7201–7206, https://doi.org/10.1021/es801116b.
[112] K. Buruga, J.T. Kalathi, K.H. Kim, Y.S. Ok, B. Danil, Polystyrene-halloysite nano tube membranes for water purification, J. Ind. Eng. Chem. 61 (2018) 169–180, https://doi.org/10.1016/j.jiec.2017.12.014.
[113] F. Medhat Bojnourd, M. Pakizeh, Preparation and characterization of a nanoclay/PVA/PSf nanocomposite membrane for removal of pharmaceuticals from water, Appl. Clay Sci. 162 (2018) 326–338, https://doi.org/10.1016/j.clay.2018.06.029.
[114] C.H. Chen, M.C. Melo, N. Berglund, A. Khan, C. de la Fuente, J.P. Ulmschneider, M.B. Ulmschneider, Understanding and modelling the interactions of peptides with membranes: from partitioning to self-assembly, Curr. Opin. Struct. Biol. 61 (2020) 160–166, https://doi.org/10.1016/j.sbi.2019.12.021.

[115] X. Wang, H. Wang, Y. Wang, J. Gao, J. Liu, Y. Zhang, Hydrotalcite/graphene oxide hybrid nanosheets functionalized nanofiltration membrane for desalination, Desalination 451 (2019) 209–218, https://doi.org/10.1016/j.desal.2017.05.012.
[116] E. Luster, D. Avisar, I. Horovitz, L. Lozzi, M. Baker, R. Grilli, H. Mamane, N-doped TiO2-coated ceramic membrane for carbamazepine degradation in different water qualities, Nanomaterials 7 (8) (2017) 206, https://doi.org/10.3390/nano7080206.
[117] W.J. Lee, Y. Bao, X. Hu, T.-T. Lim, Hybrid catalytic ozonation-membrane filtration process with CeOx and MnOx impregnated catalytic ceramic membranes for micropollutants degradation, Chem. Eng. J. 378 (2019) 121670, https://doi.org/10.1016/j.cej.2019.05.031.
[118] L. Tsapovsky, M. Simhon, V.R. Calderone, G. Rothenberg, V. Gitis, Retention of organics and degradation of micropollutants in municipal wastewater using impregnated ceramics, Clean Technol. Environ. Policy 22 (3) (2020) 689–700, https://doi.org/10.1007/s10098-020-01813-2.
[119] C. Mansas, L. Atfane-Karfane, E. Petit, J. Mendret, S. Brosillon, A. Ayral, Functionalized ceramic nanofilter for wastewater treatment by coupling membrane separation and catalytic ozonation, J. Environ. Chem. Eng. 8 (4) (2020) 104043, https://doi.org/10.1016/j.jece.2020.104043.
[120] P. Bhattacharya, D. Mukherjee, S. Dey, S. Ghosh, S. Banerjee, Development and performance evaluation of a novel CuO/TiO2 ceramic ultrafiltration membrane for ciprofloxacin removal, Mater. Chem. Phys. 229 (2019) 106–116, https://doi.org/10.1016/j.matchemphys.2019.02.094.
[121] X. Meng, Z. Liu, C. Deng, M. Zhu, D. Wang, K. Li, Y. Deng, M. Jiang, Microporous nano-MgO/diatomite ceramic membrane with high positive surface charge for tetracycline removal, J. Hazard. Mater. 320 (2016) 495–503, https://doi.org/10.1016/j.jhazmat.2016.08.068.
[122] L.P. Zhang, Z. Liu, Y. Faraj, Y. Zhao, R. Zhuang, R. Xie, X.J. Ju, W. Wang, L.Y. Chu, High-flux efficient catalytic membranes incorporated with iron-based Fenton-like catalysts for degradation of organic pollutants, J. Membr. Sci. 573 (2019) 493–503, https://doi.org/10.1016/j.memsci.2018.12.032.
[123] M. Chen, L. Zhu, J. Chen, F. Yang, C.Y. Tang, M.D. Guiver, Y. Dong, Spinel-based ceramic membranes coupling solid sludge recycling with oily wastewater treatment, Water Res. 169 (2020) 115180, https://doi.org/10.1016/j.watres.2019.115180.
[124] J.P. Fan, J.J. Luo, X.H. Zhang, B. Zhen, C.Y. Dong, Y.C. Li, J. Shen, Y.T. Cheng, H.P. Chen, A novel electrospun B-CD/CS/PVA nanofiber membrane for simultaneous and rapid removal of organic micropollutants and heavy metal ions from water, Chem. Eng. J. 378 (2019), https://doi.org/10.1016/j.cej.2019.122232.
[125] J. Geltmeyer, H. Teixido, M. Meire, T. Van Acker, K. Deventer, F. Vanhaecke, S. Van Hulle, K. De Buysser, K. De Clerck, TiO2 functionalized nanofibrous membranes for removal of organic (micro)pollutants from water, Sep. Purif. Technol. 179 (2017) 533–541, https://doi.org/10.1016/j.seppur.2017.02.037.
[126] A. Mohamed, A. Salama, W.S. Nasser, A. Uheida, Photodegradation of ibuprofen, cetirizine, and naproxen by PAN-MWCNT/TiO2–NH2 nanofiber membrane under UV light irradiation, Environ. Sci. Eur. 30 (1) (2018), https://doi.org/10.1186/s12302-018-0177-6.
[127] W.J. Weber Jr., Distributed optimal technology networks: a concept and strategy for potable water sustainability, Water Sci. Technol. 46 (6–7) (2002) 241–246, https://doi.org/10.2166/wst.2002.0685.
[128] S. Gotovac, L. Song, H. Kanoh, K. Kaneko, Assembly structure control of single wall carbon nanotubes with liquid phase naphthalene adsorption, Colloids Surf. A Physicochem. Eng. Asp. 300 (1–2) (2007) 117–121, https://doi.org/10.1016/j.colsurfa.2006.10.035.

[129] K. Yang, L. Zhu, B. Xing, Adsorption of polycyclic aromatic hydrocarbons by carbon nanomaterials, Environ. Sci. Technol. 40 (6) (2006) 1855–1861, https://doi.org/10.1021/es052208w.
[130] A. Bhatnagar, M. Sillanpää, Removal of natural organic matter (NOM) and its constituents from water by adsorption—a review, Chemosphere 166 (2017) 497–510, https://doi.org/10.1016/j.chemosphere.2016.09.098.
[131] A.S. Brady-Estévez, S. Kang, M. Elimelech, A single-walled-carbon-nanotube filter for removal of viral and bacterial pathogens, Small 4 (4) (2008) 481–484, https://doi.org/10.1002/smll.200700863.
[132] Y. Wang, H. Huang, W.-J. Lau, A.F. Ismail, A. Isloor, A. Al-Ahmed, Carbon nanotube composite membranes for microfiltration of pharmaceuticals and personal care products, in: Micro and Nano Technologies, Elsevier, 2019, pp. 183–202, https://doi.org/10.1016/B978-0-12-814503-6.00008-2 (Chapter 8).
[133] S. Kurwadkar, T.V. Hoang, K. Malwade, S.R. Kanel, W.F. Harper, G. Struckhoff, Application of carbon nanotubes for removal of emerging contaminants of concern in engineered water and wastewater treatment systems, Nanotechnol. Environ. Eng. 4 (1) (2019), https://doi.org/10.1007/s41204-019-0059-1.
[134] L. Ma, X. Dong, M. Chen, L. Zhu, C. Wang, F. Yang, Y. Dong, Fabrication and water treatment application of carbon nanotubes (CNTs)-based composite membranes: a review, Membranes 7 (1) (2017), https://doi.org/10.3390/membranes7010016.
[135] W.-L. Jiang, M.R. Haider, J.-L. Han, Y.-C. Ding, X.-Q. Li, H.-C. Wang, H.M. Adeel Sharif, A.-J. Wang, N.-Q. Ren, Carbon nanotubes intercalated RGO electro-Fenton membrane for coenhanced permeability, rejection and catalytic oxidation of organic micropollutants, J. Membr. Sci. 623 (2021) 119069, https://doi.org/10.1016/j.memsci.2021.119069.
[136] M. Masjoudi, M. Golgoli, Z. Ghobadi Nejad, S. Sadeghzadeh, S.M. Borghei, Pharmaceuticals removal by immobilized laccase on polyvinylidene fluoride nanocomposite with multi-walled carbon nanotubes, Chemosphere 263 (2021) 128043, https://doi.org/10.1016/j.chemosphere.2020.128043.
[137] E.M.M. Wanda, B.B. Mamba, T.A.M. Msagati, Nitrogen-doped carbon nanotubes/polyethersulfone blend membranes for removing emerging micropollutants, Clean Soil Air Water 45 (4) (2017), https://doi.org/10.1002/clen.201500889.
[138] S.T. Dube, R.M. Moutloali, S.P. Malinga, Hyperbranched polyethyleneimine/multiwalled carbon nanotubes polyethersulfone membrane incorporated with Fe-Cu bimetallic nanoparticles for water treatment, J. Environ. Chem. Eng. 8 (4) (2020) 103962, https://doi.org/10.1016/j.jece.2020.103962.
[139] Y. Wang, J. Zhu, H. Huang, H.H. Cho, Carbon nanotube composite membranes for microfiltration of pharmaceuticals and personal care products: capabilities and potential mechanisms, J. Membr. Sci. 479 (2015) 165–174, https://doi.org/10.1016/j.memsci.2015.01.034.
[140] A. Khalid, A. Ibrahim, O.C.S. Al-Hamouz, T. Laoui, A. Benamor, M.A. Atieh, Fabrication of polysulfone nanocomposite membranes with silver-doped carbon nanotubes and their antifouling performance, J. Appl. Polym. Sci. 134 (15) (2017), https://doi.org/10.1002/app.44688.
[141] J. Kang, H. Zhang, X. Duan, H. Sun, X. Tan, S. Liu, S. Wang, Magnetic Ni-Co alloy encapsulated N-doped carbon nanotubes for catalytic membrane degradation of emerging contaminants, Chem. Eng. J. 362 (2019) 251–261, https://doi.org/10.1016/j.cej.2019.01.035.
[142] A.R. Bakr, M.S. Rahaman, Electrochemical efficacy of a carboxylated multiwalled carbon nanotube filter for the removal of ibuprofen from aqueous solutions under

acidic conditions, Chemosphere 153 (2016) 508–520, https://doi.org/10.1016/j.chemosphere.2016.03.078.
[143] Y. Lou, F.J. Tan, R. Zeng, M. Wang, P. Li, S. Xia, Preparation of cross-linked graphene oxide on polyethersulfone membrane for pharmaceuticals and personal care products removal, Polymers (Basel) 12 (9) (2020) 1921, https://doi.org/10.3390/polym12091921.

CHAPTER 10

Challenges in commercialization of sustainable membranes with FNMs

Putu Doddy Sutrisna
Department of Chemical Engineering, University of Surabaya (UBAYA), Surabaya, Indonesia

10.1 Introduction

Separation processes in industries play an important role to improve the quality of products. Gas and liquid separation processes can be performed by several techniques including distillation, chemical absorption, physical or chemical adsorption, and membrane separation process. In recent years, membrane has gained an increased interest for application in industries due to its simplicity and versatility. Membrane is a permselective barrier between feed and permeate sides, which can retain certain species and pass through other species [1–3]. The term membrane can be trailed back to the 18th century when Abbé Nolet first proposed the term 'osmosis' to define the penetration of water through a diaphragm [1]. Since that discovery, the experiments using membrane have been conducted, however, industrial application of membrane did not occur until the early 1960s when Loeb and Sourirajan discovered a technique to synthesize defect-free, high throughput, and anisotropic reverse osmosis membranes [1,4]. Following the discovery, membrane separation techniques have been applied for water and gas separation processes. The driving force for the separation process could be the difference of pressure (ΔP), temperature (ΔT), concentration (ΔC), or electrical potential (ΔE) between two phases. In general, membrane separation processes can be classified according to their driving forces and the types of phases on either side of membranes as shown in Table 10.1.

Membrane separation process is receiving a growing interest for application in separation processes on an industrial scale due to its advantages compared to conventional techniques, such as distillation, evaporation, crystallization, chemical absorption, physical adsorption, and other physical separation techniques. Membrane is considered a clean technique because membrane does not need chemicals as in absorption and adsorption. Membrane is also considered as energy saver

Membranes with Functionalized Nanomaterials
https://doi.org/10.1016/B978-0-323-85946-2.00002-3

Table 10.1 Membrane processes and their driving forces [4].

Membrane process	Feed phase	Permeate phase	Driving force
Microfiltration	Liquid	Liquid	ΔP
Ultrafiltration	Liquid	Liquid	ΔP
Nanofiltration	Liquid	Liquid	ΔP
Reverse osmosis	Liquid	Liquid	ΔP
Piezodialysis	Liquid	Liquid	ΔP
Gas separation	Gas	Gas	Δp
Vapor permeations	Gas	Gas	Δp
Pervaporation	Liquid	Gas	Δp
Electrodialysis	Liquid	Liquid	ΔE
Membrane electrolysis	Liquid	Liquid	ΔE
Dialysis	Liquid	Liquid	Δc
Diffusion dialysis	Liquid	Liquid	Δc
Membrane contactor	Liquid	Liquid	Δc
	Gas	Liquid	$\Delta c/\Delta p$
	Liquid	Gas	$\Delta c/\Delta p$
Thermo-osmosis	Liquid	Liquid	$\Delta T/\Delta p$
Membrane distillation	Liquid	Liquid	$\Delta T/\Delta p$

Note: P is the pressure difference across the membrane, p is the difference of partial pressure of component, c is the concentration difference, E is the difference of electrical potential, and T is the temperature difference.

technique compared to distillation and evaporation. In addition, membranes offer special features, such as requiring a small footprint for a large membrane area and ease of fabrication and operation [1,5,6]. However, industrial applications of membranes are impeded because of some problems encountered during membrane operation. Such problems include concentration polarization and membrane fouling, and high operating costs due to regular membrane cleaning and replacement [1,4,6]. In addition, membranes used for the gas separation process experience the trade-off between gas permeability and selectivity. Basically, the performances of membranes are determined by several parameters, including permeate flux and separation efficiency for liquid-based membranes and gas permeability or permeance as well as gas selectivity for gas-based membranes.

10.2 Overview of important problems in liquid-based and gas-based membranes

10.2.1 Liquid-based membranes

Membranes used for liquid separation processes have been applied for various applications. Such applications include water and wastewater treatments using pressure-driven membranes (microfiltration/MF, ultrafiltration/UF, nanofiltration/UF, reverse osmosis/RO); the dehydration of alcohol using pervaporation/PV, desalination process using membrane distillation/MD, and other important separation processes in industries [1,4,6–8]. Up to now, many membrane manufacturers have commercialized membranes for liquid separation using either inorganic or organic/polymeric materials [2,7,9]. In general, inorganic membranes for liquid separation processes possess higher chemical and thermal stabilities than polymeric membranes. In addition, inorganic membranes, such as ceramic, carbon, zeolite, and metal membranes, have a relatively more uniform pore size and pore size distribution than polymeric membranes [9–11]. However, for industrial applications, polymeric membranes are preferred due to their easy processability and scale-up, as well as being much cheaper than inorganic membranes. Hence, polymeric membranes have been produced and applied more widely than inorganic membranes [1,2].

In general, polymeric membranes available today suffer from several problems, such as concentration polarization and fouling phenomena that cause a trade-off between membrane flux and rejection [1,12,13]. Concentration polarization that usually precedes membrane fouling is considered a reversible process, which means that this phenomenon can be reduced by several methods, such as controlling the operational parameters of filtration and membrane surface modification. On the other hand, membrane fouling needs to be addressed by more intensive techniques, such as membrane cleaning, surface modification, and membrane replacement. Fouling can be defined as the deposition and adsorption of contaminants in feed solutions on the membrane surface and/or inside pores and the wall of the membrane's pores. Fouling phenomenon leads to the decline in permeate flux and the alteration of separation efficiency and rejection, and ultimately shortened the membrane's life. In general, fouling in liquid-based membranes can be classified based on the type of foulants that exist on the feed solution [1,4].

Recently, efforts attempted to minimize concentration polarization and fouling are directed into combining polymeric material and inorganic, especially nanomaterials into mixed matrix membranes (MMMs) and nanocomposite membranes [2,14–17]. The main reason to combine these two materials is exploiting the advantages of each material in the separation process, in addition to the modification of properties of composite membranes that eventually will reduce the concentration polarization and fouling phenomena during the separation process.

10.2.2 Gas-based membranes

The commercial application of membranes in gas separation processes was started in the 1980s inspired by the technique developed by Loeb-Sourirajan in the 1960s [18,19]. The development of asymmetric membranes with high flux has overcome the problems encountered in dense membranes with low gas flux. Membranes for gas separation processes have been applied in several applications, such as the purification of H_2, air separation to produce oxygen-enriched air, and for separation of CO_2/H_2O in the natural gas industry [18,19]. For CO_2 capture application, several properties should be acquired by membranes. The properties include high gas permeability and selectivity, high thermal and chemical stabilities, resistant to physical aging and plasticization, and cheap [19]. Research on membrane gas separation processes is therefore directed to fabricate and control the properties of gas separation membranes. Membranes for gas separation processes can be synthesized from both inorganic and organic materials. Between the polymeric and inorganic membrane materials, inorganic membranes for gas separation processes can provide good separation performance. However, their industrial application is impeded by high cost and is difficult to scale-up. Polymeric membranes offer easy processability to be scaled up and relatively cheap. Despite many advantages of gas separation membranes made of polymeric materials, there exists a gas permeability-selectivity trade-off. The trade-off upper bound was compiled by Robeson in 1991 by plotting the logarithmic of gas permeability and selectivity of different gas pairs for the performance of existing gas separation membranes [20]. The data was then updated in 2008 as presented in Figs. 10.1 and 10.2, and then the latest CO_2/CH_4 data was presented in 2015 [21,22] with the rapid development of new materials and membranes.

One way to improve the performance of the polymeric gas separation membrane beyond the upper bound is to incorporate inorganic nanofillers within the polymer matrix, which is also known as mixed matrix membranes (MMMs) and also nanocomposite membranes. Compared with an inorganic coating on a polymeric substrate such as in composite membranes, the MMMs are relatively simple to prepare thus considered a promising solution for large-scale practical application [3,23–26]. For its successful implementation, several aspects need to be carefully considered: the selection of polymer and inorganic materials for membrane preparation, the regulation of the inorganic-polymer interfaces during the membrane fabrication process, the understanding of membrane property change, and the gas transport mechanism within the MMMs [27].

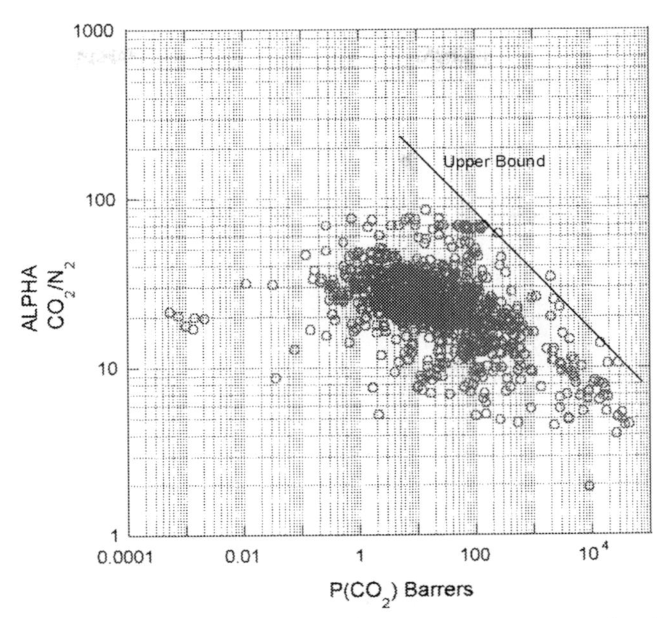

Fig. 10.1 Upper bound correlation for CO_2/N_2 separation. *(Reproduced from L.M. Robeson, The upper bound revisited, J. Membr. Sci. 320 (2008) 390–400. https://doi.org/10.1016/j.memsci.2008.04.030 with permission.)*

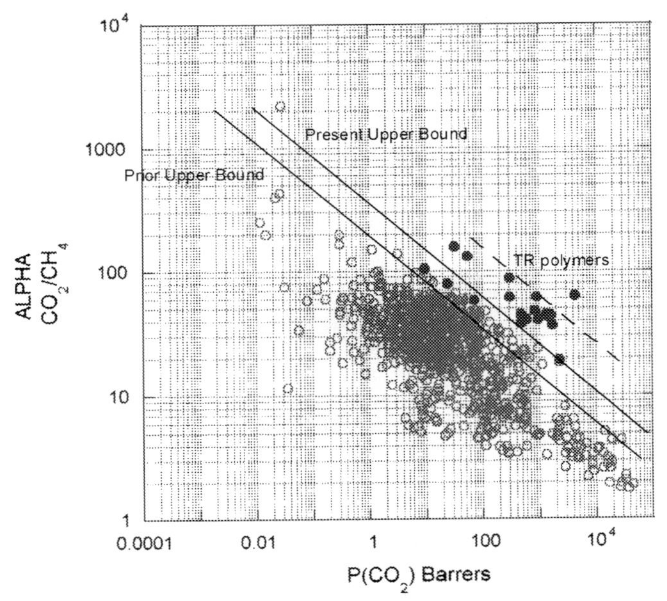

Fig. 10.2 Upper bound correlation for CO2/CH4 separation. *(Reproduced from L.M. Robeson, The upper bound revisited, J. Membr. Sci. 320 (2008) 390–400. https://doi.org/10.1016/j.memsci.2008.04.030 with permission.)*

10.3 Functionalized nanomaterials-membranes

Attempts to combine polymeric material and inorganic nanoparticles have been actively conducted during the last decade. The combination of these two different materials is usually realized as mixed matrix membranes (MMMs) and nanocomposite membranes. MMMs basically combine polymer and nanomaterials by simple mixing of two materials, while in nanocomposite membranes the mixture of polymer and inorganic particles is coated on a porous substrate. In these two types of membranes, the main components are polymer and nanomaterials. In terms of polymer material, all types of polymers such as glassy, rubbery, and copolymer, have been applied as a polymer matrix for MMMs and nanocomposite membranes. For nanomaterials, various types of nanomaterials have been mixed with polymer matrices. Such nanomaterials include halloysite nanotubes [28–30], montmorillonite, $CaCO_3$ nanoparticles [31,32], nickel-cobalt layered double hydroxide [33], Mg—Fe layered double hydroxide nanomaterials [34], silica (SiO_2) [35], iron acetate [36], zwitterionic nanoparticles-polydopamine-sulfobetaine methacrylate or P(DA-SBMA) [37], hydrose manganese oxide [38], metal organic frameworks (MOFs) [39–43], metal/metal oxide-based nanoparticles (i.e., silver nanoparticles, copper-based nanoparticles, iron oxide-based nanoparticles, aluminum oxide-based nanoparticles, titanium dioxide-based nanoparticles) [2,36,44], carbon-based nanoparticles (i.e., carbon nanotubes or CNTs, graphene, and graphene oxide or GO) [45–48], and the combination of fillers mentioned above are inorganic particles that have been used as discontinuous phase inside MMMs or nanocomposite membranes [38].

The combination of these fillers with polymers inside MMMs and nanocomposite membranes is directed to improve the performance of the membranes by altering the physical and chemical properties of the membranes. For instance, in water or wastewater treatment processes, the presence of fillers is aimed to improve the antifouling property of the membranes, as several nanomaterials (i.e., graphene oxide) possess a hydrophilic property that can alter the membrane from less hydrophilic to more hydrophilic [46,49,50]. Other studies suggested that several nanomaterials, such as silver and copper-based nanoparticles, can promote an antibacterial property to the produced membranes [2,50]. In the gas separation process, the presence of nanomaterials can change the morphology and free volume of the polymer matrix hence can promote an increase in gas permeability or gas permeance through the membranes [27].

Apart from the application of nanomaterials for the improvement of membranes performances in separation processes, many studies also reported the functionalization of pristine nanomaterials using various techniques and chemical agents [51–53]. For example, the incorporation of MOFs nanoparticles that can create aggregates inside the membranes has motivated the functionalization of nanomaterials to improve the dispersion of particles inside the polymer matrix [54–57]. One simple approach that has been attempted in recent years is the functionalization of particles using simple soaking and coating of coating agents, such as polydopamine or polydopamine/polyethylene imine [37,49,58–67]. This deposition process is relatively easy to conduct and can create a membrane with improved performance. Polydopamine as the polymer produced from the polymerization of dopamine has the ability to coat and functionalize many types of surfaces and can create a more hydrophilic surface that is very important in water and wastewater treatment processes to avoid fouling phenomenon [59,62,68]. For industrial application, research on the functionalization of nanomaterials or membranes is still attempted and optimization on operating conditions as well as efforts to scale-up the synthesis process are still required before the membranes can be applied in real applications.

10.4 Challenges in synthesis and commercialization of mixed matrix and nanocomposite membranes with functional nanomaterials

10.4.1 The obstacles in membrane synthesis

The successful application of membranes with functional nanomaterials (FNMs) for the separation of gases or liquids is mainly determined by the successful fabrication of membranes. Produced membranes should not possess voids and defects on the interface between the FNMs and the polymer matrix. Voids and defects in MMMs and composite membranes in some cases can increase permeate flux and gas permeation through the membrane but will decrease the selectivity of the membrane. The formation of void and defect-free MMMs and composite membranes is therefore growing as an interesting field of research in the membrane technology area. In general, three kinds of defects are usually found during the synthesis of MMMs and composite membranes including sieve-in-a-cage, the rigidification of polymer, and the blockage of particle pore. The illustration of the typical defects in MMMs is provided in Fig. 10.3 [23,69]. These defects can be caused by several factors, such as the incompatibility between polymer

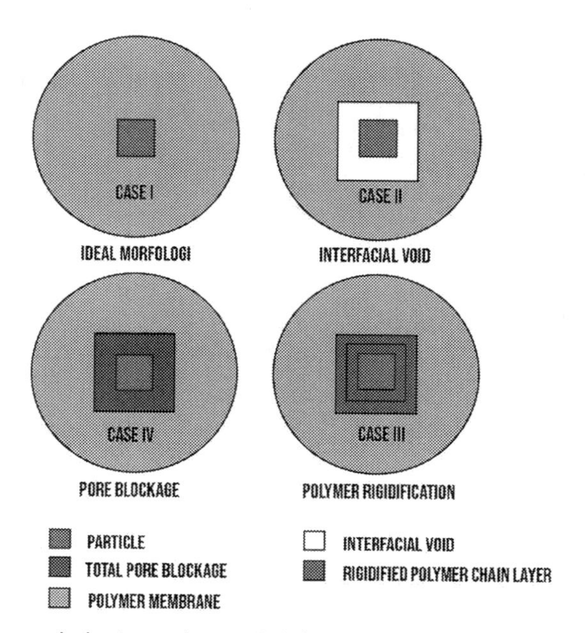

Fig. 10.3 The morphologies and typical defects in MMMs. *(Adapted from T.T. Moore, W.J. Koros, Non-ideal effects in organic–inorganic materials for gas separation membranes, J. Mol. Struct. 739 (2005) 87–98. https://doi.org/10.1016/j.molstruc.2004.05.043 with permission.)*

and particle, the evaporation of solvent during membrane formation that stresses the interface of polymer-particle, and the weak adhesion between the particle and polymer [70].

The four possible morphologies in MMMs and composite membranes can be depicted as (I) ideal morphology, (II) interfacial voids morphology, (III) rigidified polymer morphology, and (IV) reduced permeability region within sieve morphology.

Case I in MMMs is relatively difficult to obtain and the flux of gas and liquid in this membrane can be predicted by the Maxwell model that is formulated by Eq. (10.1):

$$P_{mm} = P_c\left(\frac{P_d + 2P_c - 2\O_d(P_c - P_d)}{P_d + 2P_c + \O_d(P_c - P_d)}\right) \tag{10.1}$$

where P_{mm} is the effective permeability of MMMs, Φ is volume fraction, while c and d denote the continuous and dispersed phases, respectively.

Case II or sieve-in-a-cage morphology is the easiest to diagnose. This morphology of MMMs and composite membranes is typical morphology obtained when the polymer matrix faces compatibility issues with the nanoparticles. The interfacial void results in increased permeability with essentially no change in selectivity, or in many cases, causes reduced selectivity of the membranes. The attempts to avoid the formation of this interfacial void are mainly directed to the enhancement of compatibility between polymer and nanomaterials, such as surface functionalization and the proper choice of polymer and inorganic nanomaterials.

Case III describes the presence of rigidified region on the interface between the polymer matrix and particles. The rigidified region on the interface forms because of stress experienced by two materials during the preparation of MMMs. When the rigidification of polymer matrix occurs, the ability of the membranes to discriminate and select certain components to permeate will increase, while on the other hand, the rigidification of polymer matrix might decrease the flux of permeate through the membranes. In gas separation membranes, this phenomenon can be observed by the increasing value of the glass transition temperature of the polymer [70].

Case IV depicts the decrease in permeability region near the surface of particles or the whole particles. This type of morphology might occur when the pores of inorganic nanomaterials are covered by the polymer matrix. The decrease in permeability region causes a reduction in permeate flux or gas permeability through the membranes.

Typical morphology that has been obtained by recent studies suggested the formation of voids and defects as the main morphology occurred inside MMMs and nanocomposite membranes. In order to develop MMMs and nanocomposite membranes with functional nanomaterials on an industrial scale, the problem of voids and defects should be solved as this problem decreases the performance of membrane in gas or liquid separation processes. Many attempts have been conducted to reduce the formation of voids and defects, such as controlling the mechanism of MMMs and nanocomposite membranes synthesis, and careful choice of polymer and nanomaterials because it is believed that the suitability of polymer and particle and controlling the synthesis of membranes determine the successful formation of MMMs and nanocomposite membranes [16,27]. Another technique includes maintaining flexibility. Maintaining flexibility in the course of membrane synthesis is conducted by casting the membrane at an elevated temperature to maximize stress relaxation. But this method requires the use of a solvent with very low volatility to be mixed with a polymer with a high glass transition temperature that can be operated at ambient pressure. Addition of a silane coupling agent has also been proposed to improve the performance of membranes, but the weak bond between silanated inorganic particles and polymer matrix still occurs [69,71–74].

Mahajan et al. [75] discussed the two most important factors to avoid the formation of the interphase between polymer matrix and particles, i.e., the good polymer-particle interaction and controlled evaporation of solvent during membrane formation. Another factor that can also determine the good polymer-particle interaction is the polymer's flexibility especially during the synthesis of MMMs and nanocomposite membranes. However, flexible polymer material suffers from deterioration at high pressure. Hence, the strategies employed include: (a) the annealing of produced membranes, however, this method failed to improve the dispersion of particles in the polymer matrix, (b) using plasticizer to lower the glass transition temperature (T_g) of the polymer, and applying solvent with low volatility [76]. In terms of MMM and nanocomposite membrane preparation methods, there are two different methods to disperse inorganic particles, which are claimed to be good methods for mixing polymer and particle. These methods are [27,70] as follows:

(1) Mixing of precursor particle-solvent or particle-polymer solution before the addition of bulk polymer into the mixture.

(2) Addition of the inorganic particles dispersed in a certain solvent into the polymer that has been homogenized with another solvent.

The first method, which is also known as 'priming protocol' method, is widely used as the first step technique to form void- and defect-free MMMs [27,75]. Good particle dispersion is expected when priming protocol is employed because the surface of particles will be wetted by the polymer solution.

Another method employed to produce defect-free MMMs is using as-synthesized fillers. One example is the use of as-synthesized zeolitic imidazolate framework (ZIF-8) nanoparticles (size of around 60 nm and specific surface area of 1300–1600 m^2g^{-1}) in the preparation of MMMs based on the glassy Matrimid-5218 polymer by solution mixing technique to measure the permeation of different pure gases, such as H_2, CO_2, O_2, N_2, and CH_4 [77]. Produced membranes showed a very good degree of particle dispersion in the polymer matrix and transparency. Direct dispersion of ZIF-8 particles in the polymer solution after their synthesis (as-synthesized) improved the dispersion of particles in the polymer matrix. The control of solution viscosity during solvent evaporation also determined the degree of agglomeration of ZIF-8 particles. The highest loading of ZIF-8 particles in the membrane that can give a superior result in terms of permeability was 30 wt%, while the highest selectivity of gas pairs was obtained using 5 wt% ZIF-8 loading. At higher loading, i.e., 40 wt%, the membranes became brittle.

The enhancement of particle dispersion is also observed when direct sonication of the particle is utilized during the preparation of MMMs. ZIF-8/Matrimid-5218 MMMs with improved CO_2 permeability and a slight increase in CO_2/CH_4 selectivity have been prepared and sonication provides cavitation from sonic waves, which makes a localized area or 'hot zone' and causes the dissolution of ZIF-8 constituents at the particle surfaces. This Ostwald ripening effect leads to the increase of the particle size distribution of ZIF-8 nanoparticles [78]. Another challenge in MMMs and nanocomposite membranes with FNMs synthesis is the optimum particle concentration that can be loaded inside the polymer matrix. It is well known that nanomaterials could potentially form aggregates, especially at high concentrations. Hence, many studies attempted to control the mixing of polymer and particles before casting into flat sheet membranes. The efforts to control the mixing of polymer and particles include the application of dip-coating technique with controlled parameters and also by using priming protocol and as-synthesized fillers as mentioned earlier in this section.

Despite the extensive efforts to increase the dispersion of the inorganic filler in a polymer matrix, much research focused on how to improve the dispersion of inorganic particles in the polymer matrix. Another challenge for the utilization of MMMs and nanocomposite membranes with functional nanomaterials in liquid and gas separation processes is the regulation of membrane thickness during membrane synthesis. When micrometer-sized particles are incorporated into polymer matrix, the membranes usually form a relatively thicker layer compared to membranes with nanometer size of particles or nanomaterials. Hence, the formation of MMMs and nanocomposite membranes with functional nanomaterials has a great potential to produce very thin membranes. Thicker membranes will reduce the permeate flux or permeability of the gas through the membrane. Although improved performance has been observed with the mixed matrix dense membranes, thin composite hollow fiber membranes are more competitive for large-scale industrial applications due to their higher gas permeation rate and lower consumption of expensive materials during the fabrication process. The composite membrane usually consists of a highly permeable gutter layer and a surface ultrathin selective layer, both of which are coated on porous membrane support [27,79].

The synthesis of nanocomposite membranes is proposed to reduce the mechanical resistance by forming a thin layer of polymer and inorganic fillers on the surface of the porous substrate. In this kind of membrane, the resistance for gas transport is mainly driven by the very thin layer of polymer and particles (often in the range of 200 Å to 1 μm). This thin layer is supported by a microporous layer or porous substrate [79,80]. The main challenge in the synthesis of this membrane is the production of very small particles (in nanometer size) and the production of a very thin layer on the surface of the porous substrate. The proposed structure of inorganic particles incorporating nanocomposite membrane is shown in Fig. 10.4.

The thin layer of polymer-particle in nanocomposite membranes requires a careful preparation method as it needs to be very thin and needs good dispersion of particles inside the polymer matrix. The inorganic particles dispersed in the polymer matrix are expected to function both as a selective layer and mechanical support for the nanocomposite membranes. Another important consideration during the synthesis of this membrane is the intrusion of particles into the pores of porous substrate, especially when using nanosized particles.

Poly(dimethylsiloxane) (PDMS) and poly[1-(trimethylsilyl)-1-propyne] (PTMSP) are the most commonly applied gutter layer to bridge the porous support and thin selective layer. But the formation of a thin, defect-free selective coating layer on PDMS can be challenging due to its low surface energy, leading to poor interfacial adhesion between the gutter layer and the selective layer [79,82]. To tackle this problem, researchers have developed a series of flat sheet composite membranes using surface-functionalized PDMS to introduce the addition of a thin selective layer, and the incorporation of inorganic nanoparticles, *star*-polymers, and soft nanoparticles can promote the composite membrane gas permeability [83–86]. Similarly, Li

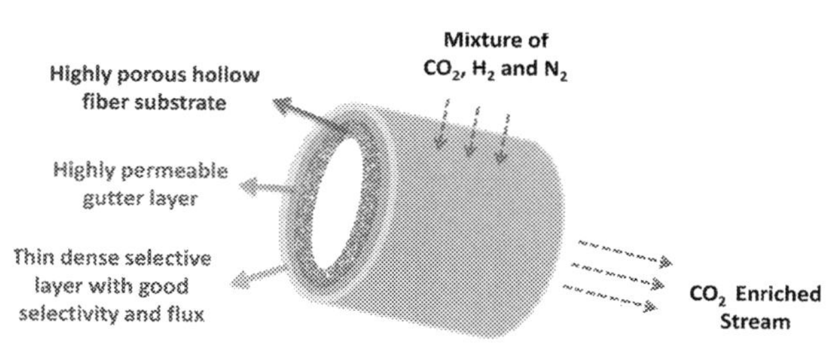

Fig. 10.4 The structure of inorganic particles incorporating nanocomposite membrane. This structure requires a thin selective layer over the porous substrate. *(Reproduced with permission from H.Z. Chen, Z. Thong, P. Li, T.-S. Chung, High performance composite hollow fiber membranes for CO2/H2 and CO2/N2 separation, Int. J. Hydrogen Energy 39 (2014) 5043–5053. https://doi.org/10.1016/j.ijhydene.2014.01.047.)*

et al. [87] applied polydopamine as an intermediate layer between PDMS and polyvinylamine (PVAm) selective layer. On the other hand, the application of PTMSP, one of the most permeable polymer materials, as the gutter layer has been limited because the gradual relaxation of its porous structure can lead to a significant loss of permeance especially for thin films [79]. For example, the dip-coated bare PTMSP thin layer can lose nearly 80% of its original permeance after 10 days at room temperature [81]. However, for the composite membrane, whether the coating of a selective polymer layer can stabilize the nonequilibrium PTMSP structure has not been fully explored.

In water and wastewater application, nanocomposite membranes with functional nanomaterials have also been synthesized and applied using different nanomaterials, such as graphene-based nanosheets, halloysite, MOFs, and CNTs. For instance, in oily wastewater treatment, current studies on MMMs and nanocomposite membranes for the oily wastewater treatment process have attempted to utilize emerging materials, such as MOFs. Up to now, ZIF and MOF-808 have been blended with a polymer matrix to form MMMs for oil–water separation and pervaporation processes [39,73]. MOF-808 was selected because it possesses very good stability, high adsorption capacity, and it is relatively easy to modify [39]. MOF-808 can be modified with EDTA and the modification and incorporation of MOF-808 into PAN matrix could improve membrane's separation efficiency, especially to adsorb selected ions presented in oily wastewater. In addition, the membrane demonstrated excellent recyclability and corrosion resistance. Overall, the membrane is highly efficient in treating wastewater.

Another challenge for MMMs and nanocomposite membranes with FNMs used in water and wastewater applications is the lack of durability of fillers inside the membranes. The wash out of fillers from polymer matrix is common especially during membrane cleaning or during operation at high temperature and pressure [50]. The highly durable and stable presence of nanomaterials inside polymer matrix can be achieved by enhancing the compatibility between nanomaterials and polymer matrix. Improving the compatibility as has been discussed in Section 10.3 can be conducted by several techniques, including surface functionalization of nanomaterials and/or polymer. The functionalization can be conducted before or after the membrane has been formed [15,52,56,72,86]. The loss of nanomaterials from the polymer matrix needs to be evaluated under various operational conditions to avoid the presence of nanomaterials in the permeate or environment, causing potential health problems to the society at large. Membranes used for liquid separation also need to be resistant to biofouling. Some nanomaterials, such as graphene and graphene oxide, have antimicrobial and antibiofouling properties, hence they are very suitable to be embedded inside the polymer matrix [50]. In addition, graphene oxide has hydrophilic functional groups that make it very important and has been applied in several studies to increase the permeate flux through the membranes because of the reduction in fouling tendency of the membranes with graphene oxide [46,48,49,64]. For further study, there is still a need to explore the enhancement of membrane surface physicochemical properties, such as hydrophilicity and electrical surface charge.

10.4.2 Challenges in operational aspects of MMMs and nanocomposite membranes with FNMs

Up to now, most studies conducted for membranes with FNMs, including MMMs and nanocomposite membranes, have been focused on dense and flat sheet configuration. This configuration is very important and relatively easy to fabricate on laboratory scale to find membranes with a suitable combination of polymer-FNMs. However, for industrial scale applications, the MMMs and nanocomposite membranes should be synthesized in other configurations, especially as hollow fiber membranes. Hollow fiber membranes offer a higher packing density than flat sheet membranes. This creates saving in plant area because hollow fiber membrane modules can hold a very high surface area of membranes in a small module, hence reducing the initial fixed capital of the plant [88,89].

In recent years, the fabrication of hollow fiber MMMs and hollow fiber nanocomposite membranes with FNMs still faces a great challenge [26,27,43,70,89]. The fabrication of hollow fiber polymeric membrane with no fillers or FNMs involves relatively complicated aspects on thermodynamics, kinetics, chemistry, and mechanical. The incorporation of FNMs inside dope solution adds another complex situation where the dope solution becomes opaque, hence the cloud point technique that is usually used to predict the ternary diagram of polymer is relatively difficult to be applied in the fabrication of hollow fiber mixed matrix membranes. Furthermore, good dispersion of fillers inside polymer matrix in hollow fiber configuration requires very careful consideration as in hollow fiber configuration, a very high amount of fillers should be mixed with the polymer solution. Fig. 10.5 illustrates the challenges on hollow fiber mixed matrix and composite membranes synthesis that need to be addressed before the hollow fiber mixed matrix membranes and nanocomposite membranes can be applied on an industrial scale. High concentration and high amount of FNMs inside mixed matrix and nanocomposite membranes could potentially create nanomaterials leakage that could lead to the formation of voids and defects in hollow fiber membranes [88–91].

In addition to the problems faced during the synthesis in MMMs and nanocomposite membranes with FNMs, the membrane stability under an elevated pressure condition is very important for the practical application, especially in gas separation processes [27,88,92]. One particular concern is the membrane plasticization in gas separation processes. Although enormous attempts have been reported for the MMMs and nanocomposite membranes preparation, only a few have studied the membrane's plasticization behavior [79,93–95]. For example, Shahid et al. incorporated three different MOFs (MIL-53, ZIF-8, and Cu_3BTC_2) into the Matrimid membrane and obtained an improved plasticization resistance [93]. In addition, two aspects are still relatively poorly understood for the MMMs yet they are crucial to the practical application. The first one is the reversibility of the plasticization, i.e., whether the membrane can restore its original separation performance after the depressurization process. The other one is the membrane plasticization behavior under high pressure mixed gas conditions. It has been demonstrated that

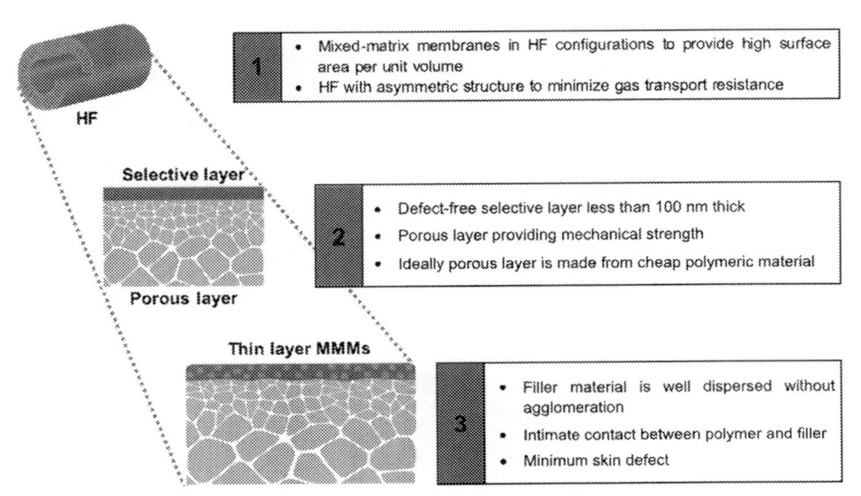

Fig. 10.5 The illustration of challenges in the synthesis of hollow fiber mixed matrix and nanocomposite membranes with FNMs. *(Reproduced with permission from M.R.A. Hamid, H.-K. Jeong, Recent advances on mixed-matrix membranes for gas separation: opportunities and engineering challenges, Korean J. Chem. Eng. 35 (2018) 1577–1600. https://doi.org/10.1007/s11814-018-0081-1.)*

the pure gas results may fail to predict the membrane performance under mixed gas conditions, which is more important for practical application [27]. In a recent study, the mixed gas and pure gas were tested for plasticization resistance of mixed matrix membranes that were fabricated using Pebax as polymer matrix and nano ZIF-8 particles in flat sheet and hollow fiber composite membranes. The presence of ZIF-8 nanomaterials could improve the antiplasticization ability of the composite membranes. This was due to the formation of a hydrogen bond between the Pebax matrix and ZIF-8 nanoparticles. The hollow fiber composite membranes also showed good antiplasticization performance when tested with both pure and mixed gases. The membranes also presented the ability to restore their original separation performance after the depressurization process, which is very important for the application that requires high feed pressure [79]. From laboratory studies mentioned above, the presence of functional nanomaterials has been able to improve the performances of membranes for various applications that require membranes with high separation performances and good stability. However, the industrial realization of MMMs and nanocomposite membranes with functionalized nanomaterials, especially for gas separation, still requires further studies to find the best membranes from a laboratory study that can be scaled up and applied in the real situation.

In addition to studies on the plasticization resistance for dense or flat sheet MMMs, the study on the plasticization effects on the stability of composite membranes with FNMs is still required and this aspect is still relatively poorly understood for composite membranes. As the thin polymer layer has inferior structural stability, the pressure-induced deformation or plasticization can be more significant. Fu et al. [84] studied the performance of the flat sheet composite membrane with a feed pressure up to 10 bar of pure CO_2, and slight plasticization was observed of the thin selective layer. But whether the self-supporting hollow fiber composite membrane can sustain high feed pressure is of great practical importance and thus is worth investigating. A study by Sutrisna et al. in 2016 showed improved performances of hollow fiber nanocomposite membrane under elevated pressures. The membranes were fabricated by dip-coating polyvinylidene fluoride (PVDF) hollow fiber support inside a mixture of Pebax copolymer and ZIF-8 particles. The membranes showed relatively stable gas permeance and gas selectivity under the pressurization-depressurization cycle [79]. The study was then expanded into different types of FNMs, such as UiO-66 and UiO-66-NH_2, which is a type of MOFs with good adsorption capacity to CO_2 gas. In this study, high loading of UiO-66 particles could be deposited on the surface of PVDF supports. The membranes also showed good CO_2 plasticization resistances, indicating promising results that can be scaled up to commercial scale [95]. However, further study and on-site trial should be conducted before the membranes can be applied in the real situation.

Membranes for gas and liquid separations are also required to be stable for long-term operation. In the gas separation process, the physical aging of membranes is still one of the major problems [300]. Recently, it has been reported that the incorporation of microporous materials such as MOFs has significantly improved the long-term performance of several polymers such as copolymer Pebax-1657 [79] so they will not suffer aging for long-term performance. This is because the incorporation of nanomaterials has been observed for being able to reduce the chain movement of the polymer resulting in performance stability as the membrane aged [79,80,95]. In addition, the selectivity of membranes for CO_2 separation was also observed to increase during the aging period because the permeation of other gases with a larger kinetic diameter such as CH_4 was more significantly reduced [79]. The improved separation performance and antiaging property are then also part of the major goals aimed for the next CO_2 separation.

The energy consumption and efficiency are still important factors in MMMs and nanocomposite membranes are used for water, wastewater, and gas treatment processes [3,24,27]. In general, membranes with high permeability require lower pressure as the driving force for the permeation process and membranes with lower fouling tendency operate for a longer time under pressure. Hence, efficient separation using MMMs and nanocomposite membranes can be achieved by optimizing the design of membrane structure and utilizing a suitable combination of polymer and functional nanomaterials [27]. In gas separation processes, for instance, in the carbon capture process, energy requirement and utilization are challenges that need to be addressed in the real situation. For a process relying on adsorption, the main energy consumption mainly comes from the regeneration process. During the regeneration process, two energies are at least required if temperature-swing adsorption is chosen, such as raising the material temperature from the adsorption process to the desorption process and endothermic desorption process. In a pressure-swing adsorption process, the energy penalty might come from continuous pressurization and depressurization of the system. Several innovative ways can be proposed to address the issue requiring less energy consumption and enhanced adsorption–desorption process. However, this regeneration energy is not required in a membrane-based CO_2 separation process. From the membrane-based separation process perspective, the concern of energy is more related to make the overall process more efficient and consume less energy. This task is usually achieved by optimizing the operating condition of the membrane. In this regard, it should be admitted that there is still a lack of details on the study to build a general strategy for the operating condition for a membrane-based CO_2 separation.

10.5 Conclusions

This chapter discusses possible challenges for the industrialization and scale-up of membranes with functionalized nanomaterials or FNMs. FNM-based membranes are generally synthesized as mixed matrix membranes (MMMs) or nanocomposite membranes. The challenges can be categorized into two different aspects, synthesis and operational challenges. From the synthesis point of view, several problems related to the fabrication of membranes need to be addressed. Such challenges correlate with the classic problem of compatibility between polymer or continuous phase and FNMs as the dispersed phase. The incompatibility between these two materials causes the formation of voids, defects, or rigidified regions near the interface between these materials. Voids, defects, and rigidified sections affect the separation performances of the membranes. Hence, many efforts have been made to handle this problem. Surface functionalization, priming protocol, and controlling the concentration of FNMs inside polymer matrix are several techniques that have been successful in minimizing the problems of incompatibility of two materials. Surface functionalization using silane and polydopamine has been conducted and considered as a potential solution for improving the dispersion of FNMs inside the polymer matrix and both of these functionalized agents have been reviewed in this chapter.

In terms of operational aspects, hollow fiber membrane configuration offers a more efficient and effective separation process than flat sheet membranes. However, the scale-up process of MMMs and nanocomposite membranes from flat sheet configuration into hollow fiber configuration still faces challenges. For the hollow fiber MMMs to be commercially attractive, the membranes should be synthesized as the asymmetric structure that has a thin selective layer of the mixed matrix of FNMs and polymer. This is a very challenging research prospect worthy of further investigation. In addition, for the fabrication of hollow fiber membrane, a higher amount of FNMs needs to be dispersed uniformly on the mixed matrix layer, and this requires optimization as the possibility of particle aggregation and wash out are possible during the synthesis. Furthermore, some issues on membrane plasticization, especially in gas separation processes, and long-term performance of produced membranes need to be investigated as these two phenomena are very important in the real situation [27,50].

References

[1] R.W. Baker, Membrane Technology and Applications, second ed., John Wiley & Sons, Ltd, 2004.

[2] M. Bassyouni, M.H. Abdel-Aziz, M.S. Zoromba, S.M.S. Abdel-Hamid, E. Drioli, A review of polymeric nanocomposite membranes for water purification, J. Ind. Eng. Chem. 73 (2019) 19–46, https://doi.org/10.1016/j.jiec.2019.01.045.

[3] P. Bernardo, E. Drioli, G. Golemme, Membrane gas separation: a review/state of the art, Ind. Eng. Chem. Res. 48 (2009) 4638–4663, https://doi.org/10.1021/ie8019032.

[4] M. Mulder, Basic Principles of Membrane Technology, second ed., Kluwer Academic Publishers, Netherlands, 1996.

[5] H.K. Lonsdale, The growth of membrane technology, J. Membr. Sci. 10 (1982) 81–181, https://doi.org/10.1016/S0376-7388(00)81408-8.

[6] A.K. Pabby, S.S.H. Rizvi, A.M.S. Requena (Eds.), Handbook of Membrane Separations, CRC Press, 2015, https://doi.org/10.1201/b18319.

[7] N.U. Barambu, M.R. Bilad, M.A. Bustam, K.A. Kurnia, M.H.D. Othman, N.A.H.M. Nordin, Development of membrane material for oily wastewater treatment: a review, Ain Shams Eng. J. (2020), https://doi.org/10.1016/j.asej.2020.08.027. S2090447920302355.

[8] B. Bolto, J. Zhang, X. Wu, Z. Xie, A review on current development of membranes for oil removal from wastewaters, Membranes 10 (2020) 65, https://doi.org/10.3390/membranes10040065.

[9] P.S. Goh, A.F. Ismail, A review on inorganic membranes for desalination and wastewater treatment, Desalination 434 (2018) 60–80, https://doi.org/10.1016/j.desal.2017.07.023.

[10] A. Kayvani Fard, G. McKay, A. Buekenhoudt, H. Al Sulaiti, F. Motmans, M. Khraisheh, M. Atieh, Inorganic Membranes: Preparation and Application for Water Treatment and Desalination, Materials, Vol. 11, 2018, p. 74, https://doi.org/10.3390/ma11010074.

[11] Y.S. Lin, Inorganic membranes for process intensification: challenges and perspective, Ind. Eng. Chem. Res. 58 (2019) 5787–5796, https://doi.org/10.1021/acs.iecr.8b04539.

[12] S. Adham, A. Hussain, J. Minier-Matar, A. Janson, R. Sharma, Membrane applications and opportunities for water management in the oil and gas industry, Desalination 440 (2018) 2–17, https://doi.org/10.1016/j.desal.2018.01.030.

[13] N.A. Ahmad, P.S. Goh, L.T. Yogarathinam, A.K. Zulhairun, A.F. Ismail, Current advances in membrane technologies for produced water desalination, Desalination 493 (2020), https://doi.org/10.1016/j.desal.2020.114643, 114643.

[14] A. Elrasheedy, N. Nady, M. Bassyouni, A. El-Shazly, Metal organic framework based polymer mixed matrix membranes: review on applications in water purification, Membranes 9 (2019) 88, https://doi.org/10.3390/membranes9070088.

[15] W.-C. Chong, C.-H. Koo, W.-J. Lau, Mixed-matrix membranes incorporated with functionalized nanomaterials for water applications, in: Handbook of Functionalized Nanomaterials for Industrial Applications, Elsevier, 2020, pp. 15–51, https://doi.org/10.1016/B978-0-12-816787-8.00002-8.

[16] J. Dechnik, J. Gascon, C.J. Doonan, C. Janiak, C.J. Sumby, Mixed-matrix membranes, Angew. Chem. Int. Ed. 56 (2017) 9292–9310, https://doi.org/10.1002/anie.201701109.
[17] S. Al Aani, C.J. Wright, M.A. Atieh, N. Hilal, Engineering nanocomposite membranes: addressing current challenges and future opportunities, Desalination 401 (2017) 1–15, https://doi.org/10.1016/j.desal.2016.08.001.
[18] W.J. Koros, G.K. Fleming, Membrane-based gas separation, J. Membr. Sci. 83 (1993) 1–80, https://doi.org/10.1016/0376-7388(93)80013-N.
[19] C.E. Powell, G.G. Qiao, Polymeric CO2/N2 gas separation membranes for the capture of carbon dioxide from power plant flue gases, J. Membr. Sci. 279 (2006) 1–49, https://doi.org/10.1016/j.memsci.2005.12.062.
[20] L.M. Robeson, Correlation of separation factor versus permeability for polymeric membranes, J. Membr. Sci. 62 (1991) 165–185, https://doi.org/10.1016/0376-7388(91)80060-J.
[21] L.M. Robeson, The upper bound revisited, J. Membr. Sci. 320 (2008) 390–400, https://doi.org/10.1016/j.memsci.2008.04.030.
[22] H. Lin, M. Yavari, Upper bound of polymeric membranes for mixed-gas CO2/CH4 separations, J. Membr. Sci. 475 (2015) 101–109, https://doi.org/10.1016/j.memsci.2014.10.007.
[23] T.-S. Chung, L.Y. Jiang, Y. Li, S. Kulprathipanja, Mixed matrix membranes (MMMs) comprising organic polymers with dispersed inorganic fillers for gas separation, Prog. Polym. Sci. 32 (2007) 483–507, https://doi.org/10.1016/j.progpolymsci.2007.01.008.
[24] S. Najari, S. Saeidi, F. Gallucci, E. Drioli, Mixed matrix membranes for hydrocarbons separation and recovery: a critical review, Rev. Chem. Eng. 0 (2019) 20180091, https://doi.org/10.1515/revce-2018-0091.
[25] D. Bastani, N. Esmaeili, M. Asadollahi, Polymeric mixed matrix membranes containing zeolites as a filler for gas separation applications: a review, J. Ind. Eng. Chem. 19 (2013) 375–393, https://doi.org/10.1016/j.jiec.2012.09.019.
[26] M. Galizia, W.S. Chi, Z.P. Smith, T.C. Merkel, R.W. Baker, B.D. Freeman, 50th Anniversary perspective : polymers and mixed matrix membranes for gas and vapor separation: a review and prospective opportunities, Macromolecules 50 (2017) 7809–7843, https://doi.org/10.1021/acs.macromol.7b01718.
[27] G. Dong, H. Li, V. Chen, Challenges and opportunities for mixed-matrix membranes for gas separation, J. Mater. Chem. A 1 (2013) 4610, https://doi.org/10.1039/c3ta00927k.
[28] S.N. Wan Ikhsan, N. Yusof, F. Aziz, N. Misdan, A.F. Ismail, W.-J. Lau, J. Jaafar, W.N. Wan Salleh, N.H., Hayati Hairom, efficient separation of oily wastewater using polyethersulfone mixed matrix membrane incorporated with halloysite nanotube-hydrous ferric oxide nanoparticle, Sep. Purif. Technol. 199 (2018) 161–169, https://doi.org/10.1016/j.seppur.2018.01.028.
[29] S.N. Wan Ikhsan, N. Yusof, N.I. Mat Nawi, M.R. Bilad, N. Shamsuddin, F. Aziz, A.F. Ismail, Halloysite nanotube-Ferrihydrite incorporated Polyethersulfone mixed matrix membrane: effect of nanocomposite loading on the antifouling performance, Polymers 13 (2021) 441, https://doi.org/10.3390/polym13030441.
[30] Z. Bai, L. Wang, C. Liu, C. Yang, G. Lin, S. Liu, K. Jia, X. Liu, Interfacial coordination mediated surface segregation of halloysite nanotubes to construct a high-flux antifouling membrane for oil-water emulsion separation, J. Membr. Sci. 620 (2021), https://doi.org/10.1016/j.memsci.2020.118828, 118828.
[31] J.S.B. Melbiah, D. Nithya, D. Mohan, Surface modification of polyacrylonitrile ultrafiltration membranes using amphiphilic Pluronic F127/CaCO3 nanoparticles for oil/water emulsion separation, Colloids Surf. A Physicochem. Eng. Asp. 516 (2017) 147–160, https://doi.org/10.1016/j.colsurfa.2016.12.008.

[32] S. Saki, N. Uzal, Preparation and characterization of PSF/PEI/CaCO3 nanocomposite membranes for oil/water separation, Environ. Sci. Pollut. Res. 25 (2018) 25315–25326, https://doi.org/10.1007/s11356-018-2615-9.

[33] J. Cui, Z. Zhou, A. Xie, Q. Wang, S. Liu, J. Lang, C. Li, Y. Yan, J. Dai, Facile preparation of grass-like structured NiCo-LDH/PVDF composite membrane for efficient oil–water emulsion separation, J. Membr. Sci. 573 (2019) 226–233, https://doi.org/10.1016/j.memsci.2018.11.064.

[34] D. Makwana, V. Polisetti, J. Castaño, P. Ray, H.C. Bajaj, Mg-Fe layered double hydroxide modified montmorillonite as hydrophilic nanofiller in polysulfone-polyvinylpyrrolidone blend ultrafiltration membranes: separation of oil-water mixture, Appl. Clay Sci. 192 (2020), https://doi.org/10.1016/j.clay.2020.105636, 105636.

[35] W. Qing, X. Li, Y. Wu, S. Shao, H. Guo, Z. Yao, Y. Chen, W. Zhang, C.Y. Tang, In situ silica growth for superhydrophilic-underwater superoleophobic Silica/PVA nanofibrous membrane for gravity-driven oil-in-water emulsion separation, J. Membr. Sci. 612 (2020), https://doi.org/10.1016/j.memsci.2020.118476, 118476.

[36] H.M. Mousa, H. Alfadhel, M. Ateia, G.T. Abdel-Jaber, Polysulfone-iron acetate/polyamide nanocomposite membrane for oil-water separation, Environ. Nanotechnol. Monitor. Manage. 14 (2020), https://doi.org/10.1016/j.enmm.2020.100314, 100314.

[37] M.R. De Guzman, C.K.A. Andra, M.B.M.Y. Ang, G.V.C. Dizon, A.R. Caparanga, S.-H. Huang, K.-R. Lee, Increased performance and antifouling of mixed-matrix membranes of cellulose acetate with hydrophilic nanoparticles of polydopamine-sulfobetaine methacrylate for oil-water separation, J. Membr. Sci. 620 (2021), https://doi.org/10.1016/j.memsci.2020.118881, 118881.

[38] G.S. Lai, M.H.M. Yusob, W.J. Lau, R.J. Gohari, D. Emadzadeh, A.F. Ismail, P.S. Goh, A.M. Isloor, M.R.-D. Arzhandi, Novel mixed matrix membranes incorporated with dual-nanofillers for enhanced oil-water separation, Sep. Purif. Technol. 178 (2017) 113–121, https://doi.org/10.1016/j.seppur.2017.01.033.

[39] X. Chen, D. Chen, N. Li, Q. Xu, H. Li, J. He, J. Lu, Modified-MOF-808-loaded Polyacrylonitrile membrane for highly efficient, simultaneous emulsion separation and heavy metal ion removal, ACS Appl. Mater. Interfaces 12 (2020) 39227–39235, https://doi.org/10.1021/acsami.0c10290.

[40] L. Shu, L.-H. Xie, Y. Meng, T. Liu, C. Zhao, J.-R. Li, A thin and high loading two-dimensional MOF nanosheet based mixed-matrix membrane for high permeance nanofiltration, J. Membr. Sci. 603 (2020), https://doi.org/10.1016/j.memsci.2020.118049, 118049.

[41] Y. Cai, D. Chen, N. Li, Q. Xu, H. Li, J. He, J. Lu, Nanofibrous metal–organic framework composite membrane for selective efficient oil/water emulsion separation, J. Membr. Sci. 543 (2017) 10–17, https://doi.org/10.1016/j.memsci.2017.08.047.

[42] M.S. Denny, S.M. Cohen, In situ modification of metal-organic frameworks in mixed-matrix membranes, Angew. Chem. Int. Ed. 54 (2015) 9029–9032, https://doi.org/10.1002/anie.201504077.

[43] V. Muthukumaraswamy Rangaraj, M.A. Wahab, K.S.K. Reddy, G. Kakosimos, O. Abdalla, E.P. Favvas, D. Reinalda, F. Geuzebroek, A. Abdala, G.N. Karanikolos, Metal organic framework-based mixed matrix membranes for carbon dioxide separation: recent advances and future directions, Front. Chem. 8 (2020) 534, https://doi.org/10.3389/fchem.2020.00534.

[44] L. Zhu, M. Chen, Y. Dong, C.Y. Tang, A. Huang, L. Li, A low-cost mullite-titania composite ceramic hollow fiber microfiltration membrane for highly efficient separation of oil-in-water emulsion, Water Res. 90 (2016) 277–285, https://doi.org/10.1016/j.watres.2015.12.035.

[45] A. Alammar, S.-H. Park, C.J. Williams, B. Derby, G. Szekely, Oil-in-water separation with graphene-based nanocomposite membranes for produced water treatment, J. Membr. Sci. 603 (2020), https://doi.org/10.1016/j.memsci.2020.118007, 118007.
[46] N.F.D. Junaidi, N.H. Othman, N.S. Fuzil, M.S. Mat Shayuti, N.H. Alias, M.Z. Shahruddin, F. Marpani, W.J. Lau, A.F. Ismail, N.D. Aba, Recent development of graphene oxide-based membranes for oil–water separation: a review, Sep. Purif. Technol. 258 (2021) 118000, https://doi.org/10.1016/j.seppur.2020.118000.
[47] N.F.D. Junaidi, N.H. Othman, M.Z. Shahruddin, N.H. Alias, F. Marpani, W.J. Lau, A.F. Ismail, Fabrication and characterization of graphene oxide–polyethersulfone (GO–PES) composite flat sheet and hollow fiber membranes for oil–water separation, J. Chem. Technol. Biotechnol. 95 (2020) 1308–1320, https://doi.org/10.1002/jctb.6366.
[48] Y. Peng, Z. Yu, F. Li, Q. Chen, D. Yin, X. Min, A novel reduced graphene oxide-based composite membrane prepared via a facile deposition method for multifunctional applications: oil/water separation and cationic dyes removal, Sep. Purif. Technol. 200 (2018) 130–140, https://doi.org/10.1016/j.seppur.2018.01.059.
[49] A. Alkhouzaam, H. Qiblawey, Novel polysulfone ultrafiltration membranes incorporating polydopamine functionalized graphene oxide with enhanced flux and fouling resistance, J. Membr. Sci. 620 (2021), https://doi.org/10.1016/j.memsci.2020.118900, 118900.
[50] M.R. Esfahani, S.A. Aktij, Z. Dabaghian, M.D. Firouzjaei, A. Rahimpour, J. Eke, I.C. Escobar, M. Abolhassani, L.F. Greenlee, A.R. Esfahani, A. Sadmani, N. Koutahzadeh, Nanocomposite membranes for water separation and purification: fabrication, modification, and applications, Sep. Purif. Technol. 213 (2019) 465–499, https://doi.org/10.1016/j.seppur.2018.12.050.
[51] M. Delfi, M. Ghomi, A. Zarrabi, R. Mohammadinejad, Z.B. Taraghdari, M. Ashrafizadeh, E.N. Zare, T. Agarwal, V.V.T. Padil, B. Mokhtari, F. Rossi, G. Perale, M. Sillanpaa, A. Borzacchiello, T. Kumar Maiti, P. Makvandi, Functionalization of polymers and nanomaterials for biomedical applications: antimicrobial platforms and drug carriers, Prosthesis 2 (2020) 117–139, https://doi.org/10.3390/prosthesis2020012.
[52] P. Makvandi, Functionalization of polymers and nanomaterials for water treatment, food packaging, textile and biomedical applications: a review, Environ. Chem. Lett. 29 (2021) 583–611, https://doi.org/10.1007/s10311-020-01089-4.
[53] S. Palit, C.M. Hussain, Functionalization of nanomaterials for industrial applications: recent and future perspectives, in: Handbook of Functionalized Nanomaterials for Industrial Applications, Elsevier, 2020, pp. 3–14, https://doi.org/10.1016/B978-0-12-816787-8.00001-6.
[54] B. Ghalei, K. Sakurai, Y. Kinoshita, K. Wakimoto, A.P. Isfahani, Q. Song, K. Doitomi, S. Furukawa, H. Hirao, H. Kusuda, S. Kitagawa, E. Sivaniah, Enhanced selectivity in mixed matrix membranes for CO2 capture through efficient dispersion of amine-functionalized MOF nanoparticles, Nat. Energy 2 (2017) 17086, https://doi.org/10.1038/nenergy.2017.86.
[55] M. Jia, Y. Feng, J. Qiu, X.-F. Zhang, J. Yao, Amine-functionalized MOFs@GO as filler in mixed matrix membrane for selective CO2 separation, Sep. Purif. Technol. 213 (2019) 63–69, https://doi.org/10.1016/j.seppur.2018.12.029.
[56] H. Molavi, A. Shojaei, S.A. Mousavi, Improving mixed-matrix membrane performance via PMMA grafting from functionalized NH2–UiO-66, J. Mater. Chem. A 6 (2018) 2775–2791, https://doi.org/10.1039/C7TA10480D.
[57] B. Zornoza, A. Martinez-Joaristi, P. Serra-Crespo, C. Tellez, J. Coronas, J. Gascon, F. Kapteijn, Functionalized flexible MOFs as fillers in mixed matrix membranes for highly

selective separation of CO2 from CH4 at elevated pressures, Chem. Commun. 47 (2011) 9522–9524, https://doi.org/10.1039/C1CC13431K.
[58] M.B.M.Y. Ang, C.R.M. Macni, A.R. Caparanga, S.-H. Huang, H.-A. Tsai, K.-R. Lee, J.-Y. Lai, Mitigating the fouling of mixed-matrix cellulose acetate membranes for oil–water separation through modification with polydopamine particles, Chem. Eng. Res. Des. 159 (2020) 195–204, https://doi.org/10.1016/j.cherd.2020.04.015.
[59] T.G. Barclay, H.M. Hegab, S.R. Clarke, M. Ginic-Markovic, Versatile surface modification using Polydopamine and related Polycatecholamines: chemistry, structure, and applications, Adv. Mater. Interfaces 4 (2017) 1601192, https://doi.org/10.1002/admi.201601192.
[60] J. Cui, Z. Zhou, A. Xie, M. Meng, Y. Cui, S. Liu, J. Lu, S. Zhou, Y. Yan, H. Dong, Bio-inspired fabrication of superhydrophilic nanocomposite membrane based on surface modification of SiO2 anchored by polydopamine towards effective oil-water emulsions separation, Sep. Purif. Technol. 209 (2019) 434–442, https://doi.org/10.1016/j.seppur.2018.03.054.
[61] H. Hu, B. Yu, Q. Ye, Y. Gu, F. Zhou, Modification of carbon nanotubes with a nanothin polydopamine layer and polydimethylamino-ethyl methacrylate brushes, Carbon 48 (2010) 2347–2353, https://doi.org/10.1016/j.carbon.2010.03.014.
[62] Q. Huang, J. Chen, M. Liu, H. Huang, X. Zhang, Y. Wei, Polydopamine-based functional materials and their applications in energy, environmental, and catalytic fields: state-of-the-art review, Chem. Eng. J. 387 (2020), https://doi.org/10.1016/j.cej.2020.124019, 124019.
[63] Y. Lv, H.-C. Yang, H.-Q. Liang, L.-S. Wan, Z.-K. Xu, Nanofiltration membranes via co-deposition of polydopamine/polyethylenimine followed by cross-linking, J. Membr. Sci. 476 (2015) 50–58, https://doi.org/10.1016/j.memsci.2014.11.024.
[64] X. Wang, X. Peng, Y. Zhao, C. Yang, K. Qi, Y. Li, P. Li, Bio-inspired modification of superhydrophilic iPP membrane based on polydopamine and graphene oxide for highly antifouling performance and reusability, Mater. Lett. 255 (2019), https://doi.org/10.1016/j.matlet.2019.126573, 126573.
[65] H. Wei, J. Ren, B. Han, L. Xu, L. Han, L. Jia, Stability of polydopamine and poly(DOPA) melanin-like films on the surface of polymer membranes under strongly acidic and alkaline conditions, Colloids Surf. B. Biointerfaces 110 (2013) 22–28, https://doi.org/10.1016/j.colsurfb.2013.04.008.
[66] Y. Xiang, F. Liu, L. Xue, Under seawater superoleophobic PVDF membrane inspired by polydopamine for efficient oil/seawater separation, J. Membr. Sci. 476 (2015) 321–329, https://doi.org/10.1016/j.memsci.2014.11.052.
[67] H.-C. Yang, W. Xu, Y. Du, J. Wu, Z.-K. Xu, Composite free-standing films of polydopamine/polyethyleneimine grown at the air/water interface, RSC Adv. 4 (2014) 45415–45418, https://doi.org/10.1039/C4RA04549A.
[68] J. Liebscher, R. Mrówczyński, H.A. Scheidt, C. Filip, N.D. Hădade, R. Turcu, A. Bende, S. Beck, Structure of Polydopamine: a never-ending story? Langmuir 29 (2013) 10539–10548, https://doi.org/10.1021/la4020288.
[69] T.T. Moore, W.J. Koros, Non-ideal effects in organic–inorganic materials for gas separation membranes, J. Mol. Struct. 739 (2005) 87–98, https://doi.org/10.1016/j.molstruc.2004.05.043.
[70] M.A. Aroon, A.F. Ismail, T. Matsuura, M.M. Montazer-Rahmati, Performance studies of mixed matrix membranes for gas separation: a review, Sep. Purif. Technol. 75 (2010) 229–242, https://doi.org/10.1016/j.seppur.2010.08.023.
[71] F. Ahmadijokani, S. Ahmadipouya, H. Molavi, M. Arjmand, Amino-silane-grafted NH2-MIL-53(Al)/polyethersulfone mixed matrix membranes for CO2/CH4 separation, Dalton Trans. 48 (2019) 13555–13566, https://doi.org/10.1039/C9DT02328C.

[72] H.R. Amedi, M. Aghajani, Aminosilane-functionalized ZIF-8/PEBA mixed matrix membrane for gas separation application, Microporous Mesoporous Mater. 247 (2017) 124–135, https://doi.org/10.1016/j.micromeso.2017.04.001.
[73] S. Li, Z. Chen, Y. Yang, Z. Si, P. Li, P. Qin, T. Tan, Improving the pervaporation performance of PDMS membranes for n-butanol by incorporating silane-modified ZIF-8 particles, Sep. Purif. Technol. 215 (2019) 163–172, https://doi.org/10.1016/j.seppur.2018.12.078.
[74] J. Zhang, Q. Xin, X. Li, M. Yun, R. Xu, S. Wang, Y. Li, L. Lin, X. Ding, H. Ye, Y. Zhang, Mixed matrix membranes comprising aminosilane-functionalized graphene oxide for enhanced CO2 separation, J. Membr. Sci. 570–571 (2019) 343–354, https://doi.org/10.1016/j.memsci.2018.10.075.
[75] R. Mahajan, W.J. Koros, Factors controlling successful formation of mixed-matrix gas separation materials, Ind. Eng. Chem. Res. 39 (2000) 2692–2696, https://doi.org/10.1021/ie990799r.
[76] R. Mahajan, R. Burns, M. Schaeffer, W.J. Koros, Challenges in forming successful mixed matrix membranes with rigid polymeric materials, J. Appl. Polym. Sci. 86 (2002) 881–890, https://doi.org/10.1002/app.10998.
[77] Q. Song, S.K. Nataraj, M.V. Roussenova, J.C. Tan, D.J. Hughes, W. Li, P. Bourgoin, M.A. Alam, A.K. Cheetham, S.A. Al-Muhtaseb, E. Sivaniah, Zeolitic imidazolate framework (ZIF-8) based polymer nanocomposite membranes for gas separation, Energ. Environ. Sci. 5 (2012) 8359–8369, https://doi.org/10.1039/C2EE21996D.
[78] J.A. Thompson, K.W. Chapman, W.J. Koros, C.W. Jones, S. Nair, Sonication-induced Ostwald ripening of ZIF-8 nanoparticles and formation of ZIF-8/polymer composite membranes, Microporous Mesoporous Mater. 158 (2012) 292–299, https://doi.org/10.1016/j.micromeso.2012.03.052.
[79] P.D. Sutrisna, J. Hou, H. Li, Y. Zhang, V. Chen, Improved operational stability of Pebax-based gas separation membranes with ZIF-8: a comparative study of flat sheet and composite hollow fibre membranes, J. Membr. Sci. 524 (2017) 266–279, https://doi.org/10.1016/j.memsci.2016.11.048.
[80] J. Hou, P.D. Sutrisna, Y. Zhang, V. Chen, Formation of ultrathin, continuous metal–organic framework membranes on flexible polymer substrates, Angew. Chem. Int. Ed. 55 (2016) 3947–3951, https://doi.org/10.1002/anie.201511340.
[81] H.Z. Chen, Z. Thong, P. Li, T.-S. Chung, High performance composite hollow fiber membranes for CO2/H2 and CO2/N2 separation, Int. J. Hydrogen Energy 39 (2014) 5043–5053, https://doi.org/10.1016/j.ijhydene.2014.01.047.
[82] Y. Wang, H. Li, G. Dong, C. Scholes, V. Chen, Effect of fabrication and operation conditions on CO2 separation performance of PEO–PA block copolymer membranes, Ind. Eng. Chem. Res. 54 (2015) 7273–7283, https://doi.org/10.1021/acs.iecr.5b01234.
[83] Q. Fu, J. Kim, P.A. Gurr, J.M.P. Scofield, S.E. Kentish, G.G. Qiao, A novel cross-linked nano-coating for carbon dioxide capture, Energ. Environ. Sci. 9 (2016) 434–440, https://doi.org/10.1039/C5EE02433A.
[84] Q. Fu, E.H.H. Wong, J. Kim, J.M.P. Scofield, P.A. Gurr, S.E. Kentish, G.G. Qiao, The effect of soft nanoparticles morphologies on thin film composite membrane performance, J. Mater. Chem. A 2 (2014) 17751–17756, https://doi.org/10.1039/C4TA02859G.
[85] A. Halim, Q. Fu, Q. Yong, P.A. Gurr, S.E. Kentish, G.G. Qiao, Soft polymeric nanoparticle additives for next generation gas separation membranes, J. Mater. Chem. A 2 (2014) 4999–5009, https://doi.org/10.1039/C3TA14170E.
[86] J. Kim, Q. Fu, K. Xie, J.M.P. Scofield, S.E. Kentish, G.G. Qiao, CO2 separation using surface-functionalized SiO2 nanoparticles incorporated ultra-thin film composite mixed matrix membranes for post-combustion carbon capture, J. Membr. Sci. 515 (2016) 54–62, https://doi.org/10.1016/j.memsci.2016.05.029.

[87] P. Li, Z. Wang, W. Li, Y. Liu, J. Wang, S. Wang, High-performance multilayer composite membranes with mussel-inspired Polydopamine as a versatile molecular bridge for CO2 separation, ACS Appl. Mater. Interfaces 7 (2015) 15481–15493, https://doi.org/10.1021/acsami.5b03786.
[88] N. Prasetya, N.F. Himma, P.D. Sutrisna, I.G. Wenten, B.P. Ladewig, A review on emerging organic-containing microporous material membranes for carbon capture and separation, Chem. Eng. J. 391 (2020), https://doi.org/10.1016/j.cej.2019.123575, 123575.
[89] M.R.A. Hamid, H.-K. Jeong, Recent advances on mixed-matrix membranes for gas separation: opportunities and engineering challenges, Korean J. Chem. Eng. 35 (2018) 1577–1600, https://doi.org/10.1007/s11814-018-0081-1.
[90] P.D. Sutrisna, N. Prasetya, N.F. Himma, I.G. Wenten, A mini-review and recent outlooks on the synthesis and applications of zeolite imidazolate framework-8 (ZIF-8) membranes on polymeric substrate, J. Chem. Technol. Biotechnol. 95 (2020) 2767–2774, https://doi.org/10.1002/jctb.6433.
[91] M.R.A. Hamid, T.C.S. Yaw, M.Z.M. Tohir, W.A.W.A.K. Ghani, P.D. Sutrisna, H.-K. Jeong, Zeolitic imidazolate framework membranes for gas separations: current state-of-the-art, challenges, and opportunities, J. Ind. Eng. Chem. 98 (2021) 17–41, https://doi.org/10.1016/j.jiec.2021.03.047.
[92] S.R. Venna, M.A. Carreon, Metal organic framework membranes for carbon dioxide separation, Chem. Eng. Sci. 124 (2015) 3–19, https://doi.org/10.1016/j.ces.2014.10.007.
[93] S. Shahid, K. Nijmeijer, High pressure gas separation performance of mixed-matrix polymer membranes containing mesoporous Fe(BTC), J. Membr. Sci. 459 (2014) 33–44, https://doi.org/10.1016/j.memsci.2014.02.009.
[94] S. Shahid, K. Nijmeijer, Performance and plasticization behavior of polymer–MOF membranes for gas separation at elevated pressures, J. Membr. Sci. 470 (2014) 166–177, https://doi.org/10.1016/j.memsci.2014.07.034.
[95] P.D. Sutrisna, J. Hou, M.Y. Zulkifli, H. Li, Y. Zhang, W. Liang, D.M. D'Alessandro, V. Chen, Surface functionalized UiO-66/Pebax-based ultrathin composite hollow fiber gas separation membranes, J. Mater. Chem. A 6 (2018) 918–931, https://doi.org/10.1039/C7TA07512J.

CHAPTER 11

Membranes with FNMs for sustainable development

Deepshikha Datta[a], K.S. Deepak[b], and Bimal Das[b]
[a]Department of Chemical Engineering, GMR Institute of Technology, Rajam, India
[b]Department of Chemical Engineering, National Institute of Technology Durgapur, Durgapur, India

11.1 Introduction

The world population has seen a quadrupled growth in the 20th century and would cross 8 billion marks by 2050 [1]. With such unprecedented numbers comes the scarcity of resources. Today's world is dwelling to find an intricate balance between providing adequate living standards to all and mitigating the consequences and aftereffects on the environment. Due to the rapid increase in population, the demands for basic facilities such as water, food, electricity, and health care are increasing, which in turn stirs up the production of such services, impacting the environment. Sustainable membranes are the need of generation to provide acceptable, environmentally effective, innovative, and high-performance substitute for various applications [2]. Nanotechnology has proved to be one of the best feasible way out owing to its immense ability to provide technical solution in the most prominent way to resolve various issues regarding the modern sustainable technological requirement. Besides the versatility of the nano-enabled membranes proves it to be a suitable product for the present era.

The notion of "sustainable development" was introduced by the Brundtland Commission created by the UN, which stated as those aspects required to meet the demands of present generations without compromising the futuregeneration's ability to meet their needs [3]. Thus, the notion of sustainability is the delicate balance between the parameters shown in Fig. 11.1. It also placed sustainability at the conjuncture of social, economic, and environmental factors, as shown in Fig. 11.2.

The National Nanotechnology Initiative was molded with expectations of nanotechnology being the panacea of challenges faced by society. For example, Roco, at his presentation at the Cornell Nanofabrication Center on September 2001, showcased ways in which nanotechnology could help

Membranes with Functionalized Nanomaterials
https://doi.org/10.1016/B978-0-323-85946-2.00006-0

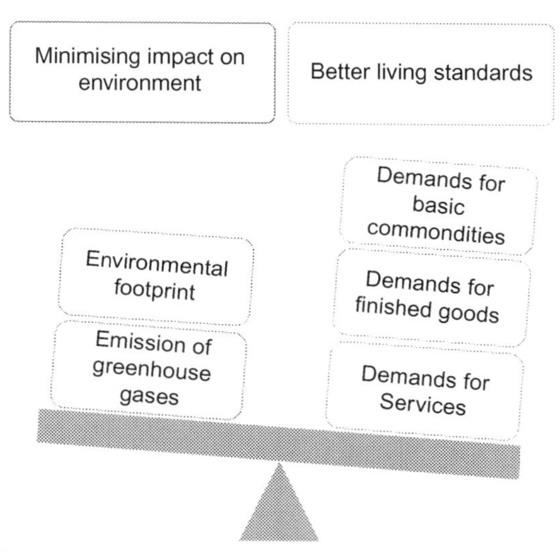

Fig. 11.1 The balance for sustainability.

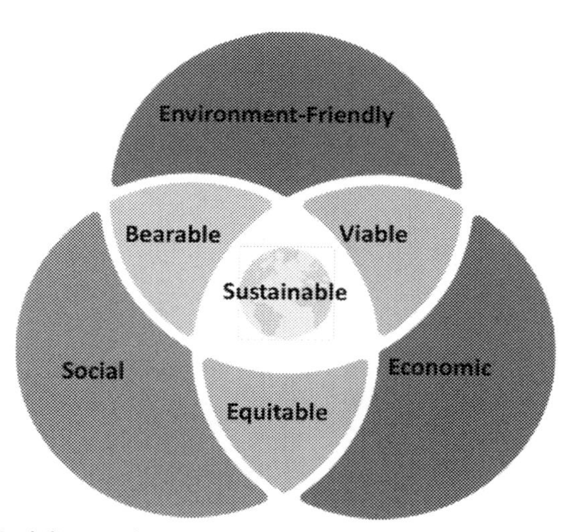

Fig. 11.2 Sustainability at the conjuncture of social economic and environmental perspective.

build a sustainable environment by creating more efficient water purification technologies and clean energy sources [4].

A change in approach from one-size-fits-all is essential so that specialized innovations and policy reforms are done to tackle issues pertaining to a particular application and then enhancing its sustainability [5]. In that direction, nanotechnology can be a torch bearer, as most of the processes specifically related to membranes occurs at nanoscale. Understanding the nuances and exploiting them in developing new and optimizing existing membranes would be a boom in attaining sustainability. With stringent environmental regulations and increased demand of water, membranes are expected to bear these changing dynamics and for which nanotechnology plays a pivotal role. Nano-based membranes can cope up with the societal and industrial requirements by maintaining better productivity and better rejection of pollutants [6]. The quantum leap in more advanced and sustainable products is a proof of the work and potential bestowed with nanotechnology and membrane technology. Tons of research have been done to develop new and improvised materials that meet the societal and industrial expectations with the help of membrane science and nanotechnology in the area of desalination, air filtration, and energy sector. Evaluating the exact amount of contribution from nanotechnology would require understanding the rapidity, affordability, sustainability, and commerciality. Each of this parameter evaluation is exhaustive. For instance, to evaluate the sustainability, we must evaluate the current position of a technology with various funnel questions, which have been mentioned in the subsequent sections in this chapter.

This book chapter looks at the role of membranes with FNMs in sustainable development through various case studies and applications by looking at them from generalized as well as eagle-eyed perspectives. Readers must understand that understanding sustainability is a complex process with various interconnected variables and direct and indirect parameters and hence the procedure would vary with each technology. What readers must gain is the holistic approach and the strategy utilized by industries across the globe in tackling sustainable problems.

11.2 Functionalized nanomaterial acts as a silver bullet in the arsenal

The state of sustainability cannot be reached only by focusing on the aspects on whose conjure it is defined. It can only be attained by studying the complex interactions between social systems, global systems, and human systems. The social system takes into account the factors that affect support our life such as industry, technology, and politics and human systems cover the factors such as health care, lifestyle, and well-being, whereas global systems are set of those factors linked to earth's ecosystem like the climate and energy. To understand the potential of FNMs toward sustainability, it essential to see the big picture taking into account the possible backlashes and any possible cost, which the current or the future generation would be forced to bear.

11.2.1 Reasons behind the failure of achievement of a successful sustainable development

At this stage, it is pivotal to consider answering why haven't we been successful in the endeavor so far. This can be answered by understanding the stigma and fame associated with the solutions, which are often claimed to the panacea of all problems that we face. The understanding of potential technology can be established by the interaction of the direct effects onto other systems mentioned earlier. The complex interactions within the various systems associated with sustainable development are schematically represented in Fig. 11.3.

Following may be the reasons which may have hindered the possibility of sustainable development:

1. Researchers often mix up the issues pertaining to sustainability as purely associated with global systems without taking into account the interactions the system may have with human systems [7].
2. The sustainable solutions are often considered as the "end-of-pipe" solutions without entirely eradicating the root causes, thus shifting the

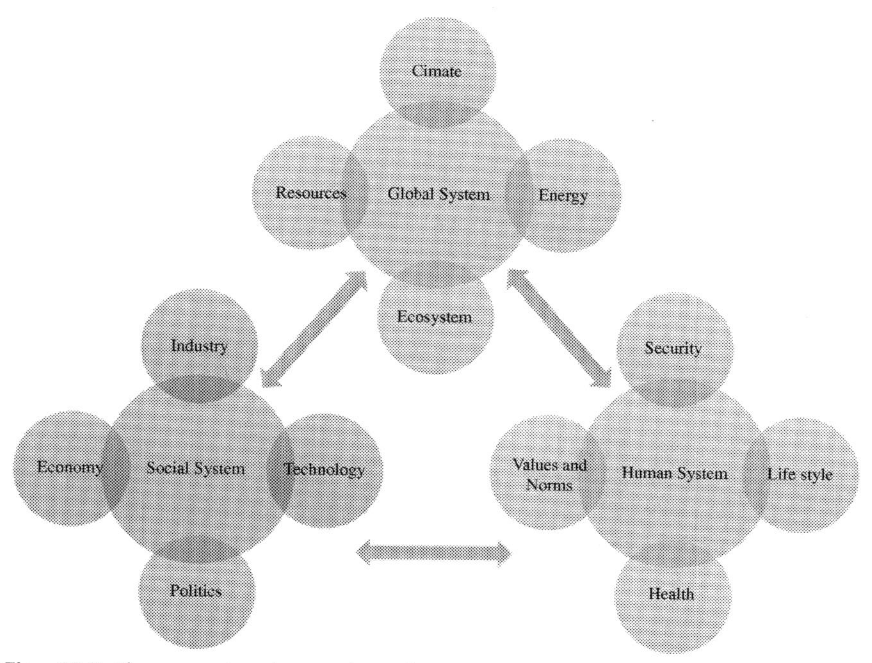

Fig. 11.3 The complex interactions between various systems associated with sustainable development.

society's focus toward mitigation rather than assessing how to prevent them in the first place [8].

3. It is fallacious to believe that the solutions provided by FNMs are the best. Such a conclusion is built on the corroded belief arising from the non availability of the comparative assessment of the next-in-line solutions to that problem [9].
4. As mentioned earlier, we must not provide a mere partial solution to a problem that may, in turn, cause adverse aftereffects. For this, a complete picture must be made by assessing the negative aspects the provided solution may have on other systems [10].
5. Most research focus on providing the potential solutions to a sustainability issue rather than working on the real solutions, which are the need of the hour [11].
6. With the notion of sustainable solutions comes the aspect of "greenwashing," which must be prevented at all costs [11].

11.2.2 Critical questions in the arena of sustainability

As research toward the search for sustainability has intensified over the years, the following questions may help to strip off any smoke screens associated with a provided solution:

1. Can FNM solutions be the answer to all sustainability issues?
2. How well can FNM-based solutions help in tacking the complex interactions between the systems?
3. How far have we succeeded in converting the potential solutions to real solutions?
4. Do FNM-based solutions have the potential to maximize benefits on the human system while having the least possible impact on the global system?

11.2.3 Understanding the potential vested in FNM-based solutions: A case study

Usually, researchers prefer to choose typical incidents of past claim making and then build a hypothetical arena to investigate a technology's assertions as hyperbole aimed to demonstrate its potential. Here, an alternate path was chosen; this case study answers the questions mentioned above in a given setting, specifically, the urban setting about sustainability. Moreover, half of the world's population lives in cities. They are faced with a pressing need for long-term sustainability. Cities have begun to respond to these problems on their own.

Phoenix has been given the dubious honor of being named the world's least sustainable city [12], is an excellent example of intervention research problems with urban sustainability. Several studies have been reported on the Gateway Corridor Community in Phoenix, Arizona, and we would be tracing the potential of FNMs using the same research methodology used by Wiek et al. [13]. The research considered the sustainability problem as a supply-demand with FNMs-based solutions placed on the supply side and the sustainability problems on the demand side. Another point of concern was to consider whether the proposed technology would solve the problems and review whether alternate technologies have an upper hand during the comparative assessment, and understand the underlying root causes and constraints.

To understand the extent of the potential of FNMs to solve sustainability problems, let us first list down the issues reported in the literature and then look onto the areas, where FNMs could be a viable solution. Deprived economy, lack of green energy systems, automobile mobility, polluted air and water, water scarcity, lack of social cohesion, child obesity and health care, waste management, and Island heat effect are the problems the gateway corridor community is left to face. Among these sustainable problems, those being solved using FNMs-based solutions are listed in Table 11.1.

The underlying causes are also a point of concern as if left unresolved, new problems would start to spur, and our resources and scientific focus would further drift onto the mitigation efforts rather than preventing the reasons for those in the first place. In most of the urban communities' reasons such as lack of attention from concerned authorities, lack of consumer activism, and potholes in the enforcement of policies are the culprits for the policy being restricted to the book and not seen on the ground. Hence, one can understand the potential within FNM-based solutions to drive an economy toward sustainability but at the same time would not consider a single solution as a panacea for all problems. Only through the amalgamation of resources and potential vested in various technologies can we turn our day-dream of a sustainable economy into ground reality.

11.3 Role of sustainable membranes with FNMs in modern society

To further elucidate the possible applications of functionalized nanomaterials-based membranes, let us look into the beneficiaries of membrane technology in critical areas such as water, energy, air, and food. Understanding the role of

Table 11.1 List of sustainable problems in gateway corridor community in Phoenix and their possible solutions related to FNMs.

Gateway Corridor Community in Phoenix, Arizona, USA		
Sustainability problems (demand)	**Few proposed solutions (supply)**	**Reference**
Energy systems	TiO_2 incorporated proton exchange membranes	[14]
	SiO_2-doped proton exchange membrane	[15]
	Pt nanoparticles on CNTs as improvised catalysts for PEMs	[16]
Air pollution and GHG management	Nanofilters for filtration of particulate matter	[17]
	FNs incorporated membranes for air filtration	[18]
	Remediation of organic pollutants from air	[19]
	Thermally remodulated membrane for CO_2 removal	[20]
Water scarcity and pollution	Incorporation of FNs in MMMs for removal of heavy metals	[21]
	Antimicrobial functionalization of membranes using NMs	[22]
	Photocatalytically functionalized membranes	[23]
Social system	Nanotechnology policies in Korea	[24]

membranes and assessing their potential is sine qua non in building the bigger picture of its role in attaining sustainability particularly in a modern society.

11.3.1 Water

11.3.1.1 Desalination and water filtration

Clean and safe drinking water is one of the cornerstones of sustainability. With population increased fourfold in the past century, the water demands have boomed sevenfold. According to a recent estimate by the World Water Council, a whopping number of 3.9 billion population would live in areas with acute scarcity of water [25]. In such a grave situation, the available water resources must be preserved and reused, and other viable resources for drinking water must be found. Thus, an aggressive thrust on water treatment technologies is required. Among the available water treatment processes, the use of membranes had been preferably owing to the nonrequirement of chemicals and its tuneable efficiency.

Historical overview

Tracing back to 150 years, Maxwell introduced his "sorting demon" to the world [26] (Fig. 11.4). A being capable of separating molecules selectively. However, even after its long journey to our present scientific space, we have not successfully synthesized an "ideal membrane."

For water treatment processes, pressure-driven membrane processes are preferred. The inclusion of nanotechnology into the picture of the water treatment process has provided researchers with varied improvisations such as better permeability, sterner resistance to fouling, and selectivity owing to the functionalization provided by the nanoparticles added in the membrane.

Pendergast et al. have reviewed the performance of FNMs in their study by taking into account two indicators, namely performance and commerciality. Using a simple marking scheme where the positive improvements were marked "+," degraded ones "−"and no changes "0," they marked most of the pressure-driven membranes used in the water treatment process. The performance enhancement indicator consisted of subsidiary indicators such as permeability, selectivity, and stability (i.e., chemical, mechanical, and thermal stability), whereas the commerciality indicator was composed of capital cost, ability to manufacture using current infrastructure, and ability to expand commercially.

The biologically inspired membranes like aquaporins showed the maximum performance improvement, whereas the inorganic-organic membranes had the highest commerciality points. It also showed that none of the FNMs were having the optimum indicators, as shown in Fig. 11.5.

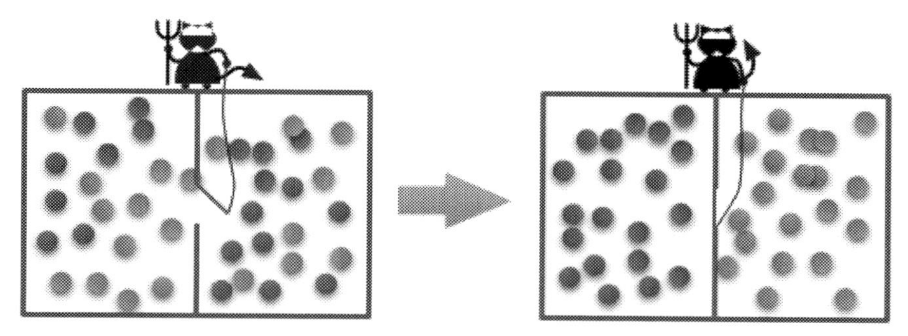

Fig. 11.4 Schematic representation of Maxwell's sorting demon.

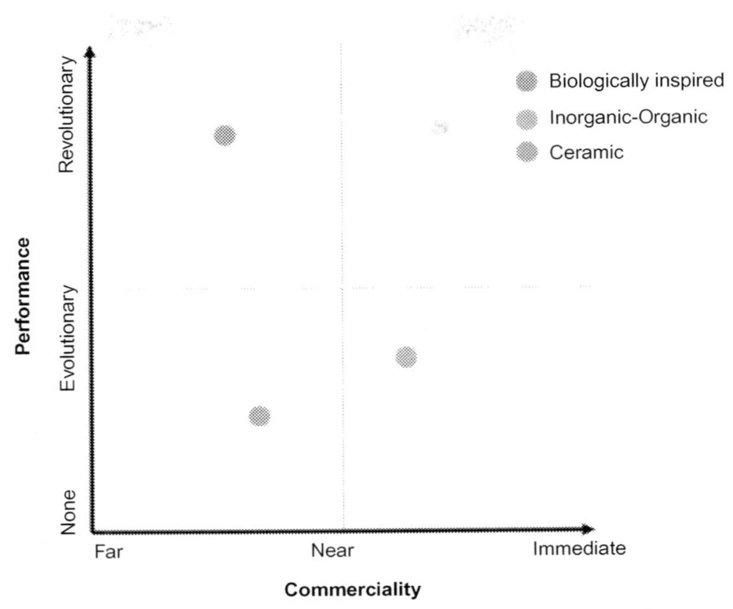

Fig. 11.5 Comparison of commerciality and performance indicators of various FNMs.

Nevertheless, researchers cautioned the study's limitations stating a "snap of time" and captures only the picture frame of scientific improvements in that time period in which the author reviewed [27].

Environmental concerns related to desalination

Desalination technologies have raised a lot of questions about their environmental impact. These are closely connected to the source and the process of energy needed to power the desalination plant, as well as the design and management of the process. Fig. 11.6 provides us with points that must be considered to evaluate the impact of a desalination technology and to analyze the sustainability of the technology.

However, it should be noted that research on the ecological implications of desalination, particularly the long-term effects on the marine biota, is limited. But considerable improvements have been there in the past decade to identify the adverse effects associated with desalination technology through evaluation of ecosystem surrounding operating plants. Table 11.2 points out some concerning points related to desalination technology majority of which had been successfully controlled and regulated owing to the strict environmental regulations implied onto such plants.

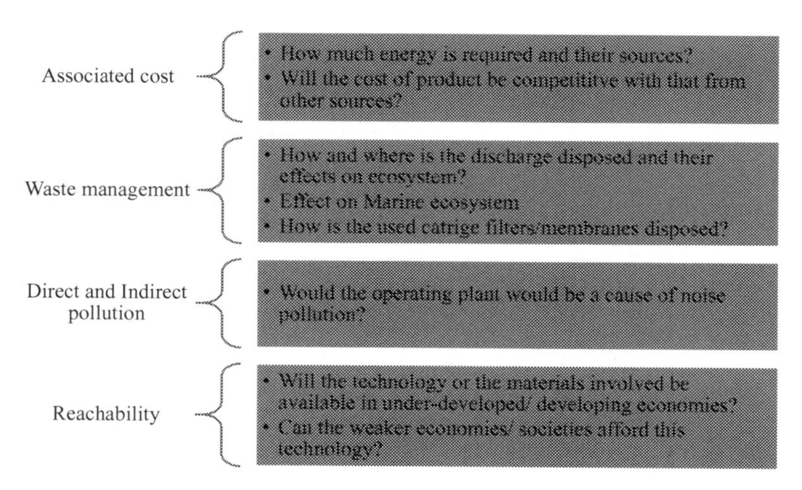

Fig. 11.6 Points to consider to evaluate the sustainability of a desalination technology [28].

Table 11.2 Points of concern related to desalination technologies' effect on environment.

Points of concern		Effect on environment	Reference
Uptake of raw material (seawater) for desalination		Small aquatic organisms are drawn into the intake systems leading to entrapment	[29]
Management of discharge	Salinity	Highly saline discharge can have adverse effect on the marine ecosystem	[30]
	Heavy metals	Detrimental for marine biota	[31]
	Chemicals	Chemicals used for antiscaling, antifoaming and cleaning purposes can lead to algal bloom	[32]

Role of nanomaterials in enhancing the sustainability of membrane technology

Desalination has been the viable option to counter the water scarcity, which the human race has started to experience and would be aggravated in the coming years. Membrane-based processes have been the pillar of desalination over the years. Nanotechnology in membranes has helped researchers in developing membranes have both high productivity and rejection [33].

Goh et al. had presented a list of questions that they stated sine qua non to understand the role of nanotechnology for the sustainability of membrane-based desalination technology, and it can be extended by understanding the actual stake FNMs have in the sustainable development of membrane-based desalination technology and how the boundaries of current technologies can be expanded while keeping sustainability in hand. Nowadays, the push is on developing membranes that favor minimal power consumption and provide high rejection simultaneously, which is why the FNMs are being explored as an alternative for conventional RO membranes. The inclusion of nanomaterials such as metals and their oxides [34,35], zeolite [36,37], silica [38–40], CNTs [41–43], and graphene-based nanomaterials [44–46] have found to be extremely advantageous. Owing to the hydrophobic and frictionless channels of functionalized CNTs, membranes possessing them have impressible water flux and salt rejection [47]. Some studies have reported a feat of increased water flux by 10-folds using functionalized graphene monolayers in membranes [48,49]. Functionalized biologically inspired nanomaterials-based membranes are an emerging area of research. Studies have shown deprived fragility [50], improvised water flux [51], and better salt rejection [52,53] in membranes incorporated with aquaporins.

Industrial plants employ thin-film composites-based membranes for the RO process. These membranes come with a set of limitations that can be countered by bringing nanotechnology into the picture. The first limitation is the exclusion of neutral solutes, which may account for the diminished quality of water [54]. FNMs can provide a solution to this by opting for size exclusion from diffusion as the transport phenomena in membranes [55]. The second flaw is the biofouling which can be attributed to the surface morphology and chemistry of the TFC membranes [56]. By inducting FNMs in operation, the Achilles' heel can be countered. The third major limitation is the intolerance of TFCs toward oxidants, thus ruling out the option of using chlorine against fouling [56]. Interestingly, FNMs provide a solution to this intolerance by introducing functionalized nanomaterials with antimicrobial and antiadhesive characteristics [57].

Thus, an amalgamation of functionalized nanomaterials and membrane technology can be the door opener for sustainable water treatment processes. FNMs can be used as mainstream membranes in the further course due to the tunability of various properties such as wettability and resistance toward fouling. Along with those, FNMs can help us create cost-efficient treatment processes without compromising their efficiency. This delicate balance would be of enormous help for third world countries, which will not be in a position to bear the cost of highly efficient but costly technologies.

11.3.1.2 Oily water

The increased energy demand craves for increased production, which has been done majorly with the help of fossil fuels. Today, even though countries have started encouraging green transportation systems, a majority of automobile industry and other industries run on fossil fuels. With soaring demands of fossil fuels, the stress on water-based transportation has increased considerably. This has led to several mismanagements and accidents such as oil spill and discharge, and the chemical nature of oils possessing toxic hydrocarbons pose a huge risk on environment [58]. A list of major oil spills in the history is provided in Fig. 11.7.

Over the coming years, the worldwide value of water purification is anticipated to rise. As a result, oily wastewater separation is both necessary and useful. Researchers and industry are being directed by tight laws and growing environmental consciousness to develop innovative technologies for separating oils from residential and industrial wastewater, marine and ocean water, and oil spill mixes [59].

Long-term environmental harm, health risks, and energy loss are all consequences of oil pollution. To separate wastewater laden with oil, many techniques have been developed such as flotation, microbial treatment, and membrane technology. A brief overview of the promising technologies in this area and their associated limitations have been summed up in Table 11.3.

Membrane technology has been the most viable option of treating wastewater containing oil but owing to several challenges such as fouling, flux, and selectivity researchers are searching for ways in enhancing their performance capabilities.

Among the pressure-driven membrane processes, ultrafiltration (UF) is one of the most efficient membrane technologies for separating oil-water mixture. UF offers a superior removal efficiency than conventional separation techniques, with no need for chemical additives and minimal energy expenses [65].

In addition to UF, RO and NF forward osmosis (FO) is a newer treatment for separation of oil laden wastewater. This method, being an osmotically driven process, uses relatively little hydraulic pressure and has numerous potential benefits, including a decreased fouling, simpler fouling removal, and greater water recovery capacity [66,67].

Excessive energy use resulted in more demand for oil, which has a number of negative environmental consequences. Membrane separation is one of the most promising separation processes for oily water. The importance of

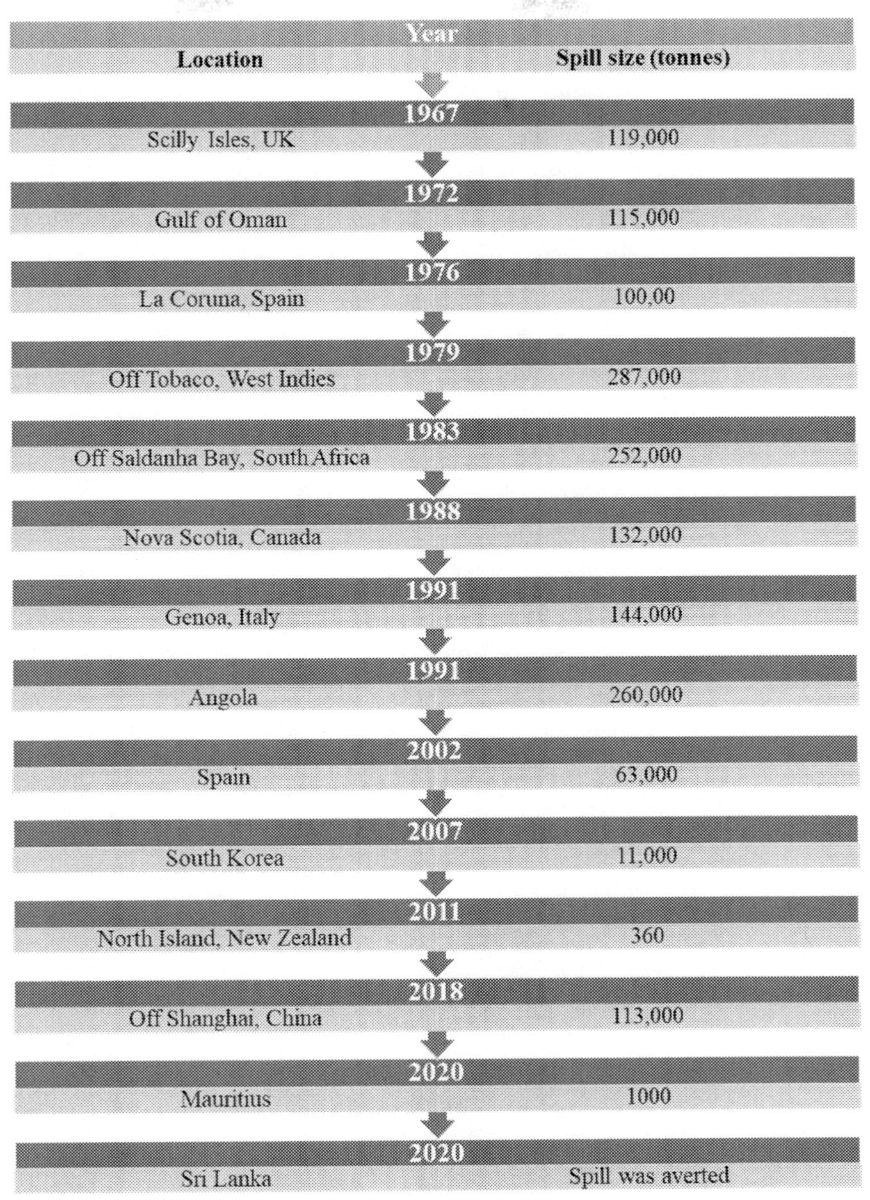

Fig. 11.7 List of major oil spills in history with their locations and the extent of the spill [58,60].

Table 11.3 List of technologies used for separating oil-water mixture with their challenges.

Technology	Separation caused by	Challenges	Reference
Flotation	Density difference	Energy consumption Operation time	[61]
Coagulation	Surface charge reduction	Cost Operation time	[62]
Microbial treatment	Microbial metabolism	Cost Oxygen-intensive	[62]
Adsorption	Physical adhesion	Frequent replacement of adsorbent	[63]
Membrane technology	Permeability of membrane	Fouling Frequent replacement of membrane	[62,64]

water for living creatures enhances the researcher's obligation to offer pure water at a low cost. Membrane fouling is the most major hindrance to the use of membrane processes. Due to interactions with numerous components in the wastewater, an irreversible decrease in flow over time occurs which in turn, increases in hydraulic resistance and thus hindering the whole process [68]. Many studies have focused on functional design membranes to decrease fouling to this point [62,69–71]. Membranes with surface modifications provide excellent antifouling characteristics. However, the homogeneity of modifications over a broad membrane area, the durability of the materials, preservation of the functionalized properties of the materials, efficiency of the fabrication method, expense of chemicals, and the stability and sustainability of the modified materials incorporate huge challenges on membrane modification (Fig. 11.8). An ideal surface alteration approach will introduce desirable functionalized nanoparticles onto the membrane surface with preferable properties like antifouling while posing minimal risks.

11.3.2 Energy

11.3.2.1 Photocatalysis

Chemicals like chlorine and other disinfectants have been taken for granted in water treatment. The resulting residuals from, which can produce carcinogenic products when they react with water. This process facilitates decontamination by using high-energy low wavelength UV protons, which is detrimental by the cell walls and DNA (Fig. 11.9) [72].

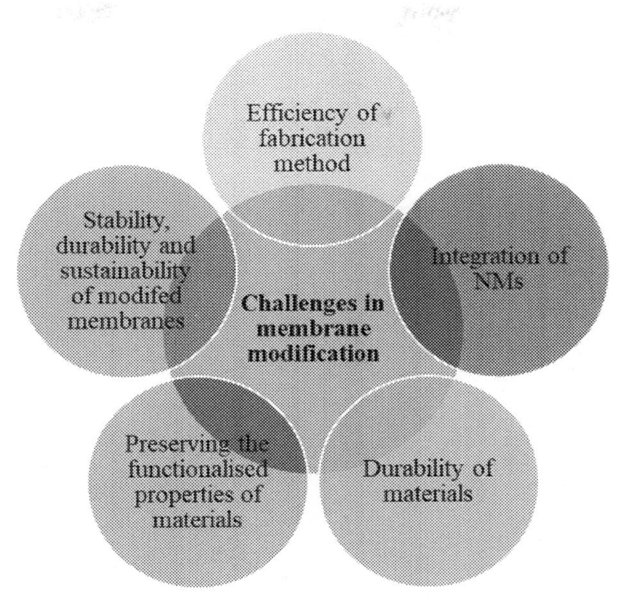

Fig. 11.8 Challenges associated with membrane modification [6].

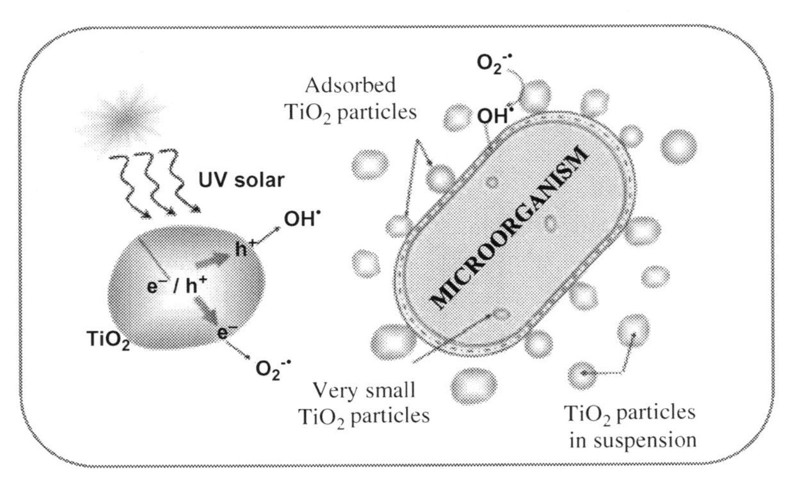

Fig. 11.9 Antibacterial property exhibited by TiO_2 [72].

Thus, removing pollutants via semiconductor-based photocatalysis with superior photoactivity, chemical stability, nontoxicity, and low cost was seen as a better alternative [73]. Titania is most commonly used in the photocatalysis industry, owing to its low cost and nontoxicity (Fig. 11.10). However, one hurdle in using photocatalysis in the decontamination of

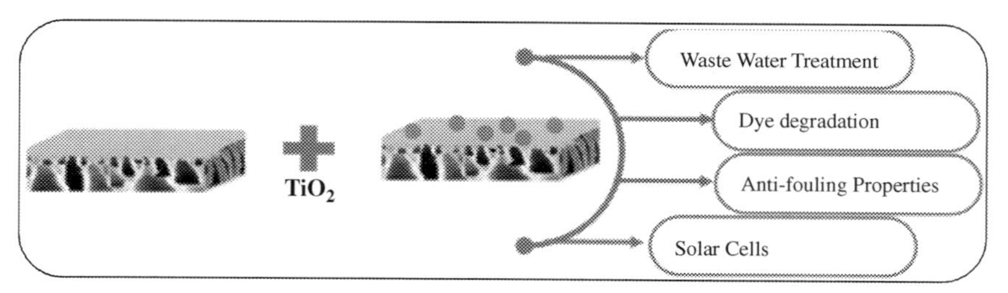

Fig. 11.10 TiO_2 incorporated membrane for photocatalysis [75].

wastewater is its inability to handle large amounts of pollutants. This inability is due to protons' constrained flux, which renders this method infeasible for commercialization [74].

11.3.2.2 Fuel cells

Another field where NPs can have huge potential is in fuel cells. However, irrespective of the field, researchers have shown that the electrochemical properties of NPs can be altered. Fuel cells are devices that, by electrocatalysis, convert chemical to electrical energy. The proton exchange membrane fuel cell is a subcategory of the fuel cell in which the fuel splits into protons and electrons, which then move through a polymer electrolyte membrane placed in between the electrodes (Fig. 11.11) [76].

Owing to the ability of PEM to allow protons by vehicle mechanism and Grotthuss mechanism and reject electrons to reach cathode avoiding short circuiting [78]. It is in this membrane where the NPs incorporated membranes find colossal potential. This potential of the membrane-nanomaterial composite can be credited to the strong interactions between the support

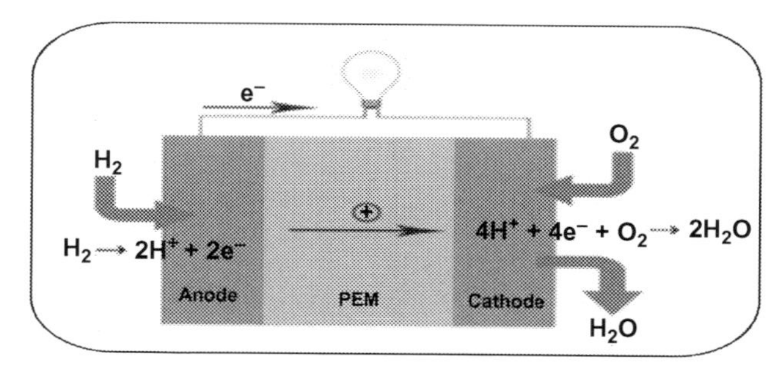

Fig. 11.11 Schematic showing the mechanism of electricity generation in PEM fuel cells [77].

matrix and the NPs. In addition to the strong interfacial interactions, the electromagnetic and barrier resistance makes nanomaterials such as TiO_2, SiO_2, and zeolite, an ideal choice for property enhancement.

11.3.3 Air

11.3.3.1 Gas separation

Membranes' gas separation characteristics have been known for more than a century. The first documented accounts are from the 19th century research studies of Mitchell, Fick [79], and Graham [80]. Then with development of composite membranes, membrane technology went on to become one of the viable alternatives in gas separation. A brief overview of historical milestones in this area is shown in Fig. 11.12.

Membrane-based separation of gaseous mixtures is a pressure-driven technique that involves forcing a gas mixture through a membrane under high pressure. The partial pressure differential between the gases in the feed and permeate drives the separation. Because it is reliant on the relative permeation rate of the gas mixture, membrane separation is regarded a nonequilibrium mechanism [80].

Fossil fuel-based power plants

Among the most difficult environmental challenges connected to global warming and climate change is the management of emissions of GHGs such

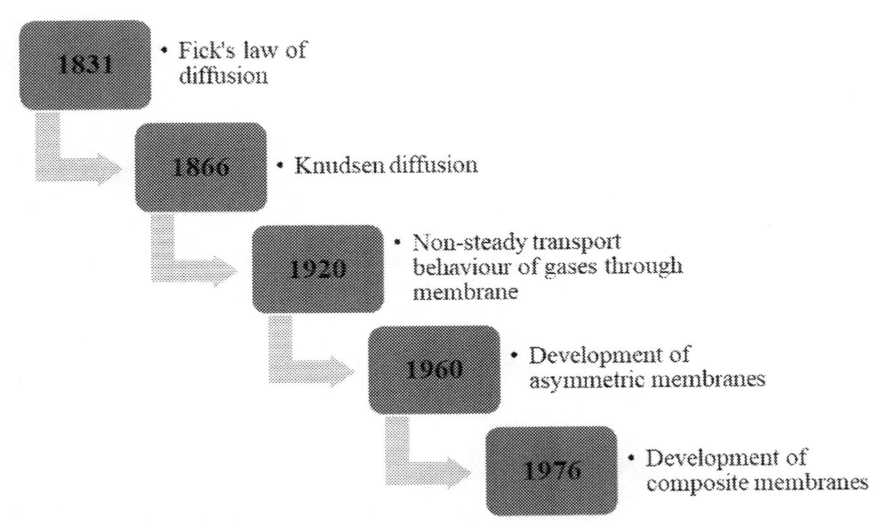

Fig. 11.12 Historical breakthroughs in membrane technology for gas separation [80].

as CO_2 and hydrocarbons [81]. CO_2 capture and sequestration (CCS) technology-based reduction of CO_2 emissions from major CO_2 producing anthropogenic activities, particularly fossil fuel-based power stations, might be a possible strategy for combating global warming [82]. A significant problem is determining a capture technique that meets the objectives of desired separation performances while imposing the smallest possible energy cost. Membrane processes are currently being investigated for CO_2 collection from power plant emissions due to their basic technical and economic benefits over rival separation methods [83–85]. Renewable energy resources such as wind, solar, and hydrogen may become another viable alternative for lowering CO_2 emissions [84]. Nonetheless, in order to meet current and future energy demands, existing alternative energy generation technologies must be pushed beyond their current boundaries and challenges, and new sustainable sources of energy must be identified.

Due to the implications of CO_2 emissions in global warming and climate change, current fossil fuel-based power plants for energy production should involve a CO_2 capture method without which averting the problems of global warming would be problematic. Developing an efficient CO_2 collection technique that can be combined in operational plants is one way. Research and past developments have shown that such a system could be imbibed into the currently operational plant at either the precombustion, oxy-combustion or postcombustion stage as shown in Fig. 11.13 [86,87].

All these stages offer their own challenges when it comes to membrane technology [89]. The postcombustion stage requires membranes with excellent thermal and chemical stability. Also, the membrane must be capable to

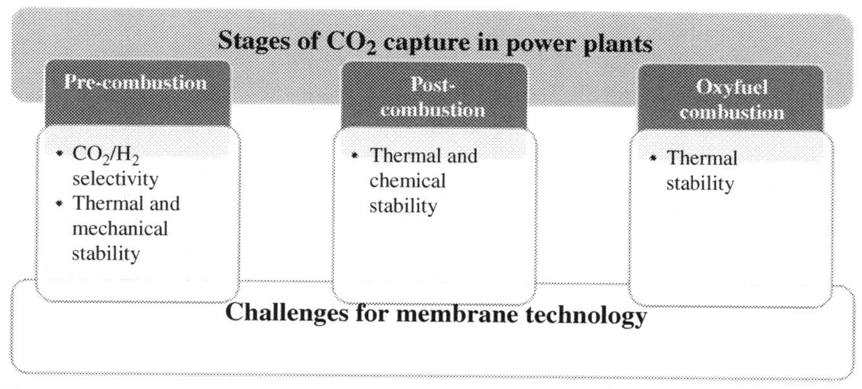

Fig. 11.13 Stages of CO_2 and their associated challenges for membrane technology [88].

undergo large amounts of gas separation and that too with appropriate selectivity of CO_2. Liquid membranes [90–93] and hollow fiber carbon membranes [94–97] have emerged as an promising membrane technology for CO_2 separation at postcombustion stage but the stability offered by these membranes still remains a concern that must be solved prior to commercialization of this technology. Whereas, retrofitting capturing system at precombustion stage demands membranes selective toward either CO_2 or H_2. Moreover, the membrane must be capable to handle high pressure and temperature. Several membranes like Pd incorporated membranes [98–100] and several thermally stable polymers coated ceramic membranes [101,102] have been developed which are hydrogen selective whereas the latter class is rare, but their capabilities at presented operating conditions like high pressure and temperature still needs to be improved.

11.3.4 Food

According to the ancient papyrus rolls, the Egyptians utilized a sort of ceramic clay mesh for wine clarification years ago. These ceramic filters could be regarded as the first food-processing membranes [103]. Commercialization of membrane production and processing has its roots in the 20th century with the commercialization of microfilters before, which membrane technology-based processing was restricted to laboratories.

The use of membrane techniques in sterilizing and clarifying is also motivated by the desire to improve the organoleptic characteristics of foods. However, owing to the intricacy of the system, selective extraction is challenging which could be attributed to the mixtures involved. As a result, extremely selective membranes are required. Fig. 11.14 shows the various applications of membrane technology in modern food-processing industries.

11.3.4.1 Diary industry

Diary industry has been the greatest beneficiary of membrane technology. Due to its nature, milk is a suitable fluid for membrane processing. Membrane technology has been used in the dairy sector since the 19th century, and it ranks second only to water and wastewater treatment as feasible alternative to more conventional technologies used in diary industry such as distillation and extraction [105–108]. The preference of membrane technology is because of its lesser energy requirement, ability to separate unwanted substances, better selectivity, and lower maintenance cost [109,110].

Membranes are used in modern dairy processing to clarify milk, enhance the concentration of chosen components, and separate certain components

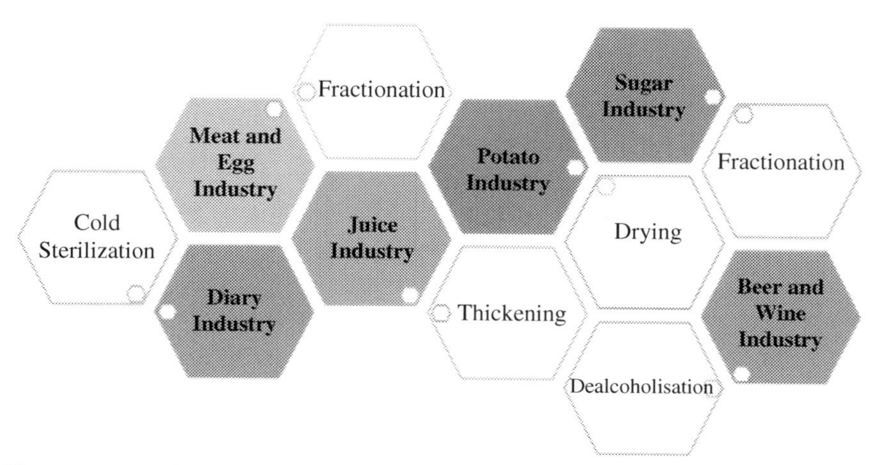

Fig. 11.14 Applications of membrane technology in specific food-processing industries [104].

from milk whose monetary values are more in nature as shown in Fig. 11.15 [111–113]. The membranes are also used in enhancing the shell life of dairy products by reducing the microbial counts using processes like microfiltration keeping the flavors intact [114]. Interestingly, studies have shown that it is possible to make milk almost completely devoid of any bacterial presence without having a significant impact on the protein content of the milk [115,116].

Membrane technology in diary industry too isn't devoid of challenges like fouling [121–124]. Membranes for dairy operations currently have a poor capacity due to a rapid flux drop caused by fouling, and procedures are energy intensive owing to the high cross-flow velocity that must be maintained to counter fouling [125]. Development of innovative applications based on the sustainable technology is thus required.

11.3.4.2 Juice industry

Juice industry is also one of the beneficiaries of membrane technology. Traditionally, thermal processing was used to enhance the shell life of products in food industry but owing to the detrimental impact of such processing on nutritional value, smell, flavor, and color of the product, membranes have been encouraged. Over the ages, studies have endeavored to come up with new ways to keep the flavor, fragrance, look, and creamy mouthfeel of

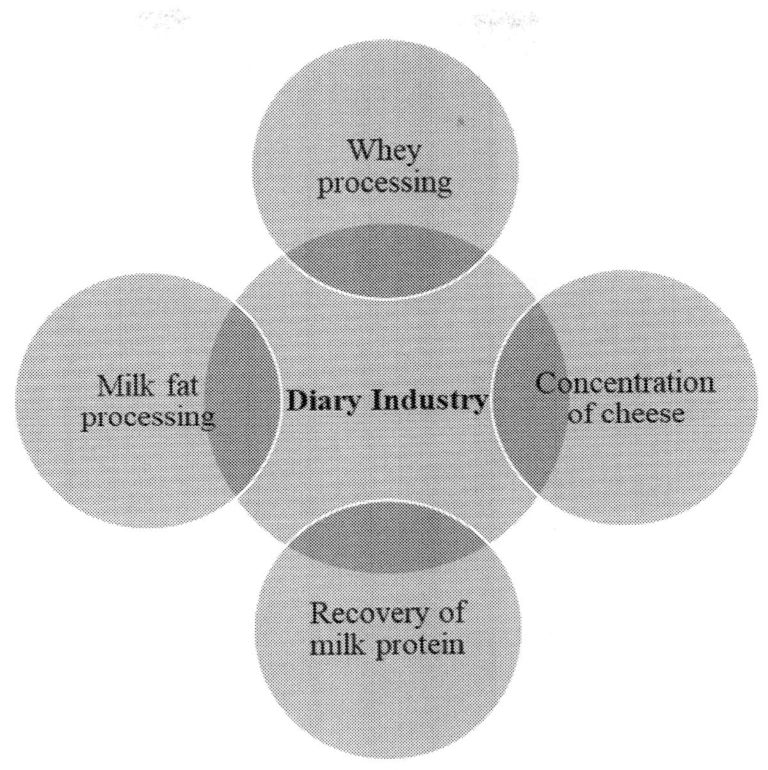

Fig. 11.15 Areas of application for membranes in diary industry [117–120].

freshly squeezed juices in the concentrate and, eventually, the regenerated juice [126]. Researchers have made significant progress in creating aroma preservation, novel process control, and product mixing procedures that will result in a high-quality concentrate that will appeal to consumers. Membrane technology has been proclaimed as the best alternative to thermal processing [127]. Emerging membrane technologies, such as membrane and osmotic distillation, as well as the integration entwinement of various techniques, may help to enhance quality and make fruit juice preparation more economically viable on a large scale. The application of membranes in juice industry can be categorized vaguely as concentration, fractionation, clarification, and restoration of aroma in juice (Fig. 11.16).

Membrane fouling is one of the primary difficulties preventing the growth of membrane operations for juice industry in the clarification of fruit juices (Table 11.4). Fouling causes a reduction in permeate flux as well as a

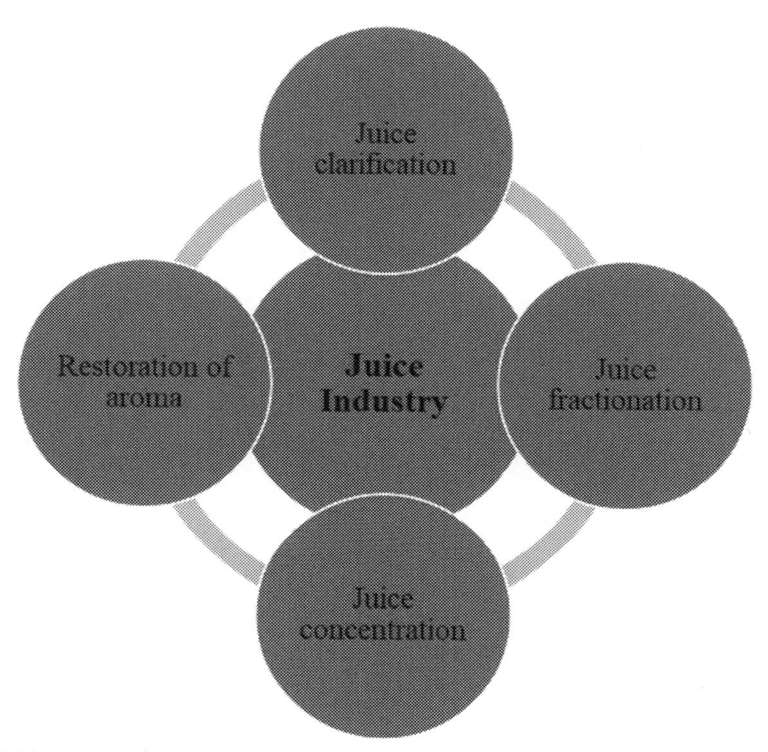

Fig. 11.16 Areas of application of membranes in juice industry [127].

Table 11.4 List of membrane processes with their applications in juice industry.

Membrane process	Application	Reference
Microfiltration	Removal of suspended solids and specific proteins	[128]
Ultrafiltration	Product concentration	[129]
Nanofiltration	Retaining compounds beneficial for health	[130]
Reverse osmosis	Purification/concentration of product prune to thermal degradation	[131]
Electrodialysis	Demineralization of product	[132]
Pervaporation	Retention of aroma	[133]

shortening of membrane life span. Cleaning solutions that are innovative can assist to broaden their use in the juice business. For the long-term expansion of membrane operations in the processing of fruit juices and drinks, new antifouling components for membranes, module design, and innovative cleaning technologies are all significant areas of research.

11.4 Road map for sustainable development through FNM-based membranes

With current water and wastewater systems, energy systems, and other conventional resources in developing nations nearing the end of their life span, there is a vital window of opportunity for FNMs to modulate the future path toward sustainability.

11.4.1 Aspects for developing new technologies in this arena

With current water and wastewater systems in developing nations nearing the end of their life span, there is a vital window of opportunity for FNMs to modulate the future path toward sustainability. Therefore, the benefit-risk trade-off is a critical parameter that needs to be analyzed with utmost caution before launching a technology into the market (Fig. 11.17).

Even with researchers including such trade-offs in their studies, we still have a considerable pothole persistent in such evaluations as the effects are trimmed only to the specific technical perspective with primary aftereffects.

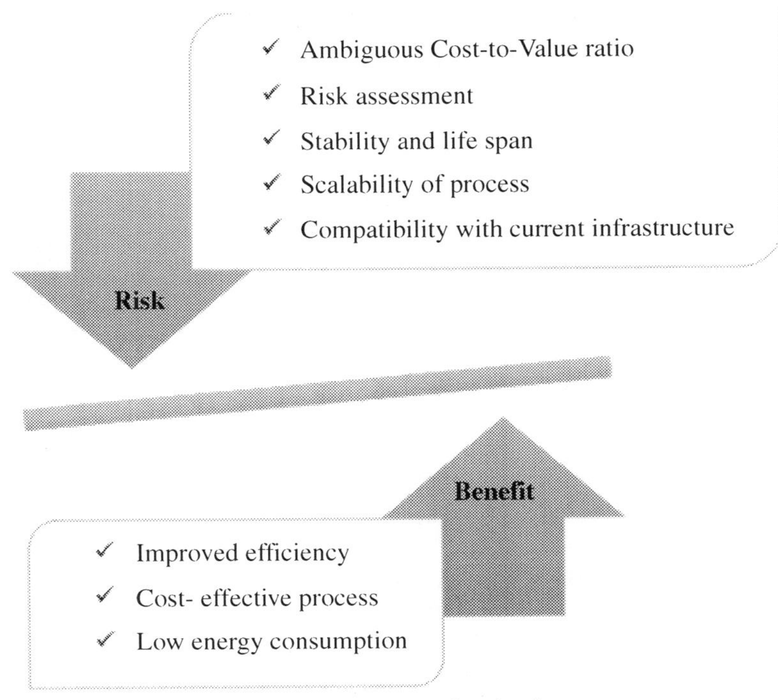

Fig. 11.17 Benefit-risk trade-off of the proposed technology.

Seldom do the studies involve secondary and indirect aftereffects into the discussion, which puts the responsibility of looking into them onto the shoulders of regulatory agencies and the end-service providers. Thus, a paradigm shift in proactively taking cross-media effects into account at the development phase itself is the need of the hour. Moreover, it must be borne in the minds of researchers that the developed technology should be able to adapt to diverse geographical, economic, and sociological barriers. Technology should not be viable to the wealthy nations but also to those sections of society who are suffering the most. Therefore, overcoming the economic perspective solely would not be enough for the future technologies but stripping off the stigmas associated with it. Furthermore, the involvement of functional nanomaterials would bring health-related stigmas hand in hand. Thus, trust-building measures and transparency toward risk assessment should be the pillars on which any technology should be developed in the coming future.

Given the varied social, cultural, and regulatory settings in which these systems are expected to be deployed, the future research on establishing FNMs-based technologies must focus on platforms that encourage usability, dependability, and clear accountability for failure.

11.5 Conclusions

Process intensification and environmental consideration has focused on the development of functionalized material for the production of membrane to offer innovative solution to resolve the scarcity in fresh water, food and energy. The proper process of optimum extraction of valuable product and conversion of waste product into useful materials has been a challenge for the environment and economy. This chapter substantiates on the understanding of the basic sustainable problems in the field of energy systems, food processing, air pollution, greenhouse gas problem, and water scarcity issues. The functionalized nanomaterial-based sustainable membranes play an important part in the modern society in the process of resolving the water crisis by enhancing the process of desalination owing to the advantage of creation of impressible water flux and salt rejection by the use of functionalized CNTs. Besides this, the effective utilization of nanomaterials for the treatment of wastewater containing oil has paved way for the development of environmentally efficient technology of sustainable membranes. The use of TiO_2-incorporated membrane for photocatalysis has improvised the photoactivity and chemical stability of the membrane with an added advantage of production of membrane with low toxicity and cost. The electromagnetic and barrier resistance makes nanomaterials such as TiO_2, SiO_2, and zeolite an ideal choice for property enhancement in fuel cells. Functionalized nanomaterial membrane also plays a significant role in the separation of valuable products from milk and for reducing the microbial counts using antimicrobial functionalized material in the matrix. Food-processing sectors depend substantially on functionalized material-enhanced membrane technology for the separation of valuable contents during its extraction process along with the consideration of economy and environment.

References

[1] M. Diallo, C.J. Brinker, Nanotechnology for sustainability: environment, water, food, minerals, and climate, in: Nanotechnology Research Directions for Societal Needs in 2020, Springer Netherlands, 2011, pp. 221–259.

[2] P.S. Goh, A.F. Ismail, Review: is interplay between nanomaterial and membrane technology the way forward for desalination? J. Chem. Technol. Biotechnol. 90 (6) (2015) 971–980, https://doi.org/10.1002/jctb.4531.

[3] H. Brundtland, Towards Sustainable Development, 1987, [Online]. Available: http://www.un-documents.net/ocf-02.htm. (Accessed 20 June 2021).

[4] M.C. Roco, From vision to the implementation of the U.S. national nanotechnology initiative, J. Nanopart. Res. 3 (1) (2001) 5–11, https://doi.org/10.1023/A:1011429917892.

[5] H. March, D. Saurí, A.M. Rico-Amorós, The end of scarcity? Water desalination as the new cornucopia for Mediterranean Spain, J. Hydrol. 519 (PC) (2014) 2642–2651, https://doi.org/10.1016/J.JHYDROL.2014.04.023.

[6] S.S. Ray, et al., Recent developments in nanomaterials-modified membranes for improved membrane distillation performance, Membranes (Basel) 10 (7) (2020) 140, https://doi.org/10.3390/MEMBRANES10070140.

[7] A. Jerneck, et al., Structuring sustainability science, Sustain. Sci. 6 (1) (2011) 69–82, https://doi.org/10.1007/s11625-010-0117-x.

[8] S. Lerner, Sacrifice Zones: The Front Lines of Toxic Chemical Exposure in the United States, The MIT Press, 2010.

[9] D. Sarewitz, R. Nelson, Three rules for technological fixes, Nature 456 (7224) (2008) 871–872, https://doi.org/10.1038/456871a.

[10] R. Seager, et al., Model projections of an imminent transition to a more arid climate in southwestern North America, Science 316 (5828) (2007) 1181–1184, https://doi.org/10.1126/science.1139601.

[11] R. Jones, Can nanotechnology ever prove that it is green? Nat. Nanotechnol. 2 (2) (2007) 71–72, https://doi.org/10.1038/nnano.2007.12.

[12] A. Ross, Bird on Fire: Lessons from the World's Least Sustainable City, Oxford University Press Inc., New York, 2011.

[13] A. Wiek, R.W. Foley, D.H. Guston, Nanotechnology for sustainability: what does nanotechnology offer to address complex sustainability problems? J. Nanopart. Res. 14 (9) (2012) 1–20, https://doi.org/10.1007/s11051-012-1093-0.

[14] A. Asla, A. Bozkurt, Preparation of proton conducting membranes containing bifunctional titania nanoparticles, in: Nanotechnology for Sustainable Development, first ed., Springer International Publishing, 2014, pp. 235–244.
[15] B. Lin, et al., Imidazolium-functionalized SiO2 nanoparticle doped proton conducting membranes for anhydrous proton exchange membrane applications, Fuel Cells 13 (1) (2013) 72–78, https://doi.org/10.1002/fuce.201200096.
[16] W. Yuan, S. Lu, Y. Xiang, S.P. Jiang, Pt-based nanoparticles on non-covalent functionalized carbon nanotubes as effective electrocatalysts for proton exchange membrane fuel cells, RSC Adv. 4 (86) (2014) 46265–46284, https://doi.org/10.1039/c4ra05120c.
[17] S. Wang, X. Zhao, X. Yin, J. Yu, B. Ding, Electret polyvinylidene fluoride nanofibers hybridized by polytetrafluoroethylene nanoparticles for high-efficiency air filtration, ACS Appl. Mater. Interfaces 8 (36) (2016) 23985–23994, https://doi.org/10.1021/acsami.6b08262.
[18] S. Chaudhary, P. Sharma, P. Chauhan, R. Kumar, A. Umar, Functionalized nanomaterials: a new avenue for mitigating environmental problems, Int. J. Environ. Sci. Technol. 16 (9) (2019) 5331–5358, https://doi.org/10.1007/s13762-019-02253-2.
[19] M. Rani, U. Shanker, Remediation of organic pollutants by potential functionalized nanomaterials, in: Handbook of Functionalized Nanomaterials for Industrial Applications, Elsevier, 2020, pp. 327–398.
[20] S. Kim, Y.M. Lee, Thermally rearranged (TR) polymer membranes with nanoengineered cavities tuned for CO2 separation, in: Nanotechnology for Sustainable Development, first ed., Springer International Publishing, 2014, pp. 265–276.
[21] W.-C. Chong, C.-H. Koo, W.-J. Lau, Mixed-matrix membranes incorporated with functionalized nanomaterials for water applications, in: Handbook of Functionalized Nanomaterials for Industrial Applications, Elsevier, 2020, pp. 15–51.
[22] J. Zhu, et al., Polymeric antimicrobial membranes enabled by nanomaterials for water treatment, J. Membr. Sci. 550 (2018) 173–197, https://doi.org/10.1016/j.memsci.2017.12.071.
[23] A. Ayral, Ceramic membranes photocatalytically functionalized on the permeate side and their application to water treatment, Membranes (Basel) 9 (5) (2019) 64, https://doi.org/10.3390/membranes9050064.
[24] D.S. So, C.W. Kim, P.S. Chung, M.S. Jhon, Nanotechnology policy in Korea for sustainable growth, in: Nanotechnology for Sustainable Development, first ed., Springer International Publishing, 2014, pp. 391–401.
[25] World Water Council, Water Security, Sustainability and Resilience, 2018, [Online]. Available: www.worldwatercouncil.org. (Accessed 24 June 2021).
[26] M. Mulder, Basic Principles of Membrane Technology, second ed., Kluwer Academic Publishers, 1996.
[27] M.M. Pendergast, E.M.V. Hoek, A review of water treatment membrane nanotechnologies, Energy Environ. Sci. 4 (6) (2011) 1946–1971, https://doi.org/10.1039/c0ee00541j.
[28] R. Baten, K. Stummeyer, How sustainable can desalination be? Desalin. Water Treat. 51 (1–3) (2013) 44–52, https://doi.org/10.1080/19443994.2012.705061.
[29] S. Miller, H. Shemer, R. Semiat, Energy and environmental issues in desalination, Desalination 366 (2015) 2–8, https://doi.org/10.1016/J.DESAL.2014.11.034.
[30] S.T. Panicker, P.K. Tewari, Safety and reliability aspects of seawater reverse osmosis desalination plant of nuclear desalination demonstration project, Int. J. Nucl. Desalin. 2 (3) (2007) 244–252, https://doi.org/10.1504/IJND.2007.013548.
[31] D. Hasson, H. Shemer, A. Sher, State of the art of friendly 'green' scale control inhibitors: a review article, Ind. Eng. Chem. Res. 50 (12) (2011) 7601–7607, https://doi.org/10.1021/IE200370V.

[32] T. Mezher, H. Fath, Z. Abbas, A. Khaled, Techno-economic assessment and environmental impacts of desalination technologies, Desalination 266 (1–3) (2011) 263–273, https://doi.org/10.1016/J.DESAL.2010.08.035.
[33] P. Wang, J. Ma, F. Shi, Y. Ma, Z. Wang, X. Zhao, Behaviors and effects of differing dimensional nanomaterials in water filtration membranes through the classical phase inversion process: a review, Ind. Eng. Chem. Res. 52 (31) (2013) 10355–10363, https://doi.org/10.1021/ie303289k.
[34] M. Ben-Sasson, X. Lu, S. Nejati, H. Jaramillo, M. Elimelech, In situ surface functionalization of reverse osmosis membranes with biocidal copper nanoparticles, Desalination 388 (2016) 1–8, https://doi.org/10.1016/j.desal.2016.03.005.
[35] M. Ben-Sasson, K.R. Zodrow, Q. Genggeng, Y. Kang, E.P. Giannelis, M. Elimelech, Surface functionalization of thin-film composite membranes with copper nanoparticles for antimicrobial surface properties, Environ. Sci. Technol. 48 (1) (2014) 384–393, https://doi.org/10.1021/es404232s.
[36] H. Dong, L. Zhao, L. Zhang, H. Chen, C. Gao, W.S. Winston Ho, High-flux reverse osmosis membranes incorporated with NaY zeolite nanoparticles for brackish water desalination, J. Membr. Sci. 476 (2015) 373–383, https://doi.org/10.1016/j.memsci.2014.11.054.
[37] J. Duan, Y. Pan, F. Pacheco, E. Litwiller, Z. Lai, I. Pinnau, High-performance polyamide thin-film-nanocomposite reverse osmosis membranes containing hydrophobic zeolitic imidazolate framework-8, J. Membr. Sci. 476 (2015) 303–310, https://doi.org/10.1016/j.memsci.2014.11.038.
[38] A. Peyki, A. Rahimpour, M. Jahanshahi, Preparation and characterization of thin film composite reverse osmosis membranes incorporated with hydrophilic SiO2 nanoparticles, Desalination 368 (2015) 152–158, https://doi.org/10.1016/j.desal.2014.05.025.
[39] M. Zargar, Y. Hartanto, B. Jin, S. Dai, Understanding functionalized silica nanoparticles incorporation in thin film composite membranes: interactions and desalination performance, J. Membr. Sci. 521 (2017) 53–64, https://doi.org/10.1016/j.memsci.2016.08.069.
[40] A.M.A. Abdelsamad, A.S.G. Khalil, M. Ulbricht, Influence of controlled functionalization of mesoporous silica nanoparticles as tailored fillers for thin-film nanocomposite membranes on desalination performance, J. Membr. Sci. 563 (2018) 149–161, https://doi.org/10.1016/j.memsci.2018.05.043.
[41] R. Das, M.E. Ali, S.B.A. Hamid, S. Ramakrishna, Z.Z. Chowdhury, Carbon nanotube membranes for water purification: a bright future in water desalination, Desalination 336 (1) (2014) 97–109, https://doi.org/10.1016/j.desal.2013.12.026.
[42] S. Roy, M. Bhadra, S. Mitra, Enhanced desalination via functionalized carbon nanotube immobilized membrane in direct contact membrane distillation, Sep. Purif. Technol. 136 (2014) 58–65, https://doi.org/10.1016/j.seppur.2014.08.009.
[43] W.F. Chan, et al., Zwitterion functionalized carbon nanotube/polyamide nanocomposite membranes for water desalination, ACS Nano 7 (6) (2013) 5308–5319, https://doi.org/10.1021/nn4011494.
[44] P.S. Goh, A.F. Ismail, Graphene-based nanomaterial: the state-of-the-art material for cutting edge desalination technology, Desalination 356 (2015) 115–128, https://doi.org/10.1016/j.desal.2014.10.001.
[45] Y. Yuan, et al., Enhanced desalination performance of carboxyl functionalized graphene oxide nanofiltration membranes, Desalination 405 (2017) 29–39, https://doi.org/10.1016/j.desal.2016.11.024.
[46] A.K. Mishra, S. Ramaprabhu, Functionalized graphene sheets for arsenic removal and desalination of sea water, Desalination 282 (2011) 39–45, https://doi.org/10.1016/j.desal.2011.01.038.

[47] H.J. Kim, et al., High-performance reverse osmosis CNT/polyamide nanocomposite membrane by controlled interfacial interactions, ACS Appl. Mater. Interfaces 6 (4) (2014) 2819–2829, https://doi.org/10.1021/am405398f.
[48] M. Hu, B. Mi, Enabling graphene oxide nanosheets as water separation membranes, Environ. Sci. Technol. 47 (8) (2013) 3715–3723, https://doi.org/10.1021/es400571g.
[49] W. Choi, J. Choi, J. Bang, J.H. Lee, Layer-by-layer assembly of graphene oxide nanosheets on polyamide membranes for durable reverse-osmosis applications, ACS Appl. Mater. Interfaces 5 (23) (2013) 12510–12519, https://doi.org/10.1021/am403790s.
[50] G. Sun, T.S. Chung, K. Jeyaseelan, A. Armugam, A layer-by-layer self-assembly approach to developing an aquaporin-embedded mixed matrix membrane, RSC Adv. 3 (2) (2013) 473–481, https://doi.org/10.1039/c2ra21767h.
[51] C.Y. Tang, Y. Zhao, R. Wang, C. Hélix-Nielsen, A.G. Fane, Desalination by biomimetic aquaporin membranes: review of status and prospects, Desalination 308 (2013) 34–40, https://doi.org/10.1016/j.desal.2012.07.007.
[52] Y. He, H. Hoi, C.D. Montemagno, S. Abraham, Functionalized polymeric membrane with aquaporin using click chemistry for water purification application, J. Appl. Polym. Sci. 135 (35) (2018) 46678, https://doi.org/10.1002/app.46678.
[53] X. Li, R. Wang, F. Wicaksana, C. Tang, J. Torres, A.G. Fane, Preparation of high performance nanofiltration (NF) membranes incorporated with aquaporin Z, J. Membr. Sci. 450 (2014) 181–188, https://doi.org/10.1016/j.memsci.2013.09.007.
[54] M.S. Mauter, I. Zucker, F. Perreault, J.R. Werber, J.H. Kim, M. Elimelech, The role of nanotechnology in tackling global water challenges, Nat. Sustain. 1 (4) (2018) 166–175, https://doi.org/10.1038/s41893-018-0046-8.
[55] J.R. Werber, C.O. Osuji, M. Elimelech, Materials for next-generation desalination and water purification membranes, Nat. Rev. Mater. 1 (5) (2016) 1–15, https://doi.org/10.1038/natrevmats.2016.18.
[56] M. Elimelech, W.A. Phillip, The future of seawater desalination: energy, technology, and the environment, Science 333 (6043) (2011) 712–717, https://doi.org/10.1126/science.1200488.
[57] C. Liu, J. Lee, C. Small, J. Ma, M. Elimelech, Comparison of organic fouling resistance of thin-film composite membranes modified by hydrophilic silica nanoparticles and zwitterionic polymer brushes, J. Membr. Sci. 544 (2017) 135–142, https://doi.org/10.1016/j.memsci.2017.09.017.
[58] SAFETY4SEA, 1978–2020: list of major oil spills, in: Green, Pollution, 2020. https://safety4sea.com/1978-2020-list-of-major-oil-spills/?__cf_chl_jschl_tk__=pmd_e7d6f22e81a72bb12ce9365902fe58bc1d6d3bad-1627150788-0-gqNtZGzNAfijcnBszQki. (Accessed 25 July 2021).
[59] B. Jing, H. Wang, K.Y. Lin, P.J. McGinn, C. Na, Y. Zhu, A facile method to functionalize engineering solid membrane supports for rapid and efficient oil–water separation, Polymer (Guildf) 54 (21) (2013) 5771–5778, https://doi.org/10.1016/J.POLYMER.2013.08.030.
[60] ITOPF, Oil Tanker Spill Statistics 2020, January 2020, [Online]. Available: https://www.itopf.org/fileadmin/uploads/itopf/data/Documents/Company_Lit/Oil_Spill_Stats_publication_2020.pdf. (Accessed 25 July 2021).
[61] A. Fakhru'l-Razi, A. Pendashteh, L.C. Abdullah, D.R.A. Biak, S.S. Madaeni, Z.Z. Abidin, Review of technologies for oil and gas produced water treatment, J. Hazard. Mater. 170 (2–3) (2009) 530–551, https://doi.org/10.1016/J.JHAZMAT.2009.05.044.
[62] F. Yalcinkaya, E. Boyraz, J. Maryska, K. Kucerova, A review on membrane technology and chemical surface modification for the oily wastewater treatment, Materials 13 (2) (2020) 493, https://doi.org/10.3390/MA13020493.

[63] A.J. Wu, X.D. Li, J. Yang, J.H. Yan, Synthesis and characterization of a plasma carbon aerosol coated sponge for recyclable and efficient separation and adsorption, RSC Adv. 7 (15) (2017) 9303–9308, https://doi.org/10.1039/C6RA26275A.
[64] M. Padaki, et al., Membrane technology enhancement in oil–water separation. A review, Desalination 357 (2015) 197–207, https://doi.org/10.1016/J.DESAL.2014.11.023.
[65] Y. He, Z.W. Jiang, Technology review: treating oilfield wastewater, Filtr. Sep. 45 (5) (2008) 14–16, https://doi.org/10.1016/S0015-1882(08)70174-5.
[66] B. Mi, M. Elimelech, Organic fouling of forward osmosis membranes: fouling reversibility and cleaning without chemical reagents, J. Membr. Sci. 348 (1–2) (2010) 337–345, https://doi.org/10.1016/J.MEMSCI.2009.11.021.
[67] M. Cheryan, N. Rajagopalan, Membrane processing of oily streams. Wastewater treatment and waste reduction, J. Membr. Sci. 151 (1) (1998) 13–28, https://doi.org/10.1016/S0376-7388(98)00190-2.
[68] Y. Zhu, D. Wang, L. Jiang, J. Jin, Recent progress in developing advanced membranes for emulsified oil/water separation, NPG Asia Mater. 6 (5) (2014) e101, https://doi.org/10.1038/am.2014.23.
[69] B. Cheng, Z. Li, Q. Li, J. Ju, W. Kang, M. Naebe, Development of smart poly(vinylidene fluoride)-graft-poly(acrylic acid) tree-like nanofiber membrane for pH-responsive oil/water separation, J. Membr. Sci. 534 (2017) 1–8, https://doi.org/10.1016/J.MEMSCI.2017.03.053.
[70] A. Venault, et al., Low-biofouling membranes prepared by liquid-induced phase separation of the PVDF/polystyrene-b-poly (ethylene glycol) methacrylate blend, J. Membr. Sci. 450 (2014) 340–350, https://doi.org/10.1016/J.MEMSCI.2013.09.004.
[71] Y. Fatma, A. Siekierka, M. Bryjak, J. Maryska, Preparation of various nanofibrous composite membranes using wire electrospinning for oil-water separation, IOP Conf. Ser. Mater. Sci. Eng. 254 (10) (2017) 102011, https://doi.org/10.1088/1757-899X/254/10/102011.
[72] J. Blanco-Galvez, P. Fernández-Ibáñez, S. Malato-Rodríguez, Solar photo catalytic detoxification and disinfection of water: recent overview, J. Sol. Energy Eng. Trans. ASME 129 (1) (2007) 4–15, https://doi.org/10.1115/1.2390948.
[73] M.M.A. Shirazi, A. Kargari, M. Tabatabaei, Evaluation of commercial PTFE membranes in desalination by direct contact membrane distillation, Chem. Eng. Process. Process Intensif. 76 (2014) 16–25, https://doi.org/10.1016/j.cep.2013.11.010.
[74] D.F. Ollis, E. Pelizzetti, N. Serpone, Destruction of water contaminants, Environ. Sci. Technol. 25 (9) (1991) 1522–1529, https://doi.org/10.1021/es00021a001.
[75] E. Bet-Moushoul, Y. Mansourpanah, K. Farhadi, M. Tabatabaei, TiO2 nanocomposite based polymeric membranes: a review on performance improvement for various applications in chemical engineering processes, Chem. Eng. J. 283 (2016) 29–46, https://doi.org/10.1016/j.cej.2015.06.124.
[76] N. Cele, S.S. Ray, Recent progress on nafion-based nanocomposite membranes for fuel cell applications, Macromol. Mater. Eng. 294 (11) (2009) 719–738, https://doi.org/10.1002/mame.200900143.
[77] Y. Ren, G.H. Chia, Z. Gao, Metal-organic frameworks in fuel cell technologies, Nano Today 8 (6) (2013) 577–597, https://doi.org/10.1016/j.nantod.2013.11.004.
[78] S. Chandra, Fast proton transport in solids, Mater. Sci. Forum 1 (1984) 153–169.
[79] D.B. Jaynes, A.S. Rogowski, Applicability of Fick's law to gas diffusion, Soil Sci. Soc. Am. J. 47 (3) (1983) 425–430, https://doi.org/10.2136/SSSAJ1983.03615995004700030007X.
[80] A. Tabe-Mohammadi, A review of the applications of membrane separation technology in natural gas treatment, Sep Sci Technol 34 (10) (2007) 2095–2111, https://doi.org/10.1081/SS-100100758.

[81] A. Brunetti, F. Scura, G. Barbieri, E. Drioli, Membrane technologies for CO2 separation, J. Membr. Sci. 359 (1–2) (2010) 115–125, https://doi.org/10.1016/J.MEMSCI.2009.11.040.
[82] N.W. Ockwig, T.M. Nenoff, Membranes for hydrogen separation, Chem. Rev. 107 (10) (2007) 4078–4110, https://doi.org/10.1021/CR0501792.
[83] R. Bredesen, K. Jordal, O. Bolland, High-temperature membranes in power generation with CO2 capture, Chem. Eng. Process. Process Intensif. 43 (9) (2004) 1129–1158, https://doi.org/10.1016/J.CEP.2003.11.011.
[84] A. Evans, V. Strezov, T.J. Evans, Assessment of sustainability indicators for renewable energy technologies, Renew. Sust. Energ. Rev. 13 (5) (2009) 1082–1088, https://doi.org/10.1016/J.RSER.2008.03.008.
[85] M.-B. Hägg, A. Lindbråthen, CO2 capture from natural gas fired power plants by using membrane technology, Ind. Eng. Chem. Res. 44 (20) (2005) 7668–7675, https://doi.org/10.1021/IE050174V.
[86] H. Yang, et al., Progress in carbon dioxide separation and capture: a review, J. Environ. Sci. 20 (1) (2008) 14–27, https://doi.org/10.1016/S1001-0742(08)60002-9.
[87] E. Favre, Membrane processes and postcombustion carbon dioxide capture: challenges and prospects, Chem. Eng. J. 171 (3) (2011) 782–793, https://doi.org/10.1016/J.CEJ.2011.01.010.
[88] X. He, M.-B. Hägg, Membranes for environmentally friendly energy processes, Membranes 2 (4) (2012) 706–726, https://doi.org/10.3390/MEMBRANES2040706.
[89] M.A. Habib, et al., A review of recent developments in carbon capture utilizing oxy-fuel combustion in conventional and ion transport membrane systems, Int. J. Energy Res. 35 (9) (2011) 741–764, https://doi.org/10.1002/ER.1798.
[90] D.H. Kim, I.H. Baek, S.U. Hong, H.K. Lee, Study on immobilized liquid membrane using ionic liquid and PVDF hollow fiber as a support for CO2/N2 separation, J. Membr. Sci. 372 (1–2) (2011) 346–354, https://doi.org/10.1016/J.MEMSCI.2011.02.025.
[91] F.F. Krull, C. Fritzmann, T. Melin, Liquid membranes for gas/vapor separations, J. Membr. Sci. 325 (2) (2008) 509–519, https://doi.org/10.1016/J.MEMSCI.2008.09.018.
[92] L.A. Neves, J.G. Crespo, I.M. Coelhoso, Gas permeation studies in supported ionic liquid membranes, J. Membr. Sci. 357 (1–2) (2010) 160–170, https://doi.org/10.1016/J.MEMSCI.2010.04.016.
[93] W. Zhao, G. He, F. Nie, L. Zhang, H. Feng, H. Liu, Membrane liquid loss mechanism of supported ionic liquid membrane for gas separation, J. Membr. Sci. 411–412 (2012) 73–80, https://doi.org/10.1016/J.MEMSCI.2012.04.016.
[94] T.C. Merkel, H. Lin, X. Wei, R. Baker, Power plant post-combustion carbon dioxide capture: an opportunity for membranes, J. Membr. Sci. 359 (1–2) (2010) 126–139, https://doi.org/10.1016/J.MEMSCI.2009.10.041.
[95] V.Y. Dindore, D.W.F. Brilman, P.H.M. Feron, G.F. Versteeg, CO2 absorption at elevated pressures using a hollow fiber membrane contactor, J. Membr. Sci. 235 (1–2) (2004) 99–109, https://doi.org/10.1016/J.MEMSCI.2003.12.029.
[96] X. He, J.A. Lie, E. Sheridan, M.-B. Hägg, Preparation and characterization of hollow fiber carbon membranes from cellulose acetate precursors, Ind. Eng. Chem. Res. 50 (4) (2011) 2080–2087, https://doi.org/10.1021/IE101978Q.
[97] X. He, J. Arvid Lie, E. Sheridan, M.B. Hägg, CO2 capture by hollow fibre carbon membranes: experiments and process simulations, Energy Procedia 1 (1) (2009) 261–268, https://doi.org/10.1016/J.EGYPRO.2009.01.037.
[98] T.A. Peters, T. Kaleta, M. Stange, R. Bredesen, Development of thin binary and ternary Pd-based alloy membranes for use in hydrogen production, J. Membr. Sci. 383 (1–2) (2011) 124–134, https://doi.org/10.1016/J.MEMSCI.2011.08.050.

[99] G. Huiyuan, Y.S. Lis, L. Yongdan, B. Zhang, Chemical stability and its improvement of palladium-based metallic membranes, Ind. Eng. Chem. Res. 43 (22) (2004) 6920–6930, https://doi.org/10.1021/IE049722F.
[100] C.A. Scholes, K.H. Smith, S.E. Kentish, G.W. Stevens, CO2 capture from pre-combustion processes—strategies for membrane gas separation, Int. J. Greenhouse Gas Control 4 (5) (2010) 739–755, https://doi.org/10.1016/J.IJGGC.2010.04.001.
[101] D. Grainger, M.B. Hägg, Techno-economic evaluation of a PVAm CO2-selective membrane in an IGCC power plant with CO2 capture, Fuel 87 (1) (2008) 14–24, https://doi.org/10.1016/J.FUEL.2007.03.042.
[102] G. Krishnan, D. Steele, K. O'Brien, R. Callahan, K. Berchtold, J. Figueroa, Simulation of a process to capture CO2 from IGCC syngas using a high temperature PBI membrane, Energy Procedia 1 (1) (2009) 4079–4088, https://doi.org/10.1016/J.EGYPRO.2009.02.215.
[103] C. Gelman, Microporous membrane technology. I. Historical development and applications, Anal. Chem. 37 (1965) 29A–37A. Accessed 25 July 2021. [Online]. Available: http://www.ncbi.nlm.nih.gov/pubmed/14292759.
[104] F.P. Cuperus, H.H. Nijhuis, Applications of membrane technology to food processing, Trends Food Sci. Technol. 4 (9) (1993) 277–282, https://doi.org/10.1016/0924-2244(93)90070-Q.
[105] B. Balannec, M. Vourch, M. Rabiller-Baudry, B. Chaufer, Comparative study of different nanofiltration and reverse osmosis membranes for dairy effluent treatment by dead-end filtration, Sep. Purif. Technol. 42 (2) (2005) 195–200, https://doi.org/10.1016/J.SEPPUR.2004.07.013.
[106] A. Saxena, B.P. Tripathi, M. Kumar, V.K. Shahi, Membrane-based techniques for the separation and purification of proteins: an overview, Adv. Colloid Interf. Sci. 145 (1–2) (2009) 1–22, https://doi.org/10.1016/J.CIS.2008.07.004.
[107] G. Daufin, J.P. Escudier, H. Carrère, S. Bérot, L. Fillaudeau, M. Decloux, Recent and emerging applications of membrane processes in the food and dairy industry, Food Bioprod. Process. 79 (2) (2001) 89–102, https://doi.org/10.1205/096030801750286131.
[108] M. Vourch, B. Balannec, B. Chaufer, G. Dorange, Nanofiltration and reverse osmosis of model process waters from the dairy industry to produce water for reuse, Desalination 172 (3) (2005) 245–256, https://doi.org/10.1016/J.DESAL.2004.07.038.
[109] S.-Y. Kim, A. Lalor, J.W. Siebert, The commercial potential of new dairy products from membrane technology, J. Food Distrib. Res. 32 (3) (2001) 24–33, https://doi.org/10.22004/AG.ECON.27582.
[110] G. Brans, C.G.P.H. Schroën, R.G.M. Van Der Sman, R.M. Boom, Membrane fractionation of milk: state of the art and challenges, J. Memb. Sci. 243 (1–2) (2004) 263–272, https://doi.org/10.1016/J.MEMSCI.2004.06.029.
[111] Z. Caplan, D.M. Barbano, Shelf life of pasteurized microfiltered milk containing 2% fat, J. Dairy Sci. 96 (12) (2013) 8035–8046, https://doi.org/10.3168/JDS.2013-6657.
[112] S. Wallen, et al., G83-678 producing milk with a low bacteria count, Hist. Mater. From Univ. Nebraska-Lincoln Ext. (1983). Accessed 25 July2021. [Online]. Available: https://digitalcommons.unl.edu/extensionhist/482.
[113] I. Pafylias, M. Cheryan, M.A. Mehaia, N. Saglam, Microfiltration of milk with ceramic membranes, Food Res. Int. 29 (2) (1996) 141–146, https://doi.org/10.1016/0963-9969(96)00007-5.
[114] G. Rysstad, J. Kolstad, Extended shelf life milk—advances in technology, Int. J. Dairy Technol. 59 (2) (2006) 85–96, https://doi.org/10.1111/J.1471-0307.2006.00247.X.
[115] J. Fritsch, C.I. Moraru, Development and optimization of a carbon dioxide-aided cold microfiltration process for the physical removal of microorganisms and somatic cells from skim milk, J. Dairy Sci. 91 (10) (2008) 3744–3760, https://doi.org/10.3168/JDS.2007-0899.

[116] M. Rosenberg, Current and future applications for membrane processes in the dairy industry, Trends Food Sci. Technol. 6 (1) (1995) 12–19, https://doi.org/10.1016/S0924-2244(00)88912-8.
[117] B. Chackravorty, D.P. Singh, Concentration and purification of gelatin liquor by ultrafiltration, Desalination 78 (2) (1990) 279–286, https://doi.org/10.1016/0011-9164(90)80047-F.
[118] A. Caron, D. St-Gelais, Y. Pouliot, Coagulation of milk enriched with ultrafiltered or diafiltered microfiltered milk retentate powders, Int. Dairy J. 7 (6–7) (1997) 445–451, https://doi.org/10.1016/S0958-6946(97)00024-1.
[119] H. Goudédranche, J. Fauquant, J.-L. Maubois, Fractionation of globular milk fat by membrane microfiltration, Lait 80 (1) (2000) 93–98, https://doi.org/10.1051/LAIT:2000110.
[120] M. Greiter, S. Novalin, M. Wendland, K.D. Kulbe, J. Fischer, Desalination of whey by electrodialysis and ion exchange resins: analysis of both processes with regard to sustainability by calculating their cumulative energy demand, J. Memb. Sci. 210 (1) (2002) 91–102, https://doi.org/10.1016/S0376-7388(02)00378-2.
[121] B.J. James, Y. Jing, X.D. Chen, Membrane fouling during filtration of milk—a microstructural study, J. Food Eng. 60 (4) (2003) 431–437, https://doi.org/10.1016/S0260-8774(03)00066-9.
[122] H. Ma, L.F. Hakim, C.N. Bowman, R.H. Davis, Factors affecting membrane fouling reduction by surface modification and backpulsing, J. Memb. Sci. 189 (2) (2001) 255–270, https://doi.org/10.1016/S0376-7388(01)00422-7.
[123] A. Makardij, X.D. Chen, M.M. Farid, Microfiltration and ultrafiltration of milk: some aspects of fouling and cleaning, Food Bioprod. Process. 77 (2) (1999) 107–113, https://doi.org/10.1205/096030899532394.
[124] G. Gesan, G. Daufin, U. Merin, J.-P. Labbe, A. Quemerais, Microfiltration performance: physicochemical aspects of whey pretreatment, J. Dairy Res. 62 (2) (1995) 269–279, https://doi.org/10.1017/S0022029900030971.
[125] P. Kumar, N. Sharma, R. Ranjan, S. Kumar, Z.F. Bhat, D.K. Jeong, Perspective of membrane technology in dairy industry: a review, Asian Australas. J. Anim. Sci. 26 (9) (2013) 1347, https://doi.org/10.5713/AJAS.2013.13082.
[126] C. Bhattacharjee, V.K. Saxena, S. Dutta, Fruit juice processing using membrane technology: a review, Innov. Food Sci. Emerg. Technol. 43 (2017) 136–153, https://doi.org/10.1016/J.IFSET.2017.08.002.
[127] C. Conidi, E. Drioli, A. Cassano, Perspective of membrane technology in pomegranate juice processing: a review, Foods 9 (7) (2020) 889, https://doi.org/10.3390/FOODS9070889.
[128] R.C. De Oliveira, R.C. Docê, S.T.D. De Barros, Clarification of passion fruit juice by microfiltration: analyses of operating parameters, study of membrane fouling and juice quality, J. Food Eng. 111 (2) (2012) 432–439, https://doi.org/10.1016/J.JFOODENG.2012.01.021.
[129] H.K. Vyas, P.S. Tong, Process for calcium retention during skim milk ultrafiltration, J. Dairy Sci. 86 (9) (2003) 2761–2766, https://doi.org/10.3168/JDS.S0022-0302(03)73872-7.
[130] N.A. Arriola, G.D. dos Santos, E.S. Prudêncio, L. Vitali, J.C.C. Petrus, R.D.M.C. Amboni, Potential of nanofiltration for the concentration of bioactive compounds from watermelon juice, Int. J. Food Sci. Technol. 49 (9) (2014) 2052–2060, https://doi.org/10.1111/IJFS.12513.
[131] R.L. Merson, G. Paredes, D.B. Hosaka, Concentrating fruit juices by reverse osmosis, Polym. Sci. Technol. 13 (1980) 405–413, https://doi.org/10.1007/978-1-4613-3162-9_27.

[132] L.J. Andrés, F.A. Riera, R. Alvarez, Skimmed milk demineralization by electrodialysis: conventional versus selective membranes, J. Food Eng. 26 (1) (1995) 57–66, https://doi.org/10.1016/0260-8774(94)00042-8.
[133] T. Schafer, G. Bengtson, H. Pingel, K.W. Boddeker, J.P.S. Crespo, Recovery of aroma compounds from a wine-must fermentation by organophilic pervaporation, Biotechnol. Bioeng. 62 (4) (2000) 412–421, https://doi.org/10.1002/(SICI)1097-0290(19990220)62:4%3C412::AID-BIT4%3E3.0.CO;2-R (Accessed 25 July 2021).

CHAPTER 12

Future prospects of sustainable membranes

Deepshikha Datta[a], K.S. Deepak[b], and Bimal Das[b]
[a]Department of Chemical Engineering, GMR Institute of Technology, Rajam, India
[b]Department of Chemical Engineering, National Institute of Technology Durgapur, Durgapur, India

12.1 Introduction

Many countries have made progress in economic growth over the last two decades. The current global energy crisis stems not just from dwindling fossil fuel sources, but also from the environmental consequences of the whole energy life cycle and from processes of extraction to emissions, waste management, and recycling. Furthermore, water and energy are critical resources for the economic, social, and cultural fabric of any economy. These resources have long been assumed to be plentiful. Their demand rose as a result of the rise in population and the advancements brought about by the industrial revolution, and scarcity is now an evident effect. Various obstacles, such as geographical disparities in natural resource availability, sensitivity to the environmental consequences of fossil fuel-based energy, rising water shortages, and varying economic policies, all limit the process of sustainable growth. Even though certain sustainable strategies are being implemented, many have failed to be commercialized. Although the efficiency of the process is still a significant priority, nanotechnology entwined with membrane technology is a promising combination.

Membrane technologies have recently begun to play an increasingly important role in the development of infrastructure for sustainable energy, particularly in the water and energy sectors [1]. Membrane technology has considerably improved our ability to reorganize manufacturing processes, safeguard the environment and public health, and develop new technologies for long-term growth. Membrane technology is a new technology that is becoming more prevalent in our lives. Although the oldest known research of membrane phenomena may date back to the middle of the 18th century, a significant breakthrough for commercial uses of synthetic membranes began in the 1960s [2]. The development of innovative or enhanced membrane materials and membranes with superior chemical, thermal, and mechanical

Membranes with Functionalized Nanomaterials
https://doi.org/10.1016/B978-0-323-85946-2.00012-6

stabilities, higher permeability and tunability, along with increasing affordability and accessibility, continues to expand the spectrum of membrane technology applications in various processes such as desalination [3], food processing [4], energy harnessing [5], energy storage [6], and air purification [7]. Even after such developments, there is a need to reduce prices and increase accessibility, which can be accomplished through advancements in membrane-based technology.

12.2 Changing dynamics of the world

For a generation with an exploding population, with the poor struggling to make a living by treasure hunting for basic fundamental facilities like food, clean air and water, and affordable health care, sustainable development is the focal point. Sustainability is the only way of both understanding and solving the problems we face in this century.

12.2.1 The Post Brundtland World

The gigantic global economy is accruing at a rate of 3%–4% per year [8]. But at the same time, this economy is uneven and unequal with some sections enjoying a rich economic status, whereas others are struggling to live by coping up with the chits and bits of fundamental facilities. An estimation of a billion people living in grave poverty with no safe water, air, shelter, or health care should be a lesson taught to all and an eye-opener for gearing up our efforts for sustainable development [8]. This gargantuan economy is not only irregular, but it is also acting as a knot to the bottlenecks in the environmental resources. With the growing population, the irregularities are turning more hostile and pulling more people into the trenches of poverty.

Today, we are facing a multifaced trouble/problem with the changing earth's climate, greenhouse gas emissions [9], increasing energy requirements [10], water and air pollution [11], and expensive health care. What impact they would have is a debatable question with thousands of known and unknown interactions in place, but the seriousness of that gruesome impact is a universally accepted theory. What we lack is implementing or catering in proportion to the seriousness associated with them. Thus came the need for sustainable development which tries to make sense of the complex interactions between the social, human, and global systems. In addition to these systems came the aspect of governance in the picture. Such a complex problem with each of these systems having its own inherent turbulence requires the need for a collaborative approach [12]. The mere idea of being able to strip down the problem related to sustainability to a panacea is vague and erroneous. For ages, sustainability had been associated with global economic growth till the Kyoto agreement which brought the environmental perspective into the picture [13].

12.2.2 Growing volatility in the world: A ticking bomb

Despite the widespread acceptance of globalization in nearly all aspects of social, political, and economic life, various stakeholders like the policymakers, academicians, and the community leadership remain narrow-minded [15]. Our old mindset of acknowledging the human and environment as isolated spheres has been the biggest smokescreen leadership is facing in their fight for sustainability [16]. The aspect of sustainable development was brought to take into account the interactions between these systems which had been wrongly thought to be nonintegrating. The pressure built up on the depleting resources by the exploding population represented by infographics in Fig. 12.1 has made the world more volatile and has further grimed the chances for a sustainable future.

The groundbreaking publication of "Our Common Future" made sustainable development a common word in our vocabulary which was then followed by lakhs of publications and soon became the focal point of policymaking [17]. A boost to sustainable development came with the United Nations Conference on Environment and Development in Rio de Janeiro in 1992 [18] and World Summit on Sustainable Development in Johannesburg in 2002 [19]. A world commission came up with the definition of sustainable development in a report in 1987 [20]. But its holistic and generalized definition became both its strengths and its weakness with various stakeholders coming up with their own versions plagued with profits [21]. Regional differences in resources, a varied sense of seriousness toward the issues, and the narrow-minded perspectives of various stakeholders have added to the burning fire of unsustainability.

12.2.3 Sustainable development is a cul-de-sac or a clue to a Pandora's box

Our world has been blindfolded with widening differences in equality, income, availability of basic facilities, snollygosters, and the ostensible implementation of sustainable policies. In such a dog-eat-dog world with growing volatility, the feasibility of sustainable development seems to be a daydream.

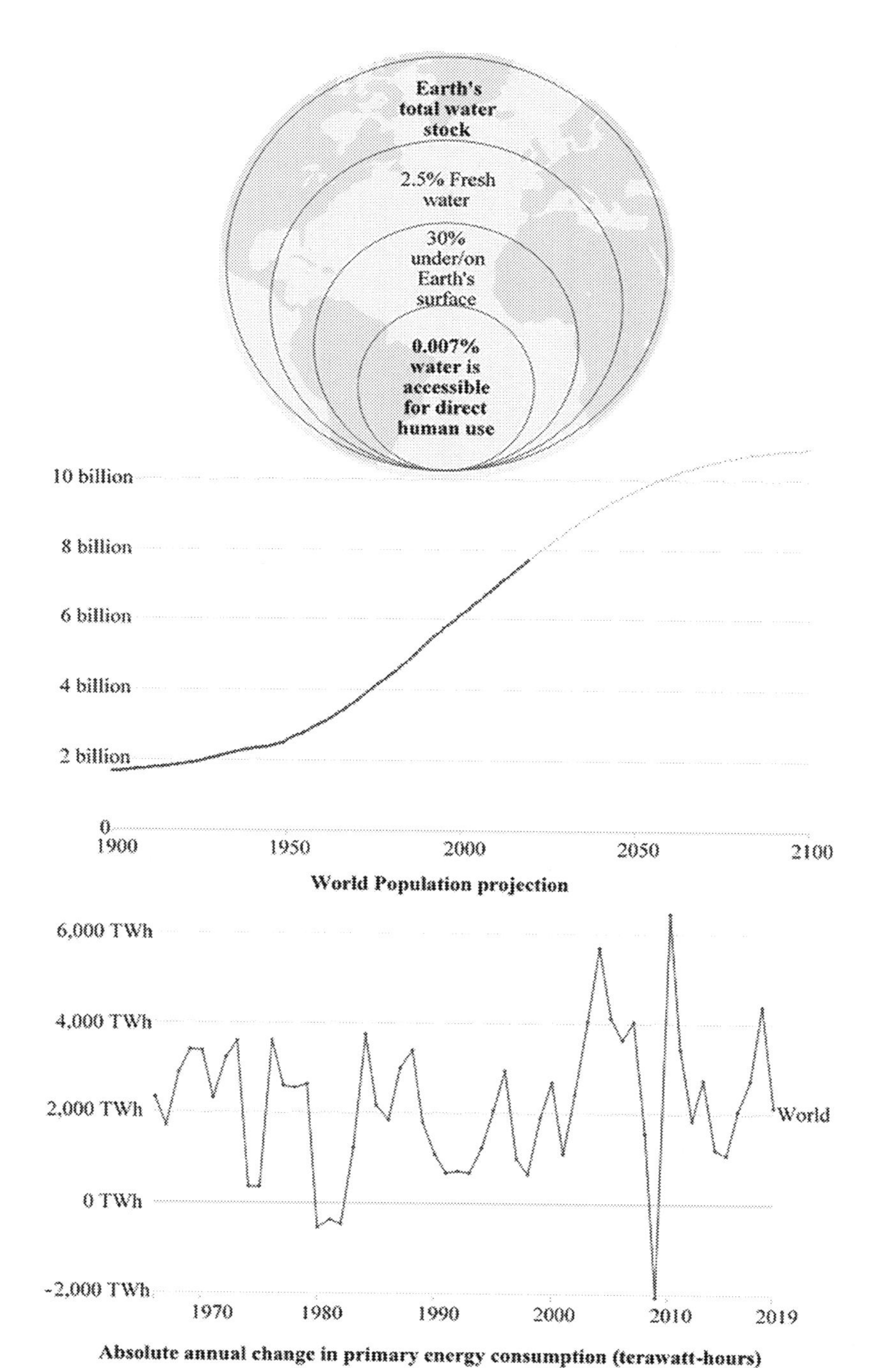

Fig. 12.1 Graphical and schematic representation of water scarcity, increasing world population, and energy consumption over the years [14].

Several thinkers have penned their pessimism toward the success of this topic [22–24]. Given the complex interactions between the systems, exploding population, and deteriorating resources, sustainable development can be referred to as one of the most complex problems humanity has ever faced and hence requires a paradigm shift in the strategies from the normal course as shown in Fig. 12.2. These are complicated issues with scientific roots that lack the global public literacy required to address them. In chaotic, non-linear, complex systems, these are concerns of enormous uncertainty. This is an intergenerational issue for which we are unprepared due to tradition. It touches on vital aspects of our economy, such as energy, transportation, infrastructure, and food supply, all of which require significant technical upgrades.

12.3 The path ahead

One of the challenges faced today is to refer to the problems and potential in the energy and water sectors as separate entities. The entwinement of relations between these sectors has to be kept in mind while developing future strategies or deciding the fate of any novel technology.

The interdependence of the water, food, and energy sectors has recently been the focus of research (Fig. 12.3). One of the major issues policymakers face is to consider them as isolated sectors. For instance, the energy sector is the largest user of water, whereas the water sector utilizes 0.2% of the total energy consumption in the United States [27]. Although the energy utilized in the water sector is relatively small, it would still be a huge boost in energy efficiency if we could reduce the same.

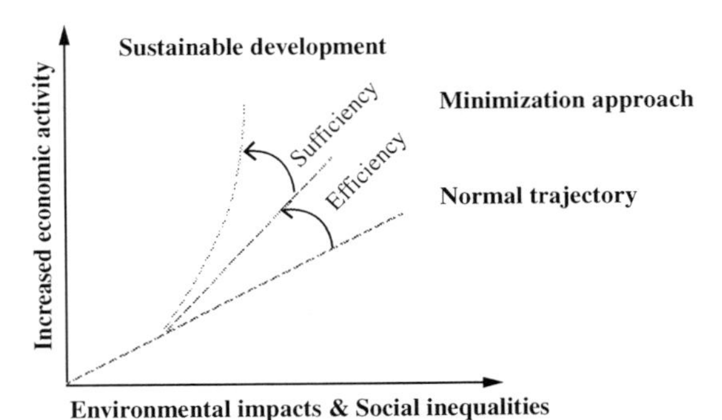

Fig. 12.2 Shift in approaches required for sustainable development [25].

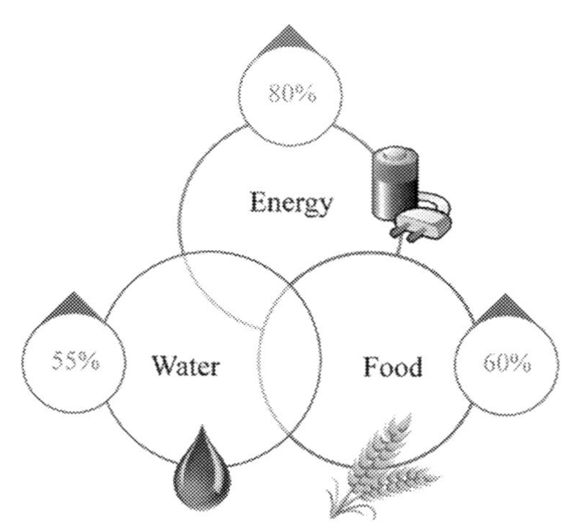

Fig. 12.3 Energy-water-food interrelations and their projected increase in demand by 2050 [26].

The realm of scientific discovery and scientific validation difficulties is vast and enthralling. Science and technological advancements are the vision of the future. The current direction and scope of membrane science research is quite broad, with an umbilical cord connecting it to the global water effort and drinking water treatment. Developed and developing nations are preparing for new challenges and visions in order to address global water shortages and crises. The characteristics of a larger holistic vision of providing clean drinking water to the general public are built on environmental sustainability and global water initiatives. In the picture of membrane science, several authors have outlined the major challenges researchers have to face in their ultimate goal toward sustainability as shown in Fig. 12.4.

12.3.1 Sustainable membranes against water scarcity

The incessant population expansion as well as fast industrialization is continually exerting pressure on global water consumption, culminating in an unsettling scenario in which the available freshwater supply no longer matches these trends. Viewing the current scenario of water scarcity and ongoing water pollution, an extensive amount of research has been put into membrane-based technologies. Currently, reverse osmosis, forward osmosis, ultra-filtration, nanofiltration, and membrane distillation are being used in desalination and other treatment processes.

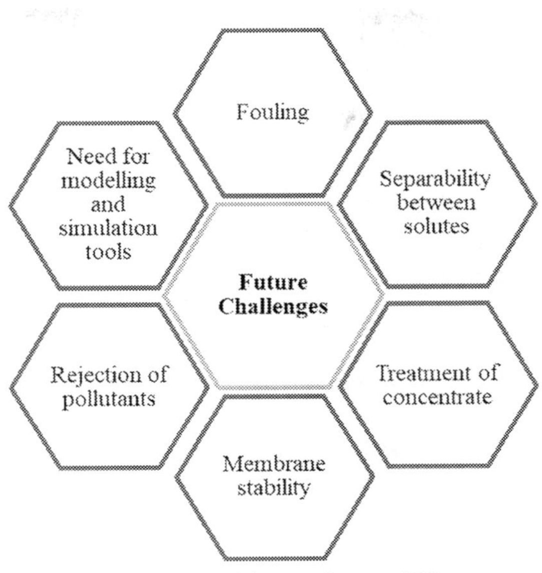

Fig. 12.4 Future challenges in sustainable membranes [28].

12.3.1.1 Desalination

In places with scarce drinkable water resources like coastal regions and isolated places, desalination has long been a viable alternative for providing clean drinking water. With 39% of the world's population living within 100 km from the sea, desalination has a vibrant market ahead [29,30]. Its undependability toward climatic conditions and other uncontrollable parameters has added to its frame and is now having a global growth rate of 55% [31]. The growth of desalination capacity is not solely because of the increased water demand but also due to the considerable reduction in desalination cost as a consequence of the major technological breakthroughs that have resulted in curbing the cost of desalinated water to make the cost comparable with other water sources. In certain select regions, desalination has now been able to successfully compete with conventional water resources [32].

Although desalination is more expensive than traditional freshwater treatment, the cost of desalination, particularly RO, is falling, while the expense of finding new freshwater sources of potable supply is rising or becoming impossible. Membrane prices have decreased dramatically in recent decades. Furthermore, as technical advancements lower the cost of equipment, the overall relative plant costs are projected to fall. Desalination, which was formerly a costly option to provide drinkable water, has become a feasible solution as a result of this trend [33,34]. Despite its capability in dealing with water shortage, desalination has been erroneously framed as the most energy-intensive water treatment technique, having negative effects in terms of high economic costs and energy demands [35]. While thermal distillation desalination methods have been limited in their uses due to their high power consumption, which makes them prohibitively expensive, membrane-based desalination has been actively pushed as the primary and mainstream desalination technique [36]. As nanotechnology advances, the focal point among the scientific fraternity has shifted from basic discovery and characterization to exploring the possibilities offered by nanotechnology and its resulting nanomaterials as a panacea to addressing the issues of world sustainability, specifically in a wide range of key areas pertaining to desalination of water [37].

Membrane fouling is a significant and unavoidable problem in every membrane process. Lower membrane fouling translates to better water production, less cleaning, and a longer membrane life span, as well as lower capital and operating expenses. The uses of pretreatment technology, operation optimization, and periodic membrane cleaning are all traditional remedies for membrane fouling. Apart from the conventional remedies, membrane modification is considered to be the most viable option as shown in Fig. 12.5. Tons of research which had been poured into identifying the

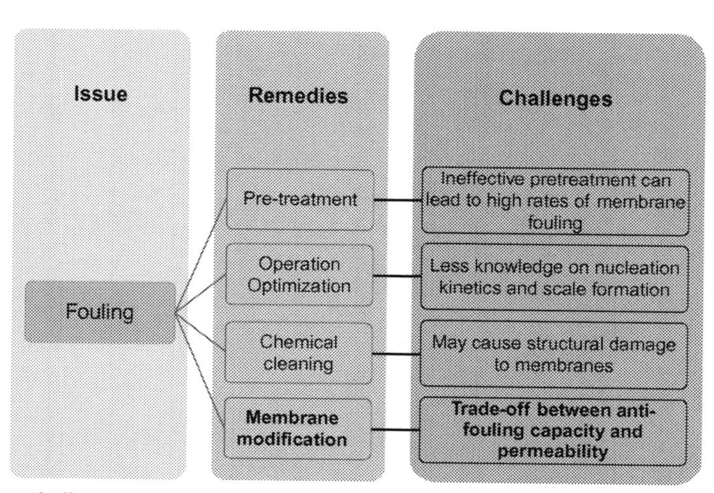

Fig. 12.5 Challenges associated with remedies for membrane fouling.

parameters that affect fouling have provided us with their chemical structure and their morphology being the significant properties.

Even though preliminary studies have shown that alterations can significantly improve antifouling capabilities, they consequently have a negative effect on permeability. Furthermore, many of them are expensive, complex, and restricted to laboratory scale. Moreover, while most methods are effective in delaying the early phases of fouling, they are not able to prevent it in the long run.

Although nanotechnology has been a relief for years of research in overcoming the hurdles possessed by membranes-based processes like permeation, rejection, and fouling, eliminating such issues would require the fabrication and synthesis of newly advanced materials capable of tackling them in an efficient and yet pocket friendly manner. Modulating the pore sizes, transport mechanism, and the functionalities associated with such materials would truly open up new doors of potential by enhancing their separability and recoverability which is the sine qua non in our fight for sustainability.

The usage of traditional RO membranes has been severely limited due to poor water recovery and high-pressure operations, which in turn generate significant volumes of liquid waste. Industry and municipal suppliers, as the end users of the desalination business, are anxiously seeking solutions from the scientific community to fill these gaps. Thus, to provide a feasible method for long-term membrane desalination, synthesis of membranes with improvised characteristics like minimal energy consumption and cost and enhanced salt rejection capabilities is necessary. The amalgamation of nanomaterials and polymers has had a kick effect onto further research. But developing new materials with better stability and tunability would be a huge success as we require materials that could adapt to the diverse atmospheric conditions where they would be used.

By considering membrane desalination technology, achieving sustainability entails addressing the exciting transition with the fewest negative effects on the environment, which means less energy waste, fewer materials, and less chemicals. Integrating the newly developed membrane components into the existing desalination system might provide the finest options for achieving the maximum sustainability advantages and optimal cost savings. To guarantee that the nanotechnology-based desalination business sectors grow and expand, a suitable climate must be created that allows new commercial entities to take the required risks to turn these new inventions into marketable goods. To create such an environment, public leadership, private sectors, and research should work hand in hand by framing a holistic pathway that would guide the stakeholders in developing newer technologies and implementing them keeping the societal needs in mind and respecting the regulatory standards.

12.3.1.2 Wastewater treatment

Water is one of the most important resources that maintains and nurtures human existence on this planet, and it is readily available in our surroundings. Agriculture, industries, and households are the three major consumers of water with the former two sectors accounting for 100 times more water than the latter [38–40]. In many parts of the globe, desalination and wastewater reuse are two feasible options for ensuring water security. To be considered "sustainable," they must be cost-effective, ecologically friendly, and socially acceptable. The membrane-based technology for water desalination and wastewater treatment has grown in popularity since it satisfies sustainability standards.

The membrane-based technology has been evolving with newer advanced technological approaches toward sustainability which has been showcased in Fig. 12.6. Zero discharge and reusability of membranes have been the most viable options to achieve circular economy. The present linear nature of this technology can be attributed to the materials which are disposed of after their life cycle and the wastes that are generated as part

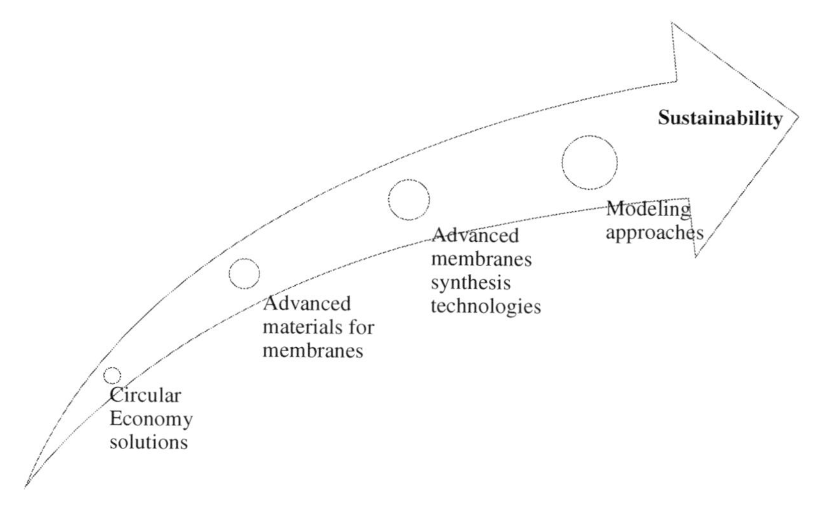

Fig. 12.6 Budding pathways and approaches in membrane-based technology for sustainability [6].

of the process. In membrane-based desalination technologies, future circular economy trends must assure waste minimization. Although membrane reuse has been found as a feasible option, it is not without its own set of problems that must be addressed. Biofouling is one of the elements that makes membrane reuse difficult and has been dealt with in detail in the previous section.

Brine treatment and energy consumption

Seawater and brackish water desalination facilities generate brine, which is made up of chemicals, salts, and precipitates. Desalination plants frequently discharge brine into the environment at high temperatures. It has been noted that unsustainable brine discharge can have significant environmental consequences [41,42].

Cost is another factor which must be considered while developing a sustainable solution. Since a sustainable technology would also be used in developing nations with a poor economy, such solutions should be affordable for the governments to implement them. Desalination, especially from saltwater resources, continues to be a very energy-intensive process for the production of drinkable water, despite the fact that its power needs have been cut in half over the previous two decades. The two main technologies that drive the desalination industry are thermal and membrane procedures. In the thermal desalination process, heat is used to evaporate pure water from its salt mixes, whereas in commercial membrane desalination methods, electric energy is employed to power high-pressure feed pumps that filter away dissolved particles. Membrane technologies, particularly RO, are favored over other systems because of their reduced energy demands. The specific energy consumption (SEC) of various methods varies greatly, and this number changes even more for a given technology depending on the process. An overview of the SEC of various technologies is shown in Fig. 12.7 and the share of electricity among the total cost associated with membrane technology is shown in Fig. 12.8. The salinity and cost of the operation is interconnected as a higher concentration would require more pressure and hence more power consumption.

12.3.1.2.1 Reutilization of wastewater

The reutilization strategy has been the blue-eyed approach for attaining sustainability. The water produced as a by-product of industrial processes or used in industries can be reutilized after treatment in various areas like agriculture and mining as shown in Fig. 12.9.

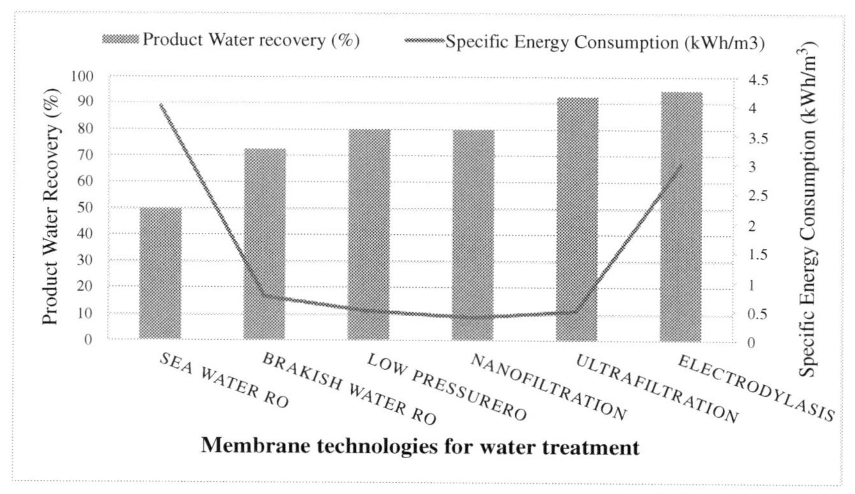

Fig. 12.7 Product water recovery (%) and specific energy consumption (kWh/m^3) of prominent membrane technologies in water treatment [43].

The management of produced water is one of the hurdles that the oil and gas industry has in terms of safeguarding human health and the environment. The amount of water produced by the oil and gas industry is three times the amount of oil produced by it [45]. The treatment of generated water for reuse and recycling is a viable alternative for its management, since it has the potential to become a useful and harmless product. Owing to the complex physicochemical composition of produced water, which can alter over

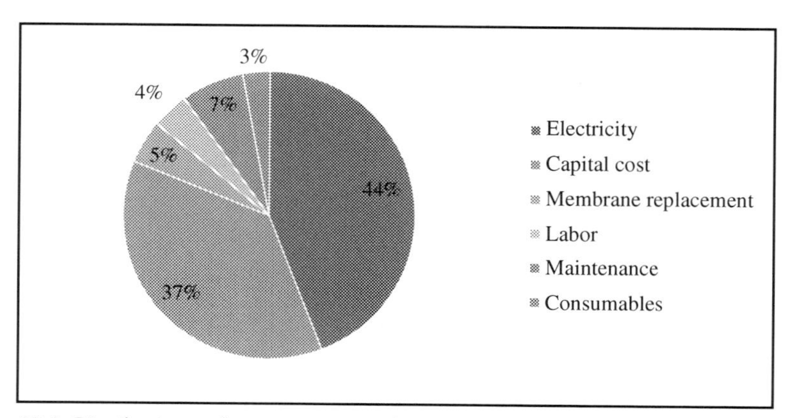

Fig. 12.8 Distribution of cost associated with membrane technology for water treatment [44].

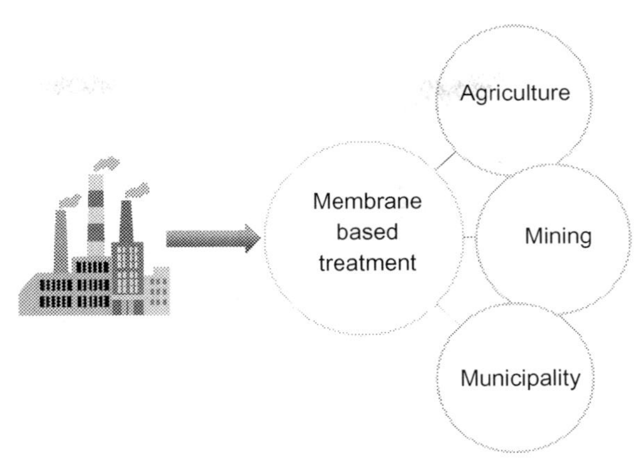

Fig. 12.9 Possible reutilization areas of treated wastewater.

time and from well to well, produced water is difficult to reuse. Organics and solids are dissolved and suspended in produced water which makes membrane technology increasingly essential in the treatment of wastewater. The high TDS content limits the types of desalination methods that may be used. Because of the enormous hydraulic pressure necessary to overcome the osmotic pressure of high-salinity generated water, RO is not a viable solution for desalinating saltwater [43,46].

As a pretreatment step, membrane processes such as microfiltration and ultrafiltration have been integrated and used effectively in separating suspended particles, macromolecules, and oil, while the amalgamation of nanofiltration and reverse osmosis has been effective in treating produced water for higher water quality standards that are potable and can be used in agriculture [47–49]. Membrane fouling is a key stumbling block for membrane filtration operations, especially when the raw water includes high levels of natural organic matter, and hence is unable to be treated with a single membrane process. Thus, hybrid/integrated membrane processes are being encouraged in the water treatment plant as they provide superior characteristics such as less fouling proficiency and the ability to remove a high concentration of organic content. A schematic representation of the challenges faced by single/conventional membrane processes, which has caused a shift toward integrated/hybrid membrane processes and the improvements or challenges these membranes could solve in the future, is shown in Fig. 12.10.

Direct supplies of fresh water were once inexpensive, making the use of more expensive desalinated water in agriculture unnecessary. Desalinated

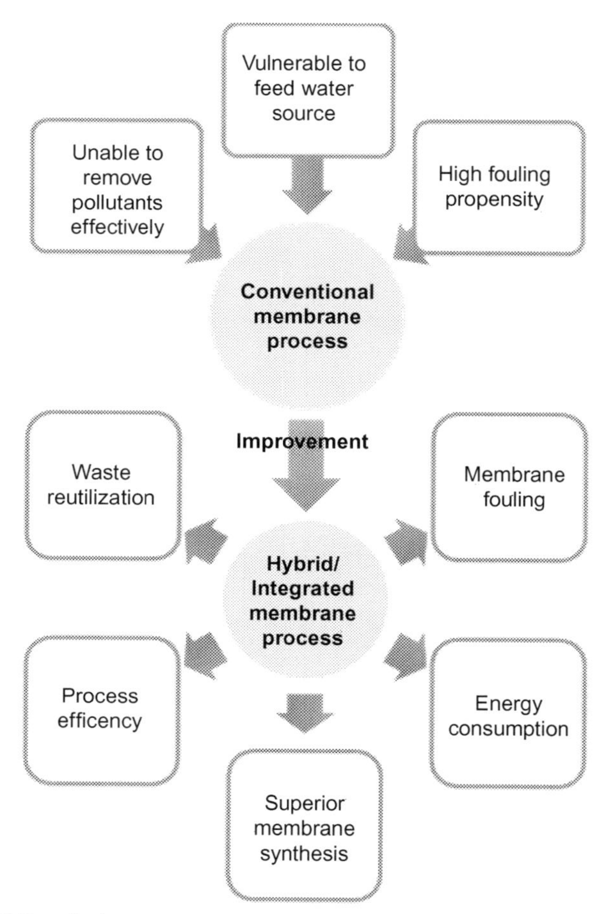

Fig. 12.10 Shift of the water treatment process from conventional to hybrid membranes and their underlying reasons along with the suggested improvements of hybrid membrane processes in the future [44].

water, on the other hand, is becoming more competitive as its cost falls, while the costs of surface water and groundwater rise [50]. Even though tagging desalinated water as affordable for agriculture may be debatable, it not only has its own benefits like decreasing the irrigation dependibility on unpredictable weather patterns, but also helps in better and more efficient management of water resources. Remineralization and concentration of pollutants such as boron is a major issue faced by agriculturalists opting for desalinated water as evident by a case study of the same in Spain [51,52]. Both of these can be eliminated by using additional treatment steps but would add to the cost factor making them unaffordable processes, and hence a benign way of tackling these issues is needed.

12.3.2 Sustainable membranes for energy sector

Renewable energy sources, particularly for power generation, are becoming a bigger part of the global energy picture. Their share in the total energy consumption as shown in Fig. 12.11 has seen a hike over the years. The weightage over the energy sources increases mainly due to the greenhouse gas emissions (GHG) associated with their production. Currently, around three-fourths of the GHG emissions are from energy production. With such vast numbers, it is pivotal that we opt for energy sources with lower emissions.

Membrane processes have become more important in a variety of industrial applications due to their low energy needs. Fig. 12.12 shows the ways in which membrane technology can be helpful in attaining sustainability, particularly in the energy sector. Energy storage devices are not only becoming more viable as alternatives for sustainable technologies in energy production, but they are also projected to play a significant part in storage too. Due to their high energy densities, batteries and fuel cells have become the centerpiece of energy storage technologies [54].

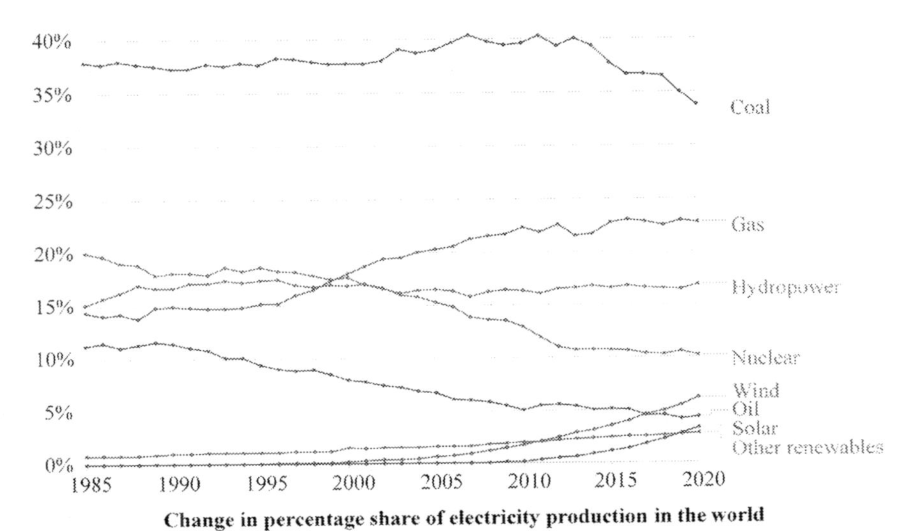

Fig. 12.11 Graphical representation of share of electricity production by varied energy sources and amount of energy generated by varied renewable energy sources over the years [53].

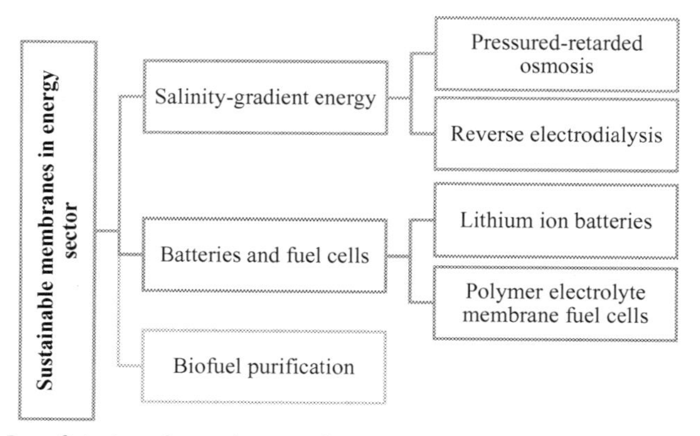

Fig. 12.12 Beneficiaries of membrane science in attaining sustainability [5].

One major hurdle is the scalability of such technologies which depends on the efficiency and the membrane costs. Furthermore, owing to the discrepancies in fuel cells, voltage decreases with an increase in current [55]. Thus, the target of an efficient, scalable, and cost-effective storage technology is still miles away. Biofuels are another beneficiary of membrane science. In biodiesel production, they are used in the separation of glycerol from unreacted lipids, whereas in production of bioethanol they are used to separate the algal biomass. But similar to the challenges associated with membrane-based solution in other sectors, fouling has been the pothole which researchers are trying to fill up. Moreover, membranes in biofuels are yet an emerging field with tons of more research to be done on exploiting the effects of membranes on those processes.

12.3.2.1 Harnessing energy

Blue energy

In 1954, Pattle came up with the theory of blue energy which is fundamentally salinity gradient energy generated by virtue of the difference in osmotic pressure in solutions. Owing to this difference in osmotic pressure, membrane electromotive force can be summed up to generate an electromotive force that can be produced [56]. Researchers have shown the immense potential vested in this form of energy, and the chemical potential generated was equal to the power produced from a 270 m high waterfall [57]. It is estimated to be second largest marine-based energy source. The preference of this form of energy can be attributed to its clean and sustainable nature.

Two emerging technologies in this arena are pressure-retarded osmosis (PRO) and reverse electrodialysis (RED), of which both are membrane-based. Both these processes are in their early stages of commercialization with currently being in the pilot stage. The economic viability of these projects is still a debatable topic as a large section of their cost is associated with the membranes. An overview of their fundamental, economical, and future prospects is listed in Table 12.1.

Low-grade waste heat energy

Out of the total energy produced by various sources, 50% of them is waste heat energy [75]. Depending on the temperature of the heat source, waste energy can be divided into three categories, namely high-grade energy having a temperature of 650°C or more, medium-grade energy having a temperature between 650°C and 230°C, and low-grade energy having a temperature less than 230°C [76]. Extensive research has been done over the years for harvesting this form of energy. The challenge faced by researchers in this area is due to the low source temperature which is related to the efficiency of the Carnot cycle making the conversion of this type of energy not affordable to useful forms of energy using traditional devices [77,78].

Membrane distillation is an interesting technology which can harness the low-grade waste heat energy present in seawater and was first patented by B.R. Bordel in 1968 [79]. Owing to several advantages like its capability to produce desalinated water along with production of energy, independency over concentration of brine, and less foot space required for the process, it is being tried to be implemented on a commercial scale [80]. Moreover, researchers have shown its ability to be customized with renewable sources of energy like solar energy [81,82] and geothermal energy [83,84]. In membrane distillation, the difference in vapor pressure is used to spin the turbine which then generates electricity. It employs a hydrophobic membrane which transfers the condensed vapor formed by the energy source to water, thus causing a volume change which translates to a hydrostatic pressure which is used to drive the turbine. Furthermore, membrane distillation can be used in nutrients recovery too as it can concentrate them at a supersaturation level [85]. The commerciality of this process is still in its nascent state and whether it can truly become a path for high quality water production, mineral recovery and energy generation through renewable sources needs to be seen.

Table 12.1 Overview of fundamental, economical and future prospects of PRO and RED technology WRT Salinity-gradient energy.

Comparison of PRO and RED technologies				
	PRO	Reference	RED	Reference
Fundamentals				
Foundation	S. Loeb, 1976	[58]	R. Pattle, 1954	[56]
Favorable Feed	Concentrated brines	[59]	Seawater	[59]
Electricity generation	Spinning turbine	[60]	Flow of ions	[61]
Type of membrane	Semipermeable	[62]	Ion-exchange membrane	[63]
Economics				
Share of capital cost in total cost	60%	[64]	80%	[65]
Highest power density reported in literature	$2.3\,W/m^2$	[66,67]	$2.2\,W/m^2$	[68]
Future prospects				
Crucial parameter	Trade-off between permeability and structural factor	[69]	Trade-off between permselectivity and Ionic resistance	[70]
Challenges	Specific membrane fabrication	[71]	Specific membrane fabrication	[63]
	Reduction of capital cost	[72]	Cost of membranes	[73]
	Optimization of process design	[74]	–	–

12.3.2.2 Energy storage

Energy storage devices are projected to play a significant role in electricity networks as a result of the extraordinary growth of sustainable technologies for power generation. Because of their high energy density, batteries and fuel cells have gotten a lot of interest among energy storage technologies [86]. Fundamentally, both convert the chemical energy stored by virtue of the components in them to electrical energy. Internally, batteries store chemical energy and, as a result, the batteries must be replenished when this energy is depleted, whereas Fuel cells generate electricity from externally stored reactants and, as a result, will last as long as fuel is available. Both these competitors of energy storage are beneficiaries of membrane technology as it acts as a barrier between the charged rods present in them from short-circuiting [87]. A separator's primary role is to avoid physical contact between the positive and negative electrodes while allowing free ion movement. Although the separator is not involved in any cell processes, its shape and characteristics have a significant impact on battery performance. Proton exchange membrane fuel cells are polymer electrolyte membrane fuel cells that are frequently utilized in automotive and transport applications because of their fast start-up and shutdown times and ability to tolerate thermal shock and high temperature corrosion.

Lithium-ion batteries (LIBs)

Separators for lithium-Ion Batteries are of three types, namely microporous membrane, nonwoven mat, composite membrane and electrolyte membrane separators. Apart from the electrolyte separators, all other types of separators require the addition of electrolyte into their matrixes to facilitate the movement of lithium ions through them. There is no one separator that can satisfy all of the criteria for various battery applications stated in Table 12.2. For rechargeable lithium-ion batteries, microporous polyolefin membranes are now the most often utilized separator. Several microporous multilayer membranes have been utilized in commercial lithium-ion batteries to overcome the difficulties presented by microporous monolayer membranes' shortcomings. Ionically conductive electrolyte membranes, on the other hand, can serve as both a separator and an electrolyte between the two electrodes. They are suited for flexible batteries and/or high-safety solid-state batteries. However, further effort is needed to enhance the membranes' ionic conductivity and mechanical characteristics.

Table 12.2 Overview of required characteristics of membranes of LIBs and PEMFCs.

Overview of membrane requirements of LIBs and PEMFCs			
LIBs	**Reference**	**PEMFCs**	**Reference**
Required characteristics of membrane			
Excellent chemical stability	[88]	Excellent chemical stability	[89]
Fast wettability	[90]	Excellent proton conductivity	[91]
Superior mechanical stability	[88]	Superior mechanical stability	[92]
Good puncture strength	[93]	Excellent dimensional stability	[89]
Appropriate and uniform membrane thickness	[90]	Negligible electronic conductivity	[94]
Small pore size and uniform distribution	[95]	Appropriate water transportation capabilities	[96]
Excellent thermal stability	[97]	Excellent thermal stability	[92]
High porosity	[88]	Small permeability of fuels	[98]
Shutdown capacity at runaway temperatures	[97]	Long lifespan	[99]
Minimal cost	[95]	Minimal cost	[96]

Proton exchange membrane fuel cells (PEMFCs)

PEMFCs have been in the focus of researchers of energy sustainability. PEMFCs use hydrogen as their fuel either in its pure state or mixed with other gases like CO_2, CO, and H_2O [100]. The current surge in interest in this arena can be attributed to the increased demand for energy and the depleting fossil fuels whose environmental impacts are also a major concern today [14]. Owing to their freedom from limitations possessed by the Carnot cycle, they can be operated with efficiencies up to 90% provided the heat formed is also brought into use [92,101]. Moreover, their low emissions and their ability to start up at lower temperatures have added points to their fame and applicability in areas like automobiles and portable devices [92].

Furthermore, their power density is on par with batteries making them a viable competitor of energy storage [102]. Even after such benefits, they have still not been successful in full commercialization [96]. One major hurdle this technology faces is with the proton exchange membrane which

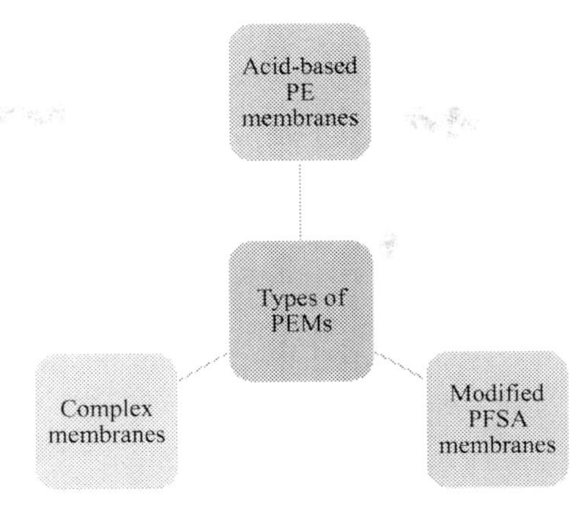

Fig. 12.13 Types of membranes used in PEMFCs [106].

facilitates the movement of ions. Some of the commonly used membranes in PEMFCs are mentioned in Fig. 12.13 To maintain this movement, the conductivity of a membrane is crucial for which they must be hydrated regularly [103]. A fall in hydration would increase the resistance by many folds and since the resistance offered by membranes accounts for 75% of the total resistance, conductivity remains an absolutely indispensable parameter [104]. Poisoning is another area of concern as it causes degradation and thinning of the membrane which is crucial for the operation of PEMFCs. Over time, due to corrosion, metal ions like Fe^{2+} accumulate starting a series of reaction resulting in membrane degradation [105].

Operational limitations are another category of hurdles faced by PEMFCs. Studies have shown that their operability at a higher temperature reduces the humidity, which in turn decreases the life span of membranes [99]. Hence, the chemical and mechanical stability of the membrane is a major area craving for developments [106]. All such limitations must be eradicated keeping their economic aspect in mind. Only by making a sustainable technology that is affordable to develop, developing and underdeveloped countries can exploit its full potential.

12.3.3 Sustainable membranes for biomedical applications

The current pandemic with a tight slap has made us realize the importance of biomedical research and the medical infrastructure. Membranes entwined with the potential of nanotechnology can play a major role in various biomedical applications like filtration and drug delivery.

12.3.3.1 Air filtration

Nonwoven fibrous membranes are widely utilized in a variety of air filtration applications, ranging from interior air filters to personal protective equipment (PPE) like N95 masks. Filtration efficiency and permeability are two parameters often correlated with the quality of the filters.

Both these parameters are interdependent, as shown in Fig. 12.14, making it difficult for researchers to enhance both simultaneously. Enhancement of filtration efficiency can be implemented by making the fibrous membrane finer, but that would result in a decrease in permeability and hence increase the pressure drop substantially [107]. Coating nanoscale fibers on a supporting medium is found to be an effective method of curbing these limitations to an extent [108].

In a world hassled by COVID-19 pandemic, antimicrobial efficiency is an mushrooming area of concern particularly when it comes to the PPEs. An example of a face mask with antimicrobial properties has been depicted in Fig. 12.15. Microorganisms can accumulate over these membranes posing a health risk. Also, the growth of a colony of microorganisms on the membranes would decrease its filtration efficiency [109]. Studies have shown that functionalization of the nanofibers can be a viable solution for enhanced antimicrobial property at lowered cost compared to the traditional way of impregnating membranes with nanoparticles of antimicrobial property [108,110,111].

12.4 Conclusion

Membranes can offer an important contribution in the attainment of sustainable technologies in various sectors by their efficient and innovative

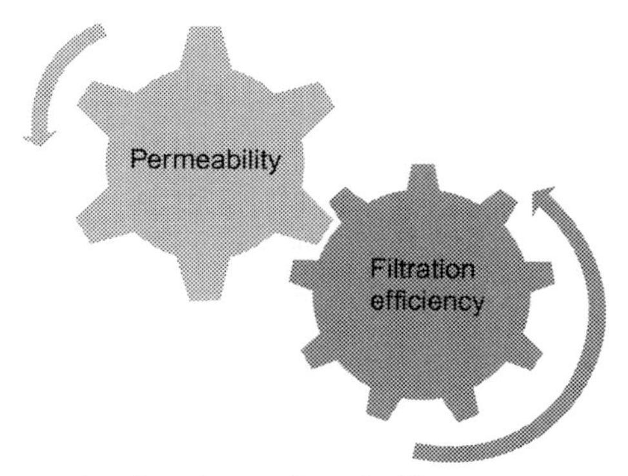

Fig. 12.14 Parameters related to the quality of a filter [107].

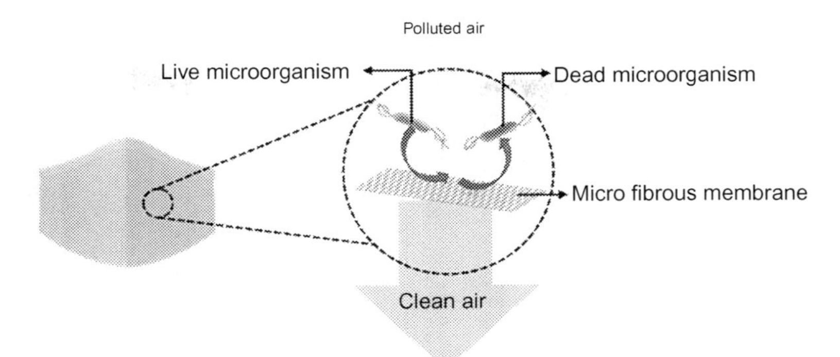

Fig. 12.15 Antimicrobial properties of nanofibrous membrane-based masks.

application minimizing the risk and effectively mitigating the problems along with the enhancement in the recovery of the value-added product. The emerging technologies need the effective utilization of membranes for an industrial revolution. The use of membranes has a significant role in the sustainability of water and energy as they are effectively used for the desalination of brine water, treatment of wastewater, production of fuel cells, and development of various efficient energy storage devices. The flexibility and high adaptability of the membranes has categorized them as an efficient material for current considerations. Besides this, the use of nanostructured advanced material for membrane fabrication has created a new generation of sustainable product having great significance on the environmental aspect along with the advantage of less land usage and easy applicability.

References

[1] S. Roy, S. Ragunath, Emerging membrane technologies for water and energy sustainability: future prospects, constraints and challenges, Energies 11 (11) (2018) 2997, https://doi.org/10.3390/EN11112997.

[2] V.T. Stannett, W.J. Koros, D.R. Paul, H.K. Lonsdale, R.W. Baker, Recent advances in membrane science and technology, Adv. Polym. Sci. 32 (1979) 69–121, https://doi.org/10.1007/3-540-09442-3_5.

[3] P.S. Goh, A.F. Ismail, Review: Is interplay between nanomaterial and membrane technology the way forward for desalination? J. Chem. Technol. Biotechnol. 90 (6) (2015) 971–980, https://doi.org/10.1002/jctb.4531.

[4] F.P. Cuperus, H.H. Nijhuis, Applications of membrane technology to food processing, Trends Food Sci. Technol. 4 (9) (1993) 277–282, https://doi.org/10.1016/0924-2244(93)90070-Q.

[5] N.L. Le, S.P. Nunes, Materials and membrane technologies for water and energy sustainability, Sustain. Mater. Technol. 7 (2016) 1–28, https://doi.org/10.1016/j.susmat.2016.02.001.
[6] A. Yusuf, et al., A review of emerging trends in membrane science and technology for sustainable water treatment, J. Clean. Prod. 266 (2020), https://doi.org/10.1016/J.JCLEPRO.2020.121867, 121867.
[7] N. Galka, A. Saxena, High efficiency air filtration: the growing impact of membranes, Filtr. Sep. 46 (4) (2009) 22–25, https://doi.org/10.1016/S0015-1882(09)70157-0.
[8] J.D. Sachs, The Age of Sustainable Development, Columbia University Press, 2015.
[9] H. Hondo, Life cycle GHG emission analysis of power generation systems: Japanese case, Energy 30 (11–12 SPEC. ISS) (2005) 2042–2056, https://doi.org/10.1016/j.energy.2004.07.020.
[10] I. Dincer, Environmental impacts of energy, Energy Policy 27 (14) (1999) 845–854, https://doi.org/10.1016/S0301-4215(99)00068-3.
[11] A. Ali, R.A. Tufa, F. Macedonio, E. Curcio, E. Drioli, Membrane technology in renewable-energy-driven desalination, Renew. Sust. Energ. Rev. 81 (2018) 1–21, https://doi.org/10.1016/j.rser.2017.07.047.
[12] C. Böhringer, C. Vogt, Economic and environmental impacts of the Kyoto Protocol, Can. J. Econ. 36 (2) (2003) 475–496, https://doi.org/10.1111/1540-5982.t01-1-00010.
[13] B. Bolin, A History of the Science and Politics of Climate Change: The Role of the Intergovernmental Panel on Climate Change, U.S Department of Energy Office of Scientific and Technical Information, Argonne, IL (United States), 2007.
[14] Our World in Data, Absolute Annual Change in Primary Energy Consumption (Online), 2020, Available from: https://ourworldindata.org/grapher/abs-change-energy-consumption?tab=chart&country=~OWID_WRL. (Accessed 01 July 2021).
[15] J. Agyeman, Sustainable Communities and the Challenge of Environmental Justice, New York University Press, 2005.
[16] L. Busch, Bringing nature back in: principles for a new social science of nature, Centennial Rev. 40 (1996).
[17] B.R. Keeble, The Brundtland Report: 'Our Common Future', Taylor & Francis Group, 1988, https://doi.org/10.1080/07488008808408783.
[18] A.L.T. McCammon, United nations conference on environment and development, held in Rio de Janeiro, Brazil, during 3–14 June 1992, and the '92 global forum, Rio de Janeiro, Brazil, 1–14 June 1992, Environ. Conserv. 19 (4) (1992) 372–373, https://doi.org/10.1017/s0376892900031647.
[19] I. Von Frantzius, World Summit on sustainable development Johannesburg 2002: a critical analysis and assessment of the outcomes, Env. Polit. 13 (2) (2004) 467–473, https://doi.org/10.1080/0964401041000l689214.
[20] WCED, World Commission on Environment and Development, January 1987.
[21] S. May, G. Cheney, J. Roper (Eds.), The Debate over Corporate Social Responsibility, Oxford University Press, Inc., 2007.
[22] J. Lovelock, The revenge of Gaiga: Why the Earth is fighting back and how we can still save humanity, vol. 36, Penguin UK, 2007.
[23] J. Jacobs, Dark Age Ahead, vol. 1, Vintage Canada, 2010.
[24] M.J. Rees, Our Final Century: Will the Human Race Survive the twenty-first century?, BasicBooks, 2003.
[25] P. Koltun, Materials and sustainable development, Prog. Nat. Sci. Mater. Int. 20 (1) (2010) 16–29, https://doi.org/10.1016/s1002-0071(12)60002-1.
[26] IRENA, Renewable Energy in the Water, Energy & Food Nexus (Online), January 2015. Available from: /publications/2015/Jan/Renewable-Energy-in-the-Water-Energy–Food-Nexus (Accessed 03 July 2021).

[27] K. Hussey, J. Pittock, The energy-water nexus: managing the links between energy and water for a sustainable future, Ecol. Soc. 17 (1) (2012), https://doi.org/10.5751/ES-04641-170131.
[28] S. Palit, Application of nanotechnology, nanofiltration, and drinking and wastewater treatment—a vision for the future, Water Purif. (2017) 587–620.
[29] N. Ghaffour, The challenge of capacity-building strategies and perspectives for desalination for sustainable water use in MENA, New Pub. Balaban 5 (1–3) (2012) 48–53, https://doi.org/10.5004/DWT.2009.564.
[30] J.R. Ziolkowska, Is desalination affordable?—Regional cost and price analysis, Water Resour. Manag 29 (5) (2014) 1385–1397, https://doi.org/10.1007/S11269-014-0901-Y.
[31] N. Ghaffour, T.M. Missimer, G.L. Amy, Technical review and evaluation of the economics of water desalination: current and future challenges for better water supply sustainability, Desalination 309 (2013) 197–207, https://doi.org/10.1016/J.DESAL.2012.10.015.
[32] M. Kurihara, Seawater reverse osmosis desalination, Membranes 11 (4) (2021) 243, https://doi.org/10.3390/MEMBRANES11040243.
[33] T.M. Missimer, S.A. Stroves, Feedwater Quality Variation and the Economics of Water Plant Operation: The Island Water Association Brakish Water RO Treatment Facility, Sanibel Island, Florida, 2002.
[34] F.H. Kiand, Supply of Desalinated Water by the Private Sector: 30 MGD Singapore Seawater Desalination Plant, December 2004.
[35] Global Water Intelligence, Desal returns to strength, Glob. Water Intell. 21 (10) (2020). (Online). Available from: https://www.globalwaterintel.com/global-water-intelligence-magazine/22/5/general/boosting-off-grid-desalination-with-better-control-systems. (Accessed 08 July 2021).
[36] J. Schallenberg-Rodríguez, J.M. Veza, A. Blanco-Marigorta, Energy efficiency and desalination in the Canary Islands, Renew. Sust. Energ. Rev. 40 (2014) 741–748, https://doi.org/10.1016/J.RSER.2014.07.213.
[37] A.K. Hussein, Applications of nanotechnology in renewable energies—a comprehensive overview and understanding, Renew. Sust. Energ. Rev. 42 (2015) 460–476, https://doi.org/10.1016/J.RSER.2014.10.027.
[38] M. Qadir, B.R. Sharma, A. Bruggeman, R. Choukr-Allah, F. Karajeh, Non-conventional water resources and opportunities for water augmentation to achieve food security in water scarce countries, Agric. Water Manag. 87 (1) (2007) 2–22, https://doi.org/10.1016/J.AGWAT.2006.03.018.
[39] AQUASTAT, Agriculture Water management, 2019, (Online). Available from: http://www.fao.org/aquastat/en/data-analysis/documents/. (Accessed 08 July 2021).
[40] UN-Water, Water Analytical Brief on Unconventional Water Resources, Switzerland, June 2020, (Online). Available from: https://www.unwater.org/publications/un-water-analytical-brief-on-unconventional-water-resources/. (Accessed 08 July 2021).
[41] R. Riera, F. Tuya, A. Sacramento, E. Ramos, M. Rodríguez, Ó. Monterroso, The effects of brine disposal on a subtidal meiofauna community, Estuar. Coast. Shelf Sci. 93 (4) (2011) 359–365, https://doi.org/10.1016/J.ECSS.2011.05.001.
[42] D.A. Roberts, E.L. Johnston, N.A. Knott, Impacts of desalination plant discharges on the marine environment: a critical review of published studies, Water Res. 44 (18) (2010) 5117–5128, https://doi.org/10.1016/J.WATRES.2010.04.036.
[43] N. Hankins, R. Singh, Emerging Membrane Technology for Sustainable Water Treatment, Elsevier, 2016.
[44] W.L. Ang, A.W. Mohammad, N. Hilal, C.P. Leo, A review on the applicability of integrated/hybrid membrane processes in water treatment and desalination plants, Desalination 363 (2015) 2–18, https://doi.org/10.1016/J.DESAL.2014.03.008.

[45] Z. Khatib, P. Verbeek, Water to value—produced water management for sustainable field development of mature and green fields, in: SPE International Conference on Health, Safety and Environment in Oil and Gas Exploration and Production, March 2002, pp. 91–94, https://doi.org/10.2118/73853-MS.

[46] M.A. Engle, et al., U.S. Geological Survey National Produced Waters Geochemical Database v2.3, 2019, https://doi.org/10.5066/F7J964W8 (Online). (Accessed 09 July 2021).

[47] P. Xu, J.E. Drewes, D. Heil, Beneficial use of co-produced water through membrane treatment: technical-economic assessment, Desalination 225 (1–3) (2008) 139–155, https://doi.org/10.1016/J.DESAL.2007.04.093.

[48] Y. He, Z.W. Jiang, Technology review: treating oilfield wastewater, Filtr. Sep. 45 (5) (2008) 14–16, https://doi.org/10.1016/S0015-1882(08)70174-5.

[49] B. Buranaj Hoxha, R. Ettehadi, J. Porter, U. Prasad, S.H. Ong, Improving Shale Wellbore Integrity through Membrane Efficiency and Geomechanics Solutions, November 2020, https://doi.org/10.2118/202388-MS.

[50] O. Barron, et al., Feasibility assessment of desalination application in Australian traditional agriculture, Desalination 364 (2015) 33–45, https://doi.org/10.1016/J.DESAL.2014.07.024.

[51] D. Zarzo, E. Campos, P. Terrero, Spanish experience in desalination for agriculture, New Pub. Balaban 51 (1–3) (2013) 53–66, https://doi.org/10.1080/19443994.2012.708155.

[52] M.A. García-Rubio, J. Guardiola, Desalination in spain: a growing alternative for water supply, Int. J. Water Resour. Develop. 28 (1) (2012) 171–186, https://doi.org/10.1080/07900627.2012.642245.

[53] H. Ritchie, M. Roser, Renewable Energy, Our World Data, December 2020. (Online). Available from: https://ourworldindata.org/renewable-energy. (Accessed 05 July 2021).

[54] A. Kraytsberg, Y. Ein-Eli, Review of advanced materials for proton exchange membrane fuel cells, Energy Fuels 28 (12) (2014) 7303–7330, https://doi.org/10.1021/ef501977k.

[55] V. Das, S. Padmanaban, K. Venkitusamy, R. Selvamuthukumaran, F. Blaabjerg, P. Siano, Recent advances and challenges of fuel cell based power system architectures and control – a review, Renew. Sustain. Energy Rev. 73 (2017) 10–18, https://doi.org/10.1016/j.rser.2017.01.148.

[56] R.E. Pattle, Production of electric power by mixing fresh and salt water in the hydroelectric pile, Nature 174 (4431) (1954) 660, https://doi.org/10.1038/174660a0.

[57] T. Thorsen, T. Holt, The potential for power production from salinity gradients by pressure retarded osmosis, J. Memb. Sci. 335 (1–2) (2009) 103–110, https://doi.org/10.1016/J.MEMSCI.2009.03.003.

[58] S. Loeb, Production of energy from concentrated brines by pressure-retarded osmosis: I. Preliminary technical and economic correlations, J. Membr. Sci. 1 (C) (1976) 49–63, https://doi.org/10.1016/S0376-7388(00)82257-7.

[59] J.W. Post, et al., Salinity-gradient power: evaluation of pressure-retarded osmosis and reverse electrodialysis, J. Membr. Sci. 288 (1–2) (2007) 218–230, https://doi.org/10.1016/J.MEMSCI.2006.11.018.

[60] S. Loeb, F. Van Hessen, D. Shahaf, Production of energy from concentrated brines by pressure-retarded osmosis: II. Experimental results and projected energy costs, J. Membr. Sci. 1 (C) (1976) 249–269, https://doi.org/10.1016/S0376-7388(00)82271-1.

[61] B.E. Logan, M. Elimelech, Membrane-based processes for sustainable power generation using water, Nature 488 (7411) (2012) 313–319, https://doi.org/10.1038/nature11477.

[62] K.L. Lee, R.W. Baker, H.K. Lonsdale, Membranes for power generation by pressure-retarded osmosis, J. Membr. Sci. 8 (2) (1981) 141–171, https://doi.org/10.1016/S0376-7388(00)82088-8.

[63] J.G. Hong, et al., Potential ion exchange membranes and system performance in reverse electrodialysis for power generation: a review, J. Memb. Sci. 486 (2015) 71–88, https://doi.org/10.1016/J.MEMSCI.2015.02.039.

[64] Y. Emami, S. Mehrangiz, A. Etemadi, A. Mostafazadeh, S. Darvishi, A brief review about salinity gradient energy, Int. J. Smart Grid Clean Energy 2 (2) (2013) 295–300. (Online). Available from: http://citeseerx.ist.psu.edu/viewdoc/download?doi=10.1.1.1061.7748&rep=rep1&type=pdf. (Accessed 11 July 2021).

[65] A. Zhu, P.D. Christofides, Y. Cohen, On RO membrane and energy costs and associated incentives for future enhancements of membrane permeability, J. Membe. Sci. 344 (1–2) (2009) 1–5, https://doi.org/10.1016/J.MEMSCI.2009.08.006.

[66] K. Gerstandt, K.V. Peinemann, S.E. Skilhagen, T. Thorsen, T. Holt, Membrane processes in energy supply for an osmotic power plant, Desalination 224 (1–3) (2008) 64–70, https://doi.org/10.1016/J.DESAL.2007.02.080.

[67] G. Han, J. Zuo, C. Wan, T.-S. Chung, Hybrid pressure retarded osmosis–membrane distillation (PRO–MD) process for osmotic power and clean water generation, Environ. Sci. Water Res. Technol. 1 (4) (2015) 507–515, https://doi.org/10.1039/C5EW00127G.

[68] D.A. Vermaas, M. Saakes, K. Nijmeijer, Doubled power density from salinity gradients at reduced intermembrane distance, Environ. Sci. Technol. 45 (16) (2011) 7089–7095, https://doi.org/10.1021/ES2012758.

[69] Y. Xu, X. Peng, C.Y. Tang, Q.S. Fu, S. Nie, Effect of draw solution concentration and operating conditions on forward osmosis and pressure retarded osmosis performance in a spiral wound module, J. Membr. Sci. 348 (1–2) (2010) 298–309, https://doi.org/10.1016/J.MEMSCI.2009.11.013.

[70] G.M. Geise, M.A. Hickner, B.E. Logan, Ionic resistance and permselectivity tradeoffs in anion exchange membranes, ACS Appl. Mater. Interfaces 5 (20) (2013) 10294–10301, https://doi.org/10.1021/AM403207W.

[71] G. Han, S. Zhang, X. Li, T.S. Chung, Progress in pressure retarded osmosis (PRO) membranes for osmotic power generation, Prog. Polym. Sci. 51 (2015) 1–27, https://doi.org/10.1016/J.PROGPOLYMSCI.2015.04.005.

[72] F. Helfer, C. Lemckert, Y.G. Anissimov, Osmotic power with pressure retarded osmosis: theory, performance and trends – a review, J. Membr. Sci. 453 (2014) 337–358, https://doi.org/10.1016/J.MEMSCI.2013.10.053.

[73] M. Elimelech, W.A. Phillip, The future of seawater desalination: energy, technology, and the environment, Science 333 (6043) (2011) 712–717, https://doi.org/10.1126/science.1200488.

[74] G.Z. Ramon, B.J. Feinberg, E.M.V. Hoek, Membrane-based production of salinity-gradient power, Energy Environ. Sci. 4 (11) (2011) 4423–4434, https://doi.org/10.1039/C1EE01913A.

[75] C. Forman, I.K. Muritala, R. Pardemann, B. Meyer, Estimating the global waste heat potential, Renew. Sust. Energ. Rev. 57 (2016) 1568–1579, https://doi.org/10.1016/J.RSER.2015.12.192.

[76] I. Johnson, W.T. Choate, A. Davidson, Waste heat recovery. Technology and opportunities in U.S., Industry (2008), https://doi.org/10.2172/1218716.

[77] H.D. Madhawa Hettiarachchi, M. Golubovic, W.M. Worek, Y. Ikegami, Optimum design criteria for an organic rankine cycle using low-temperature geothermal heat sources, Energy 32 (9) (2007) 1698–1706, https://doi.org/10.1016/J.ENERGY.2007.01.005.

[78] R.A. Kishore, D. Singh, R. Sriramdas, A.J. Garcia, M. Sanghadasa, S. Priya, Linear thermomagnetic energy harvester for low-grade thermal energy harvesting, J. Appl. Phys. 127 (4) (2020), https://doi.org/10.1063/1.5124312, 044501.
[79] B.R. Bodel, Distillation of Saline Water Using Silicone Rubber Membrane, US3361645A, August 09, 1966.
[80] E. Drioli, A. Ali, F. Macedonio, Membrane distillation: recent developments and perspectives, Desalination 356 (2015) 56–84, https://doi.org/10.1016/J.DESAL.2014.10.028.
[81] F. Banat, R. Jumah, M. Garaibeh, Exploitation of solar energy collected by solar stills for desalination by membrane distillation, Renew. Energy 25 (2) (2002) 293–305, https://doi.org/10.1016/S0960-1481(01)00058-1.
[82] F. Banat, N. Jwaied, Economic evaluation of desalination by small-scale autonomous solar-powered membrane distillation units, Desalination 220 (1–3) (2008) 566–573, https://doi.org/10.1016/J.DESAL.2007.01.057.
[83] C.S. Turchi, S. Akar, T. Cath, J. Vanneste, M. Geza, Use of Low-Temperature Geothermal Energy for Desalination in the Western United States, November 2015, (Online). Available from: www.nrel.gov/publications. (Accessed 11 July 2021).
[84] R. Sarbatly, C.K. Chiam, Evaluation of geothermal energy in desalination by vacuum membrane distillation, Appl. Energy 112 (2013) 737–746, https://doi.org/10.1016/J.APENERGY.2012.12.028.
[85] E. Drioli, A. Criscuoli, E. Curcio, Integrated membrane operations for seawater desalination, Desalination 147 (1–3) (2002) 77–81, https://doi.org/10.1016/S0011-9164(02)00579-9.
[86] H.L. Ferreira, R. Garde, G. Fulli, W. Kling, J.P. Lopes, Characterisation of electrical energy storage technologies, Energy 53 (2013) 288–298, https://doi.org/10.1016/J.ENERGY.2013.02.037.
[87] N. Balaban, Energy 2000: The Beginning of a New Millennium, 2000.
[88] S.S. Zhang, A review on the separators of liquid electrolyte Li-ion batteries, J. Power Sources 164 (1) (2007) 351–364, https://doi.org/10.1016/J.JPOWSOUR.2006.10.065.
[89] V. Neburchilov, J. Martin, H. Wang, J. Zhang, A review of polymer electrolyte membranes for direct methanol fuel cells, J. Power Sources 169 (2) (2007) 221–238, https://doi.org/10.1016/J.JPOWSOUR.2007.03.044.
[90] X. Huang, Separator technologies for lithium-ion batteries, J. Solid State Electrochem. 15 (4) (2010) 649–662, https://doi.org/10.1007/S10008-010-1264-9.
[91] A.S. Arico, S. Srinivasan, V. Antonucci, DMFCs: from fundamental aspects to technology development, Fuel Cells Fundam. Syst. 1 (2) (2001) 87–169, https://doi.org/10.1002/1615-6854(200107)1:2%3C133::AID-FUCE133%3E3.0.CO;2-5 (Online).
[92] Y. Wang, K.S. Chen, J. Mishler, S.C. Cho, X.C. Adroher, A review of polymer electrolyte membrane fuel cells: technology, applications, and needs on fundamental research, Appl. Energy 88 (4) (2011) 981–1007, https://doi.org/10.1016/J.APENERGY.2010.09.030.
[93] H. Bierenbaum, R.B. Isaacson, M.L. Druin, S.G. Plovan, Microporous polymeric films, Ind. Eng. Chem., Prod. Res. Dev. 13 (1) (1974) 19. (Online). Available from: https://pubs.acs.org/sharingguidelines. (Accessed 12 July 2021).
[94] B. Smitha, S. Sridhar, A.A. Khan, Solid polymer electrolyte membranes for fuel cell applications—a review, J. Membr. Sci. 259 (1–2) (2005) 10–26, https://doi.org/10.1016/J.MEMSCI.2005.01.035.
[95] P. Arora, Z. (John) Zhang, Battery separators, Chem. Rev. 104 (10) (2004) 4419–4462, https://doi.org/10.1021/CR020738U.
[96] O.Z. Sharaf, M.F. Orhan, An overview of fuel cell technology: fundamentals and applications, Renew. Sust. Energ. Rev. 32 (2014) 810–853, https://doi.org/10.1016/J.RSER.2014.01.012.

[97] H. Lee, M. Yanilmaz, O. Toprakci, F. Kun, X. Zhang, A review of recent developments in membrane separators for rechargeable lithium-ion batteries, Energy Environ. Sci. 7 (12) (2014) 3857–3886, https://doi.org/10.1039/C4EE01432D.
[98] J. Zhang, Y. Xiang, S. Lu, S.P. Jiang, High temperature polymer electrolyte membrane fuel cells for integrated fuel cell – methanol reformer power systems: a critical review, Adv. Sustain. Syst. 2 (8–9) (2018) 1700184, https://doi.org/10.1002/ADSU.201700184.
[99] S.J. Hamrock, M.A. Yandrasits, Proton exchange membranes for fuel cell applications, J. Macromol. Sci. Part C 46 (3) (2007) 219–244, https://doi.org/10.1080/15583720600796474.
[100] K. (Karl) Kordesch, G. Simader, Fuel cells and their applications, U.S. Department of Energy Office of Scientific and Technical Information, Germany, 1996.
[101] G. Hinds, Performance and Durability of PEM Fuel Cells: A Review, Middlesex, 2004, (Online). Available from: http://eprintspublications.npl.co.uk/id/eprint/3056. (Accessed 17 July 2021).
[102] S.J. Peighambardoust, S. Rowshanzamir, M. Amjadi, Review of the proton exchange membranes for fuel cell applications, Int. J. Hydrog. Energy 35 (17) (2010) 9349–9384, https://doi.org/10.1016/J.IJHYDENE.2010.05.017.
[103] T.A. Zawodzinski, et al., Water uptake by and transport through Nafion® 117 membranes, J. Electrochem. Soc. 140 (4) (1993) 1041, https://doi.org/10.1149/1.2056194.
[104] L. Gancs, B.N. Hult, N. Hakim, S. Mukerjee, The impact of Ru contamination of a Pt/C electrocatalyst on its oxygen-reducing activity, Electrochem. Solid-State Lett. 10 (9) (2007) B150, https://doi.org/10.1149/1.2754382.
[105] M. Inaba, T. Kinumoto, M. Kiriake, R. Umebayashi, A. Tasaka, Z. Ogumi, Gas crossover and membrane degradation in polymer electrolyte fuel cells, Electrochim. Acta 51 (26) (2006) 5746–5753, https://doi.org/10.1016/J.ELECTACTA.2006.03.008.
[106] A. Albarbar, M. Alrweq, Proton exchange membrane fuel cells: review, Prot. Exch. Membr. Fuel Cells (2018) 9–29, https://doi.org/10.1007/978-3-319-70727-3_2.
[107] C.H. Hung, W.W.F. Leung, Filtration of nano-aerosol using nanofiber filter under low Peclet number and transitional flow regime, Sep. Purif. Technol. 79 (1) (2011) 34–42, https://doi.org/10.1016/J.SEPPUR.2011.03.008.
[108] M.S. Mauter, Y. Wang, K.C. Okemgbo, C.O. Osuji, E.P. Giannelis, M. Elimelech, Antifouling ultrafiltration membranes via post-fabrication grafting of biocidal nanomaterials, ACS Appl. Mater. Interfaces 3 (8) (2011) 2861–2868, https://doi.org/10.1021/AM200522V.
[109] D. Kharaghani, Y. Kee Jo, M.Q. Khan, Y. Jeong, H.J. Cha, I.S. Kim, Electrospun antibacterial polyacrylonitrile nanofiber membranes functionalized with silver nanoparticles by a facile wetting method, Eur. Polym. J. 108 (2018) 69–75, https://doi.org/10.1016/J.EURPOLYMJ.2018.08.021.
[110] J.D. Schiffman, Y. Wang, E.P. Giannelis, M. Elimelech, Biocidal activity of plasma modified electrospun polysulfone mats functionalized with polyethyleneimine-capped silver nanoparticles, Langmuir 27 (21) (2011) 13159–13164, https://doi.org/10.1021/LA202605Z.
[111] D. Kharaghani, H. Lee, T. Ishikawa, T. Nagaishi, S.H. Kim, I.S. Kim, Comparison of fabrication methods for the effective loading of Ag onto PVA nanofibers, Text. Res. J. 89 (4) (2018) 625–634, https://doi.org/10.1177/0040517517753635.

Index

Note: Page numbers followed by *f* indicate figures and *t* indicate tables.

G

H

I

K

L

M

O

P

Printed in the United States
by Baker & Taylor Publisher Services